Springer Theses

Recognizing Outstanding Ph.D. Research

More information about this series at http://www.springer.com/series/8790

Aims and Scope

The series "Springer Theses" brings together a selection of the very best Ph.D. theses from around the world and across the physical sciences. Nominated and endorsed by two recognized specialists, each published volume has been selected for its scientific excellence and the high impact of its contents for the pertinent field of research. For greater accessibility to non-specialists, the published versions include an extended introduction, as well as a foreword by the student's supervisor explaining the special relevance of the work for the field. As a whole, the series will provide a valuable resource both for newcomers to the research fields described, and for other scientists seeking detailed background information on special questions. Finally, it provides an accredited documentation of the valuable contributions made by today's younger generation of scientists.

Theses are accepted into the series by invited nomination only and must fulfill all of the following criteria

- They must be written in good English.
- The topic should fall within the confines of Chemistry, Physics, Earth Sciences, Engineering and related interdisciplinary fields such as Materials, Nanoscience, Chemical Engineering, Complex Systems and Biophysics.
- The work reported in the thesis must represent a significant scientific advance.
- If the thesis includes previously published material, permission to reproduce this must be gained from the respective copyright holder.
- They must have been examined and passed during the 12 months prior to nomination.
- Each thesis should include a foreword by the supervisor outlining the significance of its content.
- The theses should have a clearly defined structure including an introduction accessible to scientists not expert in that particular field.

Michael L. Wall

Quantum Many-Body Physics of Ultracold Molecules in Optical Lattices

Models and Simulation Methods

Doctoral Thesis accepted by the Colorado School of Mines, Golden, CO, USA

Springer

Michael L. Wall
Colorado School of Mines
Golden, CO, USA

JILA, NIST, and University of Colorado
Boulder, CO, USA

ISSN 2190-5053 ISSN 2190-5061 (electronic)
Springer Theses
ISBN 978-3-319-14251-7 ISBN 978-3-319-14252-4 (eBook)
DOI 10.1007/978-3-319-14252-4

Library of Congress Control Number: 2015934925

Springer Cham Heidelberg New York Dordrecht London

Printed on acid-free paper

Springer International Publishing AG Switzerland is part of Springer Science+Business Media (www.springer.com)

To the memory of my grandfathers,
Richard V. Walker and Richard H. Wall, Sr.

Test everything; retain what is good.
1 Thessalonians 5:21

Foreword

The field of ultracold gases has transformed physics in myriad ways, from advancing quantum computing to testing the very foundations of quantum mechanics. As opposed to solid-state systems, in which many-body dynamics often occurs on extremely short (e.g. femtosecond) timescales, the dynamics in ultracold gas settings occurs on the order of milliseconds. Hence, the extraordinary impact of ultracold quantum gases has been especially evident in studies of non-equilibrium quantum dynamics. For example, in a landmark experiment in 2002, Greiner, Mandel, Esslinger, Hänsch, and Bloch observed a quantum phase transition in an ultracold quantum gas trapped in a crystal of light, including dynamical oscillations between phases in the form of quantum revivals. Such optical lattices have become standard tools of ultracold experiments, and have opened up quantum lattice physics in important new ways, with whole swaths of researchers now devoted to the topic. These initial experiments used atoms, which experience only short-range interactions and are scalar objects, with at most a simple internal structure given by hyperfine degrees of freedom. However, in the last 5 years, ultracold molecules have also reached the ultracold regime. Such compound objects, trapped in crystals of light and with a hierarchical set of controllable degrees of freedom and long-range dipole–dipole interactions, present a whole new area of quantum many-body physics which is only beginning to be explored. We have dubbed this nascent field *quantum complexity*. To make progress in such rich systems, one requires significant new numerical tools as well as totally new models.

Dr. Michael L. Wall's thesis provides such tools and develops with great care and attention the necessary models to found quantum complexity in ultracold molecules. In recognition of the extraordinary impact and quality of his work, in 2013 he was awarded the Nicholas Metropolis Award for Outstanding Thesis Work in Computational Physics by the American Physical Society. His thesis, contained in the following, covers four major topics on the themes of quantum many-body physics, non-equilibrium quantum dynamics and entanglement, and ultracold molecules.

First, he opens up a new area of quantum many-body physics by deriving, explaining, and exploring the *molecular Hubbard Hamiltonian*. This new Hamiltonian describes ultracold molecules in a crystal of light formed from an optical lattice. The ultracold molecules he considers in this initial work, called singlet sigma molecules, have a tunable number of internal states, from two to about one hundred and fifty, in addition to the usual spatial degrees of freedom one finds in Hubbard-like models. One can contrast this model to the well-known Hubbard models for electrons, with just two internal states (spin-1/2, or a qubit). The molecular Hubbard Hamiltonian Dr. Wall develops offers highly controllable multi-scale many-body dynamics.

Second, he solves the problem of how a gas of ultracold fermions pairs into bosons on the lattice, that is, without taking a long-wavelength continuum limit. As matter is made of fermions, and yet we observe many effective bosons, such pairing is a foundational question, and one which many people worked on but no one solved properly on the lattice until Dr. Wall. In the strongly interacting or unitary regime, the question of how to describe such strongly correlated fluids in a discrete or lattice geometry poses a challenge to holographic duality.

Third, he not only created open-source matrix product state (MPS) code used worldwide, including an implementation in the widely used Algorithms and Libraries for Physics Simulations (ALPS) package, he developed new algorithms to deal with dynamics and excited states in systems with long-range interactions. Such numerical methods are key to the new picture of Hilbert space the quantum many-body community is presently developing, in which the apparent exponential difficulty of simulating quantum systems is often illusory when answering practical questions. The difficulty of simulating quantum systems is key to justifying the quantum computing effort in the first place, and Dr. Wall's code helps us refine and direct our efforts to build effective quantum computers. In older algorithms, like density matrix renormalization group (DMRG) methods, one has to rewrite code extensively to deal with each different long-range model. As molecules interact via a dipole–dipole interaction, one has to treat long-range effects. Dr. Wall created a new, large-scale, highly functional code used worldwide. This code is over 10,000 lines long and requires only minor modifications to treat different molecules, field configurations, etc. Without these new algorithmic developments, one would struggle to treat molecules with DMRG or its time-dependent generalization (tDMRG). He also implemented proof-of-principle codes of many new MPS methods, which were proposed during the course of his degree, including minimally entangled typical thermal states (METTS), which combines DMRG with Monte Carlo. The thesis covers both the central MPS codes and these other explorations in detail, and will be very useful to those wanting to understand data compression of a quantum state on classical computers and quantum many-body entangled methods.

Finally, Dr. Wall created extensive educational materials I myself use to educate my students. His impact on my group cannot be overstated. The reader will find these materials useful to bring undergraduates up to speed rapidly and accomplish quantum many-body calculations and research without extensive graduate course work.

Department of Physics
Colorado School of Mines
Golden, CO, USA
February 2015

Lincoln D. Carr

Preface

Ultracold atoms have revolutionized the field of quantum many-body physics due to excellent understanding of their microscopic dynamics and a high degree of control over these dynamics with external fields. The next revolution in ultracold physics promises to come with ultracold molecules, whose production lies at the cutting edge of research. In this thesis, we are concerned with how ultracold molecules trapped in optical lattices may be used as resources for novel many-body physics.

There are six main parts to this thesis. The first part gives a general introduction to the topics covered in the thesis, and provides some technical details on the derivation of many-body lattice models from few-body physics. In the second part, we derive a low-energy Hamiltonian, the molecular Hubbard Hamiltonian, describing ultracold heteronuclear bialkali dimer molecules loaded into an optical lattice and elucidate its many-body properties. These molecules have large permanent electric dipole moments, which give rise to long-range and anisotropic dipole–dipole interactions in an electric field and allow access to the rich internal structure of rotational and hyperfine states in an AC microwave field. Rather than focusing on simulating models relevant to condensed matter physics, we focus on the many-body physics available to near-term experimental setups with the minimal tuning of external fields.

The third part of this thesis studies how fermions pair to make bosons in a discrete context, via a two-channel model for a Feshbach resonance in the presence of an optical lattice and possibly strong inter-channel coupling. The two-body problem is solved numerically using a scaling theory to extract the result for an infinite number of Bloch bands. The bound states of a partitioned Hamiltonian, which we call dressed molecules, are identified as the relevant short-range degrees of freedom at low density, and are chosen so as to reproduce the two-body scattering length identically. From this two-body solution, we derive a low-energy many-body Hamiltonian, which takes the form of a multichannel resonance model between unpaired fermions in the lowest Bloch band and dressed molecules. This approach is valid for arbitrary two-body scattering length and resonance width, and is systematically correctible to higher relative scattering energy.

In the fourth part of this thesis, we discuss matrix product states (MPSs), a class of entanglement-restricted states which are useful for variational calculations in one spatial dimension. This part begins with an overview of the theory of MPSs, with special emphasis on intuitive notions of their use as variational ansätze. Algorithms for finding eigenstates of 1D Hamiltonians, for time-evolution under a general time-dependent 1D Hamiltonian, and for equilibrium properties at finite temperature are presented. Furthermore, it is shown how MPS algorithms may be made generic by the identification of a class of operators known as a matrix product operators and a set of rules for constructing such operators.

The fifth part of this thesis deals with open source implementations of variational MPS algorithms and educational materials designed to facilitate the use and understanding of these methods. The open source projects include a stand-alone open source implementation of the time-evolving block decimation algorithm, open source TEBD, and an implementation of time-evolving block decimation for the widely used algorithms and libraries for physics simulations (ALPS) package, as part of the ALPS international collaboration. In the final part of the thesis, we conclude, give suggestions for future work, and provide appendices.

Golden, CO, USA Michael L. Wall

Acknowledgement

First I would like to thank my advisor, Lincoln Carr. Lincoln's enthusiasm for research and his dedication to his students were what drew me to the Colorado School of Mines in the first place. During the course of my Ph.D., he has given me every opportunity I could have wished for. Additionally, Lincoln has been very supportive of my family life over the progress of my degree, and for that I am immensely grateful.

I would also like to thank my committee, John Bohn, Mahadevan Ganesh, James McNeil, and Zhigang Wu, for their encouragement and for reading a long thesis.

Thanks are due to the members of the Algorithms and Libraries for Physics Simulations (ALPS) collaboration for maintaining great coding and pedagogical resources and allowing me to be a part of the collaboration. Special thanks go to Matthias Troyer, who was very helpful and patient as I grew accustomed to being part of a large code collaboration and learned Python in the process.

I am grateful to Dr. Tim Kaiser for managing the high performance computing facilities here at Mines and for his advice on high performance computing applications.

Thanks to the many people who contributed to what would become open source TEBD, including Charles Clark, Barry Schneider, Jamie Williams, Ippei Danshita, and Daniel Schirmer. I have also benefitted from feedback on the open source projects and their associated educational materials from many people, including Giuseppe Carleo, Meng Cheng, Stephen Clark, Juan Manuel Florez, Lukas Gamper, Joseph Glick, Martin Heimsoth, Alexander Itin, Mateusz Łacki, Hong Lu, Grégoire Misguich, Dominik Muth, Jen Patterson, Han Pu, Bryce Robbins, and Joe Whalen.

Hans Peter Büchler was very kind to discuss his work on strongly interacting fermions in optical lattices and share his computer code with me. His generosity, as well as the initial inspiration from his work, was crucial to bringing the Fermi resonance Hamiltonian to fruition.

I would like to thank J. Ignacio Cirac and his group for their hospitality during a 2-week stay in Garching in October 2009. Special thanks go to Mari Carmen Bañuls for her assistance in scheduling and for arranging a pleasant evening out in Munich.

Also special thanks to Juan José García Ripoll for sharing some Matlab-based MPS code with me during my visit.

I am grateful to Erich Hoover for maintaining a very nice LaTeX thesis template, and for modifying the template to suit some specific needs of the present thesis.

I have been fortunate to attend many great workshops and conferences during my tenure as a graduate student. I would like to thank the Institute for Pure and Applied Mathematics at the University of California, Los Angeles and the organizers for the opportunity to attend the program "Numerical Approaches to Quantum Many-Body Systems"; the organizers of the 2010 Boulder summer school for condensed matter and materials physics for allowing me to participate in a wonderful and very insightful summer school; the organizers of the physics of quantum electronics conference for their invitation to speak in 2012; the organizers of the first quantum technologies conference for a nice conference and a pleasant week in Toruń; and Ana Maria Rey for an invitation to speak about the Fermi resonance Hamiltonian at her group meeting.

I have benefitted from discussions with many researchers about molecules and matrix product states, including Jesus Aldegunde, Jeremy Hutson, Roman Krems, Salvatore Manmana, Ian McCulloch, Ulrich Schollwöck, Miles Stoudenmire, Frank Verstraete, and Steve White. I would especially like to acknowledge Silke Ospelkaus and the groups of Jun Ye and Debbie Jin at the University of Colorado, Boulder for showing me around their experimental labs.

I would like to thank all of the members of the Carr theoretical physics research group over the years for their friendship and insightful discussions. Special thanks are due to Laith Haddad for his humor, kindness, and empathy as we finished our degrees simultaneously; to Scott Strong for the good times and for watching the cat while I was away; and to Erman Bekaroglu for his careful readings of my manuscripts and discussions about molecules. I would also like to thank Ryan Mishmash for his friendship, for sharing his knowledge of condensed matter physics, and for always giving advice with his characteristic acerbic wit.

I would not be where I am today without the support and love of my parents, Gail and Rick, and my brother, John. Finally, my deepest thanks to my wife, Teresa, and our two little girls, Maren and Juniper, for their love, patience, caring support, and perspective.

Contents

Part I Introduction

1 General Introduction ... 3
1.1 Feshbach Resonances ... 7
1.2 Production of Ultracold Molecules ... 9
1.3 Classifications and Few-Body Physics of Ultracold Molecules ... 10
1.4 Simulation Methods ... 15
1.5 Related Work by Other Groups ... 18
1.6 Outline ... 21
References ... 24

2 Models for Strongly Correlated Lattice Physics ... 37
2.1 The Single-Particle Problem: Bloch States and Wannier Functions ... 38
2.1.1 Single-Particle Hubbard Parameters ... 42
2.2 Two-Particle Hubbard Parameters ... 43
2.2.1 Short-Range Interactions ... 44
2.2.2 Long-Range Interactions ... 46
References ... 51

Part II The Molecular Hubbard Hamiltonian

3 Emergent Timescales in Entangled Quantum Dynamics of Ultracold Molecules in Optical Lattices ... 55
3.1 Introduction ... 55
3.2 The MHH ... 57
3.2.1 Derivation of the MHH ... 57
3.2.2 Tunneling ... 59
3.2.3 Interaction with an Optical Lattice ... 60
3.2.4 Dipole–Dipole Interactions ... 65
3.2.5 Energy Scales ... 67
3.2.6 Novel Features of the MHH ... 68

3.3 Methods 69
3.3.1 Time-Evolving Block Decimation 69
3.3.2 Quantum Measures 71
3.4 Case Study: Hard-Core Bosonic Molecules at Half-Filling 72
3.5 Conclusions 81
Appendix 1: Single-Molecule Physics 82
1 Relationship Between Operators in Space-Fixed and Molecule-Fixed Coordinate Systems 82
2 Rotational Hamiltonian 82
3 DC Field Term 83
4 AC Field Term 86
Appendix 2: Convergence 87
1 Single-Molecule Considerations 87
2 Many-Body Considerations 88
References 90

4 Hyperfine Molecular Hubbard Hamiltonian 93
4.1 Introduction 93
4.2 Statement of the Hamiltonian and Experimental Consequences 95
4.2.1 Quantum Dephasing 96
4.2.2 Internal State Dependence of Phase Diagram 96
4.2.3 Tunable Complexity 98
4.3 Derivation of the HMHH 99
4.3.1 Tunneling Energies 102
4.3.2 Two-Molecule Interactions 103
4.3.3 Interactions with Static External Fields 105
4.3.4 Interaction with an ac Microwave Field 106
4.4 Conclusions 108
Appendix 1: The Internal Hamiltonian 108
Appendix 2: Interactions with Static External Fields 111
Appendix 3: Explicit Values for the Single-Molecule Matrix Elements 115
References 118

Part III The Fermi Resonance Hamiltonian

5 Microscopic Model for Feshbach Interacting Fermions in an Optical Lattice with Arbitrary Scattering Length and Resonance Width 123
5.1 Exact Solution for Two-Particles 125
5.2 Fermi Resonance Hamiltonian 127
5.3 Supplemental Material: Derivation of the Nonlinear Eigenequation 130
5.4 Supplemental Material: Classification of the Two-Particle Bound States 133
5.5 Supplemental Material: Finite Width Resonances 136
References 137

Part IV Matrix Product States

6 Matrix Product States: Foundations 141
6.1 Bird's Eye View of MPSs 141
6.2 Definitions 147
6.3 Canonical Forms for MPSs 152
6.4 Examples of MPSs 156
6.4.1 Product State 156
6.4.2 GHZ State 156
6.4.3 W State 157
6.4.4 AKLT State 158
6.5 Correlations Within MPSs and the Transfer Operator 160
6.6 Symmetry-Adapted MPSs 164
References 173

7 Out-of-Equilibrium Dynamics with Matrix Product States 177
7.1 Introduction 178
7.2 Brief Review of Matrix Product Formalism 180
7.2.1 Matrix Product States 180
7.2.2 Matrix Product Operators 185
7.3 Variational Ground State Search 191
7.4 Variational Excited State Search 194
7.5 Calculation of Observables 198
7.6 Time Evolution with MPSs 199
7.7 Commutator-Free Magnus Expansions 200
7.8 Krylov Subspace Propagation 203
7.9 Simulation Protocol 207
7.10 Case Studies 208
7.10.1 Case Study: Ising Model in a Transverse Field 208
7.10.2 Case Study: Dipolar Ising Chain 212
7.11 Conclusions 214
References 218

8 The Infinite Size Variational Matrix Product State Algorithm 223
8.1 The Orthogonality Fidelity 227
8.2 Orthogonalization of MPSs in the Thermodynamic Limit 228
8.3 Expectation Values 232
8.4 Examples 237
References 239

9 Finite Temperature Matrix Product State Algorithms and Applications 241
9.1 Introduction 241
9.2 Methodology 242
9.2.1 Matrix Product States 242
9.2.2 The Ancilla Method 246
9.2.3 Minimally Entangled Typical Thermal States 247

9.3 Validity Issues 249
9.4 Application: Specific Heat of the Hard-Core Extended Bose–Hubbard Model 249
References 250

Part V Open Source Code and Educational Materials

10 Open Source Code Development 255
10.1 OSTEBD Overview 256
10.1.1 Parallel Extensions in OSTEBD 257
10.2 The ALPS TEBD Routines 261
10.2.1 Numerical Optimizations of the ALPS Code 262
10.2.2 The Python Front End 265
10.2.3 Integration with the VisTrails Workflow Provenance System 268
References 270

11 Educational Materials 273
11.1 Materials Distributed with the Open Source Packages 273
11.2 Materials Intended for CTPRG Use 277
References 280

Part VI Conclusions and Appendices

12 Conclusions and Suggestions for Future Research 285
References 288

A Documentation for ALPS v2.0 TEBD Code 291
A.1 Documentation: TEBD 291
A.1.1 Time-Evolving Block Decimation 291
A.1.2 References 292
A.1.3 TEBD-Specific Parameters 292
A.2 Tutorials: TEBD-01 bhquench 294
A.2.1 The Hardcore Boson Model 294
A.2.2 Linear Quench 295
A.2.3 Linear Quench with Hold 298
A.2.4 Nonlinear Quenches 300
A.3 Tutorials: TEBD-02 kink 303
A.3.1 Evolution of a Domain Wall 303
A.3.2 Exact Solution for the Case of the XX Model 303
A.3.3 Error Analysis of TEBD 1: Time Step Error 308
A.3.4 Error Analysis of TEBD 2: Entanglement Cutoff Error 310
A.3.5 Solution in the Case of the XXZ Model 313

B Educational Materials: A Gentle Introduction to Time Evolving Block Decimation ... 317
B.1 Pre-test ... 317
B.1.1 Exercise 1 ... 317
B.1.2 Exercise 2 ... 317
B.1.3 Exercise 3 ... 318
B.2 The Limitations of Exact Diagonalization ... 318
B.2.1 Exercise 4: The Limitations of Exact Diagonalization ... 318
B.3 Preliminaries ... 319
B.3.1 Mathematical Preliminaries ... 319
B.3.2 Entanglement ... 323
B.4 Conceptual Basis of TEBD ... 324
B.5 TEBD in Summary ... 326
B.6 Using OSTEBD ... 326
B.6.1 Exercise 8: Using TEBD: Non-number Conserving Version ... 328
B.6.2 Exercise 9: Using TEBD: Number Conserving Version .. 328
B.6.3 Exercise 10: The Tight-Binding Chain ... 328
B.6.4 Exercise 11: Fidelity Susceptibility and the Superfluid-Mott Insulator Quantum Phase Transition ... 330

C Educational Materials: Introduction to MPS Algorithms ... 337
C.1 MPSs and Their Canonical Forms ... 337
C.1.1 Step 1: MPS Structure ... 338
C.1.2 Step 2: MPS Initialization ... 338
C.1.3 Step 3: Canonical Form ... 340
C.2 Matrix Product Operators ... 340
C.2.1 Step 4: MPO Structure ... 342
C.2.2 Step 5: MPO Representation of the Ising Hamiltonian ... 343
C.2.3 ∗ Long-Range Interactions with MPO ... 343
C.3 Construction of the Effective Hamiltonian ... 345
C.3.1 Step 6: LR Array ... 347
C.3.2 Step 7: Effective Hamiltonian ... 347
C.3.3 Step 8: Putting It All Together ... 348
C.3.4 ∗ Sparse Eigensolvers: The Lanczos Algorithm ... 349
C.4 Observables ... 350
C.5 Excited States ... 351
C.5.1 ∗ Sparse Solution of the Projected Eigenproblem ... 353
C.6 Time Evolution with TEBD ... 354
C.7 Source Code ... 356
C.7.1 Module: LinearAlgebraBackend.f90 ... 356
C.7.2 Module: MPSOps.f90 ... 359
C.7.3 Module: MPOOps.f90 ... 361
C.7.4 Module: EffHamiOps.f90 ... 368
C.7.5 Main Program: OrthTestMain.f90 ... 372
C.7.6 Main Program: IsingGSMain.f90 ... 373

List of Figures

Fig. 1.1 Intrinsic energy scales of alkali metal dimers. In descending order, the energy scales of a $^1\Sigma$ diatomic molecule are excitations of the electronic state, vibrational modes of the nuclei about the equilibrium internuclear separation, rotation of the molecule about its center of mass, and couplings between the nuclear spins and the other angular momenta of the molecule. These energy scales span nearly twelve orders of magnitude 11

Fig. 3.1 Dependence of the field-free tunneling (hopping) coefficient on rotational state and lattice height 65
Fig. 3.2 Dependence of site-averaged number on lattice size L 75
Fig. 3.3 Dependence of structure factors within and between rotational states J on the number of lattice sites 76
Fig. 3.4 Dependence of the asymptotic behavior of rotational state populations on the lattice height η 78
Fig. 3.5 Dependence of the asymptotic behavior of structure factors on the lattice height η.. 79
Fig. 3.6 Dependence of emergent timescales on number of lattice sites 80
Fig. 3.7 Dependence of emergent timescales on lattice height 81
Fig. 3.8 Dressed state dipole moments and energies 85
Fig. 3.9 Compositions of dressed states vs. scaled rotational energy........ 85
Fig. 3.10 Resonant AC field induced population cycling in the dressed and field-free bases 86
Fig. 3.11 Convergence with respect to DC dressing rotational state cutoff 87
Fig. 3.12 Convergence with respect to local dimension cutoff................ 88
Fig. 3.13 Convergence with respect to entanglement cutoff parameter 89

Fig. 4.1 Quantum dephasing in the HMHH.................................. 97
Fig. 4.2 Tunneling matrix elements in a dc electric field..................... 97

Fig. 4.3 Geometry of the HMHH 98
Fig. 4.4 Gigahertz scale view of the Stark effect for KRb 104
Fig. 4.5 Distribution of dipolar character 106
Fig. 4.6 Kilohertz scale view of the Stark effect for $^{40}K^{87}Rb$, N=0 110
Fig. 4.7 Induced dipoles for KRb in an electric field 113
Fig. 4.8 Zeeman effect for $^{40}K^{87}Rb$, $N = 0$ 114
Fig. 4.9 Zeeman effect for $^{40}K^{87}Rb$, $N=1$ 115

Fig. 5.1 Schematic of the FRH transformation 124
Fig. 5.2 Exact two-particle band structures for various a_s in a strong optical lattice 126
Fig. 5.3 Hubbard parameters for the FRH 129
Fig. 5.4 Classification of two-particle bound states 135
Fig. 5.5 Bound state energies for finite width resonances 136

Fig. 6.1 Schematic of the MPS construction 145
Fig. 6.2 Examples of basic tensor operations in diagrammatic notation 149
Fig. 6.3 Examples of tensor networks 150
Fig. 6.4 Correlation structure of PBC states 152
Fig. 6.5 Approximating algebraic decay by sums of exponentials 163
Fig. 6.6 Structure of charge flow through an MPS in diagrammatic notation 169

Fig. 7.1 Diagrammatic notation for tensor networks 182
Fig. 7.2 Gauge conditions for MPSs 183
Fig. 7.3 MPO in diagrammatic notation 185
Fig. 7.4 Variational ground state search in diagrammatic notation 192
Fig. 7.5 Linear forms enforcing orthogonality 195
Fig. 7.6 Statics of the Ising model 209
Fig. 7.7 Dynamics of the Ising model 211
Fig. 7.8 Statics of the dipolar Ising model 212
Fig. 7.9 Dynamics of the dipolar Ising model 213
Fig. 7.10 Demonstration of Bogoliubov–de Gennes method convergence 218

Fig. 8.1 Normalization conditions for infinite MPSs 229
Fig. 8.2 Expectation of a finitely supported operator within an infinite MPS 232
Fig. 8.3 Expectation of an infinitely supported operator within an infinite MPS 234
Fig. 8.4 Fixed point relations for the dominant matrices of the Hamiltonian MPO transfer matrix, Eq. (8.37) 235
Fig. 8.5 Correlation functions for the spin-1/2 and spin-1 Heisenberg models computed with iMPS 238
Fig. 8.6 Scaling of the correlation length with the bond dimension χ in iMPS 239

Fig. 9.1 Tensor network representation of full four site wavefunction 242
Fig. 9.2 Tensor network representation of the scalar product procedure 244
Fig. 9.3 Tensor network representation of the quadratic form 245
Fig. 9.4 Specific heat of the hard-core extended Bose–Hubbard model..... 250

Fig. 10.1 Open source time-evolving block decimation blog statistics 256
Fig. 10.2 Performance of intrinsically parallel OSTEBD routines........... 261
Fig. 10.3 Example VisTrail for ALPS TEBD tutorial1a....................... 269

Fig. 11.1 Results of ALPS TEBD tutorial 2 275
Fig. 11.2 Results of ALPS TEBD tutorial 1 278

Fig. B.1 SVD representation of an image 325
Fig. B.2 Single particle density matrix, noninteracting system of $L = 50$ lattice sites and $N = 50$ particles 331
Fig. B.3 Fidelity susceptibility for (*bottom to top*) $L = 3, 5, 10, 20, 40$..... 335

Fig. C.1 Algorithm to put site j into left-canonical form 339
Fig. C.2 Algorithm to put site j into right canonical form 340
Fig. C.3 Construction of the effective hamiltonian in tensor network diagram representation....................................... 345
Fig. C.4 Pseudocode to find the ground state variationally.................. 346
Fig. C.5 Construction of linear forms for projection in tensor network diagram representation....................................... 352

List of Tables

Table 1.1 Coupling constants defining a $^1\Sigma$ molecular species 13

Table 2.1 Fit parameters for the tunneling at the nearest, next-nearest, and third nearest neighbor distances together with their asymptotic standard errors 43

Table 3.1 Values of the polarizabilities for LiCs in different rotational states $|JM\rangle$ 64

Table 3.2 Comparison of energy and length scales for the molecular hubbard hamiltonian of Eq. (3.1) 68

Table 3.3 Emergent timescales τ and τ_Q and their fit asymptotic standard errors for various lattice heights and system sizes 77

Table 4.1 Table of energy scales of the hyperfine molecular Hubbard Hamiltonian 96

List of Symbols

Molecular state with total electronic spin and total projection of electronic orbital angular momentum along the internuclear axis equal to zero	$^{1}\Sigma$
Dynamical polarizability tensor	$\tilde{\alpha}$
DC electric field	$\mathbf{E}_{\mathrm{DC}}$
AC electric field	$\mathbf{E}_{\mathrm{AC}}$
Optical lattice field	$\mathbf{E}_{\mathrm{opt}}$
Unnormalized spherical harmonic of order k and projection q	$C_q^{(k)}$
Permanent dipole moment (Part II). Local dimension (Part IV)	d
Projection of dipole operator along spherical direction q	$\hat{d}_q$
Rotational constant	B_N (B in Chap. 3)
Rotational angular momentum operator	$\hat{\mathbf{N}}$ ($\hat{\mathbf{J}}$ in Chap. 3)
Hyperfine spin operator of the ith nucleus	$\hat{\mathbf{I}}_i$
Lattice spacing	a
Characteristic lattice momentum π/a	k_l
Recoil energy	E_R
Number of lattice sites	L
3D Bloch function	$\psi_{\sigma\mathbf{nq}}$
1D Bloch function	$\psi_{\sigma nq}$
3D Wannier function	$w_{\sigma\mathbf{n}}(\mathbf{r}-\mathbf{r}_i)$
s-wave scattering length	a_s
Effective range	r_B
Interchannel coupling constant	g (Part III)
Interchannel detuning	ν (Part III)
Total quasimomentum	$\mathbf{K}$
von Neumann entropy of entanglement in region A	$S_{\mathrm{vN}}(A)$
Kronecker product of two indices a and b	(ab)
Bond dimension of a matrix product object	χ

List of Abbreviations

1D/2D/3D	One dimension/two dimensions/three dimensions
AKLT	Affleck, Lieb, Kennedy, and Tasaki
ALPS	Algorithms and libraries for physics simulations
BCS	Bardeen–Cooper–Schrieffer (theory of superconductivity)
BEC	Bose–Einstein condensate
BHH	Bose–Hubbard Hamiltonian
BZ	Brillouin zone
CDW	Charge-density wave
CFME	Commutator-free Magnus expansion
CPS	Classical product state
CTPRG	Carr theoretical physics research group
DFT	Density functional theory
DMFT	Dynamical mean-field theory
DMRG	Density-matrix renormalization group
ED	Exact diagonalization
FFT	Fast Fourier transform algorithm
FLOP	Floating-point operation
FRH	Fermi resonance Hamiltonian
FSA	Finite state automaton
GAXPY	Generalized A times X plus Y
GMRES	Generalized minimal residual algorithm
HMHH	Hyperfine molecular Hubbard Hamiltonian
iMPS	Infinite size variational matrix product state algorithm
MCQDT	Multichannel quantum defect theory
MERA	Multiscale entanglement renormalization ansatz
METTS	Minimally entangled typical thermal states
MHH	Molecular Hubbard Hamiltonian
MI	Mott insulator
MPI	Message passing interface
MPO(s)	Matrix product operator(s)
MPS(s)	Matrix product state(s)

OA	Operator alphabet
OBC	Open boundary conditions
OpenMP	Open multiprocessing
OSTEBD	Open source time-evolving block decimation
PBC	Periodic boundary conditions
PEPS	Projected entangled-pair states
QMC	Quantum Monte Carlo
RG	Renormalization group
SF	Superfluid
SPDM	Single particle density matrix
STIRAP	Stimulated Raman adiabatic passage
SVD	Singular value decomposition
TEBD	Time-evolving block decimation
ZGESVD	Complex double precision singular value decomposition

Part I
Introduction

Chapter 1
General Introduction

According to the postulates of quantum mechanics, a quantum system is completely specified by the state $|\psi(t)\rangle$ whose evolution is provided by the Schrödinger equation

$$i\hbar\frac{\partial}{\partial t}|\psi(t)\rangle = \hat{H}|\psi(t)\rangle. \tag{1.1}$$

Here, $\hat{H}$ is the Hamiltonian describing the interactions of all of the microscopic degrees of freedom in the system under study. Unfortunately, the dimension of the Hilbert space in which the state $|\psi(t)\rangle$ lives grows exponentially with the number of constituents in a many-body system, rendering Eq. (1.1) essentially useless for extracting physically relevant information from systems with more than a few particles. Practical concerns aside, there is a more fundamental reason why Eq. (1.1) does not enable us to answer all relevant questions in many-body physics. This reason is put succinctly by P.W. Anderson in his now famous article "More is different" [1] when he says that "The ability to reduce everything to simple fundamental laws does not imply the ability to start from those laws and reconstruct the universe." That is to say, many-body systems can display very different, *emergent*, behavior from their microscopic constituents. In particular, the ground state of a many-body system need not have the same symmetry as its governing Hamiltonian due to the phenomenon of *spontaneous symmetry breaking*.

A powerful method for studying weakly interacting[1] many-body systems is provided by the principle of *adiabatic continuity* [2] which allows the eigenstates of the interacting system to be connected to the eigenstates of the non-interacting system. The key assumption of this notion is that levels do not change symmetry as

[1]The precise sense in which we mean the system is weakly interacting is that it is not strongly correlated, the latter of which will be defined below. In particular, by weakly interacting we do not mean that the interactions lie within the radius of convergence of a perturbation series.

M.L. Wall, *Quantum Many-Body Physics of Ultracold Molecules in Optical Lattices*, Springer Theses, DOI 10.1007/978-3-319-14252-4_1

interactions are introduced, and so levels corresponding to the same symmetry have no crossings with increasing interactions. Hence, the ordering of quantum numbers in the interacting and non-interacting systems will be roughly the same. The most prominent many-body theory resulting from adiabatic continuity is Landau's Fermi liquid theory, which applies to weakly interacting fermions in three dimensions. Here, the elementary excitations of the interacting system are quasiparticles which stand in one-to-one correspondence with those of a free Fermi gas, albeit with renormalized physical parameters [3]. Adiabatic continuity clearly fails if the system changes its symmetry, as it does near a point of non-analyticity known as a *quantum phase transition* (QPT) [4]. We define the class of systems which cannot be adiabatically connected to the non-interacting counterparts of their microscopic degrees of freedom as being *strongly correlated.*

The degrees of freedom which are relevant to the low-energy theory of a strongly correlated system may be difficult to identify, as we have no reference non-interacting state to which they may be related. An example of this is the fractional quantum Hall effect, in which the relevant degrees of freedom are quasiparticles which carry rational fractions of the elementary charge [5] and also obey fractional exchange statistics [6, 7]. The identification of the relevant macroscopic degrees of freedom in a many-body system is provided, at least in principle, by the *renormalization group* (RG) procedure [8]. An RG analysis involves studying the behavior of a system under a scaling transformation in which some set of degrees of freedom are integrated out to yield an effective description of the system in fewer variables. This procedure is called *coarse graining*. The main classical focus of RG was in studying critical systems. Critical systems have no length scales due to the diverging of the correlation length ξ describing exponential decay of equal-time order parameter correlations in the ground state.[2] Thus, provided that we have coarse grained to a scale large compared to the microscopic scales, further coarse graining should not produce any significant effect on our description of the system. That is, critical points represent fixed points of the RG iteration. More generally, the fixed points of the RG iteration correspond to the possible macroscopic states of the system. Here the remarkable feature of *universality* naturally arises, in which microscopic details of the system are not relevant to its macroscopic behavior.

However powerful the RG idea, one still must choose an appropriate coarse graining procedure. A particularly powerful procedure for one-dimensional (1D) systems is *White's rule* [9], which posits that the states which should be kept when coarse graining a system in real space are those which have the largest weight in the reduced density matrix obtained by tracing out all degrees of freedom not being coarse grained. The real-space RG procedure utilizing White's rule is formulated algorithmically as the *density-matrix renormalization group* (DMRG), which has been the method of choice for strongly correlated 1D systems for nearly 20 years. Theoretical analysis of DMRG has revealed [10, 11] that it may be

[2] In the absence of exponential decay, the correlation length may be taken to be a length scale on which correlations qualitatively shift to a long-distance behavior.

formulated as a variational method within a class of quantum states known as *matrix product states* (MPSs). This realization together with insights from quantum information theory, particularly from entanglement theory, led to a generalization of the DMRG procedure to time evolution for short-range interacting systems [12–14] and to the proper entanglement structure to describe states with periodic boundary conditions efficiently via DMRG [15]. Also, research along these lines has led to proposals of real-space RG schemes based on entanglement decimation in higher dimensions such as projected entangled-pair states (PEPS) [16, 17] and the multiscale entanglement renormalization algorithm (MERA) [18]; this is still an area of intense research activity.

Using MPSs as a variational ansatz builds upon a long history of variational methods for strongly correlated systems. Many of the most successful methods in many-body theory, for example the Bardeen–Cooper–Schrieffer (BCS) theory of superconductivity [19], the theory of the quantum Hall effect [5], and density functional theory (DFT) [20, 21], are all variational in nature. The availability of open source packages for DFT such as SIESTA [22], ABINIT [23], and OCTOPUS [24] has allowed for great progress in materials physics. MPSs and their generalizations provide the best hope of stimulating such progress for strongly correlated systems. Hence, one of the goals of this thesis is to provide flexible MPS algorithms which are applicable to a wide array of systems, and open source implementations of MPS algorithms which are simultaneously high performance and easily modified to meet user's needs.

In addition to the theoretical hurdles to studying strongly correlated systems, it is difficult in a condensed matter setting to predict which systems will be strongly correlated. Even when a strongly correlated system has been discovered, often its parameters are difficult to control, and its microscopic dynamics may be too fast to be reliably studied. The crossover of atomic physics into the field of strongly correlated condensed matter began circa 1995 when advances in laser cooling [25] and evaporative cooling [26, 27] led to the creation of Bose–Einstein condensates (BEC) of the alkali metal species Rb [28], Na [29], and Li [30, 31]. As opposed to liquid Helium, the only other known elemental quantum liquid at the time, the interactions in these highly dilute gases were weak, enabling more detailed analysis both theoretically and experimentally. Following the success of bosonic atoms, fermionic atoms were also brought to quantum degeneracy [32–34]. While Pauli exclusion prevents fermions in the same internal state from interacting through s-wave interactions and hence drastically slows evaporative cooling at low temperatures [35], Fermi gases may be cooled either through sympathetic cooling of the gas with a BEC or through evaporative cooling when multiple species of fermions are present.

Ultracold[3] atomic gases have many advantages over traditional condensed matter systems for studying many-body physics. For one, the timescales which are relevant to an atomic many-body system are on the order of milliseconds to minutes,

[3]By ultracold, we mean temperatures less than 1 μK.

several orders of magnitude longer than typical timescales of condensed matter experiments. This has enabled for the study of slow many-body dynamics of a QPT [36] and the collapse and revival of matter wave coherence [37]. Additionally, ultracold atoms are extremely well isolated from their environment, and sources of decoherence such as spontaneous emission events can be controlled such that lifetimes are on the order of seconds to more than a minute. Finally, using tools such as optically or magnetically tunable Feshbach resonances [38] allows for precise tunability of the interactions in ultracold gases. We will discuss Feshbach resonances in more detail in Sect. 1.1.

Another essential tool for strongly correlated many-body physics with ultracold atoms is provided by *optical lattices*. An optical lattice [39] is a standing wave array of light formed by counter-propagating laser beams in three dimensions. The light couples to the dynamical polarizability of the object, and the resulting AC Stark shift induces a periodic trapping potential. By altering the geometry or phase of the beams forming the lattice, one can induce a wide variety of geometries [40], including confinement to a quasi-1D geometry [41, 42], as well as time-dependence [43]. Such a lattice mimics the effects of the lattices common in solid state systems, but without the difficulties provided by disorder and phonons which are inevitable in solid state systems at finite temperature. A major avenue of research stemming from this capability is in tuning the parameters of an atomic gas trapped in an optical lattice such that the governing Hamiltonian reproduces a model relevant to condensed matter physics. Such specialized experimental systems represent *quantum simulators* [44], which are essentially single-program quantum computers first envisioned by Feynman [45]. The first atomic quantum simulator to be both proposed [46] and built [36] was the Bose–Hubbard model [47]

$$\hat{H} = -t \sum_{\langle i,j \rangle} \left[\hat{b}_i^{\dagger} \hat{b}_j + \text{h.c.} \right] + \frac{U}{2} \sum_i \hat{n}_i \left(\hat{n}_i - 1 \right). \tag{1.2}$$

Here i and j label sites in a lattice, $\langle i, j \rangle$ denotes all pairs which are nearest-neighbors in the lattice, $\hat{b}_i$ destroys a bosonic particle at site i, and $\hat{n}_i = \hat{b}_i^{\dagger} \hat{b}_i$ counts the number of bosonic particles on site i. The first term represents quantum mechanical tunneling of bosons between neighboring sites with an associated tunneling energy t, and the latter term is an energetic penalty for two or more bosons to occupy the same site due to interactions. While it is remarkable that ultracold atoms are able to realize a model of great relevance to condensed matter physics, Eq. (1.2) is also important in that it represents a minimal, natural Hamiltonian for ultracold atoms in an optical lattice. That is, if one were to load an ultracold bosonic alkali gas into an optical lattice without fine-tuning of fields, Eq. (1.2) would govern its properties. A topic which forms much of the bulk of this thesis is determining the corresponding natural Hamiltonian for molecules, which have more a complex internal structure than atoms. We will discuss the microscopic derivation of Hamiltonians such as Eq. (1.2) in further detail in Chap. 2.

The remainder of the introduction is organized as follows. In Sect. 1.1 we review Feshbach resonances, an indispensable experimental tool for tuning the interactions of ultracold atomic gases and the production of ultracold molecules. In Sect. 1.2 we discuss experimental production of ultracold molecules. In Sect. 1.3 we provide a basic review of the structure and few-body properties of the molecules relevant to this thesis. In Sect. 1.4 we present a digest of modern numerical techniques for the quantum many-body problem; with a focus on methods applicable to strongly correlated systems. In Sect. 1.5 we overview work by other groups which is related to the results of this thesis. Finally, in Sect. 1.6, we outline the layout of the body of the thesis.

1.1 Feshbach Resonances

Feshbach resonances [38, 48] are an essential component both of the tunability of interactions in ultracold atomic gases and of the production of ultracold molecules. The basic physics of a Feshbach resonance can be explained through a two-channel model. In this model, we partition our Hilbert space into open and (energetically) closed channels, with the asymptotic limit of the open channel potential corresponding to two free atoms. The closed channel potential is assumed to support a bound molecular state near the threshold of the open channel potential. In the presence of a phenomenological interchannel coupling g, the bound state of the closed channel is no longer a true bound state but becomes a resonance due to its mixing with the open channel.[4] Now, a Feshbach resonance occurs when one of the bound states in the closed channel becomes near degenerate with the scattering state in the open channel. Even a weak coupling g is sufficient to cause strong mixing of the two channels when these energies nearly coincide, and this causes a drastic change in the scattering properties. Expressed in terms of the detuning ν between the open and closed channels, the two-channel scatting amplitude is [49]

$$f(\mathbf{k}) = -\frac{2\mu}{4\pi\hbar^2}\frac{g^2}{\epsilon_{\mathbf{k}} - \nu + \frac{\mu g^2}{2\pi\hbar^2}ik}, \tag{1.3}$$

where μ is the reduced mass, $\mathbf{k}$ the incident momentum, and $\epsilon_{\mathbf{k}}$ the incident energy. We may write this scattering amplitude as

$$f(\mathbf{k}) = -\frac{1}{1/a_s + ik + r_b k^2}, \tag{1.4}$$

by identifying the s-wave scattering length $a_s = -2\mu g^2/4\pi\hbar^2\nu$ and the effective range $r_B = \pi\hbar^4/\mu^2 g^2$, both of which are experimentally measurable. Although

[4] This is what differentiates a Feshbach resonance from a shape resonance. In the latter no bound state exists in the absence of the coupling.

this is the asymptotic form of the scattering amplitude arising from scattering of low-momentum particles with $k\,|r_B| \ll 1$ [50], this form is exact for the two-channel model.[5] Hence, at higher relative energy the two-channel model breaks down in its ability to describe the full momentum dependence of the scattering amplitude.

Feshbach resonances in ultracold gases are most frequently provided by hyperfine couplings between atoms whose valence electrons reside in singlet and triplet configurations.[6] The singlet potential is generally much deeper than the triplet potential, and also generally appears above the threshold of the triplet potential. Hence, the singlet forms the energetically closed channel. Because of the difference in magnetic moment between the singlet and triplet states, the energetic difference between their scattering thresholds may be tuned with a magnetic field. Denoting the detuning ν between a bound state of the closed channel and the scattering threshold of the open channel in terms of the difference in magnetic moment $\delta\mu \equiv \mu_{\text{closed}} - \mu_{\text{open}}$, the magnetic field strength B, and the critical magnetic field strength B_c where the detuning vanishes,

$$\nu\,(B) = \delta\mu\,(B - B_c)\,, \tag{1.5}$$

we may parameterize the dependence of the s-wave scattering length on B as [38]

$$a_s\,(B) = a_{\text{bg}}\left(1 - \frac{\Delta}{B - B_0}\right). \tag{1.6}$$

Here, $\Delta = \Gamma_0/\delta\mu$ with Γ_0 defining the strength of the resonance, $B_0 = B_c + \delta B$ where $\delta B = \delta E/\delta\mu$ is an interchannel interaction-induced shift, a_{bg} is the background scattering length the open channel would have in the absence of the closed channel bound state, and we have neglected any inelastic processes.

A common parameterization of the interactions in dilute atomic gases is provided by the regularized s-wave pseudopotential [51, 52]

$$U\,(\mathbf{r}) = \frac{2\pi\hbar^2 a_s}{\mu}\delta\,(\mathbf{r})\,\partial_r r. \tag{1.7}$$

This pseudopotential is exact in the low-energy limit $E \to 0$, and provides an excellent approximation provided that $k\,|a_s| \ll 1$ and $kr_B \ll 1$ [38]. Hence, the controllability of the scattering length translates quite readily into controllability of the two-body interactions of the gas. We will return to this parameterization in Chap. 2, when we discuss Hubbard models.

[5] Note that the effective range given in the context of scattering of slow particles, $r_\star$, is related to the effective range defined here as $r_\star = -2r_B$.

[6] The spin states of real atoms are never purely singlet or triplet, but rather singlet-dominated or triplet-dominated.

1.2 Production of Ultracold Molecules

Molecules have a complex internal structure of vibrational and rotational energy levels which has no counterpart in alkali atoms.[7] Transitions between the internal states of a molecule are not typically governed by strict selection rules as in atoms, but rather by the square modulus of the wave function overlap between the states in question. These overlaps are known as *Franck–Condon factors*. This complicated internal structure makes direct laser cooling of a molecule challenging, as it is very difficult to devise a closed pumping cycle. Several laser cooling schemes have been proposed for specific molecules [53–56], and enhancement of laser cooling has been predicted when molecules are placed within an optical resonator cavity [57–59], with laser-cooled ions [60], or with atoms in Rydberg states [61, 62]. Experimental progress in direct laser cooling of molecules has been made in rare cases, most notably Strontium Fluoride [63]. Other *direct* methods of cooling which start from preformed molecules and attempt to extract energy from them include buffer gas cooling [64] and deceleration [65] by means of electric [66–68], magnetic [69, 70], or optical [71–73] fields. In the buffer gas cooling technique, hot molecules of the desired species are loaded into a chamber containing a cryogenic noble gas, most commonly Helium or Neon, and allowed to equilibrate. The latter technique amounts to essentially running a particle accelerator in reverse. That is, translational energy is extracted by imposing a field gradient on the molecule. The advantage of direct means such as buffer gas cooling or deceleration is that they are applicable to a wide array of species. However, direct methods to date have only produced molecules in the cold regime of temperatures $T \sim 10\,\mathrm{mK}$–$1\,\mathrm{K}$ which does not allow access to the fully quantum degenerate regime.

The most successful methods for producing ultracold, high-density samples of molecules have been *indirect* methods which form molecules from atoms which have themselves been cooled to ultracold temperatures. Because of the large effort in cooling alkali atoms [28–30], most of the molecules formed in this fashion are bialkali molecules. Early attempts at the indirect production of molecules focused on *photo-associating* two atoms scattering in S states into a bound state of the $S + P$ excited potential [74, 75]. Molecules can transition to the ground (electronic) state by spontaneous decay [76, 77]. However, the most successful method for creating ultracold molecular samples in very deeply bound levels is by *magneto-association* of ultracold atoms into bound *Feshbach molecules* [38, 78] by sweeping across a Feshbach resonance. Generally, the magneto-association process creates molecules which are weakly bound but translationally ultracold, and the process has been optimized so as to have nearly unit efficiency. Magneto-association has been achieved for heteronuclear species [79–86] as well as two-component fermionic species [87–93] and single-component bosons [94, 95]. In the latter two cases, these molecules have been observed to condense and form a molecular BEC. This

[7] An overview of molecular degrees of freedom is provided in Sect. 1.3.

tunability enables the study [91–93, 96–98] of the crossover from a BEC of diatomic molecules to a Cooper paired dilute Fermi gas, a phenomenon known as the *BEC-BCS crossover* [99–103]. Part III of this thesis focuses on solving the crossover problem in an optical lattice at the two-particle level using numerically exact techniques, and then using this numerical solution to derive an effective many-body model for the crossover in the lattice. Although the continuum limit of the BEC-BCS crossover problem has been solved, the lattice problem has been the subject of considerable debate. The low-energy effective theory describing a two-channel model in the lattice takes the form of a multi-channel model between unpaired atoms in the lowest band and a set of *dressed molecules* which form an effective closed channel. However, as with the relevant degrees of freedom of the other strongly correlated systems mentioned in the introductory paragraphs, the effective closed channel bears little resemblance to the microscopic closed channel, instead consisting of an infinite summation over Bloch bands from both the open and closed microscopic channels.

While the molecules which are produced through Feshbach association are translationally ultracold, they are often very highly internally excited. For ultracold atoms, the presence of excitations in the internal degrees of freedom does not cause concern, as these degrees of freedom are usually hyperfine states which can be manipulated via optical pumping. For molecules, the presence of a rich internal structure of rotational and vibrational levels, together with the absence of strict selection rules for transitions, makes the isolation of a single molecular state a more daunting task. Remarkable progress has been made in transferring a collection of Feshbach molecules to a low-lying internal state via the stimulated Raman adiabatic passage (STIRAP) procedure [82, 86, 95, 104–107]. In this procedure, Feshbach molecules are transferred coherently to a much more deeply bound state via a two-photon process chosen such that the Franck–Condon factors of the intermediate state with both the target deeply bound state and the Feshbach molecular state are large. This process has been demonstrated to have 90 % one-way efficiency in the case of KRb [106]. Finally, by using the mixing of hyperfine and rotational states induced by a nuclear quadrupole coupling, one can transfer the molecules into the lowest hyperfine state at a given magnetic field, resulting in absolute ground state molecules [108].

1.3 Classifications and Few-Body Physics of Ultracold Molecules

STIRAP has been the most successful method for producing ultracold, high phase-space density samples of molecules. Because STIRAP requires that the constituent atoms are already ultracold before they are assembled into molecules, ultracold molecules to date have been mostly alkali metal dimers [82, 85, 86, 109, 110], with mixed alkali metal-alkaline earth molecules on the horizon. The natural energy scales governing the various degrees of freedom of an alkali metal dimer molecule

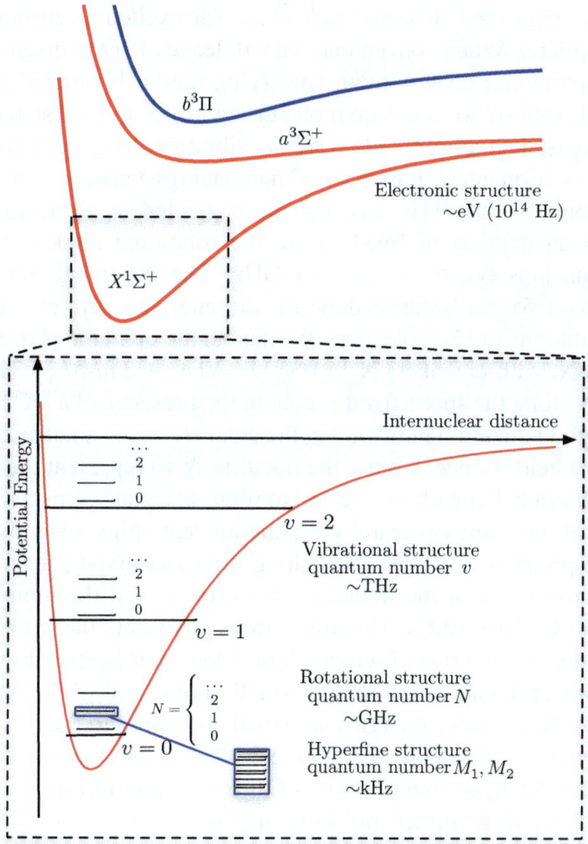

Fig. 1.1 Intrinsic energy scales of alkali metal dimers. In descending order, the energy scales of a ${}^1\Sigma$ diatomic molecule are excitations of the electronic state, vibrational modes of the nuclei about the equilibrium internuclear separation, rotation of the molecule about its center of mass, and couplings between the nuclear spins and the other angular momenta of the molecule. These energy scales span nearly twelve orders of magnitude

span nearly twelve orders of magnitude, and can be classified as in Fig. 1.1. The largest energy scales are set by the electronic degrees of freedom and are of order a few electron-volts. The electronic degrees of freedom are involved in the coupling of the molecule to the optical lattice via the polarizability tensor [111], but these fields are far detuned from any resonances so that the molecule remains in its electronic ground state. The molecular term symbol denoting the electronic state is of the form $Q^{2S+1}\Lambda^{\pm}$, where S is the total electronic spin, Λ is the absolute value of the projection of the total orbital angular momentum along the internuclear axis, and $\pm$ denotes the parity under reflections in a plane containing the internuclear axis [50]. The term $Q = X$ for the ground electronic potential and then $A, B, C, \ldots$ in energetically ascending order for excited electronic potentials of the same S, Λ

multiplicity as the ground state and $a, b, c, \ldots$ for excited electronic potentials of different multiplicity. When convenient, we will leave off Q to discuss the multiplet structure of a particular level without specifying how it is ordered with respect to the others. In this thesis we focus on molecules with $^1\Sigma$ ground states.

The next largest degrees of freedom are the vibrational degrees of freedom, which are classified by a quantum number v. The spacings between vibrational energy levels are of order a few THz, and can be neglected at ultracold temperatures. The next relevant degrees of freedom are the rotational modes of the molecule which have spacings on the order of a GHz. The rotational modes will be of particular interest for us because they are the energetically lowest-lying dipole-accessible excitations of $^1\Sigma$ molecules. We denote the operator of rotational angular momentum as $\hat{\mathbf{N}}$, with eigenkets $\hat{\mathbf{N}}^2|NM_N\rangle = N(N+1)|NM_N\rangle$. Here M_N is the projection of $\mathbf{N}$ along the space-fixed z-axis. In the presence of a DC electric field N is no longer a good quantum number, but the eigenstates are adiabatically connected to states in zero field. Hence, we use the notation $\bar{N}$ to represent the eigenstate in a DC field which is adiabatically connected to rotational state N in zero field. Finally, at the bottom of the energy hierarchy are the nuclear spins given by the operators $\hat{\mathbf{I}}_1$ and $\hat{\mathbf{I}}_2$ with projections M_1 and M_2 along the space-fixed z axis. Here, 1 and 2 refer to the constituents of the molecule in the order that the molecule is named, e.g. Rb is 1 and Cs 2 for RbCs. The interactions governing the nuclear spins are of order 100 Hz–1 kHz. In terms of temperature, 1 kHz corresponds to roughly 50 nK, and so hyperfine structure is typically thermally populated while the other internal structure is not. While these energies are small compared to the other scales of the problem, the precise, state-selective nature of the STIRAP process requires us to take into account the hyperfine structure of the molecules [108].

Explicitly, a $^1\Sigma$ alkali dimer molecule may be characterized by the parameters displayed in Table 1.1 [108, 111–113]. The numerical values of these parameters for the three most experimentally relevant species at the time of the writing of this thesis [82, 109, 110] are also collected in Table 1.1.[8] The physical origins of these terms and their role in the microscopic Hamiltonian are given in Chaps. 3 and 4.

What are the new features of ultracold molecules relevant to many-body physics? The dipole–dipole interaction

$$\hat{H}_{\mathrm{DD}}(\mathbf{R}) = \frac{\hat{\mathbf{d}}\cdot\hat{\mathbf{d}} - 3\left(\hat{\mathbf{d}}\cdot\mathbf{e}_R\right)\left(\mathbf{e}_R\cdot\hat{\mathbf{d}}\right)}{R^3}, \tag{1.8}$$

where $\mathbf{R}$ is the vector connecting the two dipoles, is the source of much interest in polar molecules as it provides interactions which are both long range and anisotropic. The anisotropy has been observed in studying the stereodynamics of ultracold collisions [114, 115]. In addition, because of the anisotropy, angular momentum is not conserved during low-energy scattering. This adds a short-range

[8]The dynamic polarizability of LiCs has not yet been calculated, to our knowledge. We estimate that the ratio of the perpendicular and parallel polarizabilities will be similar to that of KRb and RbCs.

Table 1.1 Coupling constants defining a $^1\Sigma$ molecular species

Description	Symbol	^{87}Rb^{133}Cs	^{40}K^{87}Rb	^{7}Li^{133}Cs
Rotational constant	B_N (GHz)	0.488	1.1139	5.636
Permanent dipole moment	d (Debye)	1.25	0.566	5.52
Nuclear spin 1	I_1	3/2	4	3/2
Nuclear spin 2	I_2	7/2	3/2	7/2
Nuclear g-factor 1	g_1	1.834	−0.324	2.171
Nuclear g-factor 2	g_2	0.738	1.834	0.738
Nuclear quadrupole coupling 1	$(eqQ)_1$ (MHz)	−0.872	0.45	0.0185
Nuclear quadrupole coupling 2	$(eqQ)_2$ (MHz)	0.051	−1.41	0.188
Rotation-nuclear spin coupling 1	c_1 (Hz)	98.4	−24.1	32
Rotation-nuclear spin coupling 2	c_2 (Hz)	194.1	420.1	3014
Tensor nuclear spin–spin coupling	c_3 (Hz)	192.4	−48.2	140
Scalar nuclear spin–spin coupling	c_4 (Hz)	17345.4	−2030.4	1610
Rotational g-factor	g_r	0.0062	0.014	0.0106
Nuclear shielding factor 1	σ_1 (ppm)	3531	1321	108.2
Nuclear shielding factor 2	σ_2 (ppm)	6367	3469	6242.5
Parallel polarizability	$\alpha_\parallel$ (au)	3033.97	2116.77	?
Perpendicular Polarizability	$\alpha_\perp$ (au)	675.966	471.61	$\sim 0.2\alpha_\parallel$

Most of the values in this table have been obtained by DFT or other ab-initio theoretical means [111–113] and have not yet been verified by experiment. The exceptions are the rotational constant, dipole moment, and nuclear quadrupole couplings for KRb [108]

contribution in the s-wave channel [116–118] which gives rise to a weak shape resonance in all even-ℓ channels. The given form of the dipole–dipole potential breaks down at short distances where dispersion and chemical effects become relevant. We account for this by imposing a short distance cutoff b of order r_{vdW}, the van der Waals length, on the dipole–dipole potential and account for the short-distance behavior by adding a contribution [119, 120]

$$\frac{4\pi\hbar^2 a\left(d\right)}{m}\delta\left(\mathbf{r}\right) \tag{1.9}$$

to the short-range pseudopotential. Here $a\left(d\right)$ is the dipole-dependent scattering length which is provided by the low-energy limit of the scattering matrix. Such a pseudopotential has been shown [119] to reproduce the correct physics away from any shape resonances.

In addition to the dipole–dipole interaction, molecules have a rich internal state space which can be tunably accessed using AC microwave fields. Thus, in addition to the ability to fine-tune fields to produce spin models for quantum simulation [121, 122], one can also access regimes where a large number of internal degrees of freedom are interacting over a disparate range of timescales and thus build a simulator of a quantum complex system [123]. The means by which one can tune the internal state space using external fields will be covered in Chaps. 3 and 4.

Molecules also display complex short-range physics due to both the large number of internal states and large energy scales at short-range [124] and the possibility of chemical reactions [125, 126]. In order to discuss short range physics, we first clarify the scales which classify processes as being either short or long range. We define the dipole length $r_d = \left(md^2/\hbar^2\right)$ as the separation where the dipolar energy of two-particles is comparable to the relative collision energy of two-particles with a wavelength of r_d. We also define the van der Waals length $r_{\mathrm{vdW}} = \left[2\pi/\Gamma\left(1/4\right)^2\right]\left(2\mu C_6/\hbar^2\right)^{1/4}$, where C_6 is the coefficient of the $1/R^6$ dispersion potential with R the intermolecular separation, $\mu = M/2$ the reduced mass of two molecules of mass M, and $\Gamma\left(x\right)$ is the Gamma function, as the separation where the dipolar interaction becomes comparable to the short-range dispersive interaction [38, 127]. The scale at which chemical reactions become relevant is R_e, the bond length, which is smaller than 1 nm for the species considered here [128]. The associated energy scale is on the order of 100 THz, the chemical bond scale. The van der Waals length r_{vdW} is the next relevant length scale, ranging from 6 nm for KRb to 30 and 50 nm for LiCs and RbCs, respectively. The dipole length is very large compared to all of these, and increases with increasing dipole moment. The short-range region where chemical and state-changing collisions become relevant corresponds to $R < r_{\mathrm{vdW}}$.

A complete picture of low-energy scattering for these molecules is provided by multichannel quantum defect theory (MCQDT) [129] which uses the large separation of scales to define dimensionless parameters s and y which characterize the short range phase shift and chemical reactivity [130]. LiCs and KRb are both highly reactive [131]; there is unit probability of loss at short range. Thus, the collision rates, both elastic and inelastic, depend only on the long-range potential and there can be no scattering resonances. Universal formulae exist for such species [131–133] and agree well with experimental measurements of molecular lifetimes [125]. Also, hyperfine spin appears not to be relevant in the scattering of highly reactive molecules [125, 131]. For non-reactive species such as RbCs and NaK, the collisions are non-universal, and may have electric field dependent scattering resonances [134].

Recent work has shown that the short-range physics, both inelastic and elastic, also depends strongly on an applied electric field [114, 133, 135, 136]. In particular, it has been well established that the stability of a molecular ensemble increases with increasing dipole moment in quasi-2D traps [133, 137]. Somewhat surprisingly, the electric field does not appear to be relevant to the stability of a molecular gas in a 3D optical lattice [126]. The detailed description of the short-range scattering is outside the scope of this work. We characterize the short-range physics by a complex scattering length a_s describing both elastic and inelastic scattering, which can be computed using MCQDT. We note that large real or imaginary parts of a_s lead to an effective hard-core condition in an optical lattice where only one molecule can exist per lattice site due to strong resonant interactions or the continuous quantum Zeno effect [138], respectively.

1.4 Simulation Methods

Analytical solutions of the many-particle Schrödinger equation are rare, and so often one resorts to numerical techniques. One of the most successful methods for the quantum many-body problem is DFT, which relies on the fact that the ground state energy is a unique functional of the ground state density [20, 21]. However, this energy functional is only known exactly for a free electron gas, and so approximate functionals are used for interacting systems. The computational procedure of DFT consists of minimizing the energy generated from an approximate functional by changing the density. The density corresponding to the minimum energy is expected to most closely represent the true ground state density. While in principle the many-body wave function of the ground state is itself also a functional of the ground state density, this functional is not known, and so the predictive power of DFT is limited to observables which depend only on the density. Hence, DFT is not generically useful for detecting order in strongly correlated systems, which generally requires two-point correlation functions such as the density–density correlation function $\langle \hat{n}_i \hat{n}_j \rangle$. Also, approximate functionals are only available for Coulomb-interacting electronic systems, and so DFT is not an appropriate starting point for systems which interact through different potentials such as molecules interacting through the dipole–dipole potential. Finally, we note that no improvements made to DFT can turn it into a "black box" method applicable to any interacting electronic system, even in principle, as the existence of an efficient approximation to the universal functionals relevant to electronic systems would imply that the hardest problems for quantum computers to solve would be efficiently solvable by classical computers [139].[9] This is thought to be impossible, although no formal proof has yet been provided.

The simplest method to solve the many-body Schrödinger equation is to form a matrix representation of the Hamiltonian operator and numerically diagonalize it. This method is referred to as *exact diagonalization*. The fact that the Hilbert space of a many-body system grows exponentially with the number of constituents in the system implies that this procedure is only practical for systems with very few constituents. Some improvement can be made by considering *sparse eigensolver methods* such as the Lanczos [140] or Davidson [141] algorithms which use

[9]Explicitly, it would imply that a *Quantum Merlin–Arthur* (QMA)-complete problem lies in P, where P is the class of problems which can be solved in polynomial time by a deterministic Turing machine. QMA is the quantum analog to the classical complexity class NP.

only a procedure applying the Hamiltonian to a given vector to find solutions corresponding to extremal eigenvalues. By carefully accounting for all symmetries of the system, for example point group symmetries of the lattice, U(1) symmetry corresponding to conservation of number of particles, or SU(2) spin symmetry, the nonzero values of the Hamiltonian can be codified and calculated on-the-fly to optimize performance. Even with all of these optimizations, the cutting edge of modern exact diagonalization is limited to roughly 40 two-component spins, or qubits, or a half-filled electronic system on a square lattice with 20 sites.

Another popular class of methods for the many-body problem are *Quantum Monte Carlo* (QMC) methods. Path-integral QMC methods use the *worldline* mapping from a d dimensional quantum system to a $d+1$ dimensional classical system [4, 142] and then use the classical Metropolis algorithm [143] to generate the equilibrium expectations of observables by sampling worldline configurations. The most common path integral QMC methods for strongly correlated systems are loop QMC [144] and the worm algorithm [145]. Modern implementations of these algorithms can simulate millions of particles at low temperatures, even when strongly correlated. However, path integral QMC suffers from the *sign problem* for interacting fermionic systems in dimensions greater than one or *frustrated* systems which have an extensive classical ground state degeneracy. In these systems, there exist configuration updates of the worldlines which amount to negative probabilities in the classical Monte Carlo scheme, and so Monte Carlo can no longer be applied. If the sign is ignored in the update procedure, an exponentially growing cancellation in the sign expectation leads to an exponential growth of errors rather than a statistically limited behavior of errors as in convergent Monte Carlo [146]. A similar *phase problem* exists when applying QMC to the unitary dynamics of a many-body system. There also exist variational QMC methods such as diffusion Monte Carlo [147] or variational Monte Carlo [148] in which a trial wavefunction is optimized using a Monte Carlo procedure. These algorithms suffer from bias in the choice of the variational ansatz, and also are not generally applicable to dynamics. Diagrammatic Monte Carlo [149] samples terms[10] in a series rather than configuration updates. If the series is a perturbation series, the associated diagrams are Feynman diagrams. Bold-line diagrammatic Monte Carlo [150–152] uses re-summation techniques [153] over subsets of diagrams to improve the efficiency, and do not suffer from sign problem in some cases. Such methods may also be applicable to real-time dynamics [154].

A more recent method for strongly correlated systems is *Dynamical mean-field theory* (DMFT) [155–159], in which the full many-body problem is mapped onto an effective impurity problem for one of the constituents with a self-consistently defined path describing the coupling of this impurity to the surrounding system. The sign problem in the associated impurity problem can be better controlled due to the small size of the impurity, and the mean-field approximation can be assuaged

[10]Series expansions often employ a diagrammatic notation for the terms in the series, hence the name diagrammatic Monte Carlo.

by considering clusters of sites rather than single-site impurities and extrapolating in the size of the cluster. DMFT is most useful in dimensions $D > 2$, and becomes exact in the limit that $D \to \infty$. DMFT requires the analytical mapping from the original many-body problem to the impurity problem, and so has only been formulated for specific interaction potentials such as the those appearing in the Bose–Hubbard [160] and Fermi–Hubbard [156] models or for Bose–Fermi mixtures [159]. Hence, at present, generic applications in DMFT require new development on a case-by-case basis. Both DMFT and most flavors of QMC have the advantage that they incorporate finite temperature naturally.

The simulation methods that we apply for many-body studies in this thesis are variational methods based on MPSs, which are covered in detail in Part IV of this thesis. MPSs are a class of states which are generated through the process of a real-space renormalization group procedure known as the DMRG. The main convergence parameter χ used in an MPS simulation is a cutoff in the entanglement between any two complementary subsections. One of the greatest advantages of MPS methods is that, like exact diagonalization, MPS methods produce wave-functions, and so a vast array of properties may be computed. Furthermore, MPS methods can be formulated for any microscopic degrees of freedom and for any interactions, making them suitable for generic implementation. Finally, one can devise efficient variational MPS algorithms for the dynamics of systems. The flexibility of MPS algorithms to adapt to different degrees of freedom, a variety of interactions, and simulation of time evolution makes them best suited for our purposes. The main drawback of MPS methods is that they work best in 1D and at zero temperature. Higher dimensional generalizations of MPS algorithms exist, but their numerical conditioning is much worse than that of MPS algorithms at present. Additionally, generalizations to finite temperature exist, but they also scale worse than their zero-temperature counterparts [161–164].

With MPS methods having been used successfully for nearly 20 years, why do molecules necessitate new code and algorithm development? The first reason is that Hubbard models for dipolar molecules have long-range interactions as discussed in Chap. 2. While DMRG-type methods have been devised for long-range interacting systems [163, 165], these methods are often Hamiltonian-specialized and inefficient. A relatively new characterization of Hamiltonians based on *matrix product operators* [166–168] (MPOs) enables for long-range interactions to be accommodated efficiently. Furthermore, as discussed in Chap. 7, by using a small set of finite state automaton rules for the construction of Hamiltonian terms from local operators, a wide array of Hamiltonians can be provided as input in a consistent form to an MPS program. This completely eradicates the need for Hamiltonian-specialized implementation. The need for flexibility in an MPS program is vital for studying molecules; as more complex molecules become cooled to quantum degeneracy, a wider array of more complex many-body models are expected. Hence, flexible code prevents the need to "reinvent the wheel" as new many-body models become relevant. The MPS algorithms presented in this thesis make no reference to the microscopic constituents of the model under study or to the range or nature of their interactions. Rather, the only requirements are that the system be quasi one-dimensional and the Hamiltonian be expressed as an MPO.

Long-range interactions allow for more complex translation-breaking orders than finite-range interactions. An example of this is the "devil's staircase" of insulating phases at rational fillings of bosons interacting with $1/r^3$ interactions [169, 170]. The insulating states can be true long-range ordered crystals in 1D as opposed to, e.g., superfluids in 1D which are prevented from long range ordering by the Mermin–Wagner theorem [171, 172]. Broken translational order causes special difficulties for MPS algorithms, as these algorithms are stated most naturally for open boundary conditions. For a system with true long-range translational order, the presence of open boundaries strongly affects the bulk behavior of the system even several hundreds of sites away from the boundaries. To avoid this difficulty, it is necessary to work directly in the limit of an infinite lattice, in which we assume that the many-body state has a periodically repeating unit cell. An algorithm for variationally finding the unit cell of the ground state of an infinite system is provided in Chap. 8.

Finally, even for the simplest diatomic molecules with $^1\Sigma$ ground states the internal state space which is accessible can be significant due to a large hyperfine manifold. A powerful way of breaking a large Hilbert space into its smaller relevant components is through the explicit conservation of symmetries. In Chap. 6 the structure of symmetry-adapted MPSs is elucidated for the simplest case of Abelian symmetries, in which all of the irreducible representations are one-dimensional. Remarkably, the conservation of Abelian symmetries can be implemented in a completely generic way, and an arbitrary number of such symmetries may be simultaneously conserved. The only place in which the particular symmetry group is relevant is in determining how two quantum numbers transform under the group operation. Implementation details are provided in Chap. 6.

1.5 Related Work by Other Groups

The strongest connection of our work on polar molecules to other groups is the dipole–dipole interaction. The most detailed many-body studies of dipole–dipole interactions in ultracold atomic systems have focused on strongly magnetic atoms such as Chromium, Erbium, Europium, and Dysprosium. Chromium was the first of these atoms to be Bose-condensed [173, 174], with Dysprosium [175] and Erbium [176] only having been Bose-condensed very recently. Fermionic isotopes of Dysprosium have also been brought to quantum degeneracy [177]. The dipolar interaction of Chromium is about one-sixth its short-range interaction, and so the dipolar effects are generally perturbative in nature. Still, clear signatures of the dipole–dipole interaction have been observed, such as d-wave expansion following the collapse of a Chromium BEC [178]. For broad reviews on the physics of dipolar gases, we refer the reader to [179, 180].

More closely related to our work are studies of dipolar particles in harmonic traps [119, 120, 181–186] and lattices [170, 187–193]. Most of the work done in harmonic traps is based on the mean-field Gross–Pitaevskii formalism which assumes the presence of a condensate. In the lattice commensurability effects can drive the system out of a condensed phase and into an insulating phase with a periodicity commensurate with the lattice. Thus, a variety of techniques have been brought to bear on the lattice problem. The quantum phases of bosonic dipoles in optical lattices have been investigated for various 2D geometries, including on the square lattice using mean field techniques [189, 191] and via QMC on the square [170] and triangular lattices [192]. A wide variety of quantum phases are possible including checkerboard solid and supersolid phases in addition to the superfluid and Mott insulator phases present for short-range interacting bosons. Supersolids are characterized by coexisting translational and phase order, and have been the subject of intense study [194]. Studies beyond the ground state properties include the characterization and stability properties of metastable states [187, 188]. The presence of a lattice has also been proposed to enhance the production of molecules through indirect means [190]. The scattering properties of a dipolar gas strongly confined in two dimensions but free to move along the third have been studied [183], paralleling investigations of confinement-induced resonances appearing in the short-range case [195]. Finally, the parameters of a Luttinger liquid theory have been postulated for general power-law interactions in 1D [169], including an analysis of the Berezinskii–Kosterlitz–Thouless transition to pinned phases in the presence of a weak lattice.

The dipole–dipole interaction is not the only interesting feature of polar molecules. The presence of a permanent electric dipole moment allows for transitions between rotational states in an AC microwave field. An early suggestion based on this observation was provided by Demille [196] who proposed that polar molecules in a one-dimensional trap could be used for quantum computation. Many other groups have proposed using the internal structure to build quantum simulators of spin models [197, 198] or to study other condensed matter phenomena such as the Holstein model [199], exciton physics [200, 201], the physics of liquid crystals [202, 203], and string orders related to the Haldane phase of the antiferromagnetic spin-1 Heisenberg model [204, 205]. The internal structure can also be tuned through external fields in order to produce a desired interaction potential for a specific molecular state in a dressed-state picture [206–208]. More exotic proposals involve loading atoms into self-assembled lattices of dipolar molecules [209, 210] or tuning three-body interactions via external fields [211, 212]. A great deal of recent interest has been garnered for dipolar molecules in bilayer geometries where the dipolar interaction within a layer is repulsive, stabilizing the gases within the layers, but the interaction between molecules in different layers is attractive. This can induce exotic interlayer pairing [213, 214] and induce soliton filaments in a stack of such layers [215].

The most closely related works to our own are those which simulate quantum magnetism, particularly a long-range generalization of the t-J model, using the internal structure of polar molecules [121, 122, 216]. These works have an approach similar to ours, in which the properties of the many-body models are related to those of the few-body physics through microscopic analysis. The main difference between these works and ours is motivation. The t-J work specifically fine-tunes the external fields in order to achieve a quantum simulator of a known model. Our work, on the other hand, begins from a near-term experimental setup and asks what the naturally arising many-body Hamiltonian is and what its many-body features are.

There is also a body of work by other authors focusing on the problem addressed by the Fermi resonance Hamiltonian (FRH) which is the topic of Part III. The FRH maps a resonance model in the continuum onto a resonance model in the lattice by identifying a set of dressed molecules which form the effective closed channel in the lattice. To our knowledge the first time that an effective closed channel was used for a many-body model in a cold atoms context was in [217], although such terms had been used phenomenologically in the study of high-T_c superconductivity for many years prior [218–220]. The analogous high-T_c model, known as the "cooperon model," is still a subject of current research [221].

A proper description of the physics of the two-channel model requires a renormalization of the theory to remove divergences associated with a point-like boson [222, 223]. The exact solution for two-particles interacting via a Feshbach resonance in a (possibly anisotropic) harmonic trap was obtained in [224]. The importance of intra- as well as inter-band coupling terms was stressed, and the theory was properly renormalized to remove divergences from using a point-like boson. The authors of [225] consider a lattice two-channel model, and determine the properties of the dressed molecule by considering deep lattices and replacing the lattice with a single harmonic well. The harmonic trap approximation both artificially leads to separability of the center of mass motion from both the relative motion and internal structure and underestimates the extent of Wannier functions, often by an order of magnitude. This implies that qualitative properties of the tunneling, as well as its general order of magnitude, cannot be accounted for using this approach. Duan has derived effective two-channel models using both a projection operator formalism [226] and general symmetry considerations [227]. He considers the dressed molecule to be the exact solution of an on-site Schrödinger equation, and then couples in many-body physics using atom-molecule couplings between neighboring sites. He then considers the case where one of the on-site eigenstates is close to the scattering continuum of two-particles in a specific band n, and then projects the Hamiltonian onto the Hilbert space of empty sites, singly occupied sites with a fermion in band n, and doubly occupied sites containing a dressed molecule. Such a Hamiltonian cannot describe the full BEC-BCS crossover, as it restricts the dressed molecules to behave as hard-core bosons whereas deep in the BEC side the molecules are tightly bound, weakly interacting bosons. This work also does not give a prescription for solving the on-site problem, but references the exact solution in the harmonic trap. More recent work [228] uses the numerical solution from a double-well potential to avoid some of the shortcomings of the harmonic oscillator approximation. While this work captures some of the physics of the lattice at the nearest-neighbor level, it does not capture the full quasimomentum dependence of the lattice Hamiltonian.

Büchler was the first to give the exact solution for two fermions interacting through a zero-range Feshbach resonance in an optical lattice, properly accounting for the effects of higher bands and renormalization [229]. He then showed that when the interaction term U of the single-band Hubbard model was determined from the scattering properties of this exact two-body solution self-consistently, the Hubbard model still failed to reproduce the correct physics even for moderate s-wave scattering length, or very far from the actual pole of the Feshbach resonance. This has motivated our approach of using a two-channel model instead of a single-channel model such as the Hubbard model. His discussion of the two-body solution focused on states with zero total quasimomentum, although the theory encompasses states with arbitrary total quasimomentum.

Very recent work by von Stecher et al. [230] focuses on an effective two-channel model near a lattice resonance. Instead of solving the on-site problem exactly they project the two-body Hamiltonian outside of the scattering continuum of two fermions in bands n and m which gives rise to the resonance, solve this projected Hamiltonian exactly, and then use the eigenstates of this projected Hamiltonian as a dressed closed channel. This is approach is very similar in spirit to ours. In contrast to our work, low dimensionality is assumed from the outset and so this approach breaks down when the energy associated with the Feshbach coupling becomes larger than the energy associated with the transverse confinement. Instead, our model treats the population in transverse excited state as being fixed by the two-body solution and thus part of the dressed molecule. This allows any imposed conditions of reduced dimensionality to be controlled only by the transverse tunneling and coupling rates of the dressed molecules.

Finally, with regard to the open source work, an open source version of DMRG exists as a part of the algorithms and libraries for physics simulations project (ALPS) [231–233]. Open source versions of 1D time dependent DMRG also exist [234, 235]. To the best of our knowledge, the two open source coding projects described in this thesis, open source time-evolving block decimation [236] and the time-evolving block decimation routines included as part of the ALPS package, are the only open source codes which work directly on MPSs at the time of writing of this thesis.

1.6 Outline

The present thesis is divided broadly into six parts. Each chapter within a part represents either a publication, in which case the chapter begins with an abstract, or material which places the publications in context and provide background information. Materials relevant to the open source coding projects or educational development discussed in Part V are placed in appendices to avoid interrupting the flow of the thesis. Part I is an introduction to the topics of this thesis, and includes the present chapter. This part continues in Chap. 2, which discusses the general route from a few-body Hamiltonian to a many-body Hamiltonian appropriate for describing strongly interacting particles in a lattice.

Part II of this thesis is devoted to the *molecular Hubbard Hamiltonian*, a model for the low-energy physics of $^1\Sigma$ alkali dimer molecules loaded into an optical lattice. Chapter 3 describes the first attempt at deriving such a Hamiltonian by considering only the rotational degrees of freedom. Special focus was put on describing the dynamics of molecules following a sudden turning on of a microwave AC field coupling the lowest rotational level to higher rotational levels. Using time-dependent MPS methods, it was found that coherent Rabi oscillations between two internal states driven at a single-molecule resonance were exponentially damped with an emergent timescale. This effect was termed *quantum dephasing*. As experiments began to produce ultracold molecules in the absolute vibrational, rotational, and hyperfine ground state, this model was revised to include the hyperfine structure of the molecules. We refer to the resulting Hamiltonian as the *hyperfine* molecular Hubbard Hamiltonian, which is discussed in Chap. 4. Chapter 4 studies the strong static electric and magnetic field limit of this Hamiltonian.

In Part III of this thesis we study the problem of the pairing of two-component fermionic atoms via a Feshbach resonance in an optical lattice. As the scattering length for two-particles in the open channel diverges near resonance, the pseudopotential model for interactions, amounting to integrating out the closed channel, breaks down. Furthermore, as the strength of the pairing interaction g becomes stronger than the band gap, the restriction of the open channel to the lowest band is no longer appropriate at short distances. Hence, the Hubbard model [237] involving tunneling of two-component fermions in the lowest band with on-site s-wave intercomponent interactions fails to correctly describe the system.

In our approach, we project the complete two-channel model into the basis of Bloch functions appropriate to the lattice problem. A scaling analysis enables us to extract the bound state properties of two-particles to the limit of an infinite number of bands. By carefully partitioning the Hilbert space into low-energy and high-energy sectors and performing this numerically exact two-particle analysis on the high-energy sector, we derive the FRH, a lattice model which is applicable to resonances of any width and any scattering length. The FRH takes the form of a multichannel resonance model between fermions in the lowest band of the open channel and a closed channel consisting of "dressed molecules." The use of the full lattice solution, the extrapolation to an infinite number of bands, and the proper regularization of the Feshbach coupling are all vital to the proper quantitative and qualitative description of the physics.

Part IV of the present thesis is devoted to MPSs and variational algorithms associated with them. Chapter 6 gives an overview of MPSs. While this chapter is largely conceptual, it also contains details on symmetry adapted MPSs. Chapter 7 provides a thorough overview of algorithms using MPSs as variational ansätze for the eigenstates of arbitrary 1D Hamiltonians on finite lattices. Furthermore, it presents algorithms for simulating the dynamics of arbitrary time-dependent Hamiltonians whose error bounds do not depend on the smoothness properties of the Hamiltonian in question and do not require any Hamiltonian-specialized implementation. Chapter 8 discusses a variational algorithm for finding the ground state of an infinite system variationally using a translationally invariant MPS ansatz. Chapter 9 discusses how algorithms for time evolution of MPSs may be used to simulate MPSs as finite temperature, either through entangling MPSs to a fictitious reservoir or through sampling the characteristic states which arise at finite temperature. This chapter also provides information on how to formulate MPS algorithms for periodic boundary conditions which correctly describe the entanglement structure of periodic states.

In Part V of the thesis we discuss open source coding projects and educational materials. The open source coding projects include open source time-evolving block decimation [236], a stand-alone implementation of time-evolving block decimation, as well as an implementation of time-evolving block decimation as part of the ALPS open source package [231–233]. The educational materials include tutorials on the open source codes which are aimed at the level of graduate students performing research in strongly correlated physics and documents intended for use within the Carr theoretical physics research group for students who may have very little background in quantum mechanics or numerical methods. The educational materials are reprinted in the appendices.

Finally, the thesis concludes in Chap. 12 with suggestions for future work. This chapter together with the appendices forms Part VI of the thesis. The MPS codes which are part of this thesis are all freely available. Open source TEBD and open source MPS can be obtained from SourceForge repositories [238, 239], the ALPS code can be obtained from the ALPS website [231], and the MPS code used in the introduction to MPS algorithms (Appendix C) is reprinted in this thesis.

The publications included in this thesis are

1. Chapter 3. *Emergent Timescales in Entangled Quantum Dynamics of Ultracold Molecules in Optical Lattices*, M.L. Wall and L.D. Carr, New J. Phys. **11** 055027 (2009), doi:10.1088/1367-2630/11/5/055027. Permission is provided by the Creative Commons Attribution-Non-Commercial-ShareAlike 3.0 license according to the New Journal of Physics copyright statement.
2. Chapter 4. *Hyperfine molecular Hubbard Hamiltonian*, M.L. Wall and L.D. Carr, Physical Review A **82**, 013611 (2010), doi:10.1103/PhysRevA.82.013611. Copyright (2010) by the American Physical Society. Permission is provided by the American Physical Society according to the APS copyright policy.
3. Chapter 5. *Microscopic Model for Feshbach Interacting Fermions in an Optical Lattice with Arbitrary Scattering Length and Resonance Width*, M.L. Wall and L.D. Carr, Phys. Rev. Lett. **109**, 055302 (2012), doi:10.1103/PhysRevLett.109.055302. Copyright (2012) by the American Physical Society. Permission is provided by the American Physical Society according to the APS copyright policy.
4. Chapter 7. *Out of equilibrium dynamics with MPSs*, M.L. Wall and L.D. Carr, New J. Phys. **14**, 125015 (2012). Permission is provided by the Creative Commons Attribution-Non-Commercial-ShareAlike 3.0 license according to the New Journal of Physics copyright statement.
5. Chapter 9. *Finite Temperature MPS Algorithms and Applications*, M.L. Wall and L.D. Carr, Chapter in "Quantum Gases: Finite Temperature and

Non-Equilibrium Dynamics" (Vol. 1 Cold Atoms Series), N.P. Proukakis, S.A. Gardiner, M.J. Davis and M.H. Szymanska, eds. (Imperial College Press, 2012), arXiv:1008.4303v1. Permission is provided by the Creative Commons Attribution-Non-Commercial-ShareAlike 3.0 license according to the arXiv copyright statement.

6. Material appearing in Appendix A first appeared on the wiki pages for the ALPS collaboration hosted at http://alps.comp-phys.org and was written solely by the author of the present thesis. Permission to reprint these materials is provided by the ALPS collaboration.

References

1. Anderson, P.W.: More is different. Science **177**, 393–396 (1972)
2. Anderson, P.W.: Basic Notions of Condensed Matter Physics. Addison-Wesley, New York (1984)
3. Nozières, P., Pines, D.: The Theory of Quantum Liquids, vol. II. Addison-Wesley, New York (1990)
4. Sachdev, S.: Quantum Phase Transitions. Cambridge University Press, New York (1999)
5. Laughlin, R.B.: Anomalous quantum hall effect: An incompressible quantum fluid with fractionally charged excitations. Phys. Rev. Lett. **50**, 1395–1398 (1983). doi:10.1103/PhysRevLett.50.1395. http://link.aps.org/doi/10.1103/PhysRevLett.50.1395
6. Halperin, B.I.: Statistics of quasiparticles and the hierarchy of fractional quantized hall states. Phys. Rev. Lett. **52**, 1583–1586 (1984). doi:10.1103/PhysRevLett.52.1583. http://link.aps.org/doi/10.1103/PhysRevLett.52.1583
7. Arovas, D., Schrieffer, J.R., Wilczek, F.: Fractional statistics and the quantum hall effect. Phys. Rev. Lett. **53**, 722–723 (1984). doi:10.1103/PhysRevLett.53.722. http://link.aps.org/doi/10.1103/PhysRevLett.53.722
8. Wilson, K.G.: The renormalization group: Critical phenomena and the Kondo problem. Rev. Mod. Phys. **47**, 773–840 (1975)
9. White, S.R.: Density matrix formulation for quantum renormalization groups. Phys. Rev. Lett. **69**, 2863–2866 (1992)
10. Östlund, S., Rommer, S.: Thermodynamic limit of density matrix renormalization. Phys. Rev. Lett. **75**(19), 3537–3540 (1995). doi:10.1103/PhysRevLett.75.3537
11. Rommer, S., Östlund, S.: Class of ansatz wave functions for one-dimensional spin systems and their relation to the density matrix renormalization group. Phys. Rev. B **55**(4), 2164–2181 (1997). doi:10.1103/PhysRevB.55.2164
12. Vidal, G.: Efficient classical simulation of slightly entangled quantum computations. Phys. Rev. Lett. **91**(14), 147902 (2003). doi:10.1103/PhysRevLett.91.147902
13. Vidal, G.: Efficient simulation of one-dimensional quantum many-body systems. Phys. Rev. Lett. **93**(4), 040502 (2004). doi:10.1103/PhysRevLett.93.040502
14. Daley, A.J., Kollath, C., Schollwöck, U., Vidal, G.: Time-dependent density-matrix renormalization-group using adaptive effective Hilbert spaces. J. Stat. Mech. **2004**(04), P04005 (2004). http://stacks.iop.org/1742-5468/2004/i=04/a=P04005
15. Verstraete, F., Porras, D., Cirac, J.I.: Density matrix renormalization group and periodic boundary conditions: A quantum information perspective. Phys. Rev. Lett. **93**(22), 227205 (2004). doi:10.1103/PhysRevLett.93.227205
16. Verstraete, F., Cirac, J.I.: Renormalization algorithms for quantum many-body systems in two and higher dimensions (2004) [arXiv:cond-mat/0407066v1]

17. Verstraete, F., Murg, V., Cirac, J.I.: Matrix product states, projected entangled pair states, and variational renormalization group methods for quantum spin systems. Adv. Phys. **57**(2), 143–224 (2008). doi:10.1080/14789940801912366. http://www.tandfonline.com/doi/abs/10.1080/14789940801912366
18. Vidal, G.: Entanglement renormalization. Phys. Rev. Lett. **99**, 220405 (2007)
19. Bardeen, J., Cooper, L.N., Schrieffer, J.R.: Theory of superconductivity. Phys. Rev. **108**, 1175–1204 (1957)
20. Hohenberg, P., Kohn, W.: Inhomogeneous electron gas. Phys. Rev. **136**, B864–B871 (1964)
21. Kohn, W., Sham, L.J.: Self-consistent equations including exchange and correlation effects. Phys. Rev. **140**, A1133–A1138 (1965). doi:10.1103/PhysRev.140.A1133. http://link.aps.org/doi/10.1103/PhysRev.140.A1133
22. SIESTA package for real space density functional theory (2002). http://www.icmab.es/siesta/
23. ABINIT package for plane wave density functional theory (2004). http://www.abinit.org/
24. OCTOPUS package for time-dependent density functional theory (2002). http://www.tddft.org/programs/octopus/wiki/index.php/Main_Page
25. Metcalf, H.J., van der Straten, P.: Laser Cooling. Springer, New York (1999)
26. Ketterle, W., van Druten, N.J.: Evaporative cooling of atoms. Adv. At. Mol. Opt. Phys. **37**, 181 (1996)
27. Pethick, C.J., Smith, H.: Bose-Einstein Condensation in Dilute Gases. Cambridge University Press, Cambridge (2002)
28. Anderson, M.H., Ensher, J.R., Matthews, M.R., Wieman, C.E., Cornell, E.A.: Observation of Bose-Einstein condensation in a dilute atomic vapor. Science **269**, 198–201 (1995)
29. Davis, K.B., Mewes, M.-O., Andrews, M.R., van Druten, N.J., Durfee, D.S., Kurn, D.M., Ketterle, W.: Bose-Einstein condensation in a gas of sodium atoms. Phys. Rev. Lett. **75**, 3969–3972 (1995)
30. Bradley, C.C., Sackett, C.A., Tollett, J.J., Hulet, R.G.: Evidence of Bose-Einstein condensation in an atomic gas with attractive interactions. Phys. Rev. Lett. **75**, 1687 (1995)
31. Sackett, C.A., Stoof, H.T.C., Hulet, R.G.: Growth and collapse of a Bose-Einstein condensate with attractive interactions. Phys. Rev. Lett. **80**, 2031–2034 (1998)
32. DeMarco, B., Jin, D.S.: Onset of Fermi degeneracy in a trapped atomic gas. Science **285**, 1703–1706 (1999)
33. Roati, G., Riboli, F., Modugno, G., Inguscio, M.: Fermi-Bose quantum degenerate 40K-87Rb mixture with attractive interaction. Phys. Rev. Lett. **89**, 150403–1–4 (2002)
34. Hadzibabic, Z., Gupta, S., Stan, C.A., Schunck, C.H., Zwierlein, M.W., Dieckmann, K., Ketterle, W.: Fifty-fold improvement in the number of quantum degenerate fermionic atoms. Phys. Rev. Lett. **91**, 160401–1–4 (2003)
35. Holland, M.J., DeMarco, B., Jin, D.S.: Evaporative cooling of a two-component degenerate Fermi gas. Phys. Rev. A **61**, 53610 (2000)
36. Greiner, M., Mandel, O., Esslinger, T., Hansch, T.W., Bloch, I.: Quantum phase transition from a superfluid to a mott insulator in a gas of ultracold atoms. Nature **415**, 39–44 (2002)
37. Greiner, M., Mandel, O., Hansch, T.W., Bloch, I.: Collapse and revival of the matter wave field of a Bose-Einstein condensate. Nature **419**, 51–54 (2002)
38. Chin, C., Grimm, R., Julienne, P., Tiesinga, E.: Feshbach resonances in ultracold gases. Rev. Mod. Phys. **82**, 1225–1286 (2010). doi:10.1103/RevModPhys.82.1225. http://link.aps.org/doi/10.1103/RevModPhys.82.1225
39. Martin, P.J., Oldaker, B.G., Miklich, A.H., Pritchard, D.E.: Bragg scattering of atoms from a standing light wave. Phys. Rev. Lett. **60**, 515–518 (1988). doi:10.1103/PhysRevLett.60.515. http://link.aps.org/doi/10.1103/PhysRevLett.60.515
40. Blakie, P.B., Clark, C.W.: Wannier states and Bose-Hubbard parameters for 2D optical lattices. J. Phys. B **37**, 1391–1404 (2004)
41. Stöferle, T., Moritz, H., Schori, C., Köhl, M., Esslinger, T.: Transition from a strongly interacting 1D superfluid to a mott insulator. Phys. Rev. Lett. **92**, 130403 (2004). doi:10.1103/PhysRevLett.92.130403. http://link.aps.org/doi/10.1103/PhysRevLett.92.130403

42. Paredes, B., Widera, A., Murg, V., Mandel, O., Folling, S., Cirac, I., Shlyapnikov, G.V., Hansch, T.W.: Tonks-Girardeau gas of ultracold atoms in an optical lattice. Nature **429**, 277–281 (2005)
43. Mandel, O., Greiner, M., Widera, A., Rom, T., Hänsch, T.W., Bloch, I.: Coherent transport of neutral atoms in spin-dependent optical lattice potentials. Phys. Rev. Lett. **91**, 010407 (2003)
44. Lewenstein, M.S.A., Ahufinger, V., Damski, B., Sen De, A., Sen, U.: Ultracold atomic gases in optical lattices: Mimicking condensed matter physics and beyond. Adv. Phys. **56**, 243–379 (2007)
45. Feynman, R.P.: Simulating physics with computers. Int. J. Theor. Phys. **21**, 467 (1982)
46. Jaksch, D., Bruder, C., Cirac, J.I., Gardiner, C.W., Zoller, P.: Cold bosonic atoms in optical lattices. Phys. Rev. Lett. **81**(15), 3108–3111 (1998). doi:10.1103/PhysRevLett.81.3108
47. Fisher, M.P.A., Weichman, P.B., Grinstein, G., Fisher, D.S.: Boson localization and the superfluid-insulator transition. Phys. Rev. B **40**, 546–570 (1989)
48. Feshbach, H.: A unified theory of nuclear reactions, II. Ann. Phys. **19**(2), 287–313 (1962). doi:DOI:10.1016/0003-4916(62)90221-X. http://www.sciencedirect.com/science/article/pii/000349166290221X [ISSN 0003-4916]
49. Petrov, D.S., Salomon, C., Shlyapnikov, G.V.: Diatomic molecules in ultracold Fermi gases-novel composite bosons. J. Phys. B **38**(9), S645 (2005). http://stacks.iop.org/0953-4075/38/i=9/a=014
50. Landau, L.D., Lifshitz, E.M.: Quantum Mechanics (Non-relativistic Theory), vol. 3. Pergamon, Tarrytown (1977)
51. Lee, T.D., Huang, K., Yang, C.N.: Eigenvalues and eigenfunctions of a bose system of hard spheres and its low-temperature properties. Phys. Rev. **106**(6), 1135–1145 (1957). doi:10.1103/PhysRev.106.1135
52. Idziaszek, Z., Calarco, T.: Pseudopotential method for higher partial wave scattering. Phys. Rev. Lett. **96**, 013201 (2006). doi:10.1103/PhysRevLett.96.013201. http://link.aps.org/doi/10.1103/PhysRevLett.96.013201
53. Bahns, J.T., Stwalley, W.C., Gould, P.L.: Laser cooling of molecules: A sequential scheme for rotation, translation, and vibration. J. Chem. Phys. **104**(24), 9689–9697 (1996). doi:10.1063/1.471731 [ISSN 0021-9606]
54. Di Rosa, M.D.: Laser-cooling molecules. Eur. Phys. J. D **31**, 395–402 (2004). doi:10.1140/epjd/e2004-00167-2. http://dx.doi.org/10.1140/epjd/e2004-00167-2 [ISSN 1434-6060]
55. Stuhl, B.K., Sawyer, B.C., Wang, D., Ye, J.: Magneto-optical trap for polar molecules. Phys. Rev. Lett. **101**, 243002 (2008). doi:10.1103/PhysRevLett.101.243002. http://link.aps.org/doi/10.1103/PhysRevLett.101.243002
56. Robicheaux, F.: A proposal for laser cooling of OH molecules. J. Phys. B **42**(19), 195301 (2009). http://stacks.iop.org/0953-4075/42/i=19/a=195301
57. Morigi, G., Pinkse, P.W.H., Kowalewski, M., de Vivie-Riedle, R.: Cavity cooling of internal molecular motion. Phys. Rev. Lett. **99**, 073001 (2007). doi:10.1103/PhysRevLett.99.073001. http://link.aps.org/doi/10.1103/PhysRevLett.99.073001
58. Lev, B.L., Vukics, A., Hudson, E.R., Sawyer, B.C., Domokos, P., Ritsch, H., Ye, J.: Prospects for the cavity-assisted laser cooling of molecules. Phys. Rev. A **77**, 023402 (2008). doi:10.1103/PhysRevA.77.023402. http://link.aps.org/doi/10.1103/PhysRevA.77.023402
59. Wallquist, M., Rabl, P., Lukin, M.D., Zoller, P.: Theory of cavity-assisted microwave cooling of polar molecules. New J. Phys. **10**(6), 063005 (2008). http://stacks.iop.org/1367-2630/10/i=6/a=063005
60. Idziaszek, Z., Calarco, T., Zoller, P.: Ion-assisted ground-state cooling of a trapped polar molecule. Phys. Rev. A **83**, 053413 (2011). doi:10.1103/PhysRevA.83.053413. http://link.aps.org/doi/10.1103/PhysRevA.83.053413
61. Huber, S.D., Büchler, H.P.: Dipole-interaction-mediated laser cooling of polar molecules to ultracold temperatures. Phys. Rev. Lett. **108**, 193006 (2012). doi:10.1103/PhysRevLett.108.193006. http://link.aps.org/doi/10.1103/PhysRevLett.108.193006

62. Zhao, B., Glaetzle, A.W., Pupillo, G., Zoller, P.: Atomic rydberg reservoirs for polar molecules. Phys. Rev. Lett. **108**, 193007 (2012). doi:10.1103/PhysRevLett.108.193007. http://link.aps.org/doi/10.1103/PhysRevLett.108.193007
63. Shuman, E.S., Barry, J.F., DeMille, D.: Laser cooling of a diatomic molecule. Nature **467**(7317), 820–823 (2010). doi:10.1038/nature09443. http://dx.doi.org/10.1038/nature09443 [ISSN 0028-0836]
64. Egorov, D., Lahaye, T., Schöllkopf, W., Friedrich, B., Doyle, J.M.: Buffer-gas cooling of atomic and molecular beams. Phys. Rev. A **66**, 043401 (2002)
65. van de Meerakker, S.Y.T., Bethlem, H.L., Vanhaecke, N., Meijer, G.: Manipulation and control of molecular beams. Chem. Rev. **112**(9), 4828–4878 (2012). doi:10.1021/cr200349r. http://pubs.acs.org/doi/abs/10.1021/cr200349r
66. Bethlem, H.L., Berden, G., Meijer, G.: Decelerating neutral dipolar molecules. Phys. Rev. Lett. **83**, 1558–1561 (1999). doi:10.1103/PhysRevLett.83.1558. http://link.aps.org/doi/10.1103/PhysRevLett.83.1558
67. Bochinski, J.R., Lewandowski H.J., Sawyer, B.C., Hudson, E.R., Ye, J.: Efficient Stark deceleration of cold polar molecules. Eur. J. Phys. D **31**, 351–358 (2004)
68. Strebel, M., Spieler, S., Stienkemeier, F., Mudrich, M.: Guiding slow polar molecules with a charged wire. Phys. Rev. A **84**, 053430 (2011). doi:10.1103/PhysRevA.84.053430. http://link.aps.org/doi/10.1103/PhysRevA.84.053430
69. Hogan, S.D., Sprecher, D., Andrist, M., Vanhaecke, N., Merkt, F.: Zeeman deceleration of H and D. Phys. Rev. A **76**, 023412 (2007). doi:10.1103/PhysRevA.76.023412. http://link.aps.org/doi/10.1103/PhysRevA.76.023412
70. Hogan, S.D., Motsch, M., Merkt, F.: Deceleration of supersonic beams using inhomogeneous electric and magnetic fields. Phys. Chem. Chem. Phys. **13**, 18705–18723 (2011). doi:10.1039/C1CP21733J. http://dx.doi.org/10.1039/C1CP21733J
71. Fulton R., Bishop, A.I., Shneider, M.N., Barker P.F.: Controlling the motion of cold molecules with deep periodic optical potentials. Nat. Phys. **2**(7), 465–468 (2006). doi:10.1038/nphys339. http://dx.doi.org/10.1038/nphys339 [ISSN 1745-2473]
72. Bishop, A.I., Wang, L., Barker, P.F.: Creating cold stationary molecular gases by optical stark deceleration. New J. Phys. **12**(7), 073028 (2010). http://stacks.iop.org/1367-2630/12/i=7/a=073028
73. Barry, J.F., Shuman, E.S., Norrgard, E.B., DeMille, D.: Laser radiation pressure slowing of a molecular beam. Phys. Rev. Lett. **108**, 103002 (2012). doi:10.1103/PhysRevLett.108.103002. http://link.aps.org/doi/10.1103/PhysRevLett.108.103002
74. Band, Y.B., Julienne, L.: Ultracold-molecule production by laser-cooled atom photoassociation. Phys. Rev. A **51**, R4317–R4320 (1995). doi:10.1103/PhysRevA.51.R4317. http://link.aps.org/doi/10.1103/PhysRevA.51.R4317
75. Jones, K.M., Tiesinga, E., Lett, P.D., Julienne, P.S.: Ultracold photoassociation spectroscopy: Long-range molecules and atomic scattering. Rev. Mod. Phys. **78**, 483–535 (2006). doi:10.1103/RevModPhys.78.483. http://link.aps.org/doi/10.1103/RevModPhys.78.483
76. Takekoshi, T., Patterson, B.M., Knize, R.J.: Observation of optically trapped cold cesium molecules. Phys. Rev. Lett. **81**, 5105–5108 (1998). doi:10.1103/PhysRevLett.81.5105. http://link.aps.org/doi/10.1103/PhysRevLett.81.5105
77. Kerman, A.J., Sage, J.M., Sainis, S., Bergeman, T., DeMille, D.: Production and state-selective detection of ultracold RbCs molecules. Phys. Rev. Lett. **92**, 153001 (2004). doi:10.1103/PhysRevLett.92.153001. http://link.aps.org/doi/10.1103/PhysRevLett.92.153001
78. Köhler, T., Góral, K., Julienne, P.S.: Production of cold molecules via magnetically tunable Feshbach resonances. Rev. Mod. Phys. **78**, 1311–1361 (2006). doi:10.1103/RevModPhys.78.1311. http://link.aps.org/doi/10.1103/RevModPhys.78.1311
79. Ospelkaus, C., Ospelkaus, S., Humbert, L., Ernst, P., Sengstock, K., Bongs, K.: Ultracold heteronuclear molecules in a 3D optical lattice. Phys. Rev. Lett. **97**(12), 120402 (2006). doi:10.1103/PhysRevLett.97.120402. http://link.aps.org/abstract/PRL/v97/e120402

80. Zirbel, J.J., Ni, K.-K., Ospelkaus, S., D'Incao, J.P., Wieman, C.E., Ye, J., Jin, D.S.: Collisional stability of fermionic feshbach molecules. Phys. Rev. Lett. **100**, 143201 (2008). doi:10.1103/PhysRevLett.100.143201. http://link.aps.org/doi/10.1103/PhysRevLett.100.143201
81. Zirbel, J.J., Ni, K.-K., Ospelkaus, S., Nicholson, T.L., Olsen, M.L., Julienne, P.S., Wieman, C.E., Ye, J., Jin, D.S.: Heteronuclear molecules in an optical dipole trap. Phys. Rev. A **78**, 013416 (2008). doi:10.1103/PhysRevA.78.013416. http://link.aps.org/doi/10.1103/PhysRevA.78.013416
82. Ni, K.-K., Ospelkaus, S., de Miranda, M.H.G., Peér, A., Neyenhuis, B., Zirbel, J.J., Kotochigova, S., Julienne, P.S., Jin, D.S., Ye, J.: A high phase-space-density gas of polar molecules. Science **322**, 231–235 (2008)
83. Klempt, C., Henninger, T., Topic, O., Scherer, M., Kattner, L., Tiemann, E., Ertmer, W., Arlt, J.J.: Radio-frequency association of heteronuclear Feshbach molecules. Phys. Rev. A **78**, 061602 (2008). doi:10.1103/PhysRevA.78.061602. http://link.aps.org/doi/10.1103/PhysRevA.78.061602
84. Voigt, A.-C., Taglieber, M., Costa, L., Aoki, T., Wieser, W., Hänsch, T.W., Dieckmann, K.: Ultracold heteronuclear fermi-fermi molecules. Phys. Rev. Lett. **102**(2), 020405 (2009). doi:10.1103/PhysRevLett.102.020405
85. Heo, M.-S., Wang, T.T., Christensen, C.A., Rvachov, T.M., Cotta, D.A., Choi, J.-H., Lee, Y.-R., Ketterle, W.: Formation of ultracold fermionic NaLi Feshbach molecules (2012) [e-print arXiv.org:1205.5304]
86. Wu, C.-H., Park, J.W., Ahmadi, P., Will, S., Zwierlein, M.W.: Ultracold fermionic Feshbach molecules of ^{23}Na^{40}K (2012) [e-print arXiv.org:1206.5023]
87. Jochim, S., Bartenstein, M., Altmeyer, A., Hendl, G., Chin, C., Denschlag, J.H., Grimm, R.: Pure gas of optically trapped molecules created from fermionic atoms. Phys. Rev. Lett. **91**, 240402–1–4 (2003)
88. Jochim, S., Bartenstein, M., Altmeyer, A., Riedl, S., Chin, C., Denschlag, J.H., Grimm, R.: Bose-Einstein condensation of molecules. Science **302**, 2102–2103 (2003)
89. Greiner, M., Regal, C.A., Jin, D.S.: Emergence of a molecular Bose-Einstein condensate from a Fermi gas. Nature **426**, 437–540 (2003)
90. Xu, K., Mukaiyama, T., Abo-Shaeer, J.R., Chin, J.K., Miller, D.E., Ketterle, W.: Formation of quantum-degenerate sodium molecules. Phys. Rev. Lett. **91**, 210402–1–4 (2003)
91. Regal, C.A., Greiner, M., Jin, D.S.: Observation of resonance condensation of fermionic atom pairs. Phys. Rev. Lett. **92**, 040403–1–4 (2004)
92. Zwierlein, M.W., Stan, C.A., Schunck, C.H., Raupach, S.M.F., Kerman, A.J., Ketterle, W.: Condensation of pairs of fermionic atoms near a Feshbach resonance. Phys. Rev. Lett. **92**, 120403–1–4 (2004)
93. Bourdel, T., Khaykovich, L., Cubizolles, J., Zhang, J., Chevy, F., Teichmann, M., Tarruell, L., Kokkelmans, S.J. J.M.F., Salomon, C.: Experimental study of the BEC-BCS crossover region in lithium 6. Phys. Rev. Lett. **93**, 050401–1–4 (2004)
94. Herbig, J., Kraemer, T., Mark, M., Weber, T., Chin, C., Ngerl, H.-C., Grimm, R.: Preparation of a pure molecular quantum gas. Science **301**(5639), 1510–1513 (2003). doi:10.1126/science.1088876. http://www.sciencemag.org/content/301/5639/1510.abstract
95. Danzl, J.G., Haller, E., Gustavsson, M., Mark, M.J., Hart, R., Bouloufa, N., Dulieu, O., Ritsch, H., Nägerl, H.-C.: Quantum gas of deeply bound ground state molecules. Science **321**, 1062 (2008)
96. Bartenstein, M., Altmeyer, A., Riedl, S., Jochim, S., Chin, C., Denschlag, J.H., Grimm, R.: Crossover from a molecular Bose-Einstein condensate to a degenerate Fermi gas. Phys. Rev. Lett. **92**, 120401–1–4 (2004)
97. Kinast, J., Hemmer, S.L., Gehm, M.E., Turlapov, A., Thomas, J.E.: Evidence for superfluidity in a resonantly interacting Fermi gas. Phys. Rev. Lett. **92**, 150402–1–4 (2004)
98. Dürr, S., Volz, T., Marte, A., Rempe, G.: Observation of molecules produced from a Bose-Einstein condensate. Phys. Rev. Lett. **92**, 020406 (2004). doi:10.1103/PhysRevLett.92.020406. http://link.aps.org/doi/10.1103/PhysRevLett.92.020406

99. Eagles, D.M.: Possible pairing without superconductivity at low carrier concentrations in bulk and thin-film superconducting semiconductors. Phys. Rev. **186**, 456–463 (1969)
100. Leggett, A.J.: Diatomic molecules and cooper pairs. In: Pekalski, A., Przystawa, R. (eds.) Modern Trends in the Theory of Condensed Matter. Springer, Berlin (1980)
101. Noziéres, P., Schmitt-Rink, S.: Bose condensation in an attractive Fermion gas: from weak to strong coupling superconductivity. J. Low Temp. Phys. **59**, 195–211 (1985). doi:10.1007/BF00683774. http://dx.doi.org/10.1007/BF00683774 [ISSN 0022-2291]
102. Chen, Q., Stajic, J., Tan, S., Levin, K.: Bcs-bec crossover: From high temperature superconductors to ultracold superfluids. Phys. Rep. **412**(1), 1–88 (2005). doi:10.1016/j.physrep.2005.02.005. http://www.sciencedirect.com/science/article/pii/S0370157305001067 [ISSN 0370-1573]
103. Adams, A., Carr, L.D., Schaefer, T, Steinberg, P., Thomas, J.E.: Strongly correlated quantum fluids: Ultracold quantum gases, quantum chromodynamic plasmas, and holographic duality (2012). http://arxiv.org/abs/1205.5180
104. Winkler, K., Lang, F., Thalhammer, G., Straten, P.v.d., Grimm, R., Denschlag, J.H.: Coherent optical transfer of Feshbach molecules to a lower vibrational state. Phys. Rev. Lett. **98**(4), 043201 (2007). doi:10.1103/PhysRevLett.98.043201
105. Ospelkaus, S., Pe'er, A., Ni, K.K., Zirbel, J.J., Neyenhuis, B., Kotochigova, S., Julienne, P.S., Ye, J., Jin, D.S.: Ultracold dense gas of deeply bound heteronuclear molecules. Nat. Phys. **4**, 622–626 (2008)
106. Ospelkaus, S., Ni, K.K., de Miranda, M.H.G., Pe'er, A., Nyenhuis, B., Wang, D., Kotochigova, S., Julienne, P.S., Jin, D.S., Ye, J.: Ultracold polar molecules near quantum degeneracy. Faraday Discuss. **142**, 351–359 (2009)
107. Danzl, J.G., Mark, M.J., Haller, E., Gustavsson, M., Hart, R., Aldegunde, J., Hutson, J.M., Nägerl, H.-C.: An ultracold high-density sample of rovibronic ground-state molecules in an optical lattice. Nat. Phys. **6**, 265–270 (2010)
108. Ospelkaus, S., Ni, K.-K., Quéméner, G., Neyenhuis, B., Wang, D., deMiranda, M.H.G., Bohn, J.L., Ye, J., Jin, D.S.: Controlling the hyperfine state of rovibronic ground-state polar molecules. Phys. Rev. Lett. **104**(3), 030402 (2010). doi:10.1103/PhysRevLett.104.030402
109. Deiglmayr, J., Grochola, A., Repp, M., Mörtlbauer, K., Glück, C., Lange, J., Dulieu, O., Wester, R., Weidemüller, M.: Formation of ultracold polar molecules in the rovibrational ground state. Phys. Rev. Lett. **101**(13), 133004 (2008). doi:10.1103/PhysRevLett.101.133004
110. Cho, H.W., McCarron, D.J., Jenkin, D.L., Köppinger, M.P., Cornish, S.L.: A high phase-space density mixture of ^{87}Rb and ^{133}Cs: Towards ultracold heteronuclear molecules. Eur. Phys. J. D, 1–7. doi:10.1140/epjd/e2011-10716-1. http://dx.doi.org/10.1140/epjd/e2011-10716-1 [ISSN 1434-6060]
111. Kotochigova, S., DeMille, D.: Electric-field-dependent dynamic polarizability and state-insensitive conditions for optical trapping of diatomic polar molecules. Phys. Rev. A **82**, 063421 (2010). doi:10.1103/PhysRevA.82.063421. http://link.aps.org/doi/10.1103/PhysRevA.82.063421
112. Aldegunde, J., Rivington, B.A., Żuchowski, P.S., Hutson, J.M.: Hyperfine energy levels of alkali-metal dimers: Ground-state polar molecules in electric and magnetic fields. Phys. Rev. A **78**, 033434 (2008)
113. Ran, H., Aldegunde, J., Hutson, J.M.: Hyperfine structure in the microwave spectra of ultracold polar molecules. New J. Phys. **12**(4), 043015 (2010). http://stacks.iop.org/1367-2630/12/i=4/a=043015
114. Ni, K.K., Ospelkaus, S., Wang, D., Quemener, G., Neyenhuis, B., de Miranda, M.H.G., Bohn, J.L., Ye, J., Jin, D.S.: Dipolar collisions of polar molecules in the quantum regime. Nature (London) **464**, 1324 (2010)
115. deMiranda, M.H.G., Chotia, A., Neyenhuis, B., Wang, D., Quemener, G., Ospelkaus, S., Bohn, J.L., Ye, J., Jin, D.S.: Controlling the quantum stereodynamics of ultracold bimolecular reactions. Nat. Phys. **7**(6), 502–507 (2011). doi:10.1038/nphys1939. http://dx.doi.org/10.1038/nphys1939 [ISSN 1745-2473]

116. Marinescu, M., You, L.: Controlling atom-atom interaction at ultralow temperatures by dc electric fields. Phys. Rev. Lett. **81**, 4596–4599 (1998)
117. Yi, S., You, L.: Trapped atomic condensates with anisotropic interactions. Phys. Rev. A **61**, 041604 (2000). doi:10.1103/PhysRevA.61.041604. http://link.aps.org/doi/10.1103/PhysRevA.61.041604
118. Deb, B., You, L.: Low-energy atomic collision with dipole interactions. Phys. Rev. A **64**, 022717 (2001). doi:10.1103/PhysRevA.64.022717. http://link.aps.org/doi/10.1103/PhysRevA.64.022717
119. Bortolotti, D.C.E., Ronen, S., Bohn, J.L., Blume, D.: Scattering length instability in dipolar Bose-Einstein condensates. Phys. Rev. Lett. **97**, 160402 (2006). doi:10.1103/PhysRevLett.97.160402. http://link.aps.org/doi/10.1103/PhysRevLett.97.160402
120. Ronen, S., Bortolotti, D.C.E., Blume, D., Bohn, J.L.: Dipolar Bose-Einstein condensates with dipole-dependent scattering length. Phys. Rev. A **74**, 033611 (2006). doi:10.1103/PhysRevA.74.033611. http://link.aps.org/doi/10.1103/PhysRevA.74.033611
121. Gorshkov, A.V., Manmana, S.R., Chen, G., Ye, J., Demler, E., Lukin, M.D., Rey, A.M.: Tunable superfluidity and quantum magnetism with ultracold polar molecules. Phys. Rev. Lett. **107**, 115301 (2011). doi:10.1103/PhysRevLett.107.115301. http://link.aps.org/doi/10.1103/PhysRevLett.107.115301
122. Gorshkov, A.V., Manmana, S.R., Chen, G., Demler, E., Lukin, M.D., Rey, A.M.: Quantum magnetism with polar alkali-metal dimers. Phys. Rev. A **84**, 033619 (2011). doi:10.1103/PhysRevA.84.033619. http://link.aps.org/doi/10.1103/PhysRevA.84.033619
123. Wall, M.L., Carr, L.D.: Hyperfine molecular Hubbard Hamiltonian. Phys. Rev. A **82**(1), 013611 (2010). doi:10.1103/PhysRevA.82.013611
124. Mayle, M., Ruzic, B.P., Bohn, J.L.: Statistical aspects of ultracold resonant scattering. Phys. Rev. A **85**, 062712 (2012). doi:10.1103/PhysRevA.85.062712. http://link.aps.org/doi/10.1103/PhysRevA.85.062712
125. Ospelkaus, S., Ni, K.-K., Wang, D., deMiranda, M.H.G., Neyenhuis, B., Quéméner, G., Julienne, P.S., Bohn, J.L., Jin, D.S., Ye, J.: Quantum-state controlled chemical reactions of ultracold potassium-rubidium molecules. Science **327**(5967), 853–857 (2010). doi:10.1126/science.1184121. http://www.sciencemag.org/content/327/5967/853.abstract
126. Chotia, A., Neyenhuis, B., Moses, S.A., Yan, B., Covey, J.P., Foss-Feig, M., Rey, A.M., Jin, D.S., Ye, J.: Long-lived dipolar molecules and Feshbach molecules in a 3D optical lattice. Phys. Rev. Lett. **108**, 080405 (2012). doi:10.1103/PhysRevLett.108.080405. http://link.aps.org/doi/10.1103/PhysRevLett.108.080405
127. Gribakin, G.F., Flambaumm, V.V.: Calculation of the scattering length in atomic collisions using the semiclassical approximation. Phys. Rev. A **48**, 546–553 (1993). doi:10.1103/PhysRevA.48.546. http://link.aps.org/doi/10.1103/PhysRevA.48.546
128. Aymar, M., Dulieu, O.: Calculation of accurate permanent dipole moments of the lowest $^{1,3}\Sigma^+$ states of heteronuclear alkali dimers using extended basis sets. J. Chem. Phys. **122**(20) (2005). doi:10.1063/1.1903944 [ISSN 0021-9606]
129. Croft, J.F.E., Wallis, A.O.G., Hutson, J.M., Julienne, P.S.: Multichannel quantum defect theory for cold molecular collisions. Phys. Rev. A **84**, 042703 (2011). doi:10.1103/PhysRevA.84.042703. http://link.aps.org/doi/10.1103/PhysRevA.84.042703
130. Idziaszek, Z., Quéméner, G., Bohn, J.L., Julienne, P.S.: Simple quantum model of ultracold polar molecule collisions. Phys. Rev. A **82**, 020703 (2010). doi:10.1103/PhysRevA.82.020703. http://link.aps.org/doi/10.1103/PhysRevA.82.020703
131. Julienne, P.S., Hanna, T.M., Idziaszek, Z.: Universal ultracold collision rates for polar molecules of two alkali-metal atoms. Phys. Chem. Chem. Phys. **13**, 19114–19124 (2011). doi:10.1039/C1CP21270B. http://dx.doi.org/10.1039/C1CP21270B
132. Idziaszek, Z., Julienne, P.S.: Universal rate constants for reactive collisions of ultracold molecules. Phys. Rev. Lett. **104**, 113202 (2010). doi:10.1103/PhysRevLett.104.113202. http://link.aps.org/doi/10.1103/PhysRevLett.104.113202

133. Micheli, A., Idziaszek, Z., Pupillo, G., Baranov, M.A., Zoller, P., Julienne, P.S.: Universal rates for reactive ultracold polar molecules in reduced dimensions. Phys. Rev. Lett. **105**, 073202 (2010). doi:10.1103/PhysRevLett.105.073202. http://link.aps.org/doi/10.1103/PhysRevLett.105.073202
134. Hanna, T.M., Tiesinga, E., Mitchell, W.F., Julienne, P.S.: Resonant control of polar molecules in individual sites of an optical lattice. Phys. Rev. A **85**, 022703 (2012). doi:10.1103/PhysRevA.85.022703. http://link.aps.org/doi/10.1103/PhysRevA.85.022703
135. Quéméner, G., Bohn, J.L.: Electric field suppression of ultracold confined chemical reactions. Phys. Rev. A **81**, 060701 (2010). doi:10.1103/PhysRevA.81.060701. http://link.aps.org/doi/10.1103/PhysRevA.81.060701
136. Quéméner, G., Bohn, J.L.: Strong dependence of ultracold chemical rates on electric dipole moments. Phys. Rev. A **81**, 022702 (2010). doi:10.1103/PhysRevA.81.022702. http://link.aps.org/doi/10.1103/PhysRevA.81.022702
137. Quéméner, G., Bohn, J.L., Petrov, A., Kotochigova, S.: Universalities in ultracold reactions of alkali-metal polar molecules. Phys. Rev. A **84**, 062703 (2011). doi:10.1103/PhysRevA.84.062703. http://link.aps.org/doi/10.1103/PhysRevA.84.062703
138. Syassen, N., Bauer, D.M., Lettner, M., Volz, T., Dietze, D., García-Ripoll, J.J., Cirac, J.I., Rempe, G., Dürr, S.: Strong dissipation inhibits losses and induces correlations in cold molecular gases. Science **320**(5881), 1329–1331 (2008). doi:10.1126/science.1155309. http://www.sciencemag.org/content/320/5881/1329.abstract
139. Schuch, N., Verstraete, F.: Computational complexity of interacting electrons and fundamental limitations of density functional theory. Nat. Phys. **5**(10), 732–735 (2009). doi:10.1038/nphys1370. http://dx.doi.org/10.1038/nphys1370
140. Golub, G.H., Van Loan, C.F.: Matrix Computations, 3rd edn. Johns Hopkins Studies in Mathematical Sciences. The Johns Hopkins University Press, Baltimore (1996)
141. Davidson, E.R.: The iterative calculation of a few of the lowest eigenvalues and corresponding eigenvectors of large real-symmetric matrices. J. Comput. Phys. **17**(1), 87–94 (1975). doi:10.1016/0021-9991(75)90065-0. http://www.sciencedirect.com/science/article/pii/0021999175900650 [ISSN 0021-9991]
142. Feynman, R.P.: Atomic theory of the λ transition in helium. Phys. Rev. **91**, 1291–1301 (1953). doi:10.1103/PhysRev.91.1291. http://link.aps.org/doi/10.1103/PhysRev.91.1291
143. Metropolis, N., Rosenbluth, A.W., Rosenbluth, M.N., Teller, A.H., Teller, E.: Equation of state calculations by fast computing machines. J. Chem. Phys. **21**, 1087 (1953)
144. Evertz, H.G.: The loop algorithm. Adv. Phys. **52**(1), 1–66 (2003). doi:10.1080/0001873021000049195. http://www.tandfonline.com/doi/abs/10.1080/0001873021000049195
145. Prokof'ev, N., Svistunov, B., Tupitsyn, I.: Exact, complete, and universal continuous-time worldline Monte Carlo approach to the statistics of discrete quantum systems. J. Exp. Theor. Phys. **87**, 310–321 (1998). doi:10.1134/1.558661. http://dx.doi.org/10.1134/1.558661 [ISSN 1063-7761]
146. Troyer, M., Wiese, U.-J.: Computational complexity and fundamental limitations to fermionic quantum Monte Carlo simulations. Phys. Rev. Lett. **94**, 170201 (2005). doi:10.1103/PhysRevLett.94.170201. http://link.aps.org/doi/10.1103/PhysRevLett.94.170201
147. Grimm, R.C., Storer, R.G.: Monte-Carlo solution of Schrödinger's equation. Chin. J. Comput. Phys. **7**(1), 134–156 (1971). doi:10.1016/0021-9991(71)90054-4. http://www.sciencedirect.com/science/article/pii/0021999171900544 [ISSN 0021-9991]
148. McMillan, W.L.: Ground state of liquid He^4. Phys. Rev. **138**, A442–A451 (1965). doi:10.1103/PhysRev.138.A442. http://link.aps.org/doi/10.1103/PhysRev.138.A442
149. Prokof'ev, N.V., Svistunov, B.V.: Polaron problem by diagrammatic quantum Monte Carlo. Phys. Rev. Lett. **81**, 2514–2517 (1998). doi:10.1103/PhysRevLett.81.2514. http://link.aps.org/doi/10.1103/PhysRevLett.81.2514
150. Prokof'ev, N., Svistunov, B.: Bold diagrammatic Monte Carlo technique: When the sign problem is welcome. Phys. Rev. Lett. **99**, 250201 (2007). doi:10.1103/PhysRevLett.99.250201. http://link.aps.org/doi/10.1103/PhysRevLett.99.250201

151. Prokof'ev, N.V., Svistunov, B.V.: Bold diagrammatic Monte Carlo: A generic sign-problem tolerant technique for polaron models and possibly interacting many-body problems. Phys. Rev. B **77**, 125101 (2008). doi:10.1103/PhysRevB.77.125101. http://link.aps.org/doi/10.1103/PhysRevB.77.125101
152. VanHoucke, K., Werner, F., Kozik, E., Prokof/'ev, N., Svistunov, B., Ku, M.J.H., Sommer, A.T., Cheuk, L.W., Schirotzek, A., Zwierlein, M.W.: Feynman diagrams versus Fermi-gas Feynman emulator. Nat. Phys. **8**(5), 366–370 (2012) doi:10.1038/nphys2273. http://dx.doi.org/10.1038/nphys2273 [ISSN 1745-2473]
153. Fetter, A.L., Walecka, J.D.: Quantum Theory of Many-Particle Systems. McGraw-Hill, New York (1971)
154. Schiró, M.: Real-time dynamics in quantum impurity models with diagrammatic Monte Carlo. Phys. Rev. B **81**, 085126 (2010). doi:10.1103/PhysRevB.81.085126. http://link.aps.org/doi/10.1103/PhysRevB.81.085126
155. Metzner, W., Vollhardt, D.: Correlated lattice fermions in $d = \infty$ dimensions. Phys. Rev. Lett. **62**, 324–327 (1989). doi:10.1103/PhysRevLett.62.324. http://link.aps.org/doi/10.1103/PhysRevLett.62.324
156. Georges, A., Kotliar, G.: Hubbard model in infinite dimensions. Phys. Rev. B **45**, 6479–6483 (1992). doi:10.1103/PhysRevB.45.6479. http://link.aps.org/doi/10.1103/PhysRvB.45.6479
157. Georges, A., Kotliar, G., Krauth, W., Rozenberg, M.J.: Dynamical mean-field theory of strongly correlated fermion systems and the limit of infinite dimensions. Rev. Mod. Phys. **68**, 13–125 (1996). doi:10.1103/RevModPhys.68.13. http://link.aps.org/doi/10.1103/RevModPhys.68.13
158. Kotliar, G., Savrasov, S.Y., Haule, K., Oudovenko, V.S., Parcollet, O., Marianetti, C.A.: Electronic structure calculations with dynamical mean-field theory. Rev. Mod. Phys. **78**, 865–951 (2006). doi:10.1103/RevModPhys.78.865. http://link.aps.org/doi/10.1103/RevModPhys.78.865
159. Titvinidze, I., Snoek, M., Hofstetter, W.: Generalized dynamical mean-field theory for Bose-Fermi mixtures in optical lattices. Phys. Rev. B **79**, 144506 (2009)
160. Hu, W.-J., Tong, N.-H.: Dynamical mean-field theory for the Bose-Hubbard model. Phys. Rev. B **80**, 245110 (2009). doi:10.1103/PhysRevB.80.245110. http://link.aps.org/doi/10.1103/PhysRevB.80.245110
161. Feiguin, A.E., White, S.R.: Finite-temperature density matrix renormalization using an enlarged Hilbert space. Phys. Rev. B **72**(22), 220401 (2005). doi:10.1103/PhysRevB.72.220401
162. White, S.R.: Minimally entangled typical quantum states at finite temperature. Phys. Rev. Lett. **102**(19), 190601 (2009). doi:10.1103/PhysRevLett.102.190601
163. Stoudenmire, E.M., White, S.R.: Minimally entangled typical thermal state algorithms. New J. Phys. **12**(5), 055026 (2010). http://stacks.iop.org/1367-2630/12/i=5/a=055026
164. Wall, M.L., Carr, L.D.: Finite temperature matrix product state algorithms and applications. In: Davis, M., Gardiner, S., Nygaard, N., Proukakis, N., Szymanska, M. (eds.) Chapter in Non-equilibrium Quantum Gases at Finite Temperatures. World Scientific, Singapore (2011). e-print http://arxiv.org/abs/1008.4303
165. Schachenmayer, J., Lesanovsky, I., Micheli, A., Daley, A.J.: Dynamical crystal creation with polar molecules or Rydberg atoms in optical lattices. New J. Phys. **12**(10), 103044 (2010). http://stacks.iop.org/1367-2630/12/i=10/a=103044
166. McCulloch, I.P.: From density-matrix renormalization group to matrix product states. J. Stat. Mech. **2007**(10), P10014 (2007). http://stacks.iop.org/1742-5468/2007/i=10/a=P10014
167. Crosswhite, G.M., Doherty, A.C., Vidal, G.: Applying matrix product operators to model systems with long-range interactions. Phys. Rev. B **78**, 035116 (2008). doi:10.1103/PhysRevB.78.035116. http://link.aps.org/doi/10.1103/PhysRevB.78.035116
168. Pirvu, B., Murg, V., Cirac, J.I., Verstraete, F.: Matrix product operator representations. New J. Phys. **12**(2), 025012 (2010). http://stacks.iop.org/1367-2630/12/i=2/a=025012

169. Dalmonte, M., Pupillo, G., Zoller, P.: One-dimensional quantum liquids with power-law interactions: The luttinger staircase. Phys. Rev. Lett. **105**, 140401 (2010). doi:10.1103/PhysRevLett.105.140401. http://link.aps.org/doi/10.1103/PhysRevLett.105.140401
170. Capogrosso-Sansone, B., Trefzger, C., Lewenstein, M., Zoller, P., Pupillo, G.: Quantum phases of cold polar molecules in 2D optical lattices. Phys. Rev. Lett. **104**, 125301 (2010)
171. Mermin, N.D., Wagner, H.: Absence of ferromagnetism or antiferromagnetism in one- or two-dimensional isotropic heisenberg models. Phys. Rev. Lett. **17**, 1133–1136 (1966). doi:10.1103/PhysRevLett.17.1133. http://link.aps.org/doi/10.1103/PhysRevLett.17.1133
172. Hohenberg, P.C.: Existence of long-range order in one and two dimensions. Phys. Rev. **158**, 383–386 (1967). doi:10.1103/PhysRev.158.383. http://link.aps.org/doi/10.1103/PhysRev.158.383
173. Griesmaier, A., Werner, J., Hensler, S., Stuhler, J., Pfau, T.: Bose-Einstein condensation of chromium. Phys. Rev. Lett. **94**, 160401–1–4 (2005)
174. Beaufils, Q., Chicireanu, R., Zanon, T., Laburthe-Tolra, B., Maréchal, E., Vernac, L., Keller, J.-C., Gorceix, O.: All-optical production of chromium Bose-Einstein condensates. Phys. Rev. A **77**, 061601 (2008). doi:10.1103/PhysRevA.77.061601. http://link.aps.org/doi/10.1103/PhysRevA.77.061601
175. Lu, M., Burdick, N.Q., Youn, S.H., Lev, B.L.: Strongly dipolar Bose-Einstein condensate of dysprosium. Phys. Rev. Lett. **107**, 190401 (2011). doi:10.1103/PhysRevLett.107.190401. http://link.aps.org/doi/10.1103/PhysRevLett.107.190401
176. Aikawa, K., Frisch, A., Mark, M., Baier, S., Rietzler, A., Grimm, R., Ferlaino, F.: Bose-Einstein condensation of erbium. Phys. Rev. Lett. **108**, 210401 (2012). doi:10.1103/PhysRevLett.108.210401. http://link.aps.org/doi/10.1103/PhysRevLett.108.210401
177. Lu, M., Burdick, N.Q., Lev, B.L.: Quantum degenerate dipolar Fermi gas. Phys. Rev. Lett. **108**, 215301 (2012). doi:10.1103/PhysRevLett.108.215301. http://link.aps.org/doi/10.1103/PhysRevLett.108.215301
178. Lahaye, T., Metz, J., Fröhlich, B., Koch, T., Meister, M., Griesmaier, A., Pfau, T., Saito, H., Kawaguchi, Y., Ueda, M.: d-wave collapse and explosion of a dipolar Bose-Einstein condensate. Phys. Rev. Lett. **101**, 080401 (2008)
179. Baranov, M.A.: Theoretical progress in many-body physics with ultracold dipolar gases. Phys. Rep. **464**(3), 71–111 (2008)
180. Lahaye, T., Menotti, C., Santos, L., Lewenstein, M., Pfau, T.: The physics of dipolar bosonic quantum gases. Rep. Prog. Phys. **72**(12), 126401 (2009). http://stacks.iop.org/0034-4885/72/i=12/a=126401
181. Santos, L., Shlyapnikov, G.V., Zoller, P., Lewenstein, M.: Bose-Einstein condensation in trapped dipolar gases. Phys. Rev. Lett. **85**, 1791–1794 (2000). Erratum: Santos, L., Shlyapnikov, G.V., Zoller, P., Lewenstein, M.: Phys. Rev. Lett. **85**, 1791 (2000)
182. Yi, S., You, L., Pu, H.: Quantum phases of dipolar spinor condensates. Phys. Rev. Lett. **93**, 040403–1–4 (2004)
183. Sinha, S., Santos, L.: Cold dipolar gases in quasi-one-dimensional geometries. Phys. Rev. Lett. **99**, 140406 (2007)
184. Ronen, S., Bortolotti, D.C.E., Bohn, J.L.: Radial and angular rotons in trapped dipolar gases. Phys. Rev. Lett. **98**, 030406 (2007) doi:10.1103/PhysRevLett.98.030406. http://link.aps.org/doi/10.1103/PhysRevLett.98.030406
185. Kanjilal, K., Bohn, J.L., Blume, D.: Pseudopotential treatment of two aligned dipoles under external harmonic confinement. Phys. Rev. A **75**, 052705 (2007). doi:10.1103/PhysRevA.75.052705. http://link.aps.org/doi/10.1103/PhysRevA.75.052705
186. Wilson, R.M., Ronen, S., Bohn, J.L., Pu, H.: Manifestations of the roton mode in dipolar Bose-Einstein condensates. Phys. Rev. Lett. **100**, 245302 (2008). doi:10.1103/PhysRevLett.100.245302. http://link.aps.org/doi/10.1103/PhysRevLett.100.245302
187. Menotti, C., Trefzger, C., Lewenstein, M.: Metastable states of a gas of dipolar bosons in a 2D optical lattice. Phys. Rev. Lett. **98**, 235301 (2007)

188. Trefzger, C., Menotti, C., Lewenstein, M.: Ultracold dipolar gas in an optical lattice: The fate of metastable states. Phys. Rev. A **78**, 043604 (2008). doi:10.1103/PhysRevA.78.043604. http://link.aps.org/doi/10.1103/PhysRevA.78.043604
189. Goral, K., Santos, L., Lewenstein, M.: Quantum phases of dipolar bosons in optical lattices. Phys. Rev. Lett. **88**, 170406–1–4 (2002)
190. Damski, B., Santos, L., Tiemann, E., Lewenstein, M., Kotochigova, S., Julienne, P., Zoller, P.: Creation of a dipolar superfluid in optical lattices. Phys. Rev. Lett. **90**, 110401–1–4 (2003)
191. Danshita, I., de Melo, C.A.R.S.: Stability of superfluid and supersolid phases of dipolar bosons in optical lattices. Phys. Rev. Lett. **103**, 225301 (2009). doi:10.1103/PhysRevLett.103.225301. http://link.aps.org/doi/10.1103/PhysRevLett.103.225301
192. Pollet, L., Picon, J.D., Büchler, H.P., Troyer, M.: Supersolid phase with cold polar molecules on a triangular lattice. Phys. Rev. Lett. **104**, 125302 (2010). doi:10.1103/PhysRevLett.104.125302. http://link.aps.org/doi/10.1103/PhysRevLett.104.125302
193. Wilson, R.M., Bohn, J.L.: Emergent structure in a dipolar Bose gas in a one-dimensional lattice. Phys. Rev. A **83**, 023623 (2011). doi:10.1103/PhysRevA.83.023623. http://link.aps.org/doi/10.1103/PhysRevA.83.023623
194. Boninsegni M., Prokof'ev, N.V.: *Colloquium*. Supersolids: What and where are they? Rev. Mod. Phys. **84**, 759–776 (2012) doi:10.1103/RevModPhys.84.759. http://link.aps.org/doi/10.1103/RevModPhys.84.759
195. Olshanii, M.: Atomic scattering in the presence of an external confinement and a gas of impenetrable bosons. Phys. Rev. Lett. **81**, 938–941 (1998)
196. DeMille, D.: Quantum computation with trapped polar molecules. Phys. Rev. Lett. **88**, 067901 (2002)
197. Micheli, A., Brennen, G.K., Zoller, P.: A toolbox for lattice-spin models with polar molecules. Nat. Phys. **2**(5), 341–347 (2006). doi:10.1038/nphys287. http://dx.doi.org/10.1038/nphys287 [ISSN 1745-2473]
198. Brennen, G.K., Micheli, A., Zoller, P.: Designing spin-1 lattice models using polar molecules. New J. Phys. **9**, 138 (2007)
199. Herrera, F., Krems, R.V.: Tunable Holstein model with cold polar molecules. Phys. Rev. A **84**, 051401 (2011). doi:10.1103/PhysRevA.84.051401. http://link.aps.org/doi/10.1103/PhysRevA.84.051401
200. Pérez-Ríos, J., Herrera, F., Krems, R.V.: External field control of collective spin excitations in an optical lattice of $^2\Sigma$ molecules. New J. Phys. **12**(10), 103007 (2010). http://stacks.iop.org/1367-2630/12/i=10/a=103007
201. Herrera, F., Litinskaya, M., Krems, R.V.: Tunable disorder in a crystal of cold polar molecules. Phys. Rev. A **82**, 033428 (2010). doi:10.1103/PhysRevA.82.033428. http://link.aps.org/doi/10.1103/PhysRevA.82.033428
202. Quintanilla, J., Carr, S.T., Betouras, J.J.: Metanematic, smectic, and crystalline phases of dipolar fermions in an optical lattice. Phys. Rev. A **79**, 031601 (2009). doi:10.1103/PhysRevA.79.031601. http://link.aps.org/doi/10.1103/PhysRevA.79.031601
203. Lin, C., Zhao, E., Liu, W.V.: Liquid crystal phases of ultracold dipolar fermions on a lattice. Phys. Rev. B **81**, 045115 (2010). doi:10.1103/PhysRevB.81.045115. http://link.aps.org/doi/10.1103/PhysRevB.81.045115
204. DallaTorre, E.G., Berg, E., Altman, E.: Hidden order in 1D bose insulators. Phys. Rev. Lett. **97**, 260401 (2006). doi:10.1103/PhysRevLett.97.260401. http://link.aps.org/doi/10.1103/PhysRevLett.97.260401
205. Kestner, J.P., Wang, B., Sau, J.D., DasSarma, S.: Prediction of a gapless topological Haldane liquid phase in a one-dimensional cold polar molecular lattice. Phys. Rev. B **83**, 174409 (2011). doi:10.1103/PhysRevB.83.174409. http://link.aps.org/doi/10.1103/PhysRevB.83.174409
206. Pupillo, G., Micheli, A., Buchler, H.P., Zoller, P.: Cold Molecules: Creation and Applications, Chap. 12. Taylor & Francis, Boca Raton (2009)

207. Gorshkov, A.V., Rabl, P., Pupillo, G., Micheli, A., Zoller, P., Lukin, M.D., Büchler, H.P.: Suppression of inelastic collisions between polar molecules with a repulsive shield. Phys. Rev. Lett. **101**, 073201 (2008)
208. Micheli, A., Pupillo, G., Büchler, H.P., Zoller, P.: Cold polar molecules in two-dimensional traps: Tailoring interactions with external fields for novel quantum phases. Phys. Rev. A **76**, 043604 (2007)
209. Pupillo, G., Griessner, A., Micheli, A., Ortner, M., Wang, D.-W., Zoller, P.: Cold atoms and molecules in self-assembled dipolar lattices. Phys. Rev. Lett. **100**, 050402 (2008)
210. Ortner, M., Micheli, A., Pupillo, G., Zoller, P.: Quantum simulations of extended Hubbard models with dipolar crystals. New J. Phys. **11**, 055045 (2009)
211. Büchler, H.P., Micheli, A., Zoller, P.: Three-body interactions with cold polar molecules. Nat. Phys. **3**, 726–731 (2007)
212. Schmidt, K.P., Dorier, J., Läuchli, A.M.: Solids and supersolids of three-body interacting polar molecules on an optical lattice. Phys. Rev. Lett. **101**, 150405 (2008). doi:10.1103/PhysRevLett.101.150405. http://link.aps.org/doi/10.1103/PhysRevLett.101.150405
213. Trefzger, C., Menotti, C., Lewenstein, M.: Pair-supersolid phase in a bilayer system of dipolar lattice bosons. Phys. Rev. Lett. **103**, 035304 (2009). doi:10.1103/PhysRevLett.103.035304. http://link.aps.org/doi/10.1103/PhysRevLett.103.035304
214. Potter, A.C., Berg, E., Wang, D.-W., Halperin, B.I., Demler, E.: Superfluidity and dimerization in a multilayered system of fermionic polar molecules. Phys. Rev. Lett. **105**, 220406 (2010). doi:10.1103/PhysRevLett.105.220406. http://link.aps.org/doi/10.1103/PhysRevLett.105.220406
215. Łakomy, K., Nath, R., Santos, L.: Spontaneous crystallization and filamentation of solitons in dipolar condensates. Phys. Rev. A **85**, 033618 (2012). doi:10.1103/PhysRevA.85.033618. http://link.aps.org/doi/10.1103/PhysRevA.85.033618
216. Barnett, R., Petrov, D., Lukin, M., Demler, E.: Quantum magnetism with multicomponent dipolar molecules in an optical lattice. Phys. Rev. Lett. **96**, 190401 (2006)
217. Timmermans, E., Tommasini, P., Hussein, M., Kerman, A.: Feshbach resonances in atomic Bose-Einstein condensates. Phys. Rep. **315**(1-3), 199–230 (1999). doi:DOI:10.1016/S0370-1573(99)00025-3. http://www.sciencedirect.com/science/article/pii/S0370157399000253 [ISSN 0370-1573]
218. Ranninger, J., Robaszkiewicz, S.: Superconductivity of locally paired electrons. Phys. B **135**, 468 (1985)
219. Friedberg, R., Lee, T.D.: Gap energy and long-range order in the boson-fermion model of superconductivity. Phys. Rev. B **40**(10), 6745–6762 (1989). doi:10.1103/PhysRevB.40.6745
220. Ranninger, J., Robin, J.M.: The boson-fermion model of high-T_c superconductivity doping dependence. Phys. C **253**, 279–291 (1995)
221. Yang, K.-Y., Kozik, E., Wang, X., Troyer, M.: Diagrammatic quantum Monte Carlo solution of the two-dimensional cooperon-fermion model. Phys. Rev. B **83**, 214516 (2011). doi:10.1103/PhysRevB.83.214516. http://link.aps.org/doi/10.1103/PhysRevB.83.214516
222. Holland, M., Kokkelmans, S.J.J.M.F., Chiofalo, M.L., Walser, R.: Resonance superfluidity in a quantum degenerate Fermi gas. Phys. Rev. Lett. **87**, 120406–1–4 (2001)
223. Milstein, J.N., Kokkelmans, S.J.J.M.F., Holland, M.J.: Resonance theory of the crossover from Bardeen-Cooper-Schrieffer superfluidity to Bose-Einstein condensation in a dilute Fermi gas. Phys. Rev. A **66**, 043604–1–6 (2002)
224. Diener, R.B., Ho, T.-L.: Fermions in optical lattices swept across Feshbach resonances. Phys. Rev. Lett. **96**(1), 010402 (2006). doi:10.1103/PhysRevLett.96.010402
225. Dickerscheid, D.B.M., AlKhawaja, U., van Oosten, D., Stoof, H.T.C.: Feshbach resonances in an optical lattice. Phys. Rev. A **71**, 043604 (2005). doi:10.1103/PhysRevA.71.043604. http://link.aps.org/doi/10.1103/PhysRevA.71.043604
226. Duan, L.-M.: Effective Hamiltonian for fermions in an optical lattice across a Feshbach resonance. Phys. Rev. Lett. **95**(24), 243202 (2005)

227. Duan, L.-M.: General Hubbard model for strongly interacting fermions in an optical lattice and its phase detection. EPL **81**(2), 20001 (2008) http://stacks.iop.org/0295-5075/81/i=2/a=20001
228. Kestner, J.P., Duan, L.-M.: Effective single-band models for strongly interacting fermions in an optical lattice. Phys. Rev. A **81**(4), 043618 (2010) doi:10.1103/PhysRevA.81.043618
229. Büchler, H.P.: Microscopic derivation of Hubbard parameters for cold atomic gases. Phys. Rev. Lett. **104**(9), 090402 (2010). doi:10.1103/PhysRevLett.104.090402
230. von Stecher, J., Gurarie, V., Radzihovsky, L., Rey, A.M.: Lattice-induced resonances in one-dimensional bosonic systems. Phys. Rev. Lett. **106**(23), 235301 (2011). doi:10.1103/PhysRevLett.106.235301
231. Algorithms and libraries for physics simulations (2012). http://alps.comp-phys.org
232. Albuquerque, A.F., Alet, F., Corboz, P., Dayal, P., Feiguin, A., Fuchs, S., Gamper, L., Gull, E., Gurtler, S., Honecker, A., Igarashi, R., Korner, M., Kozhevnikov, A., Lauchli, A., Manmana, S.R., Matsumoto, M., McCulloch, I.P., Michel, F., Noack, R.M., Pawlowski, G., Pollet, L., Pruschke, T., Schollwöck, U., Todo, S., Trebst, S., Troyer, M., Werner, P., Wessel, S.: The ALPS project release 1.3: Open-source software for strongly correlated systems. J. Magn. Magn. Mater. **310**, 1187–1193 (2007)
233. Bauer, B., Carr, L.D., Evertz, H.G., Feiguin, A., Freire, J., Fuchs, S., Gamper, L., Gukelberger, J., Gull, E., Guertler, S., Hehn, A., Igarashi, R., Isakov, S.V., Koop, D., Ma, P.N., Mates, P., Matsuo, H., Parcollet, O., Pawłowski, G., Picon, J.D., Pollet, L., Santos, E., Scarola, V.W., Schollwöck, U., Silva, C., Surer, B., Todo, S., Trebst, S., Troyer, M., Wall, M.L., Werner, P., Wessel, S.: The ALPS project release 2.0: Open source software for strongly correlated systems. J. Stat. Mech. **2011**(05), P05001 (2011). http://stacks.iop.org/1742-5468/2011/i=05/a=P05001
234. Powder with power open source time-dependent DMRG code (2012). http://www.qti.sns.it/dmrg/phome.html
235. DMRG++ open source DMRG code (2012). http://www.ornl.gov/gz1/dmrgPlusPlus/
236. Open source time-evolving block decimation (2010). http://physics.mines.edu/downloads/software/tebd/
237. Esslinger, T.: Fermi-Hubbard physics with atoms in an optical lattice. Ann. Rev. Condens. Matter Phys. **1**(1), 129–152 (2010). doi:10.1146/annurev-conmatphys-070909-104059. http://www.annualreviews.org/doi/abs/10.1146/annurev-conmatphys-070909-104059
238. Open source TEBD SourceForge repository (2014). http://sourceforge.net/projects/opentebd/
239. Open source MPS SourceForge repository (2014). http://sourceforge.net/projects/openmps/

Chapter 2
Models for Strongly Correlated Lattice Physics

In this chapter we outline the general procedure of deriving an effective low-energy many-body lattice model appropriate for strong correlations from few-body physics. The replacement of the full many-body Hamiltonian with an effective model can be justified as the result of a renormalization group iteration [1]. Using a renormalization group analysis, a Hamiltonian can be projected into a suitably chosen low-energy subspace and the high-energy modes integrated out, resulting in an effective model for only the low-energy modes. Under this renormalization procedure, certain *irrelevant* interactions are suppressed relative to those which grow or stay constant.[1] The hope is that the renormalized theory contains fewer relevant single particle states or simpler interactions. Additionally, rather than integrating out the high-energy degrees of freedom, one can also *dress* them with interactions such that the new dressed high-energy degrees of freedom have a more easily understood structure or simpler couplings to the relevant low-energy degrees of freedom. This dressing is the main idea underlying Feynman diagram re-summation techniques such as the random phase approximation [2], and is also the key idea in the derivation of the Fermi resonance Hamiltonian discussed in Part III.

Generally, a renormalization group analysis of an interacting many-body system is difficult. For particles in deep lattices,[2] the presence of a band gap provides an energy scale which naturally separates the low-energy and high-energy degrees of freedom. If all other energy scales, e.g., temperature, interactions, etc., are smaller than this band gap, then the particles will remain in the lowest Bloch band, drastically simplifying the structure of the theory without the need for a more

[1]Interactions that grow and remain constant are called *relevant* and *marginal*, respectively.

[2]Deep in this context means that the strength of the lattice is comparable to or larger than the kinetic energy. For ultracold gases with lattice potential $V(x) = V_0 \sin^2(\pi x/a)$, a lattice may be considered deep in typical cases for $V_0/E_R \geq 2$, with $E_R = \hbar^2\pi^2/2ma^2$ the recoil energy.

M.L. Wall, *Quantum Many-Body Physics of Ultracold Molecules in Optical Lattices*, Springer Theses, DOI 10.1007/978-3-319-14252-4_2

complex renormalization group analysis. Lattice Hamiltonians obtained under these assumptions are known as *Hubbard models* [3, 4], in analogy with the lattice model proposed by Hubbard

$$\hat{H} = -t \sum_{\langle i,j\rangle} \sum_{\sigma\in\{\uparrow,\downarrow\}} \left[\hat{a}^{\dagger}_{i\sigma}\hat{a}_{j\sigma} + \text{h.c.}\right] + U\sum_{i} \hat{n}_{i\uparrow}\hat{n}_{i\downarrow}. \tag{2.1}$$

The first term represents tunneling of fermions with spin projection σ between nearest-neighbor lattice sites i and j with associated energy t. The latter term represents interactions between spin-up and spin-down fermions which occur only when two-particles occupy the same lattice site. For electrons in solids which interact via a screened Coulomb potential, the reduction of the interactions to a single on-site term is a drastic oversimplification, and so the Hubbard model represents a toy model which is not expected to capture the microscopic physics.[3] Additionally, the restriction to only the lowest Bloch band of the lattice does not hold for f-electron metals in which the interaction parameters are generally greater than the band splitting. However, for neutral ultracold gases in optical lattices, the interactions can be well modeled by a contact pseudopotential in many cases, and so Eq. (2.1) may represent an accurate microscopic many-body Hamiltonian. The remaining issue is to identify the parameters t and U from few-body physics.

2.1 The Single-Particle Problem: Bloch States and Wannier Functions

Let us consider a general many-body Hamiltonian consisting of a single-particle Hamiltonian $\hat{H}_1$ and two-body interactions $\hat{H}_2$, written in second quantization as

$$\hat{H} = \int d\mathbf{r}\hat{\psi}^{\dagger}(\mathbf{r})\,\hat{H}_1\hat{\psi}(\mathbf{r}) + \frac{1}{2}\int d\mathbf{r}\int d\mathbf{r}'\hat{\psi}^{\dagger}(\mathbf{r})\,\hat{\psi}^{\dagger}(\mathbf{r}')\,\hat{H}_2\hat{\psi}(\mathbf{r}')\,\hat{\psi}(\mathbf{r}). \tag{2.2}$$

For systems which include a lattice potential, we may write $\hat{H}_1$ as

$$\hat{H}_1 = \hat{H}_{\text{kin}} + \hat{H}_{\text{latt}} + \hat{H}_{\text{internal}}, \tag{2.3}$$

where $\hat{H}_{\text{kin}}$ is the kinetic energy operator, $\hat{H}_{\text{latt}}$ is the coupling of a particle to the lattice, and $\hat{H}_{\text{internal}}$ is the Hamiltonian describing the internal degrees of freedom in free space. We furthermore assume that all terms appearing in $\hat{H}_{\text{internal}}$ have no spatial dependence on the lattice scale. When this is the case then the basis

[3]For a detailed exposition of the assumptions leading to the Hubbard model in solid state systems, see [4].

diagonalizing this Hamiltonian is spatially independent, depending only on the internal degrees of freedom of a single particle in free space. We will refer to the basis diagonalizing this Hamiltonian as $|\sigma\rangle$: $\hat{H}_{\text{internal}}|\sigma\rangle = E^{\sigma}_{\text{internal}}|\sigma\rangle$.

The coupling to the lattice is provided by the dynamical polarizability tensor $\tilde{\alpha}$ of the particle in question [5]. This tensor has nine operator-valued elements which can be indexed by pairs (p, p') with p and p' running over a space fixed basis $\{x, y, z\}$.[4] Using the states $|\sigma\rangle$ we can find a representation of $\tilde{\alpha}$ within the eigenspace of $\hat{H}_{\text{internal}}$. We assume that the optical lattice consists of three independent retro-reflected[5] laser beams, each monochromatic with linear frequency ν, arranged in a simple cubic structure:

$$\mathbf{E}_{\text{opt}}(\mathbf{r}, t) = 2e^{2\pi i \nu t}\left[E_x \boldsymbol{\epsilon}^x \sin(k_l x) + E_y \boldsymbol{\epsilon}^y \sin(k_l y) + E_z \boldsymbol{\epsilon}^z \sin(k_l z)\right]. \tag{2.4}$$

Here $k_l = \pi/a$ with $a = \lambda/2$ the lattice spacing and λ the wavelength of the optical field. In addition to the intensities $I_x = E_x^2$ etc. along each spatial direction, this optical field is also described by three complex vectors $\boldsymbol{\epsilon}^\nu$, $\nu = x, y, z$ giving the polarization of the x, y, and z fields, respectively. If we use the spherical representation of the polarization vectors $\boldsymbol{\epsilon}^\nu$, we have the Stark shift

$$V_{\sigma'\sigma}(\mathbf{r}) \equiv \langle\sigma'|\hat{H}_{\text{latt}}(\mathbf{r})|\sigma\rangle = -\sum_{\nu=x,y,z}\sum_{q_\nu q_{\nu'}} |E_\nu|^2 \epsilon^{\nu\star}_{q_\nu}\langle\sigma'|\tilde{\alpha}_{q_\nu q'_\nu}|\sigma\rangle\epsilon^{\nu}_{q'_\nu}\sin^2(k_l x_\nu). \tag{2.5}$$

This shift is what provides the lattice potential for ultracold gases.

For anisotropic systems such as rotating molecules, Eq. (2.5) is not generally diagonal in the internal degrees of freedom. This is discussed for rotational eigenstates in Chap. 3, and will be discussed in full generality in a forthcoming paper [6]. In this case, the problem is complicated by the fact that the three independent beams forming $\mathbf{E}_{\text{opt}}$ generally compete for ordering of the internal state $|\sigma\rangle$. Thus, the lattice formed from three independent beams becomes non-separable due to the fact that the beams couple differently to the internal state. For simplicity, in the remainder of the present discussion we will assume that the matrix Eq. (2.5) is diagonal. In this case two simplifications occur. The first is that each $|\sigma\rangle$ obeys a Schrödinger equation which is decoupled from all other internal states. The second is that the lattice is separable in real space, and so we can solve each Cartesian direction separately.

As the single-particle Hamiltonian Eq. (2.3) is invariant under translations by a Bravais lattice vector, its eigenfunctions can be written in Bloch form [7]. We thus write the solutions of $\hat{H}_1$ in the form $\psi_{\sigma\mathbf{n}\mathbf{q}}(\mathbf{r})$. Here, the quasimomentum $\mathbf{q}$ whose components q_σ lie within the first Brillouin zone (BZ) $[-k_l, k_l)$ denote how this eigenfunction transforms under translations, and the band index $\mathbf{n}$ distinguishes

[4]Equivalently, a spherical basis $\{-1, 0, 1\}$.

[5]This accounts for the factors of 2 in front of the field strengths.

solutions with the same translational symmetry which differ in energy. Due to the separability of the lattice potential, we may write this solution as a product

$$\psi_{\sigma\mathbf{nq}}(\mathbf{r}) = \prod_{\nu=x,y,z} \psi_{\sigma n_\nu q_\nu}(r_\nu), \tag{2.6}$$

where the $\psi_{\sigma n_\nu q_\nu}(r_\nu)$ satisfy the 1D Schrödinger equations

$$\left[\frac{\hat{p}_\nu^2}{2m} + V_{\sigma\sigma}(r_\nu)\right]\psi_{\sigma n_\nu q_\nu}(x_\nu) = E_{\sigma q_\nu n_\nu}\psi_{\sigma n_\nu q_\nu}(r_\nu). \tag{2.7}$$

Here and throughout, we will use boldface to denote three-component vectors and ordinary type to denote scalar quantities, e.g. $\mathbf{q}$ refers to a 3D quasimomentum and q a 1D quasimomentum. Also, we will leave off the ν subscripts when they are unnecessary, writing instead e.g. $\psi_{\sigma nq}(x)$ for a 1D Bloch function.

We now turn to the solution of the equations governing the 1D Bloch functions Eq. (2.7). As is well known [7], Bloch functions may be written in the form

$$\psi_{\sigma nq}(x) = \frac{1}{\sqrt{La}} e^{iqx} u_{\sigma nq}(x), \tag{2.8}$$

where L is the number of (1D) unit cells and $u_{\sigma nq}(x)$ has the same periodicity as the potential. Because of this periodicity, these functions may be expanded in a Fourier series

$$u_{\sigma nq}(x) = \lim_{\ell\to\infty}\sum_{p=-\ell}^{\ell} c_{\sigma nq}^{p} e^{2\pi pix/a}. \tag{2.9}$$

Here ℓ is a finite Fourier cutoff used in numerics. A cutoff of a few tens captures the lowest few bands to machine precision. Inserting this expansion into the 1D Schrödinger equation yields an eigenvalue equation for the coefficients $c_{\sigma nq}^{p}$:

$$\lim_{\ell\to\infty}\sum_{p'=-\ell}^{\ell} H_{pp'} c_{\sigma nq}^{p'} = E_{\sigma nq} c_{\sigma nq}^{p}, \tag{2.10}$$

$$H_{pp'} = \left[(2p + q/k_l)^2 E_R + \frac{V}{2}\right]\delta_{pp'} - \frac{V}{4}\left(\delta_{p,p'+1} + \delta_{p,p'-1}\right), \tag{2.11}$$

where we have defined the recoil energy $E_R \equiv \hbar^2 k_l^2/2m$ and V is the coefficient of $\sin^2(k_l\nu)$ along the particular 1D direction. Numerically, this amounts to solving a real symmetric tridiagonal eigenproblem.

Bloch functions represent simultaneous eigenfunctions of translation and the single-particle Hamiltonian, and are hence delocalized. Typical interactions in strongly correlated systems, on the other hand, are spatially local and so Bloch functions represent an inadequate basis for describing strong interactions. A more suitable basis is provided by Wannier functions, which are the quasimomentum Fourier transform conjugates to the Bloch functions

$$w_{\sigma\mathbf{n}}(\mathbf{r}-\mathbf{r}_i)=\frac{1}{\sqrt{L^3}}\sum_{\mathbf{q}\in \mathrm{BZ}} e^{-i\mathbf{q}\cdot\mathbf{r}_i}\psi_{\sigma n\mathbf{q}}(\mathbf{r}), \tag{2.12}$$

where $\mathbf{r}_i$ represents the position of site i in the lattice. These functions are highly localized,[6] and represent a more appropriate basis in which to expand strongly interacting models. As is apparent from the notation of Eq. (2.12), a Wannier function centered around a particular lattice site $\mathbf{r}_i$ depends only on the distance from that site. We will also use the shorthand $w_{\sigma\mathbf{n}i}(\mathbf{r})\equiv w_{\sigma\mathbf{n}}(\mathbf{r}-\mathbf{r}_i)$ to simplify some expressions.

The general route to a Hubbard model is to expand the field operator $\hat{\psi}(\mathbf{r})$ in terms of the single-particle Wannier basis, resulting in a Hamiltonian of the form

$$\begin{aligned}\hat{H}=&\sum_{i,j}\sum_{\sigma n}\hat{a}^{\dagger}_{i\sigma\mathbf{n}}\langle i\sigma\mathbf{n}|\hat{H}_1|j\sigma\mathbf{n}\rangle\hat{a}_{j\sigma\mathbf{n}}\\&+\frac{1}{2}\sum_{i_1i_2i_1'i_2'}\sum_{\mathbf{n}_1\mathbf{n}_2\mathbf{n}_1'\mathbf{n}_2'}\sum_{\sigma_1\sigma_2\sigma_1'\sigma_2'}\langle i_1\sigma_1\mathbf{n}_1;i_2\sigma_2\mathbf{n}_2|\hat{H}_2|i_1'\sigma_1'\mathbf{n}_1';i_2'\sigma_2'\mathbf{n}_2'\rangle\\&\times\hat{a}^{\dagger}_{i_1\sigma_1\mathbf{n}_1}\hat{a}^{\dagger}_{i_2\sigma_2\mathbf{n}_2}\hat{a}_{i_2'\sigma_2'\mathbf{n}_2'}\hat{a}_{i_1'\sigma_1'\mathbf{n}_1'},\end{aligned} \tag{2.13}$$

where $\hat{a}^{\dagger}_{i\sigma\mathbf{n}}$ creates a particle in Wannier state $w_{\sigma\mathbf{n}}(\mathbf{r}-\mathbf{r}_i)$. The overlap integrals $\langle i\sigma\mathbf{n}|\hat{H}_1|j\sigma\mathbf{n}\rangle$ and $\langle i_1\sigma_1\mathbf{n}_1;i_2\sigma_2\mathbf{n}_2|\hat{H}_2|i_1'\sigma_1'\mathbf{n}_1';i_2'\sigma_2'\mathbf{n}_2'\rangle$ are called *Hubbard parameters*, and are the point of contact between the microscopic, few-body physics and the many-body physics. In order for this transformation to be useful, we must truncate the sums appearing in Eq. (2.13) in some form. Often, the sums over the band indices $\mathbf{n}$ and internal states σ are set by selection rules and energetics from the few-body physics. For example, for transitions in an AC microwave field with circular polarization q, only internal states for which $\langle\sigma'|\hat{d}_q|\sigma\rangle$ is nonzero will contribute, with d_q the projection of the dipole operator along the space-fixed spherical basis direction q. Similarly, interactions which transfer particles between bands on-site must preserve the parity of the bands, that is $(-1)^{n_1+n_2}=(-1)^{n_1'+n_2'}$. Additionally, it is most often assumed that the temperature, interaction scales, and all other scales of the problem are small compared to the band gap between the lowest two bands. This restricts all particles to remain in the lowest band, and so the summations

[6] More precisely, these functions can be made exponentially localized when subject to a one-dimensional inversion symmetric potential by an appropriate choice of phases on the Bloch functions [8]. The Wannier functions produced by this procedure are called maximally localized Wannier functions, and are used exclusively in this thesis unless indicated otherwise.

over band indices vanish. To avoid overly cumbersome notation, we will leave off band indices when we discuss the two-particle interactions. We will also discuss the truncation of spatial sums separately for single-particle and short- and long-ranged two-particle interactions.

2.1.1 Single-Particle Hubbard Parameters

The Hubbard parameters arising from the single-particle piece take the form

$$t_{\sigma\mathbf{n}}^{i,j} \equiv -\int d\mathbf{r} w_{\sigma\mathbf{n}}^{\star}(\mathbf{r}-\mathbf{r}_i)\left[\hat{H}_{\text{kin}}+\hat{H}_{\text{internal}}+\hat{H}_{\text{latt}}(\mathbf{r})\right] w_{\sigma\mathbf{n}}(\mathbf{r}-\mathbf{r}_j) \tag{2.14}$$

$$= -\frac{1}{L^3}\sum_{\mathbf{q}\mathbf{q}'}\int d\mathbf{r} e^{i\mathbf{q}\cdot\mathbf{r}_i}\psi_{\sigma\mathbf{n}\mathbf{q}}^{\star}(\mathbf{r})\left[\hat{H}_{\text{kin}}+\hat{H}_{\text{internal}}+\hat{H}_{\text{latt}}(\mathbf{r})\right] e^{i\mathbf{q}'\cdot\mathbf{r}_j}\psi_{\sigma\mathbf{n}\mathbf{q}'}(\mathbf{r}) \tag{2.15}$$

$$= -\frac{1}{L^3}\sum_{\mathbf{q}\mathbf{q}'} e^{i\left(\mathbf{q}\cdot\mathbf{r}_i-\mathbf{q}'\cdot\mathbf{r}_j\right)} E_{\sigma\mathbf{n}\mathbf{q}}\delta_{\mathbf{q}\mathbf{q}'} \tag{2.16}$$

$$= -\frac{1}{L^3}\sum_{\mathbf{q}} e^{i\mathbf{q}\left(\mathbf{r}_i-\mathbf{r}_j\right)} E_{\sigma\mathbf{n}\mathbf{q}}. \tag{2.17}$$

The third line used the fact that the Bloch functions diagonalize the single-particle Hamiltonian and the orthonormality of the Bloch functions. This orthogonality is also why the parameters $t_{\sigma\mathbf{n}}^{i,j}$ are diagonal in σ and $\mathbf{n}$. The parameters $t_{\sigma\mathbf{n}}^{i,j}$ with $i \neq j$ are called *tunneling* or *hopping* parameters, as they are interpreted as a particle quantum mechanically tunneling through the lattice from site i to site j. These parameters take the simple form of a quasimomentum Fourier transform of the band structure in the relative coordinate. Their contribution to the Hamiltonian is

$$\hat{H}_{\text{tunneling}} = -\sum_{\sigma\mathbf{n}}\sum_{i\neq j} t_{\sigma\mathbf{n}}^{i,j}\hat{a}_{i\sigma\mathbf{n}}^{\dagger}\hat{a}_{j\sigma\mathbf{n}}. \tag{2.18}$$

The minus sign in this definition is customary and accounts for the minus sign in the definition of $t_{\sigma\mathbf{n}}^{i,j}$. It arises because the nearest-neighbor tunneling coefficient defined as such is positive. For $i = j$, $t_{\sigma\mathbf{n}}^{i,i}$ is minus the energy of a particle in band $\mathbf{n}$ and internal state σ, and so it is customary to reverse the sign and define

$$E_{\sigma\mathbf{n}} = \frac{1}{L^3}\sum_{\mathbf{q}} E_{\sigma\mathbf{n}\mathbf{q}}. \tag{2.19}$$

This parameter is simply the average of the band structure for fixed band number $\mathbf{n}$ and internal state σ. The complete single-particle Hubbard Hamiltonian is thus

Table 2.1 Fit parameters for the tunneling at the nearest, next-nearest, and third nearest neighbor distances together with their asymptotic standard errors

Term	A	B	C
$t^{i,i+1}$	1.363 ± 0.0004	1.057 ± 0.0005	2.117 ± 0.0004
$-t^{i,i+2}$	2.491 ± 0.0023	1.957 ± 0.0011	4.361 ± 0.0012
$t^{i,i+3}$	7.294 ± 0.011	2.767 ± 0.0015	6.534 ± 0.0018

$$\hat{H}_{\text{Hsp}} = \sum_{\sigma\mathbf{n}} E_{\sigma\mathbf{n}} \sum_i \hat{n}_{i\sigma\mathbf{n}} - \sum_{\sigma\mathbf{n}} \sum_{i \neq j} t_{\sigma\mathbf{n}}^{i,j} \hat{a}_{i\sigma\mathbf{n}}^\dagger \hat{a}_{j\sigma\mathbf{n}}, \tag{2.20}$$

where $\hat{n}_{i\sigma\mathbf{n}} \equiv \hat{a}_{i\sigma\mathbf{n}}^\dagger \hat{a}_{i\sigma\mathbf{n}}$ is the number operator on site i for band $\mathbf{n}$ and internal state σ. Noting that $\hat{a}_{j\sigma\mathbf{n}}^\dagger \hat{a}_{i\sigma\mathbf{n}}$ is the Hermitian conjugate of $\hat{a}_{i\sigma\mathbf{n}}^\dagger \hat{a}_{j\sigma\mathbf{n}}$, we must have $t_{\sigma\mathbf{n}}^{j,i} = t_{\sigma\mathbf{n}}^{i,j}$ for a Hermitian Hamiltonian.[7] This allows us to write

$$\hat{H}_{\text{Hsp}} = \sum_{\sigma\mathbf{n}} E_{\sigma\mathbf{n}} \sum_i \hat{n}_{i\sigma\mathbf{n}} - \sum_{\sigma\mathbf{n}} \sum_{j>i} t_{\sigma\mathbf{n}}^{i,j} \left[\hat{a}_{i\sigma\mathbf{n}}^\dagger \hat{a}_{j\sigma\mathbf{n}} + \text{h.c.} \right]. \tag{2.21}$$

For a 1D lattice of the form $V(x) = V_0 \sin^2(\pi x/a)$, we can parameterize the dependence of the lowest band tunneling on the lattice height V_0 as [9]

$$t^{ij}/E_R = A \left(\frac{V_0}{E_R} \right)^B \exp\left(-C\sqrt{V_0/E_R} \right), \tag{2.22}$$

which is motivated by the large V_0/E_R limit of the tunneling computed using Mathieu functions [10]. A fit to numerically generated data is found to be quantitatively valid in the range $V_0/E_R > 2$. The results are summarized in Table 2.1. These expressions can be used to determine a consistent order of approximation when truncating long-range interactions.

2.2 Two-Particle Hubbard Parameters

We now turn to determining the Hubbard parameters for the two-particle interaction. Interactions can be broadly classified into short-range and long-range terms, where short-range terms are those which are well modeled by a pseudopotential of the form $U(\mathbf{r}) = g\delta(\mathbf{r})$ and long-range terms have nonlocal $\mathbf{r}$ dependence. We begin with an exposition of the short-range terms.

[7] Strictly speaking, this implies that $t_{\sigma\mathbf{n}}^{j,i} = t_{\sigma\mathbf{n}}^{i,j\star}$, but for time-reversal invariant single-particle Hamiltonians we can always choose the eigenvalues, and thus the tunnelings, to be real.

2.2.1 Short-Range Interactions

The quantum-mechanical scattering from a spherically symmetric potential can be codified in terms of partial waves ℓ, where ℓ is the relative orbital angular momentum of the scattering particles. At low energies, the dominant partial waves are the s- and p-waves, corresponding to $\ell = 0$ and $\ell = 1$, respectively. For short-range potentials, the contributions from these partial waves at low energies can be captured by pseudopotentials that depend only on a single parameter, the (s- or p-wave) *scattering length*. This is a readily calculated and measured quantity. The regularized s-wave pseudopotential is [11, 12]

$$U(\mathbf{r}) = \frac{4\pi\hbar^2 a_s}{m}\delta(\mathbf{r})\,\partial_r r, \tag{2.23}$$

where a_s is the s-wave scattering length. The term $\partial_r r$ is a regularization operator which incorporates boundary conditions on the spherically symmetric scattering problem, but is inconvenient for the lattice problem which is not naturally stated in spherical coordinates. In Cartesian coordinates, which are most natural for our simple cubic lattice, we have

$$U(\mathbf{r}) = \frac{4\pi\hbar^2 a_s}{m}\delta(x)\,\delta(y)\,\delta(z)\left[\left(x\partial_x + y\partial_y + z\partial_z\right) + 1\right]. \tag{2.24}$$

Thus, provided that our functions have sufficiently regular derivatives at $r \to 0$, the first term in brackets vanishes in this same limit. The regularity of the Wannier functions is assured by the fact that they are band-limited periodic functions, and so the pseudopotential becomes

$$U(\mathbf{r}) = \frac{4\pi\hbar^2 a_s}{m}\delta(\mathbf{r}). \tag{2.25}$$

That is, we can neglect the regularization operator.

Using this pseudopotential in the second-quantized many-body Hamiltonian and expanding in terms of lowest band Wannier functions, we have

$$\hat{H}_{s\text{-wave}} = \frac{2\pi\hbar^2}{m}\int d\mathbf{r}\int d\mathbf{r}'\hat{\psi}^\dagger(\mathbf{r})\,\hat{\psi}^\dagger(\mathbf{r}')\,a_s\delta(\mathbf{r}-\mathbf{r}')\,\hat{\psi}(\mathbf{r}')\,\hat{\psi}(\mathbf{r}) \tag{2.26}$$

$$= \frac{2\pi\hbar^2}{m}\sum_{i_1 i_2 i_1' i_2'}\sum_{\sigma_1\sigma_2\sigma_1'\sigma_2'} a_s^{\sigma_1'\sigma_2'}\hat{a}^\dagger_{i_1\sigma_1}\hat{a}^\dagger_{i_2\sigma_2'}\hat{a}_{i_2'\sigma_2'}\hat{a}_{i_1'\sigma_1'}$$

$$\times\int d\mathbf{r} w^\star_{i_1\sigma_1}(\mathbf{r})\,w^\star_{i_2\sigma_2}(\mathbf{r})\,w_{i_2'\sigma_2'}(\mathbf{r})\,w_{i_1'\sigma_1}(\mathbf{r}), \tag{2.27}$$

where $a_s^{\sigma\sigma'}$ is now the s-wave scattering length for particles in internal states σ and σ'. Note that this does not depend on the band or site indices, as it is a quantity calculated in free 3D space. All of the lattice physics is encapsulated in the Wannier overlap integral, which is essentially a geometrical factor. Each term in the summation can be interpreted as a scattering process where particles in states σ_1' and σ_2' at positions i_1' and i_2' scatter via the s-wave pseudopotential into internal states σ_1 and σ_2 at positions i_1 and i_2, respectively. We will call the primed indices incoming and the unprimed indices outgoing. Because of the localization of the Wannier functions, terms with $i_1 \neq i_2 \neq i_1' \neq i_2'$ are exponentially suppressed, and so the most common approximation is to keep only the $i_1 = i_2 = i_1' = i_2'$ term. In the case of a single internal state, this leads to

$$\hat{H}_{s\text{-wave}} = \frac{2\pi\hbar^2 a_s}{m} \int d\mathbf{r} \, |w_0(\mathbf{r})|^4 \sum_i \hat{n}_i (\hat{n}_i - 1), \tag{2.28}$$

from which we can immediately read off the value of U appearing in the Bose–Hubbard model Eq. (1.2). It is important to note that the lattice affects the strength of the interactions through the overlap of the Wannier functions. Hence, for a fixed scattering length the interactions can be increased by making the lattice deeper. By replacing a single lattice site with a harmonic oscillator with oscillator length $a_{\text{h.o.}} = a/\pi (V_0/E_R)^{1/4}$ and thus approximating the Wannier functions with harmonic oscillator eigenstates,[8] we find that these integrals all grow as roughly $(V_0/E_R)^{1/4}$ with the lattice depth V_0. Finally, in the case where the lattice potential is separable, the Wannier function overlaps also separate into three independent 1D integrals, and so can be computed numerically with high efficiency.

The situation is similar for p-wave interactions. The regularized p-wave pseudopotential is [12]

$$U(\mathbf{r}) = \frac{\pi\hbar^2 a_1^3}{\mu} \overleftarrow{\nabla} \delta(\mathbf{r}) \overrightarrow{\nabla} r \partial_{rrr} r^2, \tag{2.29}$$

where the arrows indicate the action of the gradient operators. Using the Leibnitz formula we have

$$\partial_{rrr}\left(r^2 f\right) = \left(6\partial_r + 6r\partial_{rr} + r^2\partial_{rrr}\right) f, \tag{2.30}$$

and so the only term which possibly doesn't vanish as $r \to 0$ for sufficiently regular functions f is the first. We will denote all such terms which vanish as $r \to 0$ as van.terms. We have

[8]The on-site interaction coefficient for delta-function interactions in the deep lattice limit is the only place that this approximation has any credence. For off-site interactions or tunneling parameters, the use of the approximation leads not only to quantitative errors of an order of magnitude or more, but also qualitative errors in the symmetries of the Hubbard parameters, see Chap. 5.

$$\overrightarrow{\nabla} r \partial_{rrr} \left(r^2 f\right) = 6\overrightarrow{\nabla} \left[x\partial_x + y\partial_y + z\partial_z\right] f + \text{van.terms}, \tag{2.31}$$

$$= 6\overrightarrow{\nabla} f + \text{van.terms}, \tag{2.32}$$

and so the regularized potential for our sufficiently regular Wannier functions becomes

$$U\left(\mathbf{r}\right) = \frac{6\pi\hbar^2 a_1^3}{\mu} \overleftarrow{\nabla} \delta\left(\mathbf{r}\right) \overrightarrow{\nabla}. \tag{2.33}$$

Expanding the second quantized Hamiltonian as above, we have

$$\hat{H}_{p\text{-wave}} = \frac{3\pi\hbar^2}{\mu} \int d\mathbf{r} \int d\mathbf{r}' \hat{\psi}^{\dagger}\left(\mathbf{r}\right) \hat{\psi}^{\dagger}\left(\mathbf{r}'\right) a_1^3 \overleftarrow{\nabla} \delta\left(\mathbf{r}-\mathbf{r}'\right) \overrightarrow{\nabla} \hat{\psi}\left(\mathbf{r}'\right) \hat{\psi}\left(\mathbf{r}\right), \tag{2.34}$$

$$= \frac{3\pi\hbar^2}{\mu} \sum_{i_1 i_2 i_2' i_1'} \sum_{\sigma_1 \sigma_2 \sigma_2' \sigma_1'} \left(a_1^{\sigma_1' \sigma_2'}\right)^3 \hat{a}_{i_1\sigma_1}^{\dagger} \hat{a}_{i_2\sigma_2}^{\dagger} \hat{a}_{i_2'\sigma_2'} \hat{a}_{i_1'\sigma_1'} \int d\mathbf{r}$$
$$\times \int d\mathbf{r}' w_{i_1\sigma_1}^{\star}\left(\mathbf{r}\right) w_{i_2\sigma_2}^{\star}\left(\mathbf{r}'\right) \overleftarrow{\nabla}_{\mathbf{r}-\mathbf{r}'} \delta\left(\mathbf{r}-\mathbf{r}'\right) \overrightarrow{\nabla}_{\mathbf{r}-\mathbf{r}'} w_{i_2'\sigma_2'}\left(\mathbf{r}'\right) w_{i_1'\sigma_1'}\left(\mathbf{r}\right), \tag{2.35}$$

$$= \frac{6\pi\hbar^2}{m} \sum_{i_1 i_2 i_2' i_1'} \sum_{\sigma_1 \sigma_2 \sigma_2' \sigma_1'} \left(a_1^{\sigma_1' \sigma_2'}\right)^3 \hat{a}_{i_1\sigma_1}^{\dagger} \hat{a}_{i_2\sigma_2}^{\dagger} \hat{a}_{i_2'\sigma_2'} \hat{a}_{i_1\sigma_1'} \int d\mathbf{r}$$
$$\times \int d\mathbf{r} \left[\left(\nabla w_{i_1\sigma_1}^{\star}\left(\mathbf{r}\right)\right) w_{i_2\sigma_2}^{\star}\left(\mathbf{r}\right) - w_{i_1\sigma_1}^{\star}\left(\mathbf{r}\right) \left(\nabla w_{i_2\sigma_2}^{\star}\left(\mathbf{r}\right)\right)\right]$$
$$\times \left[\left(\nabla w_{i_1'\sigma_1'}\left(\mathbf{r}\right)\right) w_{i_2'\sigma_2'}\left(\mathbf{r}\right) - w_{i_1'\sigma_1'}\left(\mathbf{r}\right) \left(\nabla w_{i_2'\sigma_2'}\left(\mathbf{r}\right)\right)\right]. \tag{2.36}$$

Here $a_1^{\sigma\sigma'}$ is the p-wave scattering length for internal states σ and σ'. We note that the derivatives can be rigorously taken, i.e., without any discretization error, using the Fourier representation of the Bloch functions, Eq. (2.9). Also, note that this integral vanishes identically if all of the incoming or outgoing indices are the same. That is to say, there is no on-site p-wave interaction between identical particles.

2.2.2 Long-Range Interactions

For alkali atoms, the dominant interaction is provided by a $1/R^6$ dispersive potential [13] and hence interactions are well modeled by a short-range pseudopotential. We now turn our attention to long-ranged interactions arising from the dipole–dipole interaction Eq. (1.8) which are relevant for molecules. We can recast the interaction Eq. (1.8) as the contraction of two rank-two spherical tensors as

$$\hat{H}_{\mathrm{DD}}(\mathbf{R}) = -\frac{\sqrt{6}}{R^3} C^{(2)}(\mathbf{R}) \cdot \left[\hat{\mathbf{d}}_1 \otimes \hat{\mathbf{d}}_2\right]^{(2)}, \tag{2.37}$$

where

$$\left[\hat{\mathbf{d}}_1 \otimes \hat{\mathbf{d}}_2\right]^{(2)}_q = \sum_m \langle 1, m, 1, q-m|2, q\rangle (d_1)_m (d_2)_{q-m}, \tag{2.38}$$

$\langle j_1 m_1 j_2 m_2 | j m\rangle$ is a Clebsch–Gordan coefficient, $(d_j)_m$ represents the mth component of the jth dipole in a space-fixed spherical basis, and $C^{(2)}_m(\mathbf{R})$ is an unnormalized spherical harmonic in the polar coordinates defined by the relative coordinate. We compute Hubbard parameters in the usual way by expanding the field operators in terms of Wannier functions, which yields the general terms

$$\frac{1}{2}\int d\mathbf{r} \int d\mathbf{r}' \hat{\psi}^{\dagger}(\mathbf{r})\, \hat{\psi}^{\dagger}(\mathbf{r}')\, \hat{H}_{\mathrm{DD}}(\mathbf{r}-\mathbf{r}')\, \hat{\psi}(\mathbf{r}')\, \hat{\psi}(\mathbf{r}) \tag{2.39}$$

$$= \frac{1}{2} \sum_{\sigma_1\sigma_2\sigma_2'\sigma_1'} \sum_{i_1 i_2 i_2' i_1'} V^{\sigma_1\sigma_2\sigma_2'\sigma_1'}_{i_1 i_2 i_2' i_1'} \hat{a}^{\dagger}_{i_1\sigma_1} \hat{a}^{\dagger}_{i_2\sigma_2} \hat{a}_{i_2'\sigma_2'} \hat{a}_{i_1'\sigma_1'},$$

$$V^{\sigma_1\sigma_2\sigma_2'\sigma_1'}_{i_1 i_2 i_2' i_1'} = \sum_q D^{\sigma_1'\sigma_2'}_{q;\sigma_1\sigma_2} W^{\sigma_1\sigma_2\sigma_2'\sigma_1'}_{q;i_1 i_2 i_2' i_1'},$$

$$D^{\sigma_1'\sigma_2'}_{q;\sigma_1\sigma_2} = \sqrt{6}(-1)^q \langle\sigma_1\sigma_2| \left[\hat{\mathbf{d}}_1 \otimes \hat{\mathbf{d}}_2\right]^{(2)}_q |\sigma_1'\sigma_2'\rangle, \tag{2.40}$$

$$W^{\sigma_1\sigma_2\sigma_2'\sigma_1'}_{q;i_1 i_2 i_2' i_1'} = -\int d\mathbf{r} F^{\sigma_1\sigma_1'}_{i_1 i_1'}(\mathbf{r}) \int d\mathbf{r}' \frac{C^{(2)}_{-q}(\mathbf{r}'-\mathbf{r})}{|\mathbf{r}'-\mathbf{r}|^3} F^{\sigma_2\sigma_2'}_{i_2 i_2'}(\mathbf{r}'). \tag{2.41}$$

In the last line we have defined $F^{\sigma\sigma'}_{ij}(\mathbf{r}) \equiv w^{\star}_{i\sigma}(\mathbf{r})\, w_{j\sigma'}(\mathbf{r})$.

Our evaluation of the Hubbard parameters now breaks into two pieces, one dealing only with the dipole moments' dependence on the internal state and the other with the spatial distribution of the Wannier functions. Here we focus only on the geometrical part of the integral. For convenience, however, we will anticipate the result that only the $q = 0$ terms are relevant for molecules, take a single internal state σ and define

$$W_{i_1 i_2 i_2' i_1'} \equiv W^{\sigma\sigma\sigma\sigma}_{0;i_1 i_2 i_2' i_1'}, \tag{2.42}$$

$$F_{ij}(\mathbf{r}) \equiv F^{\sigma\sigma}_{ij}(\mathbf{r}), \tag{2.43}$$

$$D \equiv D^{\sigma\sigma}_{0;\sigma\sigma}. \tag{2.44}$$

The integral in question,

$$W_{i_1 i_2 i_2' i_1'} = -\int d\mathbf{r} F_{i_1 i_1'}(\mathbf{r}) \int d\mathbf{r}' \frac{C_0^{(2)}(\mathbf{r}'-\mathbf{r})}{|\mathbf{r}'-\mathbf{r}|^3} F_{i_2 i_2'}(\mathbf{r}') , \tag{2.45}$$

is naively six-dimensional integral. However, noting that

$$-\int d\mathbf{r}' \frac{C_0^{(2)}(\mathbf{r}'-\mathbf{r})}{|\mathbf{r}'-\mathbf{r}|^3} F_{i_2 i_2'}(\mathbf{r}') \tag{2.46}$$

is the convolution of the dipole–dipole potential with the function $F_{i_2 i_2'}(\mathbf{r}')$, we can use the convolution theorem to find

$$W_{i_1 i_2 i_2' i_1'} = \int d\mathbf{r} F_{i_1 i_1'}(\mathbf{r}) \mathcal{F}^{-1}\left[\mathcal{F}\left[-\frac{C_0^{(2)}(\mathbf{r})}{|\mathbf{r}|^3}\right](\mathbf{k}) \mathcal{F}\left[F_{i_2 i_2'}(\mathbf{r})\right](\mathbf{k})\right](\mathbf{r}) , \tag{2.47}$$

where $\mathcal{F}[g(\mathbf{r})](\mathbf{k})$ is the Fourier transform of $g(\mathbf{r})$. The integral is now a series of two 3D Fourier transforms followed by an 3D integral in real space. The Fourier transform approach to compute dipolar integrals has also been used for dipolar Gross–Pitaevskii equations [14, 15]. Using the expansion of a plane wave in terms of spherical harmonics [16]

$$e^{-i\mathbf{k}\cdot\mathbf{r}} = 4\pi \sum_{\ell=0}^{\infty} \sum_{m=-\ell}^{\ell} (-i)^{\ell} j_{\ell}(kr) Y_m^{\ell\star}(\theta_r, \phi_r) Y_m^{\ell}(\theta_k, \phi_k) , \tag{2.48}$$

with $j_\ell(x)$ the spherical Bessel function of order ℓ and (θ_x, ϕ_x) the spherical angles in the x coordinate system defined with respect to the quantization axis, we find the Fourier transform of the dipole–dipole potential to be

$$\mathcal{F}\left[-2\frac{C_0^2(\mathbf{r})}{r^3}\right](\mathbf{k}) = \mathcal{F}\left[\frac{1-3\cos^2\theta}{r^3}\right](\mathbf{k}) = \frac{4\pi}{3}\left(\cos^2\theta_k - 1\right) , \tag{2.49}$$

with θ_k the polar angle in **k**-space. We note that the introduction of a spherical cutoff b on the lower limit of the r integration results in the function

$$4\pi\left(\cos^2\theta_k - 1\right)\left[\frac{\sin kb}{(kb^3)} - \frac{\cos kb}{(kb)^2}\right] . \tag{2.50}$$

The corrections to Eq. (2.49) scale as $(kb)^2$, and so are negligible for small b and k in the range we typically consider. We can thus safely take $b \to 0$ in computing these integrals and work with Eq. (2.49).

Numerically, the Fourier transforms can be performed in $\mathcal{O}\left(N^3 \log N\right)$ time using the fast Fourier transform (FFT) algorithm, where N is the number of grid points discretizing each $1D$ domain. More precisely, consider each 1D domain to be a symmetric finite interval $[-L/2, L/2]$ and then let $n_g - 1$ be even, with n_g the number of grid points. We then introduce the points

$$x_k = -\frac{L}{2} + \frac{(k-1)\,L}{n_g - 1}, \tag{2.51}$$

where k runs from 1 to n_g and $x_1 \equiv x_{n_g}$ because of the periodicity of the domain. The discrete Fourier domain of the $n_g - 1$ points $x_1 \ldots x_{n_g-1}$ is represented by the values $\frac{2\pi}{L}\left[0, 1, \ldots, \frac{n_g-1}{2} - 1, \frac{n_g-1}{2}, -\frac{n_g-1}{2} + 1, \ldots, -2, -1\right]$. Hence, the spacing in Fourier space is controlled by L, the length of the domain in real space, and the extent of the domain in Fourier space is controlled by n_g/L, the inverse step size. Because the Wannier functions $w(x)$ on a finite domain are periodic and band-limited, their discrete and continuous Fourier transforms are related by a scaling constant provided we sample the entire domain at a frequency of at least twice the largest frequency component of $w(x)$ [17]. This prevents us from having to consider the more advanced interpolation schemes required for Fourier integrals of general functions [18]. Furthermore, as is known for spectral methods, the calculation in Fourier space converges exponentially fast in L provided that n_g/L is large enough to capture the full support of the function in Fourier space.[9] Let us define g to be the extent of the function in the discrete Fourier space. We can then choose $n_g = 2 * g * L + 1$ to satisfy all of the above considerations. We find that for typical $g \sim 5 - 7$ which satisfy the Nyquist condition, the real space integration is of acceptable precision using a high-order Simpson integrator [18].

As the proposed calculation scales as $\mathcal{O}\left(L^3\right)$, we would like to determine the smallest L such that we obtain the value of the Hubbard parameter in the limit as $L \to \infty$ to within a desired tolerance. Consider computing W_{0dd0} on a domain of length L. There is a size-independent contribution to the dipole–dipole parameter which scales as $\sim 1/d^p$, where p is some power, but there is another contribution which scales as $\sim 1/|L - d|^{p_2}$ coming from the periodic boundary conditions. Because of the slow power law decay, this term can be quite sizable, especially when d itself is large. The solution is to compute the integral for a few L and then fit to the form

$$I\left(L\right) = a + \frac{b}{|L - d|^p}. \tag{2.52}$$

[9] The support of the function in Fourier space can be determined by using Parseval's theorem on finite Fourier subintervals to determine that the norm is unity to a desired tolerance.

a gives the true dipole–dipole parameter in the limit as $L \to \infty$. Practical experience shows that taking four values of L starting at roughly $4d$[10] produces an excellent fit.

Naively, it may be surprising that the integration over the dipole–dipole potential with its $1/r^3$ behavior yields convergent results. The key point is that the dipole–dipole potential is also proportional to a rank-two spherical harmonic, and so picks out the components of the Wannier function product with d-wave symmetry. This component necessarily vanishes at the origin, and so there is no actual divergence in the integrand. Additionally, because of the strong anisotropy of the spherical harmonic, this Hubbard parameter is quite sensitive to transverse confinement. Confinement modifies not only the strength of the potential as in the short-range case, but also the power p of the potential. This result is explored in greater detail in [19].

Thus, the result of expanding the dipole–dipole interaction in terms of Wannier functions is a term

$$\hat{H} = \sum_{i<j} U^{i,j} \hat{n}_i \hat{n}_j, \tag{2.53}$$

where $U^{i,j}$ exhibits power-law behavior at large separations $|i - j|$ and is sensitive to transverse confinement. That is, Hubbard models for molecules involve long-ranged interactions, as opposed to models for atoms. For molecules with multiple internal states which have dipole-allowed transitions, the dipole–dipole potential also gives rise to an "exchange" contribution which describes the long-ranged propagation of a rotational quantum through the system [20–22].

The general procedure outlined in this chapter is used in Part II to derive effective models for heteronuclear bialkali molecules in optical lattices. However, the procedure here is not the only way in which a Hubbard model can be derived. Part III uses a different technique, in which the relevant Wannier functions are formed from bound states of a two-particle Hamiltonian. These two-particle bound states are then coupled to fermions in the lowest Bloch band through a Wannier function overlap of a pairing Hamiltonian. In this way, higher bands are included in the Hubbard model, but only in the configurations which are determined to be physically relevant from few-body physics.

[10]Four was chosen by assuming that the decay was purely dipolar and then determining the L such that $1/d^3 > x/(L-d)^3$ for some scaling factor x. Taking $x \sim 100$ gives an estimate of $L \sim 4d$.

References

1. Shankar, R.: Renormalization-group approach to interacting fermions. Rev. Mod. Phys. **66**, 129–192 (1994). doi:10.1103/RevModPhys.66.129. http://link.aps.org/doi/10.1103/RevModPhys.66.129
2. Fetter, A.L., Walecka, J.D.: Quantum Theory of Many-Particle Systems. McGraw-Hill, New York (1971)
3. Hubbard, J.: Electron correlations in narrow energy bands. Proc. R. Soc. Lond. A **276**, 238 (1963)
4. Auerbach, A.: Interacting Electrons and Quantum Magnetism. Springer, Berlin (1994)
5. Bonin, K.D., Kresin, V.V. (eds.): Electric-Dipole Polarizabilities of Atoms, Molecules, and Clusters. World Scientific, Singapore (1988)
6. Wall, M.L., Carr, L.D.: In preparation
7. Ashcroft, N., Mermin, D.: Solid State Physics. Saunders College Publishing, Orlando (1976)
8. Kohn, W.: Analytic properties of bloch waves and wannier functions. Phys. Rev. **115**(4), 809–821 (1959). doi:10.1103/PhysRev.115.809
9. Rey, A.M.: Ultracold bosonic atoms in optical lattices. Ph.D. thesis, University of Maryland (2004)
10. Bloch, I., Dalibard, J., Zwerger, W.: Many-body physics with ultracold gases. Rev. Mod. Phys. **80**, 885 (2008)
11. Lee, T.D., Huang, K., Yang, C.N.: Eigenvalues and eigenfunctions of a bose system of hard spheres and its low-temperature properties. Phys. Rev. **106**(6), 1135–1145 (1957). doi:10.1103/PhysRev.106.1135
12. Idziaszek, Z., Calarco, T.: Pseudopotential method for higher partial wave scattering. Phys. Rev. Lett. **96**, 013201 (2006). doi:10.1103/PhysRevLett.96.013201. http://link.aps.org/doi/10.1103/PhysRevLett.96.013201
13. Pethick, C.J., Smith, H.: Bose-Einstein Condensation in Dilute Gases. Cambridge University Press, Cambridge (2002)
14. Góral, K., Rzążewski, K., Pfau, T.: Bose-Einstein condensation with magnetic dipole-dipole forces. Phys. Rev. A **61**, 051601 (2000). doi:10.1103/PhysRevA.61.051601. http://link.aps.org/doi/10.1103/PhysRevA.61.051601
15. Bortolotti, D.C.E., Ronen, S., Bohn, J.L., Blume, D.: Scattering length instability in dipolar Bose-Einstein condensates. Phys. Rev. Lett. **97**, 160402 (2006). doi:10.1103/PhysRevLett.97.160402. http://link.aps.org/doi/10.1103/PhysRevLett.97.160402
16. Gottfried, K., Yan, T.-M.: Quantum Mechanics, 2nd edn. Springer, New York (2004)
17. Brigham, E.O.: The Fast Fourier Transform and Applications. Prentice Hall, Englewood Cliffs (1988)
18. Press, W.H., Teukolsky, S.A., Vetterling, W.T., Flannery, B.P.: Numerical Recipes in C: The Art of Scientific Computing. Cambridge University Press, Cambridge (1993)
19. Wall, M.L., Carr, L.D.: Dipole-dipole interactions in optical lattices do not follow an inverse cube power law. New J. Phys. **15**(12), 123005 (2013). http://stacks.iop.org/1367-2630/15/i=12/a=123005
20. Gorshkov, A.V., Manmana, S.R., Chen, G., Ye, J., Demler, E., Lukin, M.D., Rey, A.M.: Tunable superfluidity and quantum magnetism with ultracold polar molecules. Phys. Rev. Lett. **107**, 115301 (2011). doi:10.1103/PhysRevLett.107.115301. http://link.aps.org/doi/10.1103/PhysRevLett.107.115301
21. Gorshkov, A.V., Manmana, S.R., Chen, G., Demler, E., Lukin, M.D., Rey, A.M.: Quantum magnetism with polar alkali-metal dimers. Phys. Rev. A **84**, 033619 (2011). doi:10.1103/PhysRevA.84.033619. http://link.aps.org/doi/10.1103/PhysRevA.84.033619
22. Herrera, F., Krems, R.V.: Tunable Holstein model with cold polar molecules. Phys. Rev. A **84**, 051401 (2011). doi:10.1103/PhysRevA.84.051401. http://link.aps.org/doi/10.1103/PhysRevA.84.051401

Part II
The Molecular Hubbard Hamiltonian

Chapter 3
Emergent Timescales in Entangled Quantum Dynamics of Ultracold Molecules in Optical Lattices

Abstract We derive a novel lattice Hamiltonian, the *molecular Hubbard Hamiltonian* (MHH), which describes the essential many-body physics of closed-shell ultracold heteronuclear molecules in their absolute ground state in a quasi-one-dimensional optical lattice. The MHH is explicitly time dependent, making a dynamic generalization of the concept of quantum phase transitions necessary. Using the time-evolving block decimation (TEBD) algorithm to study entangled dynamics, we demonstrate that, in the case of hard-core bosonic molecules at half-filling, the MHH exhibits emergent timescales over which spatial entanglement grows, crystalline order appears and oscillations between rotational states self-damp into an asymptotic superposition. We show that these timescales are non-monotonic functions of the physical parameters describing the lattice.

3.1 Introduction

In recent years, ultracold atomic gases have provided near perfect realizations of condensed matter Hamiltonians, acting as *quantum simulators* [1, 2] that allow the study of complex condensed matter phenomena in a clean and highly controllable environment. Ultracold polar molecular gases, which have recently been brought to the edge of quantum degeneracy in their absolute ground state [3, 4], offer additional features over atomic gases, such as a large internal Hilbert space and a greater susceptibility to external fields via a permanent electric dipole. There have been a number of proposals on how to use ultracold molecular gases for mimicking well-known Hamiltonians such as spin-1 lattice models [5]. Ultracold molecules have also been suggested as a model system for the study of strongly correlated 2D quantum phases [6] or for quantum information processing schemes [7–9]. However, these proposals frequently involve complex and yet-to-be implemented experimental techniques. In this article, we instead focus on the completely *new*

Published previously as *Emergent Timescales in Entangled Quantum Dynamics of Ultracold Molecules in Optical Lattices*, M.L. Wall and L.D. Carr, New J. Phys. **11** 055027 (2009).

M.L. Wall, *Quantum Many-Body Physics of Ultracold Molecules in Optical Lattices*, Springer Theses, DOI 10.1007/978-3-319-14252-4_3

quantum many-body physics which results naturally from the simplest quantum lattice experiments that can be performed in the immediate future with established techniques in ultracold molecular quantum gases.

Towards this end we derive a novel lattice Hamiltonian, which we refer to as the *molecular hubbard hamiltonian* (MHH). The MHH describes the physics of an ultracold polar molecular gas in a 1D optical lattice that is oriented using a DC electric field, giving rise to a resonant dipole–dipole interaction, and is driven between rotational levels using a microwave AC field. In particular, new aspects of our derivation include explicit dependence of hopping energy on the molecular polarizability tensor. This in turn allows a determination of the tensor elements, an outstanding experimental issue, from the borders of the static phase diagram of the MHH, which are identical to those of the extended Bose–Hubbard Hamiltonian [10] when a single molecular rotational level is occupied.

Beyond the statics, the MHH naturally has a dynamical component due to the AC driving fields, as well as an internal structure in terms of rotational modes which is inherently different from spinor atomic systems [11, 12]. We study this dynamical aspect with time-evolving block decimation (TEBD) [13, 14], a newly developed entangled quantum dynamics algorithm which takes spatial entanglement (specifically, Schmidt number [15]) as a cut-off. We find two emergent timescales in the case of half-filling for hard core bosonic molecules. We emphasize that a *quantum* lattice model requires low filling (average number of particles per site), in contrast to a mean field lattice model, for which the filling would typically be quite high. Thus, although experiments can most easily access the mean field regime of hundreds of molecules per site with a single pair of counter-propagating laser beams, we look slightly ahead to the quantum regime, which will require two pairs of such beams in order to create an array of quasi-1D "tubes." A third pair is then used to create the lattice in each tube. This technique is already well established for ultracold atoms [16].

Dynamical aspects of quantum phase transitions are just beginning to be considered [17, 18], and have so far been a limited area of study restricted to mean field considerations, due to lack of numerical tools. With the recent advent of entangled quantum dynamics algorithms, namely TEBD, dynamical properties of many-body systems are becoming amenable to numerical study. For example, TEBD has been used to address key questions such as the dynamics of a quantum quench [19, 20] or the speed at which correlations propagate in a lattice [21]; these are not issues which can be studied with other dynamical methods such as dynamical mean field theory (DMFT).[1] We give a brief review of TEBD in Sect. 3.3.1. The reader interested in computational details can find them in [23].

The first main contribution of this paper is to present a careful derivation of the MHH. This is done in Sect. 3.2, with some previously known aspects of molecular

[1]Time-dependent Density Functional Theory (TDFT) may be able to succeed in a partial analysis of entangled dynamics, which might even be complementary to methods such as TEBD, but this is not yet at all clear in the literature [22].

physics relegated to Appendix 1. The second main contribution is to present emergent timescales for half-filling; although we treat the case of hard-core bosons, the MHH can also be applied to fermionic molecules. To this end, in Sect. 3.3 we first give a brief explanation of TEBD and the quantum measures we use. Then, in Sect. 3.4 we present and analyze our simulations, with an accompanying convergence study in Appendix 2. Finally, in Sect. 3.5 we summarize.

3.2 The MHH

The MHH is

$$\begin{aligned}\hat{H} = &-\sum_{JJ'M} t_{JJ'M} \sum_{\langle i,i' \rangle} \left(\hat{a}^{\dagger}_{i',J'M} \hat{a}_{iJM} + \text{h.c.} \right) \\ &+ \sum_{JM} E_{JM} \sum_{i} \hat{n}_{iJM} - \pi \sin(\omega t) \sum_{JM} \Omega_{JM} \sum_{i} \left(\hat{a}^{\dagger}_{iJ,M} \hat{a}_{iJ+1,M} + \text{h.c.} \right) \\ &+ \frac{1}{2} \sum_{\substack{J_1, J_1', J_2, J_2' \\ M, M'}} U_{\text{dd}}^{\substack{J_1, J_1', J_2, J_2' \\ M, M'}} \sum_{\langle i,i' \rangle} \hat{a}^{\dagger}_{iJ_1M} \hat{a}_{iJ_1'M} \hat{a}^{\dagger}_{i'J_2M'} \hat{a}_{i'J_2'M'} ,\end{aligned} \quad (3.1)$$

where $\hat{a}_{iJM}$ destroys a bosonic or fermionic molecule in the $|\mathcal{E}; JM\rangle$ state (defined below) on the ith lattice site, and the bracket notation $\langle \ldots \rangle$ denotes that the sum is taken over nearest neighbors. The first term in Eq. (3.1) corresponds to hopping both between sites and molecular rotational states with quantum numbers J, M. The second term represents the rotational energy along with rotational state-dependent energy differences due to a DC electric field. The third term corresponds to an AC electric field, making this a driven system. The fourth term corresponds to electric dipole–dipole interactions. In the following subsections and Appendix 1 we justify Eq. (3.1) with a careful derivation and present the energy scales of each term.

3.2.1 *Derivation of the MHH*

The full molecular Hamiltonian in second quantization is

$$\begin{aligned}\hat{H} = &\int d^3r \, \hat{\psi}^{\dagger}(\mathbf{r}) \left[\hat{H}_{\text{kin}} + \hat{H}_{\text{rot}} + \hat{H}_{\text{DC}} + \hat{H}_{\text{AC}}(t) + \hat{H}_{\text{opt}}(\mathbf{r}) \right] \hat{\psi}(\mathbf{r}) \\ &+ \int d^3r d^3r' \, \hat{\psi}^{\dagger}(\mathbf{r}) \hat{\psi}^{\dagger}(\mathbf{r}') \hat{H}_{\text{dd}}(|\mathbf{r} - \mathbf{r}'|) \hat{\psi}(\mathbf{r}') \hat{\psi}(\mathbf{r}) .\end{aligned} \quad (3.2)$$

The terms on the first line correspond to single-molecule effects: kinetic energy, rotation, the DC electric field which orients the dipole, the AC microwave field which drives transitions between rotational levels, and the far off-resonant optical lattice potential, respectively. The second line is the two-molecule resonant dipolar energy. The field operators $\hat{\psi}$ can be either bosonic or fermionic. We focus on the bosonic case for brevity. There are five key assumptions underlying our derivation, as follows. We consider all five assumptions to be reasonable for present and near-future experiments.

1. We consider ultracold closed-shell polar heteronuclear diatomic molecules, characterized by permanent dipole moment d and rotational constant B. The most experimentally relevant bosonic species in this category are SrO, RbCs, and LiCs [6]. The individual molecules are assumed to be in their electronic and vibrational ground states, and it is assumed that none of these degrees of freedom can be excited at the large intermolecular separations and low temperatures/relative energies that we consider.
2. The molecule is assumed to have a ${}^1\Sigma$ ground state. The characteristic trapping potential length is chosen large enough compared to the internuclear axis to assume spherical symmetry, i.e. a locally constant potential.
3. We neglect any intramolecular interactions (e.g., hyperfine structure), as they are typically very small for ${}^1\Sigma$ molecules [24].
4. We consider only the lowest three rotational levels. All AC fields will be sufficiently weak to allow this assumption.
5. We work in the hard-core limit where at most one molecule is allowed per site. This is enforced by strong dipole–dipole interactions on-site. We consider the lattice spacing large enough to include only nearest-neighbor dipole–dipole interactions. Other short-range interactions such as exchange or chemical reactions or long-range interactions such as dispersion and quadrupole–quadrupole interactions are not considered.

We proceed to follow the usual procedure [2] of expanding the field operators of our second-quantized Hamiltonian in a Wannier basis of single-molecule states centered at a particular discrete position $\mathbf{r}_i$:

$$\hat{\psi} = \sum_i \hat{a}_i w\left(\mathbf{r} - \mathbf{r}_i\right), \tag{3.3}$$

where i is a site index and the sum is over all lattice sites. For our Wannier Basis we choose the single-molecule basis that diagonalizes the rotational and DC electric field Hamiltonians, spanned by kets $|\mathcal{E}; JM\rangle$. In this basis, which we refer to as the "dressed basis" (the DC field "dresses" the rotational basis) we have the field operator expansion

$$\hat{\psi}_{JM} = \sum_i \hat{a}_{iJM} w_{JM}\left(\mathbf{r} - \mathbf{r}_i\right) \equiv \sum_i \hat{a}_{iJM} |\mathcal{E}; JM\rangle_i . \tag{3.4}$$

We note that such a basis, while highly efficient for the hard-core limit we consider, becomes progressively worse for higher filling factors, till in the mean field limit the single-molecule basis, whether dressed or not, is so poor that many bands must be considered. Here we do not include a band index for simplicity, although the generalization of Eq. (3.1) to include multiple bands is straightforward.

This choice of Wannier basis associates the terms in Eq. (3.1) to the terms in Eq. (3.2) as follows:

$$t_{J,J',M} \equiv -\int d\mathbf{r}\, w^{\star}_{JM}\left(\mathbf{r}-\mathbf{r}_i\right)\left[H_{\text{kin}}+H_{\text{opt}}\right] w_{J'M}\left(\mathbf{r}-\mathbf{r}_{i+1}\right), \tag{3.5}$$

$$E_{JM} \equiv \int d\mathbf{r}\, w^{\star}_{JM}\left(\mathbf{r}-\mathbf{r}_i\right)\left[H_{\text{rot}}+H_{\text{DC}}\right] w_{JM}\left(\mathbf{r}-\mathbf{r}_i\right), \tag{3.6}$$

$$-\pi\Omega_{JM}\sin\left(\omega t\right) \equiv \int d\mathbf{r}\, w^{\star}_{JM}\left(\mathbf{r}-\mathbf{r}_i\right)\left[H_{\text{AC}}\right] w_{J+1,M}\left(\mathbf{r}-\mathbf{r}_i\right), \tag{3.7}$$

$$U_{\text{dd}}^{\substack{J_1,J_1',J_2,J_2'\\ M,M'}} \equiv \int d\mathbf{r}d\mathbf{r}'\, w^{\star}_{J_1'M}\left(\mathbf{r}-\mathbf{r}_i\right) w^{\star}_{J_2'M'}\left(\mathbf{r}'-\mathbf{r}_{i+1}\right) \times H_{\text{dd}}\left(\mathbf{r}-\mathbf{r}'\right) w_{J_1M}\left(\mathbf{r}-\mathbf{r}_i\right) w_{J_2M'}\left(\mathbf{r}'-\mathbf{r}_{i+1}\right), \tag{3.8}$$

where the operators H_{kin}, H_{opt}, etc., are taken to be in position space representation. For the derivation of the single-molecule terms (rotational, DC electric field, and AC electric field) and discussion of the properties of our Wannier basis, we refer the reader to Appendix 1. In the following sections we present the derivation of the tunneling (hopping) and dipole–dipole terms, which have new aspects not heretofore appearing in the literature [25].

3.2.2 Tunneling

The tunneling term represents the sum of the molecular kinetic energy with the potential energy of the lattice. After expanding in the Wannier basis of Eq. (3.4), we find the effective tunneling Hamiltonian

$$\hat{H}_t^{\text{eff}} = -\sum_{J,J',M} t_{JJ'M} \sum_{\langle i,i'\rangle}\left(\hat{a}^{\dagger}_{i,J'M}\hat{a}_{i',JM}+\text{h.c.}\right), \tag{3.9}$$

where $t_{J,J',M}$ was defined in Eq. (3.5). To understand why this operator mixes states of different J, we note that the kinetic energy and (far off-resonant) optical lattice potential do not mix rotational eigenstates. Because our Wannier basis states are dressed and therefore superpositions of rotational eigenstates with different J, the tunneling operator in the dressed basis will mix J. Although the dressed basis makes the tunneling more complex to analyze, it simplifies other terms in the MHH, such as the DC term, and is in any case a more standard basis for analysis of the diatomic molecules we study here. Comparable basis changes are sometimes made in other quantum many-body systems, where, for instance, particles and holes are mixed, or particles are paired. Note that, because we assume z-polarized fields, M is still a good quantum number. To discuss the actual form of the tunneling energies $\{t_{J,J',M}\}$ we must first examine the interaction of a diatomic molecule with the optical lattice.

3.2.3 Interaction with an Optical Lattice

The charge redistribution that occurs when a molecule is subjected to a static, spatially uniform electric field $\mathbf{E}$ is reflected in its dipole moment $\mathbf{d}$ via the polarizability series

$$d_j = d_j^{(0)} + \alpha_{jk} E_k + \tfrac{1}{2!}\beta_{jkl} E_k E_l + \tfrac{1}{3!}\Gamma_{jklm} E_k E_l E_m + \cdots , \tag{3.10}$$

where the first, second, and third order coefficients α_{jk}, β_{jkl}, and Γ_{jklm} are elements of the polarizability, hyperpolarizability, and second hyperpolarizability tensors, respectively. The polarizability tensor is a symmetric rank-two tensor with no more than six independent elements (less if molecular symmetry is greater), and characterizes the lowest order dipole moment induced by an applied electric field. From this tensor we can form the scalar invariants

$$\bar{\alpha} \equiv \frac{1}{3}\mathrm{Tr}\tilde{\alpha}, \tag{3.11}$$

$$(\Delta\alpha)^2 \equiv \frac{1}{2}\left[3\mathrm{Tr}(\tilde{\alpha}^2) - (\mathrm{Tr}\tilde{\alpha})^2\right], \tag{3.12}$$

referred to as the polarizability and the polarizability anisotropy, respectively. Note that we use the tilde to clarify that $\tilde{\alpha}$ with elements α_{jk} is a tensor, not a scalar—we reserve the accent circumflex (the "hat" symbol) for quantum operators. In linear molecules, such as diatomic molecules, the presence of only two distinct moments of inertia allows for the classification of $\tilde{\alpha}$ according to its components along and perpendicular to the internuclear axis, denoted $\alpha_{\parallel}$ and $\alpha_{\perp}$, respectively. In the presence of AC electric fields with frequency ω we speak of the dynamic polarizability tensor $\tilde{\alpha}(\omega)$, with the series of Eq. (3.10) being the zero frequency limit. The tensor $\tilde{\alpha}(\omega)$ is, in general, complex, with the real part inducing a dipole moment and the imaginary part accounting for power absorption by the dipole and out-of-phase dipole oscillation. In the case of Σ diatomic molecules in their electronic and vibrational ground states [25]

$$\tilde{\alpha}(\omega) \equiv \alpha_{\parallel}(\omega)\,\mathbf{e}'_0 \otimes \mathbf{e}'_0 + \alpha_{\perp}(\omega) \textstyle\sum_{\Lambda=\pm 1} (-1)^{\Lambda}\, \mathbf{e}'_{\Lambda} \otimes \mathbf{e}'_{-\Lambda}, \tag{3.13}$$

where the $\mathbf{e}'_q$ are molecule-fixed spherical basis vectors. The parallel and perpendicular dynamic polarizabilities are

$$\alpha_{\parallel} = \sum_{\pm} \sum_{\nu, v} \frac{\left|d_{\nu\Sigma(v)-X\Sigma(0)}\right|^2}{E_{\nu\Sigma(v)} - E_{X\Sigma(0)} \mp \hbar\omega}, \tag{3.14}$$

$$\alpha_{\perp} = \sum_{\pm} \sum_{\nu, v} \frac{\left|d_{\nu\Pi(v)-X\Sigma(0)}\right|^2}{E_{\nu\Pi(v)} - E_{X\Sigma(0)} \mp \hbar\omega}, \tag{3.15}$$

respectively. In these expressions $d_{\nu\Lambda(v)-X\Sigma(0)}$ is the transition dipole moment from the ground state to the $\nu\Lambda\ (v)$ state (following the usual diatomic molecular notation, $\Lambda \in \{\Sigma, \Pi\} \equiv \{0, 1\}$ is the quantum number associated with the projection of the total electronic orbital angular momentum along the internuclear axis, i.e., in the molecule-fixed basis) and the sum over $\mp$ accounts for the near-resonant and typically far off-resonant terms.

Transforming $\tilde{\alpha}$ from the molecule-fixed basis to the space-fixed basis using the transformation discussed in Appendix 1, we find

$$\begin{aligned} \tilde{\alpha}'(\omega_L) = &\sum_{p_1 p_2} \sum_{j=0,2} \sum_{m=-j}^{j} (2j+1) \begin{pmatrix} 1 & 1 & j \\ p_1 & p_2 & m \end{pmatrix} \sqrt{\frac{1}{(2-j)!\,(3+j)!}} \\ &\times \left[\alpha_{\parallel}(j+2)(j-1) - 4\alpha_{\perp}\right] C_m^{(j)} \mathbf{e}_{p_1} \otimes \mathbf{e}_{p_2}, \end{aligned} \tag{3.16}$$

where $C_m^{(j)}$ is an unnormalized spherical harmonic, $(\ldots)$ denotes the Wigner 3-j coefficient,[2] and the $\mathbf{e}_p$ are space-fixed spherical basis vectors.

The interaction of the lattice with the molecule is represented by the Hamiltonian

$$H_{\text{opt}}(\mathbf{x}) = -\mathbf{E}^{\star}_{\text{opt}}(\mathbf{r}) \cdot \tilde{\alpha}'(\omega_L) \cdot \mathbf{E}_{\text{opt}}(\mathbf{r}), \tag{3.17}$$

where $\mathbf{E}_{\text{opt}}(\mathbf{r})$ is the electric field of the optical lattice. If the electric field has polarization p in the space-fixed spherical basis, then we find

$$\begin{aligned} H_{\text{opt}}(\mathbf{x}) = &-\frac{\left|\mathbf{E}_{\text{opt}}(\mathbf{r})\right|^2}{3}\Big[\left(\alpha_{\parallel} + 2\alpha_{\perp}\right) C_0^{(0)} \\ &+ (-1)^p \frac{2}{(1-p)!\,(1+p)!}\left(\alpha_{\parallel} - \alpha_{\perp}\right) C_0^{(2)}\Big]. \end{aligned} \tag{3.18}$$

For light linearly polarized in the $\hat{x}$-direction we obtain

$$\begin{aligned} H_{\text{opt}} = &-\frac{\left|\mathbf{E}_{\text{opt}}(\mathbf{r})\right|^2}{6}\Big[2\left(\alpha_{\parallel} + 2\alpha_{\perp}\right) C_0^{(0)} \\ &+ \left(\alpha_{\parallel} - \alpha_{\perp}\right)\left(\sqrt{6}C_{-2}^{(2)} - 2C_0^{(2)} + \sqrt{6}C_2^{(2)}\right)\Big], \end{aligned} \tag{3.19}$$

[2] We use the conventions of Zare [43].

whereas for light linearly polarized in the $\hat{y}$-direction we find

$$H_{\text{opt}} = \frac{\left|\mathbf{E}_{\text{opt}}(\mathbf{r})\right|^2}{6}\Big[-2\left(\alpha_{\parallel}+2\alpha_{\perp}\right)C_0^{(0)} + \left(\alpha_{\parallel}-\alpha_{\perp}\right)\left(\sqrt{6}C_{-2}^{(2)}+2C_0^{(2)}+\sqrt{6}C_2^{(2)}\right)\Big]. \quad (3.20)$$

Since $C_0^{(0)} = 1$, these terms give a state-independent energy shift. The $C_q^{(2)}$ terms produce a tensor shift. Because the depth (in energy) of a typical optical lattice is much smaller than the energy of transitions between rotational levels (of order B, as defined in Appendix 2), we can ignore far off-resonant Raman coupling between different J manifolds and use only the diagonal matrix elements. The $C_2^{(2)}$ term and the $C_{-2}^{(2)}$ will both mix M in the $J \geq 2$ manifolds, but do not affect the lowest two rotational levels, again, because we neglect Raman couplings. Thus x, y, and z polarizations all have the same Hamiltonian in this approximation. We can calculate the matrix elements of $C_0^{(2)}$ in the field free basis using the Wigner–Eckart theorem to find

$$\langle J'M'|H_{\text{opt}}(\mathbf{r})|JM\rangle = -\frac{\left|\mathbf{E}_{\text{opt}}(\mathbf{r})\right|^2}{3}\delta_{JJ'}\delta_{MM'}\Big[\left(\alpha_{\parallel}+2\alpha_{\perp}\right) + (-1)^p\frac{2}{(1-p)!(1+p)!}\left(\alpha_{\parallel}-\alpha_{\perp}\right)\frac{J(J+1)-3M^2}{(2J-1)(2J+3)}\Big]. \quad (3.21)$$

In our effective Hamiltonian we choose right circular polarization for the z lattice, x polarization for the x lattice, and y polarization for the y lattice, where each lattice refers to a pair of counter-propagating laser beams used to create a standing wave. Other choices are of course also possible, and the mathematical derivation leads to a Hamiltonian similar in form to that of Eq. (3.1). However, it is possible for hopping to depend on two M indices, e.g. $t_{JJ'MM'}$.

We consider the fields making up the optical lattice to have sinusoidal spatial profiles, resulting in sine-squared intensity profiles. In addition, we assume that the y and z lattices are tight, meaning that the molecules are strongly confined at the potential minimum (for a red-detuned trap). This tight confinement allows us to approximate them via a Taylor series, e.g., $\sin^2(k_z z) \simeq k_z^2 z^2$ in the vicinity of the molecule. Using the above results, the matrix elements of the Hamiltonian for the optical lattice can be written

$$\begin{aligned}\langle J'M'|H_{\text{opt}}(\mathbf{r})|JM\rangle &= -\frac{\left|\mathbf{E}_{\text{opt}}(\mathbf{y})\right|^2 k_y^2 y^2 + \left|\mathbf{E}_{\text{opt}}(\mathbf{x})\right|^2 \sin^2(k_x x)}{3} \\ &\quad \left[\bar{\alpha} + 2\Delta\alpha\frac{J(J+1)-3M^2}{(2J-1)(2J+3)}\right]\delta_{JJ'}\delta_{MM'} \\ &\quad -\frac{\left|\mathbf{E}_{\text{opt}}(\mathbf{z})\right|^2 k_z^2 z^2}{3}\left[\bar{\alpha} - \Delta\alpha\frac{J(J+1)-3M^2}{(2J-1)(2J+3)}\right]\delta_{JJ'}\delta_{MM'}\end{aligned} \quad (3.22)$$

or, more compactly, as

$$\langle J'M'|H_{\rm opt}(\mathbf{r})|JM\rangle = \Big[-\alpha^{(t)}_{JM}\left|\mathbf{E}_{\rm opt}(\mathbf{y})\right|^2 k_y^2 y^2 - \alpha^{(t)}_{JM}\left|\mathbf{E}_{\rm opt}(\mathbf{x})\right|^2 \times \sin^2(k_x x)\Big]\delta_{JJ'}\delta_{MM'} - \left|\mathbf{E}_{\rm opt}(\mathbf{z})\right|^2 \alpha^{(z)}_{JM} k_z^2 z^2 \delta_{JJ'}\delta_{MM'} \quad (3.23)$$

by defining

$$\alpha^{(t)}_{JM} \equiv \frac{1}{3}\left[\bar{\alpha} + 2\Delta\alpha \frac{J(J+1) - 3M^2}{(2J-1)(2J+3)}\right], \quad (3.24)$$

$$\alpha^{(z)}_{JM} \equiv \frac{1}{3}\left[\bar{\alpha} - \Delta\alpha \frac{J(J+1) - 3M^2}{(2J-1)(2J+3)}\right]. \quad (3.25)$$

We now define, as is customary, the "lattice heights" in the x, y, and z directions, respectively, as

$$V_x^{(JM)} \equiv -\left|\mathbf{E}_{\rm opt}(\mathbf{x})\right|^2 \alpha^{(t)}_{JM}, \quad (3.26)$$

$$V_y^{(JM)} \equiv -\left|\mathbf{E}_{\rm opt}(\mathbf{y})\right|^2 \alpha^{(t)}_{JM}, \quad (3.27)$$

$$V_z^{(JM)} \equiv -\left|\mathbf{E}_{\rm opt}(\mathbf{z})\right|^2 \alpha^{(z)}_{JM}. \quad (3.28)$$

The tight confinement in the transverse (y and z) directions strongly suppresses tunneling in these directions, making the overall lattice effectively 1D along x.

From Eqs. (3.24), (3.25), it is apparent that different rotational levels experience different trapping frequencies and different tunneling energies. To make this clearer, we parse our full field-free tunneling matrix element as

$$\begin{aligned} t_{JM} &\equiv -\int d\mathbf{r}\, w^{\star}_{JM}(\mathbf{r}-\mathbf{r}_i)\left[H_{\rm kin} + H_{\rm opt}\right] w_{JM}(\mathbf{r}-\mathbf{r}_{i+1}) \\ &= \int d\mathbf{r}\, w^{\star}_{JM}(\mathbf{r}-\mathbf{r}_i)\left[-H_{\rm kin} + V_x^{(JM)}\sin^2\left(k_x x^2\right)\right] w_{JM}(\mathbf{r}-\mathbf{r}_{i+1}) \\ &\quad + \int d\mathbf{r}\, w^{\star}_{JM}(\mathbf{r}-\mathbf{r}_i)\left[V_y^{(JM)} k_y^2 y^2 + V_z^{(JM)} k_z^2 z^2\right] w_{JM}(\mathbf{r}-\mathbf{r}_{i+1}). \quad (3.29) \end{aligned}$$

Because we consider tight traps such that the lattice heights in the y- and z-directions are much greater than the lattice height in the x-direction, $V_y^{(JM)} \sim V_z^{(JM)} \gg V_x^{(JM)}$, the contributions from the $V_y^{(JM)}$ and $V_z^{(JM)}$ terms are exponentially suppressed compared to the $V_z^{(JM)}$ and $H_{\rm kin}$ terms, and so we neglect them. This is equivalent to the array of tubes we discussed in Sect. 3.1, where each tube is isolated from its neighbors. The matrix element becomes

$$t_{JM} \approx \int d\mathbf{r} w^{\star}_{JM} (\mathbf{r} - \mathbf{r}_i) \left[-H_{\text{kin}} + V_x^{(JM)} \sin^2 \left(k_x x^2 \right) \right] w_{JM} (\mathbf{r} - \mathbf{r}_{i+1}) \,. \quad (3.30)$$

In the evaluation of the integral, Eq. (3.30) we assume that the Bloch function of a molecule in the sinusoidal optical lattice is a Mathieu function along x. This may seem to contradict our assumption of spherical symmetry in the above derivation. However, the assumption of spherical symmetry (i.e., a locally constant potential) need only hold on the order of an internuclear axis (~5Å) near the molecule. In contrast, on the order of the characteristic lattice length $\sqrt{\hbar/\mu\omega_{\text{opt}}}$ the rigid-rotor molecule is indistinguishable from a point particle (such as an alkali atom), and so spherical symmetry is not required. With this understanding, we recognize $t^{(0)}_{JM}$ as the expression for the hopping energy for point particles in optical lattices [26] with the additional feature that the lattice height along the quasi-1D direction $V_0 = V_x^{(JM)}$ is dependent on J through the polarizability tensor. Thus, altering the expression from the theory of point particles in optical lattices, we obtain the result

$$\frac{t^{(0)}_{JM}}{E_R} \approx A \left(\frac{V_x^{(JM)}}{E_R} \right)^B \exp \left(-C \sqrt{\frac{V_x^{(JM)}}{E_R}} \right) , \quad (3.31)$$

where $A = 1.397$, $B = 1.051$, $C = 2.121$, and

$$E_R \equiv \hbar^2 k_x^2 / 2m \quad (3.32)$$

is the recoil energy.

Using tabulated values of the polarizabilities for LiCs[27] as given in Table 3.1, we find that, for a reasonable lattice height $V_x^{(00)}/E_R \simeq 10$, the tunneling term for the $|11\rangle$ state is only about 20 % of that in the $|00\rangle$ state, as shown in Fig. 3.1. For LiCs in a red-detuned optical lattice of wavelength $\lambda = 985$ nm, $E_R = 2\pi \times 1.46\hbar$ kHZ. Typical values of the lattice heights are $V_x \sim 10E_R$, $V_y, V_z \sim 25E_R$ [28].

We reiterate that the above matrix elements and tunneling energies $\{t_{JM}\}$ have been computed in the field-free basis for simplicity. To transform to the dressed basis, we use the unitary matrix with dressed eigenvectors as columns, recovering Eq. (3.9), where the tunneling matrix element is no longer diagonal in J.

Table 3.1 Values of the polarizabilities for LiCs in different rotational states $|JM\rangle$

$\|JM\rangle$	$3\alpha^{(r)}_{JM}/\bar{\alpha}$	$3\alpha^{(z)}_{JM}/\bar{\alpha}$
$\|00\rangle$	1	1
$\|10\rangle$	1.715	0.642
$\|1 \pm 1\rangle$	0.642	1.178
$\|20\rangle$	1.511	0.744
$\|2 \pm 1\rangle$	1.255	0.872
$\|2 \pm 2\rangle$	0.488	1.255

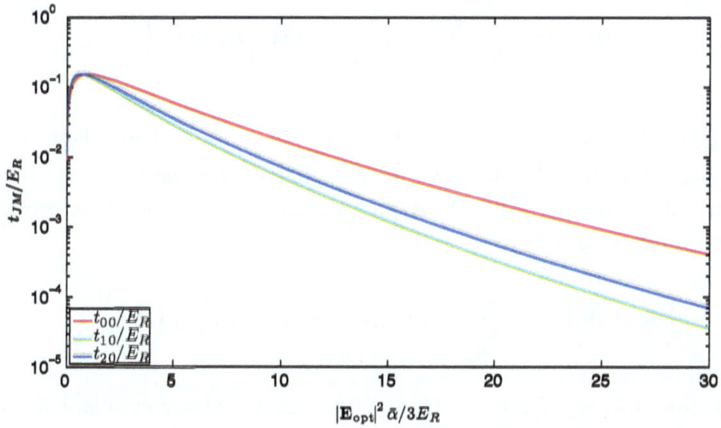

Fig. 3.1 Dependence of the field-free tunneling (hopping) coefficient on rotational state and lattice height

3.2.4 *Dipole–Dipole Interactions*

The induced dipoles from the DC field give rise to a resonant dipole–dipole interaction. The Hamiltonian for this interaction in the two-site dressed basis spanned by $|\mathcal{E}; J_1 M_1 J_2 M_2\rangle$ is

$$\hat{H}_{\rm dd} = \frac{1}{2}\sum_{J_1,J_1',J_2,J_2' \atop M,M'} U_{\rm dd}^{J_1,J_1',J_2,J_2' \atop M,M'} \sum_{\langle i,i'\rangle} \hat{a}^{\dagger}_{iJ_1M}\hat{a}_{iJ_1'M}\hat{a}^{\dagger}_{i'J_2M'}\hat{a}_{i'J_2'M'}, \tag{3.33}$$

where we have defined

$$U_{\rm dd}^{J_1,J_1',J_2,J_2' \atop M,M'} \equiv \int d\mathbf{r}d\mathbf{r}'\, w^{\star}_{J_1'M}\left(\mathbf{r}-\mathbf{r}_i\right) w^{\star}_{J_2'M'}\left(\mathbf{r}'-\mathbf{r}_{i+1}\right)$$
$$\times\, H_{\rm dd}\left(\mathbf{r}-\mathbf{r}'\right) w_{J_1M}\left(\mathbf{r}-\mathbf{r}_i\right) w_{J_2M'}\left(\mathbf{r}'-\mathbf{r}_{i+1}\right), \tag{3.34}$$

and for notational simplicity we have suppressed the $\mathcal{E}$ subscripts. Note that because of our choice of polarizations of the optical lattice and AC and DC electric fields, $M_1 = M_2 \equiv M$ and $M_1' = M_2' \equiv M'$.

The resonant dipole–dipole interaction between two permanent dipoles d_1 and d_2 whose respective centers of mass are separated by a vector $\mathbf{R}$ in the space-fixed frame is

$$\hat{H}_{\rm dd} = \frac{\hat{\mathbf{d}}_1\cdot\hat{\mathbf{d}}_2 - 3\left(\mathbf{e}_R\cdot\hat{\mathbf{d}}_1\right)\left(\hat{\mathbf{d}}_2\cdot\mathbf{e}_R\right)}{R^3}, \tag{3.35}$$

where $\mathbf{e}_R$ is a unit vector in the direction of $\mathbf{R}$. Using standard angular momentum recoupling we recast this in spherical tensor notation as

$$H_{\mathrm{dd}} = -\frac{\sqrt{6}}{R^3}\sum_{\mu}(-1)^{\mu}\, C^{(2)}_{-\mu}(\mathbf{R})\left[\hat{\mathbf{d}}_1 \otimes \hat{\mathbf{d}}_2\right]^{(2)}_{\mu}\,, \tag{3.36}$$

where $(T)^{(k)}_q$ denotes the component of the rank-k spherical tensor T that has projection q along $\mathbf{R}$, $C^{(j)}_m(\mathbf{R})$ is an unnormalized spherical harmonic in the polar coordinates defined with respect to $\mathbf{R}$, and we have defined the tensor product of the vector operators $\hat{\mathbf{d}}_1$ and $\hat{\mathbf{d}}_2$ as

$$\left[\hat{\mathbf{d}}_1 \otimes \hat{\mathbf{d}}_2\right]^{(k)}_{q} \equiv \sum_{m}\langle 1, m, 1, q-m|kq\rangle\left(\hat{\mathbf{d}}_1\right)^{(1)}_{m}\left(\hat{\mathbf{d}}_2\right)^{(1)}_{q-m}\,. \tag{3.37}$$

In the last line, $\langle j_1, m_1, j_2, m_2|J, M\rangle$ is a Clebsch–Gordan coefficient. We now take matrix elements of Eq. (3.36) in the two dressed-molecule basis $|\mathcal{E}; J_1M_1, J_2M_2\rangle$, where molecule 1 is on site i and molecule 2 is on site $i+1$, yielding

$$\begin{aligned}&\langle \mathcal{E}; J_1'M_1', J_2'M_2'|\hat{H}_{\mathrm{dd}}|\mathcal{E}; J_1M_1, J_2M_2\rangle\\ &= -\frac{\sqrt{6}}{R^3}\sum_{\mu}(-1)^{\mu}\, C^{(2)}_{-\mu}(\mathbf{R})\sum_{m}\langle 1, m, 1, \mu-m|2\mu\rangle\\ &\quad\times \langle \mathcal{E}; J_1'M_1'|\left(\hat{\mathbf{d}}_1\right)^{(1)}_{m}|\mathcal{E}; J_1M_1\rangle\langle \mathcal{E}; J_2'M_2'|\left(\hat{\mathbf{d}}_2\right)^{(1)}_{\mu-m}|\mathcal{E}; J_2M_2\rangle.\end{aligned} \tag{3.38}$$

Because our DC field is polarized along z, only $(\hat{\mathbf{d}}_1)^{(1)}_0$ and $(\hat{\mathbf{d}}_2)^{(1)}_0$ matrix elements are nonzero, enforcing $\mu = 0$, $m = 0$. With this in mind, the interaction takes the particularly simple form

$$\begin{aligned}&\langle \mathcal{E}; J_1'M_1', J_2'M_2'|\hat{H}_{\mathrm{dd}}|\mathcal{E}; J_1M_1, J_2M_2\rangle\\ &= -\frac{\sqrt{6}}{R^3}C^{(2)}_0(\mathbf{R})\\ &\quad\times \langle 1, 0, 1, 0|20\rangle\langle \mathcal{E}; J_1'M_1|\left(\hat{\mathbf{d}}_1\right)^{(1)}_{0}|\mathcal{E}; J_1M_1\rangle\langle \mathcal{E}; J_2'M_2|\left(\hat{\mathbf{d}}_2\right)^{(1)}_{0}|\mathcal{E}; J_2M_2\rangle\end{aligned} \tag{3.39}$$

$$= \langle \mathcal{E}; J_1'M_1|\left(\hat{\mathbf{d}}_1\right)^{(1)}_{0}|\mathcal{E}; J_1M_1\rangle\langle \mathcal{E}; J_2'M_2|\left(\hat{\mathbf{d}}_2\right)^{(1)}_{0}|\mathcal{E}; J_2M_2\rangle\left(\frac{1-3\cos^2\theta}{R^3}\right). \tag{3.40}$$

The intermolecular axis plays a crucial role in the sign of the interaction. Two molecules oriented along the intermolecular axis attract if their dipoles are parallel and repel if their dipoles are antiparallel. Two molecules oriented perpendicular to the intermolecular axis, on the other hand, repel if their dipoles are parallel and attract if their dipoles are antiparallel. The DC field that orients the molecules in our setup is polarized along z, perpendicular to the intermolecular quasi-1D axis x. This gives rise to repulsive interactions for positive dipole matrix elements. With this geometry the dipole potential becomes

$$\begin{aligned}&\langle\mathcal{E};J_1'M_1',J_2'M_2'|\hat{H}_{\rm dd}|\mathcal{E};J_1M_1,J_2M_2\rangle\\&=\frac{1}{R^3}\langle\mathcal{E};J_1'M_1|\left(\hat{\mathbf{d}}_1\right)_0^{(1)}|\mathcal{E};J_1M_1\rangle\langle\mathcal{E};J_2'M_2|\left(\hat{\mathbf{d}}_2\right)_0^{(1)}|\mathcal{E};J_2M_2\rangle,\end{aligned}\tag{3.41}$$

yielding

$$U_{\rm dd}^{J_1,J_1',J_2,J_2'}{}_{M,M'}=\frac{8}{\lambda^3}\langle\mathcal{E};J_1'M_1|\left(\hat{\mathbf{d}}_1\right)_0^{(1)}|\mathcal{E};J_1M_1\rangle\langle\mathcal{E};J_2'M_2|\left(\hat{\mathbf{d}}_2\right)_0^{(1)}|\mathcal{E};J_2M_2\rangle,\tag{3.42}$$

where λ is the wavelength of the optical lattice.

3.2.5 Energy Scales

We proceed to clarify the energy scales associated with each term in Eq. (3.1). Between previous discussion in Sect. 3.2 and that of Appendix 1, all terms in Eq. (3.1) are now clearly defined. The energy scales of the dressed basis are B, the rotational constant, which is roughly $60\hbar$ GHz, and $d\mathcal{E}_{\rm DC}$, which is of order 1–10 B. The DC term has no length scale associated with it because the field is uniform, and the length scale of the rotational term is the internuclear separation, on the order of angstroms. The relative contribution of the DC electric field and rotational terms in Eq. (3.1) are expressed through the dimensionless parameter

$$\beta_{\rm DC}\equiv d\mathcal{E}_{\rm DC}/B,\tag{3.43}$$

the ratio of the DC field energy to the rotational level splitting.

The energy scales of the AC term are $\hbar\omega$, where ω is the angular frequency of the driving field, and $d\mathcal{E}_{\rm AC}$. The scale $\hbar\omega$ is of order $2B$ for small $\beta_{\rm DC}\ll 1$, and of order $B\sqrt{\beta_{\rm DC}}$ for large $\beta_{\rm DC}\gg 1$. The AC field energy $d\mathcal{E}_{\rm AC}$ is of order $0.5\hbar\omega$. The single-molecule timescale associated with $d\mathcal{E}_{\rm AC}$ is the Rabi period, the time it takes for the population of a two-level system to cycle once, as seen in Fig. 3.10a. In real time, this is on the order of 10 ps for the parameters in the preceding paragraph. The timescale associated with ω is the timescale on which the small oscillations in Fig. 3.10a occur, of order 0.5 ps. The length scale of the AC field is on the order of centimeters, and so we can neglect this in light of the micron length scale of the trap.

The tunneling term has several scales. The optical lattice near the point of confinement has a length scale given by the harmonic oscillator length $l_{{\rm ho},x}^{(00)}\sim 100$ nm and an energy scale of $E_R\approx 1.4\hbar$ kHz. The energy scales of the tunneling operator

Table 3.2 Comparison of energy and length scales for the molecular hubbard hamiltonian of Eq. (3.1)

Term	Length scale	Energy scale
Rotation	Internuclear distance ~1 Å	$B \sim 60\hbar$ GHz $\approx 2\,\text{cm}^{-1}$
DC field	N/A, uniform	$d\mathcal{E}_{\text{DC}} \sim 120\hbar$ GHz $\approx 4\,\text{cm}^{-1}$
AC field	$2\pi c/\omega \sim 1$ cm	$\hbar\omega \sim 30\hbar$ GHz $\approx 1\,\text{cm}^{-1}$
Kinetic	$l_{\text{ho},x}^{(00)} \sim 100$ nm	$E_R \sim 1.46\hbar$ kHz
Tunneling	Lattice spacing ~1 μm	$\{t_{J'JM}\} \sim 100\hbar$ Hz
Resonant Dipole–dipole	Energy comparable to B at $r_B \simeq 348$ Bohr radii	$\left|\langle\mathcal{E};00|\hat{\mathbf{d}}|\mathcal{E};00\rangle\right|^2/(1\,\mu\text{m})^3 \sim 1.2\hbar$ kHz for nearest neighbors

proper are given by the $\{t_{JJ'M}\}$ which are of order 10^{-1}–$10^{-2}E_R \sim 100\hbar$ Hz for the given recoil energy.

There are also many scales for the dipole term. For the B and d specified in the first paragraph of this section and $\beta_{\text{DC}} = 1.9$, the characteristic length scale where the dipole–dipole energy becomes comparable to the rotational energy is

$$r_B \equiv \left(\left|\langle\mathcal{E};00|\hat{\mathbf{d}}|\mathcal{E};00\rangle\right|^2 \Big/ B\right)^{\frac{1}{3}}, \tag{3.44}$$

approximately 348 Bohr radii (18.4 nm). Outside this region the Born–Oppenheimer adiabatic approximation is easily fulfilled [6]. Since the length scale of our optical lattice is of order μm, we are justified in working within the Born–Oppenheimer framework. For the same parameters, the length scale where the off-resonant van der Waals potential $C_6/r^6 \approx -d^4/(6Br^6)$ becomes comparable to the dipole–dipole interaction is

$$r_{\text{vdW}} \equiv \left(2\,|C_6| \Big/ \left|\langle\mathcal{E};00|\hat{\mathbf{d}}|\mathcal{E};00\rangle\right|^2\right)^{\frac{1}{3}}. \tag{3.45}$$

This length is very small, on the order of tens to hundreds of Bohr radii. Outside of this region the resonant dipole potential dominates and the intermolecular force is repulsive. This repulsion enforces the hard-core limit. The energy scale of the dipole–dipole force is $\left|\langle\mathcal{E};00|\hat{\mathbf{d}}|\mathcal{E};00\rangle\right|^2/\lambda^3 \sim 1.2\hbar$ kHz, with higher J being an order of magnitude or so lower for small β_{DC}, and of the same order for large β_{DC} (see Fig. 3.8a).

To summarize, the scales of the problem are shown in Table 3.2.

3.2.6 Novel Features of the MHH

The MHH, Eq. (3.1), has a number of novel features which distinguish it from typical Hubbard and extended Hubbard models [2, 29]. First, the tunneling energies $\{t_{J,J'M}\}$ not only depend on the rotational level J, M but even change rotational states from J to J'. This is due both to the polarizability tensor's dependence on rotational level and to the dressed basis. This differs from other Hubbard models which consider spin degrees of freedom, as tunneling does not occur between spin states—hopping does not cause spin transitions. If we consider populating a single mode (e.g. $J = 0$, $M = 0$) in the $\Omega \to 0$ limit, then Eq. (3.1) becomes the extended Bose–Hubbard Hamiltonian, and the phase diagram is known [10, 30]. This gives ideas of how to characterize the static phases of the MHH. However, because the tunneling energy depends on J, the borders of the phase diagram will depend on the rotational state of the system. We will discuss this property and provide an application in Sect. 3.4.

Second, the Hamiltonian is fundamentally time-dependent because it is a driven system. This allows for the study of dynamic quantum phases, requiring the concept of a quantum phase diagram to be generalized to an inherently time-dependent picture. In a case study for hard-core bosonic molecules at half-filling presented in Sect. 3.4, we show that the MHH has emergent timescales.

3.3 Methods

3.3.1 Time-Evolving Block Decimation

The TEBD algorithm is a new method [13, 14] designed to study the dynamics of entangled quantum systems. The essential idea of TEBD is to provide a moving "spotlight" in Hilbert space which tracks a dynamical system. The portion of the Hilbert space so illuminated is an exponentially small fraction of the full Hilbert space; this is justified by the fact that real, physical quantum many-body systems, especially in real materials, typically explore only a small, lowly entangled part of the total Hilbert space.

In fact, TEBD moves the full quantum many-body problem from the NP-complete complexity class to the P class through an exponential reduction in the number of parameters needed to represent the many-body state. We can understand the possibility of this reduction through an analogy to image compression. Present digital cameras are capable of producing a roughly 3000×3000 array of pixels. Downloading the images from such a camera, one notices that there are far less than 10 Megapixels worth of data per image. Image compression algorithms such as JPEG produce images of remarkable quality with only a small fraction of the raw data. The reason that these algorithms are so effective is that a physical image, as opposed to a random 2D pixel array, is not the "most common" or most probable image; it contains a great deal of structure and regularity. In the same way, physical states in Hilbert space tend to be lowly entangled (by some entanglement measure), even though a general state in Hilbert space has a much larger probability of being highly entangled. There is no general proof of this fact, just as there is no guarantee that an image will come out perfectly crisp after JPEG compression; it is simply a trend observed in many-body quantum systems.

To be slightly more specific, TEBD performs a partial trace over a particular bipartite splitting of the lattice, and then keeps the χ largest eigenvalues of the resulting reduced density matrix. The cut-off parameter χ is based on the Schmidt measure [15], and so it also serves as a measure of the degree of spatial entanglement. This idea is not unique to TEBD. In fact, the density matrix renormalization group (DMRG) method first proposed by White [31] did something analogous years before. The innovation of TEBD is that at each time step it re-optimizes the truncated basis (thus the "moving spotlight"). The Schmidt number is just the number of nonzero eigenvalues in the reduced density matrix, and so is an entanglement measure natural to quantum many-body systems. The parameter χ is the number of nonzero eigenvalues in the reduced density matrix that TEBD retains. It is the principal convergence parameter of the algorithm, both in entanglement and in time. Although the time-propagation method we use is Trotter–Suzuki [32], it turns out that, due to a normalization drift, χ controls convergence at long times.

With χ interpreted as an entanglement measure, we can say that TEBD treats the system not as a wavefunction in a d^L-dimensional Hilbert space (L is the number of lattice sites), but as a collection of wavefunctions in d^2-dimensional two-site spaces that are weakly entangled with the environment created by the rest of the system. To facilitate this viewpoint, we replace the d^L coefficients of the full many-body wavefunction with L sets of $(d\chi^2 + \chi)$ coefficients corresponding to the wavefunctions of each bipartite splitting. The most computationally expensive portion of the TEBD algorithm is typically the diagonalization of these local coefficient matrices at a cost of $\mathcal{O}\left(d^3\chi^3\right)$. Looping over all $L-1$ bipartite splittings and evolving the system for a total time t_f in time steps of length δt, one obtains an asymptotic scaling of $\mathcal{O}\left(L\frac{t_f}{\delta t}d^3\chi^3\right)$.

This scaling can be greatly improved by the presence of conserved quantities. When a conserved quantity exists in the system we are able to diagonalize reduced density matrices corresponding to distinct values of this conserved quantity independently, which can result in significantly smaller reduced density matrices to diagonalize. Implementing this idea, scalings of $\mathcal{O}\left(\chi^2\right)$ have been reported for fixed d [33]. In addition, conserved quantities in the presence of selection rules can reduce the local dimension. For example, in the case of the MHH, z-polarized electric fields disallow transitions from a particular M to any other. If we begin with all molecules in a particular M state, this allows us to restrict our attention only to states with this M. In our numerics we conserve both the projection M, and the total number N. Furthermore, to match our hard-core requirement, we allow only zero or one molecules per site, so that the local dimension is $d \leq J_{\max} + 1$, $J_{\max}$ being the magnitude of the greatest angular momentum that we consider (note that the local dimension d, mentioned only here in Sect. 3.3.1, bears no relation to the permanent electric dipole moment d used throughout the rest of our treatment).

A more detailed description of TEBD can be found in [23]. We also recommend [34], besides Vidal's original papers [13, 14].

3.3.2 Quantum Measures

We use a suite of quantum measures to characterize the reduced MHH, Eq. (3.50) below. The few-body measures we use are $\langle \hat{n}_i^J \rangle$, the number in the Jth rotational state on the ith site, $E \equiv \langle \hat{H} \rangle$, the expectation of the energy, and $\frac{1}{L}\langle \hat{n}^J \rangle$, the average number in the Jth rotational state per site (L is the number of lattice sites). The latter is a J-dependent filling factor. The many-body measures we use include the density–density correlation between rotational modes J_1 and J_2 evaluated at the middle site

$$g_2^{(J_1 J_2)}\left(\left\lfloor \frac{L}{2} \right\rfloor, i\right) \equiv \left\langle \hat{n}_{\lfloor \frac{L}{2} \rfloor}^{(J_1)} \hat{n}_i^{(J_2)} \right\rangle - \left\langle \hat{n}_{\lfloor \frac{L}{2} \rfloor}^{(J_1)} \right\rangle \left\langle \hat{n}_i^{(J_2)} \right\rangle, \tag{3.46}$$

where $\lfloor q \rfloor$ is the floor function, defined as the greatest integer less than or equal to q. As an entanglement measure we use the Meyer Q-measure [35–37]

$$Q \equiv \frac{d}{d-1}\left[1 - \frac{1}{L}\sum_{k=1}^{L} \mathrm{Tr}\left(\hat{\rho}^{(k)}\right)^2\right], \tag{3.47}$$

where $\hat{\rho}^{(k)}$ is the single-site density matrix obtained by tracing over all but the kth lattice site, and the factor outside of the bracket is a normalization factor (d is the on-site dimension). This gives an average measure of the entanglement of a single site with the rest of the system. The Q-measure can also be interpreted as the average local impurity (recall that the $\mathrm{Tr}(\hat{\rho}^2) = 1$ if and only if $\hat{\rho}$ is a pure state).

To determine what measures we can use to ascertain the static phases of our model we reason by analogy with the extended Bose–Hubbard Hamiltonian where we know that the possible static phases are charge density wave, superfluid, supersolid, and Bose metal [10]. The charge density wave is an insulating phase appearing at half integer fillings which has a wavelength of two sites. Like the Mott insulating phase, it has an excitation gap and is incompressible. While the extended Bose–Hubbard Hamiltonian has only one charge density wave phase due to the presence of only one species, the MHH has the possibility of admitting several charge density wave phases due to the presence of multiple rotational states. As such, we define the structure factor

$$S_\pi^{(J_1 J_2)} = \frac{1}{N}\sum_{ij}(-1)^{|i-j|}\left\langle \hat{n}_i^{(J_1)} \hat{n}_j^{(J_2)} \right\rangle, \tag{3.48}$$

where N is the total number of molecules. We recognize this object as the spatial Fourier transform of the equal-time density–density correlation function between rotational states J_1 and J_2, evaluated at the edge of the Brillouin zone. This measure is of experimental interest because it is proportional to the intensity in many scattering experiments, e.g. neutron scattering [38]. Crystalline order between rotational states J_1 and J_2 is characterized by a nonzero structure factor $S_\pi^{(J_1 J_2)}$. The charge density wave is the phase with crystalline order but no off-diagonal long-range order as quantified by the superfluid stiffness of rotational state J

$$\rho_s^{(J)} = \lim_{\phi \to 0} L \frac{\partial^2 E^{(J)}(\phi, L)}{\partial \phi^2} \tag{3.49}$$

(note that ρ_s bears no relation to the density matrix $\hat{\rho}$). If both the structure factor and the superfluid stiffness are nonzero, the phase is called supersolid. If both the structure factor and the superfluid stiffness are zero, the phase is called Bose Metal. Finally, if the structure factor is zero and the superfluid stiffness is nonzero, the phase is superfluid. In 1D systems with short-range interactions the structure factor is zero in the thermodynamic limit and the entire superfluid phase is critical, thus there are formally no order parameters [10]. Superfluidity is instead signaled by a diverging correlation length and solid order by slow power law decay of the density–density correlator.

3.4 Case Study: Hard-Core Bosonic Molecules at Half-Filling

In the following, we consider a particular case of Eq. (3.1) for dynamical study. We choose the hard core case, which can occur naturally due to strong on-site dipole–dipole interactions, and half-filling, which is an interesting point in a number of models, including the repulsive Fermi–Hubbard Hamiltonian and the extended Bose–Hubbard Hamiltonian discussed in Sect. 3.3.2. For example, in the latter case, the charge-density-wave phase requires a minimum of half-filling [10].

If we assume that our system begins in its ground state ($J = 0$, $M = 0$), we need to only include states which have a dipole coupling to this state. For z-polarized DC and AC fields, this means we only consider $M = 0$ states, yielding the reduced Hamiltonian

$$\begin{aligned}\hat{H} = &- \sum_{JJ'} t_{JJ'} \sum_{\langle i,i' \rangle} \left(\hat{a}^{\dagger}_{i',J'} \hat{a}_{iJ} + \text{h.c.} \right) + \sum_{J} E_J \sum_{i} \hat{n}_{iJ} \\ &- \pi \sin(\omega t) \sum_{J} \Omega_J \sum_{i} \left(\hat{a}^{\dagger}_{iJ} \hat{a}_{iJ+1} + \text{h.c.} \right) \\ &+ \frac{1}{2} \sum_{J_1, J_1', J_2, J_2'} U_{\text{dd}}^{J_1, J_1', J_2, J_2'} \sum_{\langle i,i' \rangle} \hat{a}^{\dagger}_{iJ_1} \hat{a}_{iJ_1'} \hat{a}^{\dagger}_{i'J_2} \hat{a}_{i'J_2'}.\end{aligned} \tag{3.50}$$

This is the specific case of the MHH that we study using TEBD.

A matter of practical concern, as apparent in Table 3.2, is the large disparity between the timescales of the first three (Rotational, DC, and AC) and the last three (kinetic, tunneling, and Dipole–Dipole) terms. The accumulation of error resulting from truncating the Hilbert space at each TEBD timestep causes the algorithm to eventually fail after a certain "runaway time," making studies over long times intractable [39]. This invites a multiscale approach in the future [40, 41]. In our current numerics we artificially increase the recoil energy and dipole–dipole potential to be of the order of the rotational constant in order to study Eq. (3.50) using TEBD. In particular, we take

$$U_{\mathrm{dd}}^{J_1,J_1',J_2,J_2'} = \frac{10B}{d^2}\langle\mathcal{E};J_1'|\hat{\mathbf{d}}|\mathcal{E};J_1\rangle\langle\mathcal{E};J_2'|\hat{\mathbf{d}}|\mathcal{E};J_2\rangle, \tag{3.51}$$

$$t_J = 10B\left[\eta\left(1+2\frac{\Delta\alpha}{\bar{\alpha}}\frac{J(J+1)}{(2J+1)(2J+3)}\right)\right]^{1.051} \times\exp\left[-2.121\sqrt{\eta\left(1+2\frac{\Delta\alpha}{\bar{\alpha}}\frac{J(J+1)}{(2J+1)(2J+3)}\right)}\right], \tag{3.52}$$

where the dimensionless variable η becomes an ersatz "lattice height." To see the scaling more explicitly, we compare the above with the actual expressions for the MHH parameters

$$U_{\mathrm{dd}}^{J_1,J_1',J_2,J_2'} = \frac{8}{\lambda^3}\langle\mathcal{E};J_1'|\hat{\mathbf{d}}|\mathcal{E};J_1\rangle\langle\mathcal{E};J_2'|\hat{\mathbf{d}}|\mathcal{E};J_2\rangle \tag{3.53}$$

$$= \left(\frac{2mE_R d^{4/3}}{\hbar^2\pi^2}\right)^{\frac{3}{2}}\langle\mathcal{E};J_1'|\hat{\mathbf{d}}|\mathcal{E};J_1\rangle\langle\mathcal{E};J_2'|\hat{\mathbf{d}}|\mathcal{E};J_2\rangle/d^2, \tag{3.54}$$

$$t_{JM} \approx 1.397E_R\left(\frac{|\mathbf{E}_{\mathrm{opt}}|^2\bar{\alpha}}{3E_R}\left[1+2\frac{\Delta\alpha}{\bar{\alpha}}\frac{J(J+1)-3M^2}{(2J-1)(2J+3)}\right]\right)^{1.051} \tag{3.55}$$

$$\times\exp\left(-2.121\sqrt{\frac{|\mathbf{E}_{\mathrm{opt}}|^2\bar{\alpha}}{3E_R}\left[1+2\frac{\Delta\alpha}{\bar{\alpha}}\frac{J(J+1)-3M^2}{(2J-1)(2J+3)}\right]}\right). \tag{3.56}$$

If we now scale E_R to be $10B/1.397$ and set d such that $\left[2mE_Rd^{4/3}/\left(\hbar^2\pi^2\right)\right]^{\frac{3}{2}} = 10B$ for this E_R, we recover Eqs. (3.51) and (3.52) provided we make the definition

$$\eta \equiv -\left|\mathbf{E}_{\mathrm{opt}}(\mathbf{x})\right|^2\bar{\alpha}/(3E_R) = V_x^{(JM)}\bar{\alpha}/\left(3E_R\alpha_{JM}^{(t)}\right). \tag{3.57}$$

Since this dimensionless parameter plays the same role as the quasi-1D lattice height scaled to the recoil energy did in the actual MHH, we refer to it as the lattice height. For the polarizability tensor, we choose $\Delta\alpha/\bar{\alpha} = 165.8/237$, corresponding to LiCs [27]. This rescaling does not change the qualitative static and dynamical features of Eq. (3.50); it only makes Eq. (3.50) treatable directly by TEBD, without multiscale methods. In the future, we plan to apply multiscale methods to determine the emergent timescales for experimentally relevant parameters.

First, we point out that if we consider populating a single rotational state (e.g., $J = 0$, $M = 0$) in the $\Omega \to 0$ limit, then Eq. (3.50) becomes the extended Bose–Hubbard Hamiltonian, and the phase diagram is known [10, 30]. Because the tunneling energy is different for different rotational states (see Eq. (3.31)) and this difference depends only on the properties of the polarizability tensor, we can relate the borders of the phase diagram for different rotational states to properties of the polarizability tensor. The MHH thus gives a means to measure the polarizability tensor, a standing issue in experiments (Jin, 2008, JILA, NIST and University of Colorado, private communication). Our calculations in Sect. 3.2 can be used to compare directly to the phase diagram from the literature [10, 30]. In fact, this aspect of our work, unlike the simulations below, is not restricted to 1D.

However, our main focus at present is on the dynamics of the MHH. In the following numerical study, we explore dynamics as a function of the physical characteristics of the lattice, namely, number of sites L and effective lattice height η. Specifically, we study $L = 9$, 10, and 21 lattice sites with $N = 4$, 5, and 10 molecules, respectively, and η ranging from 1 to 10. We fix the dipole–dipole term as in Eq. (3.51), and fix the DC field parameter to be $\beta_{\mathrm{DC}} = 1.9$. While $\beta_{\mathrm{DC}} = 1.9$ may not correspond to a physically realizable situation, its exploration provides insight into the MHH.

The Rabi oscillations between the $J = 0$ and the $J = 1$ states damp out exponentially in the rotational time $t_r \equiv Bt/\hbar$ as

$$\langle \hat{n}_0 \rangle = a_0 - b_0\, e^{-t_r/\tau} \cos(c_0 t_r)\,, \tag{3.58}$$

$$\langle \hat{n}_1 \rangle = a_1 - b_1\, e^{-t_r/\tau} \cos(c_1 t_r)\,, \tag{3.59}$$

with some characteristic timescale τ, as seen in Fig. 3.2. We note that an exponential fit has a lower reduced chi-squared than a power-law, or algebraic fit. We also tried fit functions where the oscillations do not decay to zero, but rather persist with some asymptotic nonzero amplitude. We find that the fit functions Eqs. (3.58) and (3.59) above fit the data better as quantified by the convergence properties of the algorithms used, as discussed in Appendix 2.

The timescale τ also describes the decay of physically measurable quantities, for example the structure factors as defined in Eq. (3.48) and illustrated in Fig. 3.3. We show the emergent timescale τ for various lattice heights and systems sizes in Table 3.3.

Examining Fig. 3.2, one observes that the driven system approaches a dynamical equilibrium that is a mixture of rotational levels. The timescale with which the system relaxes to this equilibrium, τ, cannot be determined from the single-molecule physics, and so we refer to τ as an *emergent timescale*. For the low lattice height $\eta = 1$, the populations of the first two rotational states appear to oscillate around and asymptotically converge to roughly quarter filling, with $J = 1$ being lower due to contributing to population of $J = 2$ via an off-resonant AC coupling (Fig. 3.2a). For $\eta = 5$, the asymptotic equilibrium is an uneven mixture of rotational states that favors occupation of the $J = 0$ state (Fig. 3.4a), and the emergent timescale for

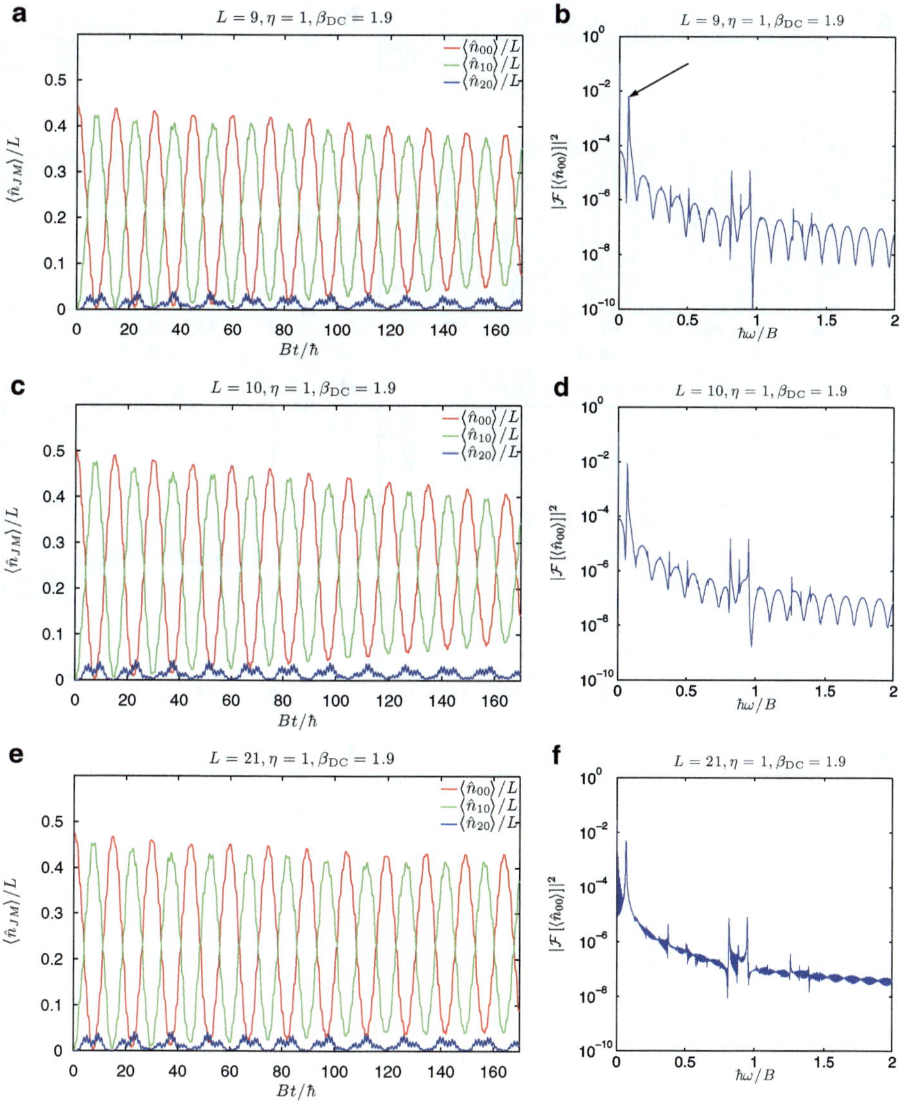

Fig. 3.2 Dependence of site-averaged number on lattice size L. For this set of parameters, the site-averaged $J = 0$ and $J = 1$ populations appear to asymptotically approach quarter filling. The $J = 2$ mode is populated slightly by off resonant AC couplings. The peak near the *left side* of the Fourier transform plots is the Rabi frequency Ω_{00}, denoted by an arrow. (**a**) Site-averaged population vs. rotational time for nine sites. Note the general theme; a gradual decrease (increase) of the maxima (minima) of oscillations. (**b**) Squared modulus of Fourier transform of site-averaged $J = 0$ population vs. rotationally scaled frequency for $L = 9$ sites. The *arrow* denotes the Rabi frequency Ω_{00}. (**c**) Site-averaged population vs. rotational time for 10 sites. Note that there is no significant difference between an odd and even number of sites. (**d**) Squared modulus of Fourier transform of site-averaged $J = 0$ population vs. rotationally scaled frequency for $L = 10$ sites. (**e**) Site-averaged population vs. rotational time for 21 sites. Note that there is no significant difference between this and the smaller system sizes. (**f**) Squared modulus of Fourier transform of site-averaged $J = 0$ population vs. rotationally scaled frequency for $L = 21$ sites

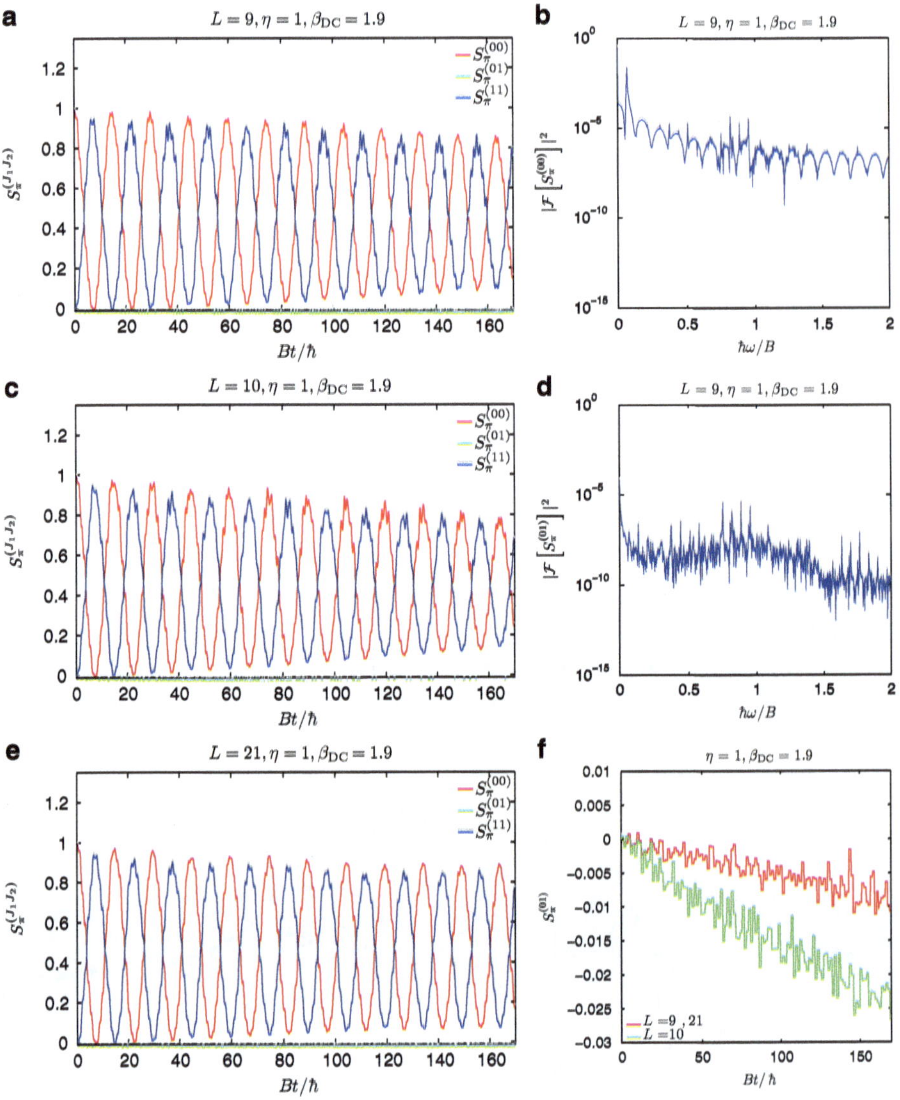

Fig. 3.3 Dependence of structure factors within and between rotational states J on the number of lattice sites. We do not consider the off-resonant $J = 2$ and higher rotational states because they have a very small occupation; $J = 2$ is shown explicitly in Fig. 3.2. (**a**) Structure factors vs. rotational time for nine sites. Note the similar asymptotic behavior to the populations in Fig. 3.2a. (**b**) Squared modulus of Fourier transform of $S_{\pi}^{(00)}$ vs. rotationally scaled frequency for $L = 9$ sites. Note the similarity with Fig. 3.2b above. (**c**) Structure factors vs. rotational time for ten sites. There is no significant difference in the $S_{\pi}^{(00)}$ and $S_{\pi}^{(11)}$ between even and odd L. For the difference in $S_{\pi}^{(01)}$, see (**f**) this figure. (**d**) Squared modulus of Fourier transform of $S_{\pi}^{(10)}$ vs. rotationally scaled frequency for $L = 9$ sites. Note the absence of the Rabi frequency. (**e**) Structure factors vs. rotational time for 21 sites. Note the lack of significant difference with the smaller odd system size. (**f**) Comparison of the $S_{\pi}^{(01)}$ correlation structure factor for odd and even numbers of sites. Note that the even site (exactly half-filling) structure factor grows faster and larger than the odd site (slightly less than half-filling) structure factor

Table 3.3 Emergent timescales τ and τ_Q and their fit asymptotic standard errors for various lattice heights and system sizes

L	η	$\tau B/\hbar$	Asymp. S.E. (%)	$\tau_Q B/\hbar$	Asymp. S.E. (%)
9	1	414.04	0.72	398.4	0.51
9	2	224.32	1.79	149.9	1.36
9	3	117.5	1.86	126.7	1.03
9	10	613.00	1.07	1079.66	14.09
10	1	259.96	0.76	240	0.6454
10	4	140.70	1.19	72.04	0.60
10	10	526.21	0.88	396.46	1.018
21	1	756.18	3.13	110.68	0.96
21	5	177.53	1.62	75.18	0.902
21	10	716.21	2.96	244.09	2.82

reaching this equilibrium is shorter than it was for $\eta = 1$ by roughly a factor of four. As the lattice height is then increased to $\eta = 10$, the populations return to the trend of $\eta = 1$, again converging to quarter filling with a timescale comparable to that of $\eta = 1$ (Fig. 3.4c). This illustrates the fact that the emergent timescale τ is not, in general, a monotonic function of the parameters of the lattice.

While the dynamics of the site-averaged rotational state populations are superficially similar for $\eta = 1$ and $\eta = 10$, the underlying physics is not identical, as can be seen by comparing Figs. 3.2f and 3.4b, d. These figures display the squared modulus of the Fourier transform of the site-averaged number in the $J = 0$ state. The only significant frequency observed for $\eta = 1$ is the Rabi frequency $\Omega \sim 0.064B/\hbar$. In contrast, the $\eta = 5$ case has numerous other characteristic frequencies. As we raise the lattice height to $\eta = 10$, the frequencies that arose for $\eta = 5$ remain, even though the overall visual trend of the site-averaged number reflects that of the single-frequency $\eta = 1$ behavior. While we do not explicitly see the new frequencies in the site-averaged number, we do see them in the structure factors. An example is Fig. 3.5b, which clearly displays the $2\,\Omega$ frequency behavior of the correlation structure factor $S_{\pi}^{(01)}$ for $\eta = 10$. This frequency, which we easily pick out in the site-averaged number's Fourier transform, can also be seen in the Fourier transform of $S_{\pi}^{(01)}$, see Fig. 3.5d.

We find that the emergent timescale τ does not depend strongly on the size of the system L, even though the distribution of molecules on the lattice is, in general, quite different for different numbers of sites, as can be seen by comparing Fig. 3.2a, e. Examining Fig. 3.2c and Table 3.3, the $L = 10$ case has a smaller τ than either of the odd L cases. We think this has to do with the filling being exactly $1/2$ and not, strictly speaking, with the number of lattice sites, as the $L = 9$ and $L = 21$ cases have fillings less than $1/2$. We see this clearly by comparing Fig. 3.6a with Fig. 3.2a, c, e. Figure 3.6a displays $\langle \hat{n}_{00} \rangle / N$, a quantity which is independent of filling but dependent, in general, on the number of lattice sites. There is a weak dependence on the number of lattice sites. On the other hand, Fig. 3.2a, c, e display $\langle \hat{n}_{00} \rangle / L$, a quantity which is independent of the number of lattice sites but dependent, in

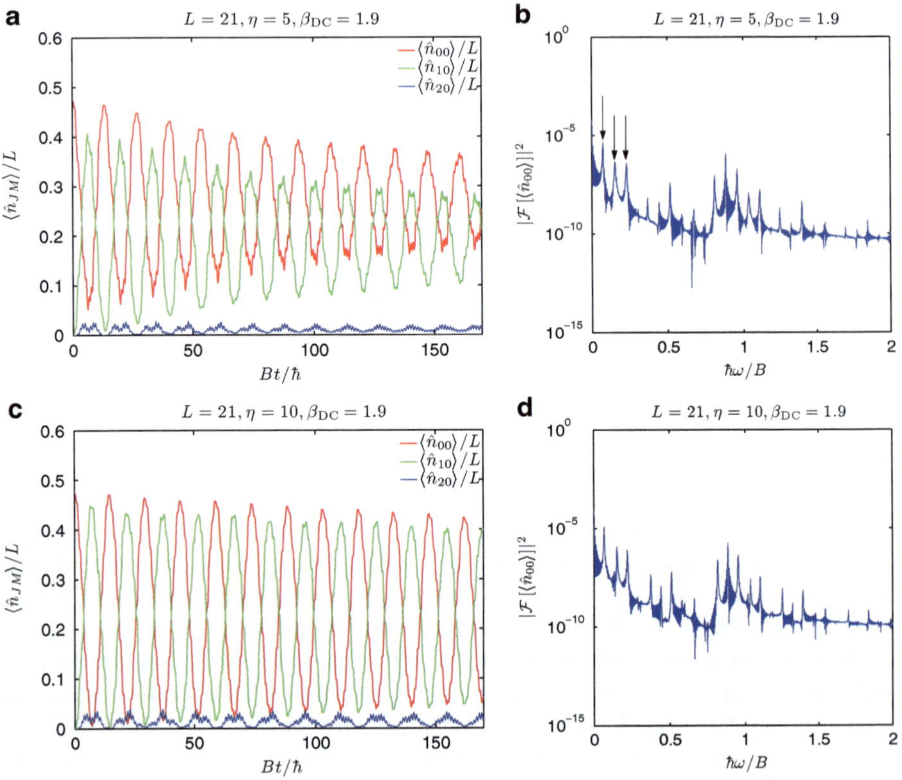

Fig. 3.4 Dependence of the asymptotic behavior of rotational state populations on the lattice height η. (**a**) Site-averaged population vs. rotational time for 21 sites with $\eta = 5$. Note that the $J = 0$ and $J = 1$ states now appear to converge to different fillings. (**b**) Squared modulus of Fourier transform of $\langle \hat{n}_{00} \rangle$ vs. rotationally scaled frequency for $L = 21$ sites and $\eta = 5$. Note the presence of several new frequencies not observed in the $\eta = 1$ case (Fig. 3.2f). In particular, Ω_{00}, $2\Omega_{00}$, and $3\Omega_{00}$ are denoted by *arrows*. (**c**) Site-averaged population vs. rotational time for 21 sites with $\eta = 10$. Note the similarity to the $\eta = 1$ case (Fig. 3.2e) and the difference from the $\eta = 5$ case [(**a**) of this figure]—the asymptotic behavior is *not* a monotonic function of the lattice height. (**d**) Squared modulus of Fourier transform of $\langle \hat{n}_{00} \rangle$ vs. rotationally scaled frequency for $L = 21$ sites and $\eta = 10$. Note that the frequencies that emerged during $\eta = 5$ have persisted

general, on the filling. There is a marked difference between $L = 10$, which has filling of $5/10 = 1/2$ and the others, which have fillings $<1/2$, but there is not a significant difference between $L = 9$ and $L = 21$, which have fillings of $4/9$ and $10/21$, respectively.

The dependence of τ on the filling is also evidenced by the correlation structure factor $S_{\pi}^{(01)}$ in Fig. 3.3f, which shows that there is a stronger correlation between the $J = 0$ and $J = 1$ states for exactly half-filling than for fillings less than half, regardless of the system size. Half-filling is known to be important in the extended Bose–Hubbard model, where it marks the introduction of the charge density wave

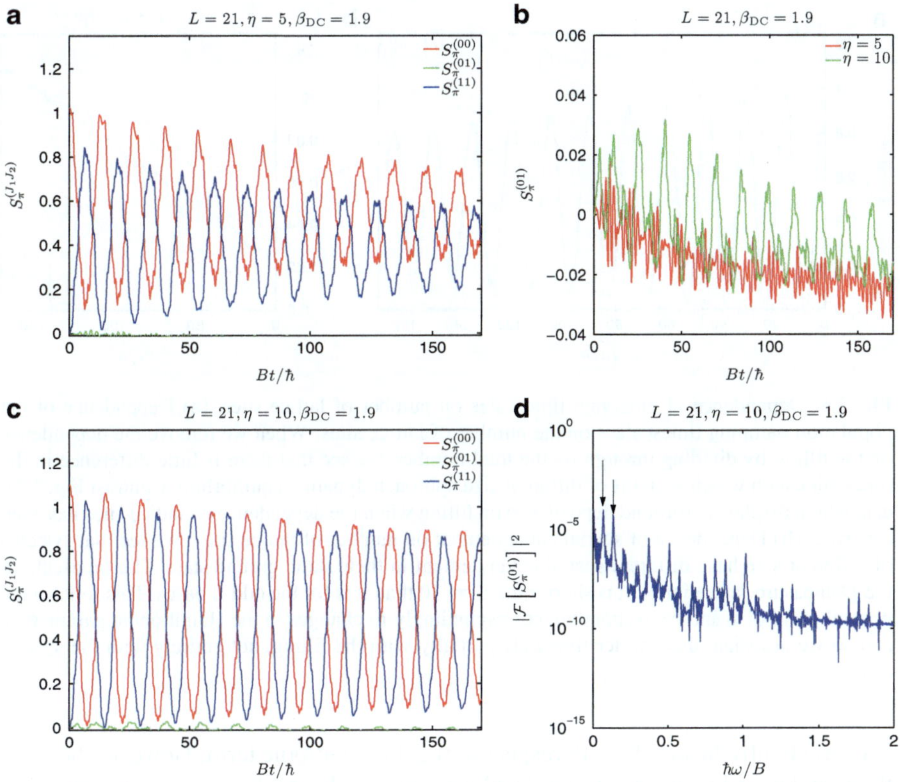

Fig. 3.5 Dependence of the asymptotic behavior of structure factors on the lattice height η. (**a**) Structure factors vs. rotational time for 21 sites with $\eta = 5$. (**b**) Correlation structure factor $S_\pi^{(01)}$ vs. rotational time for 21 sites with $\eta = 5, 10$. (**c**) Structure factors vs. rotational time for 21 sites with $\eta = 10$. Note the similarity of $S_\pi^{(00)}$ and $S_\pi^{(11)}$ to the $\eta = 1$ case (Fig. 3.3e). Note also that $S_\pi^{(01)}$ is now nonzero, and is periodic with the Rabi frequency Ω_{00} at short times and twice the Rabi frequency at long times (see also (**d**) and (**b**) of this figure). (**d**) Squared modulus of Fourier transform of $S_\pi^{(10)}$ vs. rotationally scaled frequency for $L = 21$ sites and $\eta = 10$. Many new frequencies appear, in particular the Rabi frequency and double the Rabi frequency, denoted with *arrows*

phase. We thus interpret this greater correlation structure factor as the appearance of a dynamic charge density wave phase *between* rotational states at half-filling.

This is in contrast to the usual behavior, where the structure factors $S_\pi^{(00)}$ and $S_\pi^{(11)}$ are nonzero whenever there is nonzero occupation of the particular rotational state and the structure factor $S_\pi^{(01)}$ is much smaller—essentially zero, see Fig. 3.3a, e. These results for the structure factors means that the $J = 0$ and $J = 1$ states tend to lie on top of one another, and not to "checkerboard" with a different rotational state occupying alternating sites. This is due to the fact that the Rabi flopping timescale is much shorter than the dipole–dipole timescale, meaning that the population cycles before there is sufficient time for the molecules to rearrange to a configuration which

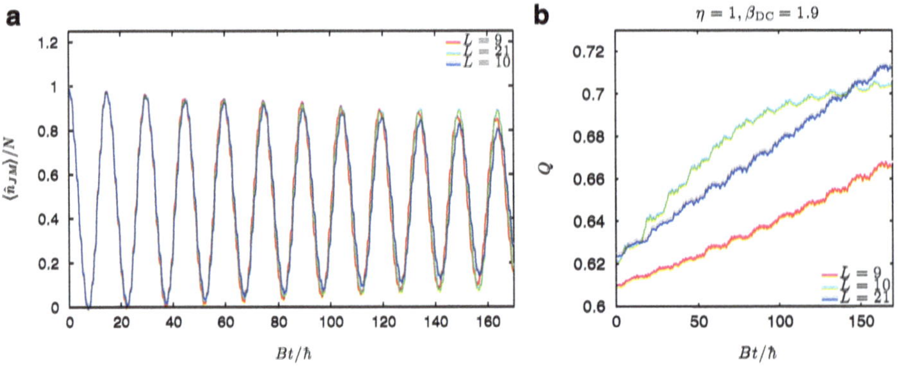

Fig. 3.6 Dependence of emergent timescales on number of lattice sites. (**a**) Dependence of the population damping timescale τ on the number of lattice sites. When we remove the dependence on the filling by dividing through by the total number, we see that there is little difference in the timescales with which systems of different size approach dynamic equilibrium. Contrast Fig. 3.2a, c, e, which display a profound dependence on filling when the dependence on lattice sites has been removed. (**b**) Dependence of spatial entanglement on number of lattice sites. We see that systems of different size have different spatial entanglement in their static ground state. The timescale of the Q-measure saturation, τ_Q, is shorter for $L = 10$ than it is for the odd L cases. This follows the general trend of τ and τ_Q responding correspondingly to changes in the Hamiltonian parameters, and so we associate this shorter timescale partially with the filling, not entirely with the system size

is energetically favorable with respect to the dipole–dipole term. However, because the population in each rotational level asymptotically reaches some nonzero value, we do see a small amount of rearrangement after many Rabi periods for any filling, corresponding to a nonzero $S_{\pi}^{(01)}$. Note that this rearrangement does not affect the site-averaged numbers, but rather the distribution of rotational states among the lattice sites. This asymptotic distribution emerges on timescales longer than we have considered, and is more prone to finite size effects than the site-averaged quantities, so we do not make a conjecture about it here.

We find that the Q-measure saturates as

$$Q = Q_{\max} - \Delta Q e^{-t_r/\tau_Q}, \tag{3.60}$$

with a different timescale τ_Q, see Fig. 3.7a and Table 3.3. We also find that the saturation timescale of the Q-measure is not, in general, a monotonic function of the lattice height η, as shown in Fig. 3.7a.

This timescale is different from the timescale τ at which the populations approach an asymptotic equilibrium, though both timescales respond similarly to changes in the Hamiltonian parameter, see Table 3.3. For example, if τ_Q gets larger as a parameter is changed, then τ also gets larger, as illustrated in Fig. 3.7a, b. The timescale τ_Q displays a stronger dependence on the number of lattice sites L than τ, as can be seen in Fig. 3.6b, a. This is because τ describes a quantity that has been averaged over sites, while τ_Q does not.

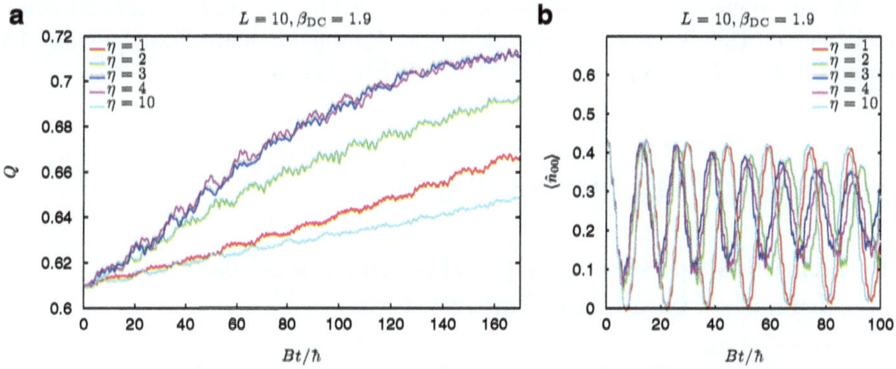

Fig. 3.7 Dependence of emergent timescales on lattice height. (**a**) Dependence of spatial entanglement on lattice height. Note that the spatial entanglement and its associated timescale are *not* monotonic functions of the lattice height. Note also that the entanglement of the static ground state appears to be largely insensitive to the lattice height. (**b**) Dependence of the site-averaged number on the lattice height. Note that the emergent timescale τ is *not* a monotonic function of the lattice height. Note also that τ responds in the same way that τ_Q does to changes in the lattice height

3.5 Conclusions

We have presented and derived a novel lattice Hamiltonian, the MHH. The MHH is a natural Hamiltonian for connecting theoretical studies of the dynamics of quantum phase transitions to near-term experimental setups using ultracold molecular gases. We presented a case study of this new Hamiltonian for hard-core bosonic molecules at half-filling. Starting from an initial condition of half-filling in the $J = 0$, $M = 0$ state, we found that initial large oscillations in the system self-damp to an asymptotic equilibrium which consists of a lattice height and filling-dependent spatially entangled superposition of dressed states. This occurs on an emergent timescale τ which cannot be predicted from the single molecule theory. We showed that τ depends non-monotonically on lattice height, weakly on lattice size, and strongly on filling (as apparent in simulations with odd and even numbers of sites). We also discovered a separate emergent timescale τ_Q which describes how quickly the many-body spatial entanglement saturates. We demonstrated that τ_Q and τ respond similarly to changes in the Hamiltonian parameters and that τ_Q depends on the filling, the lattice size, and, non-monotonically, on the lattice height. In addition to these emergent timescales, we studied the time-dependent structure factors and their frequency-domain Fourier transforms.

In future studies we will consider different filling factors, DC field strength to rotation ratios β_{DC}, and initial conditions, as well as polarized and unpolarized spin-1/2 fermionic molecules. In addition, we will use multiscale methods to study how the emergent timescale demonstrated above compares to experimental timescales for physical systems, and thereby make quantitative predictions for experiments.

We acknowledge useful discussions with Deborah Jin, Heather Lewandowski, and Jun Ye. This work was supported by the National Science Foundation under Grant PHY-0547845 as part of the NSF CAREER program.

Appendix 1: Single-Molecule Physics

1 Relationship Between Operators in Space-Fixed and Molecule-Fixed Coordinate Systems

It is well known that the representation of the angular momentum operators in a molecule-fixed coordinate frame leads to the *anomalous* commutation relations $\left[J_i, J_j\right] = -i\hbar\epsilon_{ijk}J_k$ [42]. The simplest way to avoid this trouble is to transform all expressions into the space-fixed frame where the angular momentum operators satisfy the normal commutation relations $[J_i, J_k] = i\hbar\epsilon_{ijk}J_k$ [43]. If the molecule-fixed axes are obtained by rotation of the space-fixed axes through the Euler angles $\{\phi, \theta, \chi\}$ (Zare [43]) (which we collectively abbreviate as $(\mathbf{R})$), then the component of a kth-rank spherical tensor T that has projection p along the space-fixed z axis, denoted $(T)_p^{(k)}$, can be expressed in terms of the molecule fixed components as

$$(T)_p^{(k)} = \sum_q \mathcal{D}_{pq}^{(k)}(\mathbf{R})^\star (T)_q^{(k)}, \tag{3.61}$$

where $\mathcal{D}_{pq}^{(k)}(\mathbf{R})^\star$ is the complex conjugate of the pq element of the kth-rank rotation matrix (Wigner D-matrix). To avoid confusion, we will label all space-fixed components with the letter p and all molecule-fixed components with q. From the orthogonality of the rotation matrices we have the inverse relationship

$$(T)_q^{(k)} = \sum_p \mathcal{D}_{pq}^{(k)}(\mathbf{R})(T)_p^{(k)} \tag{3.62}$$

$$= \sum_p (-1)^{p-q}\mathcal{D}_{-p,-q}^{(k)}(\mathbf{R})^\star (T)_p^k. \tag{3.63}$$

2 Rotational Hamiltonian

In the rigid rotor approximation the rotational Hamiltonian is simply

$$\hat{H}_{\rm rot} = B\hat{\mathbf{J}}^2, \tag{3.64}$$

where we have defined the rotational constant $B \equiv 1/2\mu r_e^2$, with μ the molecule's reduced mass and r_e its equilibrium internuclear separation. Typical values of B are $\sim 60\hbar$ GHz [44]. This Hamiltonian has eigenvalues $BJ(J+1)$ and eigenstates $|JM\rangle$, with J the total angular momentum and M its projection along the internuclear axis.

3 DC Field Term

The dipole moment of a polar molecule in a rotational eigenstate is zero in an average sense due to the spherical symmetry of the rotational Hamiltonian. We break this symmetry by introducing a DC electric field along the space-fixed z axis, with Hamiltonian

$$\hat{H}_{\mathrm{DC}} = -\hat{\mathbf{d}} \cdot \mathcal{E}_{\mathrm{DC}}, \tag{3.65}$$

where $\mathcal{E}_{\mathrm{DC}}$ is the electric field amplitude. The field defines the spherical space-fixed axis $p = 0$, and the molecule-fixed internuclear axis defines $q = 0$. We transform between them using a first-rank rotation matrix as outlined above:

$$\hat{H}_{\mathrm{DC}} = -\left(\hat{\mathbf{d}}\right)_0^{(1)} \mathcal{E}_{\mathrm{DC}}. \tag{3.66}$$

The matrix elements of the DC Hamiltonian in the basis which diagonalize the rotational Hamiltonian Eq. (3.65) are

$$\begin{aligned}\langle J', M'|\hat{H}_{\mathrm{DC}}|J, M\rangle &= -d\,\mathcal{E}_{\mathrm{DC}}\sqrt{(2J+1)(2J'+1)}\,(-1)^M \\ &\times \begin{pmatrix} J & 1 & J' \\ -M & 0 & M' \end{pmatrix} \begin{pmatrix} J & 1 & J' \\ 0 & 0 & 0 \end{pmatrix}\end{aligned} \tag{3.67}$$

where we use the notation $(\ldots)$ for the Wigner 3-j symbol (Zare [43]). Note that the symbol d refers to the permanent dipole moment of a molecule, and is not to be confused with the dipole operator denoted by $\hat{\mathbf{d}}$. We refer to the basis which simultaneously diagonalizes the Rotational and DC Hamiltonians as the "dressed basis," and we denote the kets that span this basis by $|\mathcal{E}; JM\rangle$, where the labels J and M are the zero field values of the corresponding quantum number and the symbol $\mathcal{E}$ is a reminder that these kets are superpositions of field free rotational states and DC field.

The effects of the DC field can be clearly seen by considering the dressed state wavefunctions, energies, and dipole moments to lowest order in perturbation theory in the dimensionless parameter $\beta_{\mathrm{DC}} \equiv d\mathcal{E}_{\mathrm{DC}}/B$, the ratio of the field energy to the rotational level splitting:

$$|\mathcal{E};J,M\rangle = |J,M\rangle - \frac{\beta_{\mathrm{DC}}}{2J}\sqrt{\frac{J^2-M^2}{4J^2-1}}|J-1,M\rangle + \frac{\beta_{\mathrm{DC}}}{2(J+1)}\sqrt{\frac{(J+1)^2-M^2}{4(J+1)^2-1}}|J+1,M\rangle, \tag{3.68}$$

$$\Delta E^{(2)}_{JM} = \frac{d^2\mathcal{E}^2_{\mathrm{DC}}}{2B}\left[\frac{J(J+1)-3M^2}{J(J+1)(2J-1)(2J+3)}\right], \tag{3.69}$$

$$\langle\mathcal{E};JM|\hat{\mathbf{d}}|\mathcal{E};JM\rangle/d = -\frac{\partial E_{JM}}{\partial\beta_{\mathrm{DC}}} = \beta_{\mathrm{DC}}\frac{3M^2/J(J+1)-1}{(2J-1)(2J+3)}, \tag{3.70}$$

where $\Delta E^{(2)}_{JM}$ is the lowest nonzero shift in the energy.

The DC field mixes states of different J, breaking the $(2J+1)$-fold degeneracy of the rotational Hamiltonian, and so J is no longer a good quantum number. In the case of a z-polarized field, M remains a good quantum number, and a degeneracy persists for all states with the same $|M|$. This mixing aligns the molecule with the field, inducing a nonzero dipole moment. This means of orienting polar molecules, known as "brute force" orientation, works well for molecules that both have a large dipole moment and can be efficiently rotationally cooled [45]. While more effective means of orienting molecules using intense laser fields are known [46], they complicate the theoretical discussion and the experimental setup, and so we do not consider them here.

In larger fields the rotational levels become deeply mixed, which allows states that are weak-field seeking in low fields to become high-field seeking in high fields [47]. The actual mixing of rotational levels vs. β_{DC} is depicted in Fig. 3.9 for the lowest three dressed levels. We note that there always exists a field $\mathcal{E}_R$ such that the lowest R dressed states' dipole moments are all positive, as this is important to ensure the stability of a collection of dipoles. The universal curve of the induced dipole moments (in units of d) vs. β_{DC} of the first two dressed rotational manifolds are shown in Fig. 3.8a. The universal curve of the dressed state energies (in units of B) vs. β_{DC} is shown in Fig. 3.8b. For reference, $\beta_{\mathrm{DC}}=1$ corresponds to a field of roughly $1.93\frac{\mathrm{kV}}{\mathrm{cm}}$ for $B\sim 60\hbar$ GHz and $d\sim 9$ D.

Expanding the field operators in Eq. (3.1) in a Wannier basis of dressed states centered at a particular discrete position $\mathbf{r}_i$ as described in Eq. (3.4), we find

$$\hat{H}_{\mathrm{rot}} + \hat{H}_{\mathrm{DC}} = \sum_J \sum_{M=-J}^{J} E_{J,M}\hat{n}_{\mathcal{E},JM}, \tag{3.71}$$

where E_{JM} is the energy of the $|\mathcal{E};J,M\rangle$ dressed state (see Fig. 3.9a) and $\hat{n}_{\mathcal{E},JM}$ is the number operator associated with this same state.

If the DC field were aligned at a small angle θ_a to the z field of the trap (say, in the xz plane), then small dipole moments mixing $M'=M\pm 1$ states would arise and the $M'=M$ dipoles would decrease slightly (we can view them as being in an effective field of $\mathcal{E}_{\mathrm{eff}}=\cos\theta_a\mathcal{E}_{\mathrm{DC}}$). Treating the new contribution perturbatively in the small parameter $\sin\theta_a\beta_{\mathrm{DC}}$, we find the lowest order couplings to the ground state

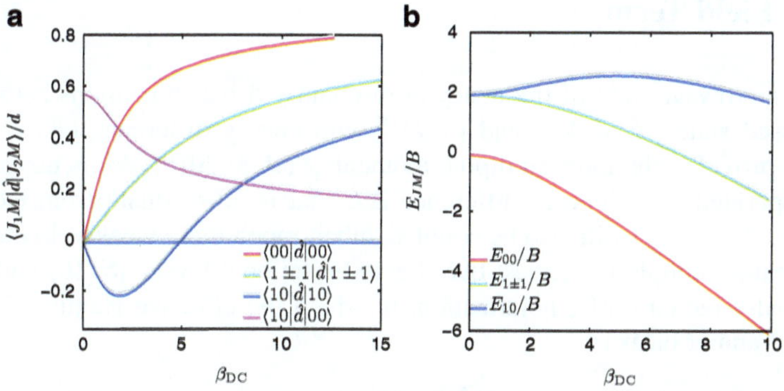

Fig. 3.8 Dressed state dipole moments and energies. Note that the $J = 1$, $M = 0$ resonant dipole moment changes from weak-field seeking to high-field seeking at $\beta_{\rm DC} \approx 5$. All rotational states have a field where this transition occurs, and the dipole tends monotonically towards unity after this field. The $\langle 10|\hat{\mathbf{d}}|00\rangle$ dipole moment (and all transition dipole moments, generically) tends towards zero monotonically as $\beta_{\rm DC}$ increases. Note also that the energetic differences between rotational levels are smallest at zero field and grow monotonically thereafter. (**a**) Scaled induced dipole moments vs. scaled DC field energy. (**b**) Scaled dressed energies vs. scaled DC field energy

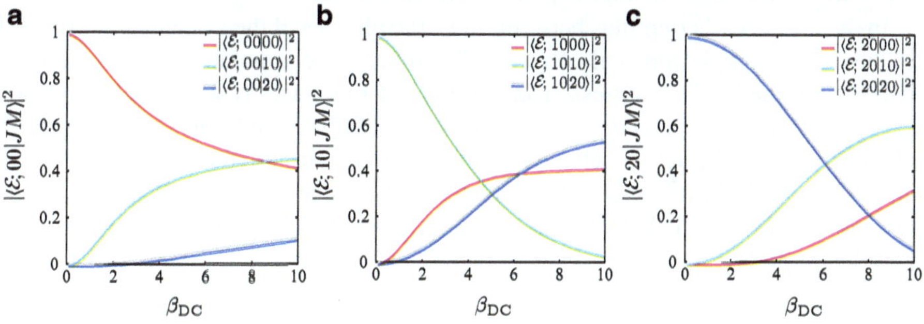

Fig. 3.9 Compositions of dressed states vs. scaled rotational energy. The states become deeply mixed in large fields, and that the dressed state $|\mathcal{E}; JM\rangle$ whose zero field value is $|JM\rangle$ does not always have the greatest overlap with $|JM\rangle$ for all $\beta_{\rm DC}$. The field strength where the first dressed state changes from weak-field to high-field seeking, $\beta_{\rm DC} = 5$, is also roughly the place where its overlap with the $|00\rangle$ field-free level is greater than the overlaps with all other field-free levels. (**a**) Composition of first dressed state. (**b**) Composition of second dressed state. (**c**) Composition of third dressed state

$$\langle \mathcal{E}; 00|\hat{H}_{\rm DC}|\mathcal{E}; 1 \pm 1\rangle \simeq \frac{\sin\theta_a d\mathcal{E}}{\sqrt{6}}\left(1 - \frac{49\sin^2\theta_a}{1440}\beta_{\rm DC}^2\right), \tag{3.72}$$

and associated timescale τ_{θ_a} for occupation of $M \neq 0$ states from the ground state,

$$\tau_{\theta_a} = \frac{\sqrt{6}\hbar}{\sin\theta_a d\mathcal{E}\left(1 - \frac{49\sin^2\theta_a}{1440}\beta_{\rm DC}^2\right)} \sim \frac{\sqrt{6}}{\beta_{\rm DC}\sin\theta_a}\frac{\hbar}{B}. \tag{3.73}$$

4 AC Field Term

An AC microwave field of frequency ω resonantly drives transitions between two DC dressed states $|\mathcal{E}; J'M'\rangle$ and $|\mathcal{E}; JM\rangle$ with energy difference $(E_{J'M'} - E_{JM}) /\hbar \approx \omega$ provided the induced dipole moment $\langle\mathcal{E}; J'M'|\hat{\mathbf{d}}|\mathcal{E}; JM\rangle$ is nonzero. Two states separated by an energy difference ΔE that is off-resonant from the driving field (i.e., $\Delta E \gg \omega$) will also be coupled, albeit much more weakly. In our system we resonantly couple the lowest two dressed rotational levels, $|\mathcal{E}; 10\rangle$ and $|\mathcal{E}; 00\rangle$. We consider the case of z polarization, in which the effective Hamiltonian in the dressed Wannier basis is

$$\hat{H}_{\mathrm{AC}}(t) = -\pi \sin(\omega t) \textstyle\sum_{JM} \Omega_{JM} \left(\hat{a}^{\dagger}_{\mathcal{E};J,M} \hat{a}_{\mathcal{E};J+1,M} + \mathrm{h.c}\right), \tag{3.74}$$

where

$$\Omega_{JM} \equiv \mathcal{E}_{\mathrm{AC}}\langle\mathcal{E}; J, M|\hat{\mathbf{d}}|\mathcal{E}; J+1, M\rangle/\hbar. \tag{3.75}$$

is the Rabi frequency. This is the frequency with which the populations of a two-level system cycles. In experiments, the AC field has spatial curvature on the order of cm which is negligible on the μm system size scale.

In the absence of couplings between sites, the physics of the system is determined by the on-site, single-molecule physics. The percentage population of each component in both the $|\mathcal{E}; J, M\rangle$ dressed and $|JM\rangle$ field-free bases is shown below for one Rabi period. In these plots only the $|\mathcal{E}; 10\rangle$ and $|\mathcal{E}; 00\rangle$ dressed states are considered, which is close to the actual behavior when all other states are far off-resonant. Each site undergoes Rabi flopping independently of the others. Figure 3.10a, b show this behavior for $\beta_{\mathrm{DC}} = 1.900$ and $\beta_{\mathrm{AC}} \equiv d\mathcal{E}_{\mathrm{AC}}/B = 0.200$, giving a Rabi period of $2\pi/\Omega_{00} = 36.5\hbar/B$.

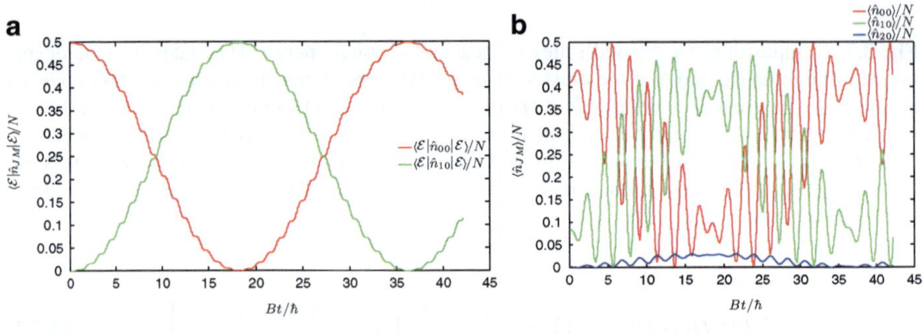

Fig. 3.10 Resonant AC field induced population cycling in the dressed and field-free bases. (**a**) Populations of the dressed states vs. rotational time. The small amplitude rapid oscillations occur on the timescale $1/\omega$, and are often averaged away via the rotating wave approximation. The large amplitude oscillations occurring on the timescale $1/\Omega_{00}$ that periodically transfer the population between $|\mathcal{E}; 00\rangle$ and $|\mathcal{E}; 10\rangle$ are the characteristic "Rabi oscillations" of a driven two-level system. (**b**) Populations of the field-free states vs. rotational time. The $|20\rangle$ state is occupied because both $|\mathcal{E}; 00\rangle$ and $|\mathcal{E}; 10\rangle$ have a nonzero projection with this state due to the mixing from the DC field, see Fig. 3.9. It is apparent from comparison with (**a**) of this figure that the dressed basis greatly simplifies the AC term in the Hamiltonian

Appendix 2: Convergence

1 Single-Molecule Considerations

Each dressed state $|\mathcal{E}; J, M\rangle$ is, in principle, an infinite linear combination of field free states

$$|\mathcal{E}; J, M\rangle = \sum_{J'=0}^{\infty} c_{J'}|J', M\rangle. \tag{3.76}$$

Numerically, we must have a finite upper bound to the sum in Eq. (3.76), which we call J_{cut}. This does not cause difficulty in practice, as the overlap of a dressed state $|\mathcal{E}; JM\rangle$ with a field-free state $|J'M\rangle$ diminishes rapidly as J' differs more greatly from J. We find the coefficients in Eq. (3.76), as well as the dressed state energies and dipole moments by simultaneously diagonalizing the rotational and DC field Hamiltonians in a basis consisting of the first J_{cut} rotational levels. Because TEBD scales poorly with the on-site dimension, we form as small an on-site basis as possible by keeping the eigenvectors corresponding to the R lowest dressed levels. To form a proper basis, we must renormalize these eigenvectors (which, for z-polarized field, does not change their orthogonality). We now demonstrate the convergence of these two procedures

To show convergence of the first procedure, we plot the difference between the energy of the Jth rotational state calculated for a particular value of $J_{\text{cut}} = i$ and one higher value, $\Delta E_J(i)$ as a function of i. The results for various field strengths are shown in Fig. 3.11a, b. We see very fast convergence for the low fields (e.g., $\beta_{\text{DC}} = 1.9$) of interest. In our numerics we use $J_{\text{cut}} = 25$, which ensures convergence for any of the β_{DC} considered.

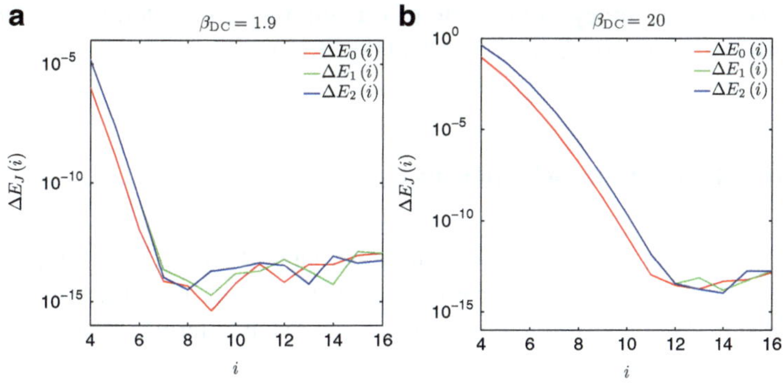

Fig. 3.11 Convergence with respect to DC dressing rotational state cutoff. As few as seven field-free levels are needed for the weak field $\beta_{\text{DC}} = 1.9$ to have the dressed state energies of interest converge to machine precision (*left panel*), and even a large DC field $\beta_{\text{DC}} = 20$ requires only 12 field-free levels for the energy to converge (*right panel*)

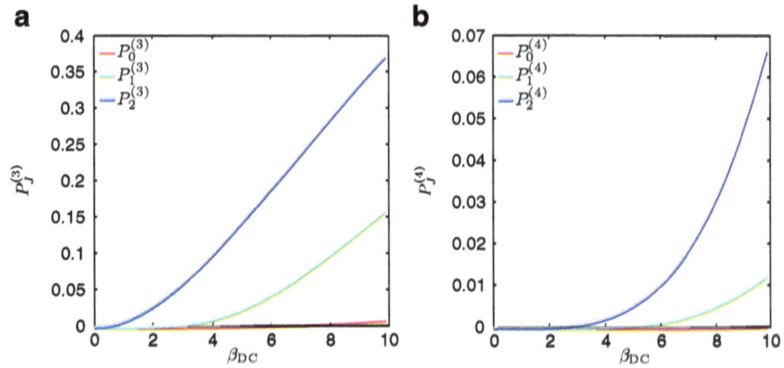

Fig. 3.12 Convergence with respect to local dimension cutoff. Dressed states with greater J lose more of their norm in truncation, as mixing occurs most strongly with adjacent J. Also, as the field is increased, the states become more deeply mixed, and so all states lose more of their norm. Truncating the local basis at the $J = 3$ dressed level incurs at most a 1 % loss of norm for any of the states that are appreciably populated during time evolution (*right panel*)

To determine convergence with respect to the second procedure, examine Fig. 3.9a, b, which show

$$P_J^{(J_{\max})} \equiv 1 - \sum_{i=0}^{J_{\max}-1} |\langle \mathcal{E}; J0|i0\rangle|^2 , \tag{3.77}$$

the amount of the total dressed wave function norm $|\langle \mathcal{E}; J0|\mathcal{E}; J0\rangle|^2$ that lies outside of the first $J_{\max}$ field-free rotational levels for $J_{\max} = 3$ and 4, respectively. For $J_{\max} = 4$ the renormalization of the first three rotational levels is a very small effect for the β_{DC} we consider, and the fourth level is not populated to any appreciable extent during time evolution for any β_{DC} (see Fig. 3.10a), so we expect that keeping the $J_{\max} = 4$ lowest levels will give sufficient accuracy. By direct simulation, we find six-digit accuracy in the suite of quantum measures defined in Sect. 3.3.2; specifically, we compare $J_{\max} = 3$ and 4 (Fig. 3.12).

2 Many-Body Considerations

There are also convergence issues that are inherent to the TEBD algorithm. The first, called the *Schmidt error*, is the error that arises from truncating the Hilbert space at each time step. We can parameterize the error per step in terms of the entanglement cutoff parameter χ as

$$\tau_l^S = 1 - \sum_{\alpha_l=1}^{\chi} \left(\lambda_{\alpha_l}^{[l]}\right)^2 \tag{3.78}$$

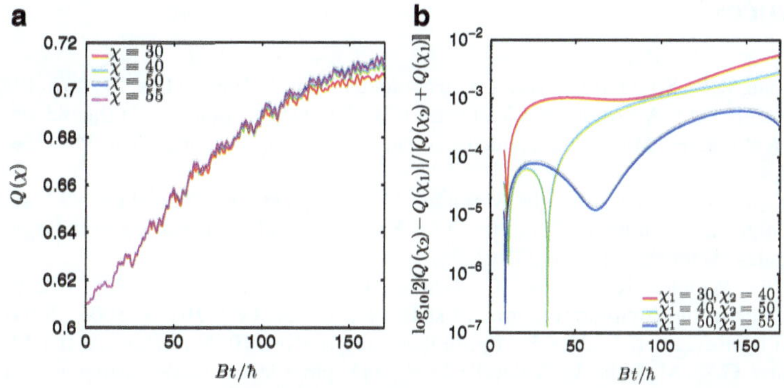

Fig. 3.13 Convergence with respect to entanglement cutoff parameter. The *left figure* shows the spatial entanglement measure Q for various values of the TEBD entanglement cutoff parameter χ. As χ is increased, Q remains close to its true value for longer. In the *right figure* we plot the log of the absolute difference in Q for two values of χ divided by its arithmetic mean. We see at least four-digit accuracy for the largest values of χ we consider. Note also that even small values of χ are accurate for short times

where $\lambda^{[l]}$ is a vector containing the eigenvalues of the reduced density matrix obtained by tracing over all sites but l, and α_l is the local index that entangles the site l with the rest of the system, with smaller α_l states having greater weight. We find that, among the measures we use, the one that is most sensitive to χ is the Q-measure, which we plot for four values of χ in Fig. 3.13. Increasing χ improves the accuracy over longer times, but there is always a time after which the measure begins to deviate. This is the normalization drift alluded to in Sect. 3.3.1. The χ-dependent time after which the Schmidt error dominates is referred to as the runaway time [39]. In the case study of Sect. 3.4, we used $\chi = 50$ for all simulations, which gives the Q-measure accurately to within four decimal places over the timescales considered.

The second intrinsic source of error in TEBD is due to the Trotter–Suzuki expansion of the propagator [32]. We parameterize this error in terms of δt, the time step. When we halve the time step from that used in the simulations above $(= 2\pi/(133\omega))$, we find no change in the measures to the ninth digit. It is clear that the Schmidt error discussed above is the chief source of error in our simulations.

To extract the emergent timescales defined in Eqs. (3.58) and (3.60), we used two different methods. The first is the nonlinear curve fitting routine "fit" in gnuplot. The second is the "NonlinearRegression" package in Mathematica 6.0. Both methods use nonlinear regression, which fits the data to a specified nonlinear function of the model parameters. The goodness of the fit is quantified by the asymptotic standard errors of the model parameters, which gives the standard deviation of each parameter. A low percent asymptotic error means that the model parameters cannot be adjusted very far without noticeably changing the goodness-of-fit. Both gnuplot and Mathematica returned the same values for the emergent timescales to within the stated asymptotic standard error.

References

1. Feynman, R.P.: Simulating physics with computers. Int. J. Theor. Phys. **21**, 467 (1982)
2. Lewenstein, M.S.A., Ahufinger, V., Damski, B., Sen De, A., Sen, U.: Ultracold atomic gases in optical lattices: Mimicking condensed matter physics and beyond. Adv. Phys. **56**, 243–379 (2007)
3. Ni, K.-K., Ospelkaus, S., de Miranda, M.H.G., Peér, A. Neyenhuis, B., Zirbel, J.J., Kotochigova, S., Julienne, P.S., Jin, D.S., Ye, J.: A high phase-space-density gas of polar molecules. Science **322**, 231–235 (2008)
4. Lang, F., Winkler, K., Strauss, C., Grimm, R., Hecker Denschlag, J.: Ultracold triplet molecules in the rovibrational ground state. Phys. Rev. Lett. **101**, 133005 (2008). doi:10.1103/PhysRevLett.101.133005. http://link.aps.org/doi/10.1103/PhysRevLett.101.133005
5. Brennen, G.K., Micheli, A., Zoller, P.: Designing spin-1 lattice models using polar molecules. New J. Phys. **9**, 138 (2007)
6. Büchler, H.P., Demler, E., Lukin, M., Micheli, A., Prokof'ev, N., Pupillo, G., Zoller, P.: Strongly correlated 2d quantum phases with cold polar molecules: Controlling the shape of the interaction potential. Phys. Rev. Lett. **98**, 060404 (2007). doi:10.1103/PhysRevLett.98.060404. http://link.aps.org/doi/10.1103/PhysRevLett.98.060404
7. DeMille, D.: Quantum computation with trapped polar molecules. Phys. Rev. Lett. **88**, 067901 (2002)
8. Goral, K., Santos, L., Lewenstein, M.: Quantum phases of dipolar bosons in optical lattices. Phys. Rev. Lett. **88**, 170406–1–4 (2002)
9. Pupillo, G., Micheli, A., Buchler, H.P., Zoller, P.: Cold Molecules: Creation and Applications, Chap. 12. Taylor & Francis, London (2009)
10. Kühner, T.D., White, S.R., Monien, H.: One-dimensional bose-hubbard model with nearest-neighbor interaction. Phys. Rev. B **61**(18), 12474–12489 (2000). doi:10.1103/PhysRevB.61.12474
11. Dickerscheid, D.B.M., Kawaguchi, Y., Ueda, M.: Heteronuclear fermionic superfluids with spin degrees of freedom. Phys. Rev. A **77**, 053605 (2008)
12. Higbie, J.M., Sadler, L.E., Inouye, S., Chikkatur, A.P., Leslie, S.R., Moore, K.L., Savalli, V., Stamper-Kurn, D.M.: Direct Nondestructive Imaging of Magnetization in a Spin-1 Bose-Einstein Gas. Phys. Rev. Lett. **95**, 050401 (2005)
13. Guifré Vidal, G.: Efficient classical simulation of slightly entangled quantum computations. Phys. Rev. Lett. **91**(14), 147902 (2003). doi:10.1103/PhysRevLett.91.147902
14. Vidal, G.: Efficient simulation of one-dimensional quantum many-body systems. Phys. Rev. Lett. **93**(4), 040502 (2004). doi:10.1103/PhysRevLett.93.040502
15. Nielsen, M., Chuang, I.: Quantum Computation and Quantum Information. Cambridge University Press, Cambridge (2000)
16. Wenger, T., Kinoshita, T., Weiss, D.S.: A quantum newton's cradle. Nature **440**, 900–903 (2006)
17. Polkovnikov, A., Sachdev, S., Girvin, S.M.: Nonequilibrium Gross-Pitaevskii dynamics of boson lattice models. Phys. Rev. A **66**, 053607–1–7 (2002)
18. Altman, E., Polkovnikov, A., Demler, E., Halperin, B., Lukin, M.D.: Superfluid-insulator transition in a moving system of interacting bosons. Phys. Rev. Lett. **9**, 020402–1–4 (2005)
19. Manmana, S.R., Wessel, S., Noack, R.M., Muramatsu, A.: Strongly correlated fermions after a quantum quench. Phys. Rev. Lett. **98**, 210405 (2007)
20. Kollath, C., Läuchli, A.M., Altman, E.: Quench dynamics and nonequilibrium phase diagram of the bose-hubbard model. Phys. Rev. Lett. **98**, 180601 (2007)
21. Läuchli, A.M., Kollath, C.: Spreading of correlations and entanglement after a quench in the one-dimensional Bose-Hubbard model. J. Stat. Mech. Theory Exp. **2008**(05), P05018 (2008). http://stacks.iop.org/1742-5468/2008/i=05/a=P05018
22. Verdozzi, C.: Time-dependent density-functional theory and strongly correlated systems: Insight from numerical studies. Phys. Rev. Lett. **101**, 166401 (2008)

23. Mishmash, R.V., Carr, L.D.: Ultracold atoms in 1D optical lattices: Mean field, quantum field, computation, and soliton formation. J. Math. Comput. Simul. **80**, 732 (2008)
24. Brown, J., Carrington, A.: Rotational Spectroscopy of Diatomic Molecules. Cambridge University Press, Cambridge (2003)
25. Micheli, A., Pupillo, G., Büchler, H.P., Zoller, P.: Cold polar molecules in two-dimensional traps: Tailoring interactions with external fields for novel quantum phases. Phys. Rev. A **76**, 043604 (2007)
26. Rey, A.M.: Ultracold Bosonic Atoms in Optical Lattices. Ph.D. thesis, University of Maryland (2004)
27. Deiglmayr, J., Aymar, M., Wester, R., Weidemüller, M., Dulieu, O.: Calculations of static dipole polarizabilities of alkali dimers: Prospects for alignment of ultracold molecules. J. Chem. Phys. **129**, 064309 (2008)
28. Jaksch, D., Bruder, C., Cirac, J.I., Gardiner, C.W., Zoller, P.: Cold bosonic atoms in optical lattices. Phys. Rev. Lett. **81**(15), 3108–3111 (1998). doi:10.1103/PhysRevLett.81.3108
29. Sachdev, S.: Quantum Phase Transitions. Cambridge University Press, New York (1999)
30. Fisher, M.P.A., Weichman, P.B., Grinstein, G., Fisher, D.S.: Boson localization and the superfluid-insulator transition. Phys. Rev. B **40**, 546–570 (1989)
31. White, S.R.: Density matrix formulation for quantum renormalization groups. Phys. Rev. Lett. **69**(19), 2863–2866 (1992)
32. Suzuki, M.: Fractal decomposition of exponential operators with applications to many-body theories and Monte Carlo simulations. Phys. Lett. A **146**, 319–323 (1990). doi:10.1016/0375-9601(90)90962-N
33. Daley, A.J.: Manipulation and Simulation of Cold Atoms in Optical Lattices. Ph.D. thesis, Leopold-Franzens-Universität Innsbruck (2005)
34. Daley, A.J., Kollath, C., Schollwöck, U., Vidal, G.: Time-dependent density-matrix renormalization-group using adaptive effective Hilbert spaces. J. Stat. Mech. Theory Exp. **2004**(04), P04005 (2004). http://stacks.iop.org/1742-5468/2004/i=04/a=P04005
35. Brennen, G.K.: An observable measure of entanglement for pure states of multi-qubit systems. Quantum Inf. Comput. **3**, 619–626 (2003)
36. Barnum, H., Knill, E., Ortiz, G., Viola, L.: Generalizations of entanglement based on coherent states and convex sets. Phys. Rev. A **68**, 032308 (2003)
37. Barnum, H., Knill, E., Ortiz, G., Somma, R., Lorenza Viola, L.: A subsystem-independent generalization of entanglement. Phys. Rev. Lett. **92**, 107902 (2004)
38. Ashcroft, N., Mermin, D.: Solid State Physics. Saunders College Publishing, Orlando (1976)
39. Gobert, D., Kollath, C., Schollwöck, U., Schütz, G.: Real-time dynamics in spin-(1/2) chains with adaptive time-dependent density matrix renormalization group. Phys. Rev. E **71**, 036102 (2005)
40. Kevorkian, J.K., Cole, J.D.: Multiple Scale and Singular Perturbation Methods, 1st edn. Springer, New York (1996)
41. Nayfeh, A.H.: Perturbation Methods, vol. 1, 1st edn. Wiley-Interscience, New York (2000)
42. Van Vleck, J.H.: The coupling of angular momentum vectors in molecules. Rev. Mod. Phys. **23**, 213–227 (1951)
43. Zare, R.: Angular momentum: Understanding spatial aspects in chemistry and physics. Wiley, New York (1988)
44. NIST: Diatomic spectral database (2008). http://physics.nist.gov/physrefdata/molspec/diatomic/
45. Stapelfeldt, H., Seideman, T.: *Colloquium*: Aligning molecules with strong laser pulses. Rev. Mod. Phys. **75**, 543–557 (2003). doi:10.1103/RevModPhys.75.543. http://link.aps.org/doi/10.1103/RevModPhys.75.543
46. Friedrich, B., Herschbach, D.: Alignment and trapping of molecules in intense laser fields. Phys. Rev. Lett. **74**, 4623–4626 (1995). doi:10.1103/PhysRevLett.74.4623. http://link.aps.org/doi/10.1103/PhysRevLett.74.4623
47. Avdeenkov, A.V., Kajita, M., Bohn, J.L.: Suppression of inelastic collisions of polar $^1\sigma$ state molecules in an electrostatic field. Phys. Rev. A **73**, 022707 (2006). doi:10.1103/PhysRevA.73.022707. http://link.aps.org/doi/10.1103/PhysRevA.73.022707

Chapter 4
Hyperfine Molecular Hubbard Hamiltonian

Abstract An ultracold gas of heteronuclear alkali dimer molecules with hyperfine structure loaded into a one-dimensional optical lattice is investigated. The hyperfine molecular Hubbard Hamiltonian (HMHH), an effective low-energy lattice Hamiltonian, is derived from first principles. The large permanent electric dipole moment of these molecules gives rise to long-range dipole–dipole forces in a dc electric field and allows for transitions between rotational states in an ac microwave field. Additionally, a strong magnetic field can be used to control the hyperfine degrees of freedom independently of the rotational degrees of freedom. By tuning the angle between the dc electric and magnetic fields and the strength of the ac field it is possible to control the number of internal states involved in the dynamics as well as the degree of correlation between the spatial and internal degrees of freedom. The HMHH's unique features have direct experimental consequences such as quantum dephasing, tunable complexity, and the dependence of the phase diagram on the molecular state.

4.1 Introduction

Ultracold molecular gases are of interest in many subfields of science ranging from precision science to quantum simulation of many-body Hamiltonians [1]. Recent success using the stimulated Raman adiabatic passage (STIRAP) method has allowed experimentalists to produce a gas of KRb molecules close to Fermi degeneracy, in the ground rovibrational state, and in a specific hyperfine level [2, 3]. Rovibonic ground state molecules have also been formed for polar LiCs [4] as well as nonpolar Cs_2 [5] and Rb_2 [6], with studies on other species currently underway [7, 8]. To reach the quantum degenerate regime one must have all molecules in the same quantum state, a task which is complicated by the rich hyperfine structure of alkali dimer molecules. Thus, a number of recent works [9–11] have investigated

Published previously as *Hyperfine molecular Hubbard Hamiltonian*, M.L. Wall and L.D. Carr, Physical Review A **82**, 013611 (2010).

M.L. Wall, *Quantum Many-Body Physics of Ultracold Molecules in Optical Lattices*, Springer Theses, DOI 10.1007/978-3-319-14252-4_4

the single-molecule microwave spectra to find a route by which all molecules are transferred to the lowest hyperfine state, yielding a gas of *absolute* ground state molecules.

From the condensed matter perspective, ultracold gases are enticing in their capacity to act as *quantum simulators* [12, 13]. Such specialized quantum computers allow for the study of complex many-body Hamiltonians in a setting where many parameters are amenable to experimental control. From this point of view, it is natural to ask how the various degrees of freedom in the quantum simulator may be controlled and used as resources. Theoretical proposals for many-body physics using ultracold molecules have so far focused only on the rotational degrees of freedom in $^1\Sigma$ molecules with external fields [14, 15] or on the hyperfine degree of freedom in $^2\Sigma$ molecules without external fields [16]. In this work we study $^1\Sigma$ molecules in strong fields including the effects of hyperfine structure and discuss how the hyperfine degrees of freedom may be controllably accessed and manipulated as a resource for generating complex quantum dynamics.

For $^1\Sigma$ molecules it has been shown that the interaction of the rotational degrees of freedom with external electric fields allows for the tuning of the strength and range of the two-molecule interaction potential [14]. Many of these results also hold for molecules with hyperfine structure, as the rotational and nuclear spin degrees are only weakly coupled in strong fields. In particular, the application of a dc field and an optical trapping potential gives rise to a purely repulsive dipole–dipole interaction between molecules in reduced geometries. Also, it has been shown that the combination of a strong uniform magnetic field and a suitably chosen microwave field allows for transitions between particular hyperfine single-molecule states, and that this may be used to transfer a collection of molecules that have been cooled to the rovibrational ground state but an excited hyperfine state to their hyperfine ground state [9–11]. This idea also works in reverse: one can select the states which are involved in many-body dynamics with the ground state by judicious choice of the field strengths and geometries. The hyperfine molecular Hubbard Hamiltonian (HMHH) reflects this fact; not only the parameters of the Hamiltonian but also the dimensionality and character of the basis are suited to experimental control.

This article is organized as follows. In Sect. 4.2 we introduce the HMHH, define its parameters, and discuss its interesting experimental consequences. This section contains the main results of the paper. In Sect. 4.2.3 we derive the HMHH from first principles and state the key assumptions underlying its derivation. Finally, in Sect. 4.4, we conclude. Some details concerning the single molecule physics are provided in the appendices in the interest of completeness.

4.2 Statement of the Hamiltonian and Experimental Consequences

The HMHH is

$$\hat{H} = \sum_{\sigma} \Delta_{\sigma} \sum_{i} \hat{n}_{i\sigma} - \sum_{\sigma} t_{\sigma} \sum_{\langle i,j \rangle} \left[\hat{a}^{\dagger}_{i\sigma} \hat{a}_{j\sigma} + \text{h.c.} \right]$$
$$+ \frac{1}{2} \sum_{\sigma,\sigma'} U_{\sigma\sigma'} \sum_{\langle i,j \rangle} \hat{n}_{i\sigma} \hat{n}_{j\sigma'}$$
$$- \frac{1}{2} \sum_{\sigma\sigma'} d_{\sigma\sigma'} E_{\text{ac}} \sum_{i} \left[\hat{a}^{\dagger}_{i\sigma} \hat{a}_{i\sigma'} + \text{h.c.} \right], \tag{4.1}$$

where $\hat{a}_{i\sigma}$ destroys a bosonic or fermionic molecule in state $|\sigma\rangle$ on the ith lattice site, and the bracket notation $\langle \ldots \rangle$ denotes that the sum is taken over nearest neighbors. The single-molecule basis $\{|\sigma\rangle\}$ takes into account the hyperfine interactions (Appendix 1) and static fields (Appendix 2) and the quantum number σ is a composite index referring to both rotational and nuclear spin degrees of freedom. The properties and dimensionality of this basis can be modified by the geometry and strength of the external fields, as will be discussed in more detail below.

The first term in the HMHH represents the energy offset of a molecule in state $|\sigma\rangle$ from a reference ground state. The second term describes the tunneling of molecules between lattice sites and depends on the rotational state. The third term describes resonant dipole–dipole interactions between molecules on neighboring sites. The final term corresponds to transitions driven between states $|\sigma\rangle$ and $|\sigma'\rangle$ by an ac microwave field. Here the transition dipole moment between two states $|\sigma\rangle$, $|\sigma'\rangle$ is $d_{\sigma\sigma'} \equiv \langle \sigma | \hat{d}_1 | \sigma' \rangle$, where $\hat{d}_1 \equiv \hat{\mathbf{d}} \cdot \mathbf{e}_1$ is the projection of the dipole operator along the space-fixed spherical basis direction $\mathbf{e}_1 = -\left(\mathbf{e}_x + i\mathbf{e}_y\right)/\sqrt{2}$.

For $^{40}K^{87}Rb$, which is the most experimentally relevant species, the energy scales of the various terms are summarized in Table 4.1. The detunings Δ_{σ} are determined chiefly by the linear Zeeman effect, and so are tunable by the dc magnetic field, and will be similar for other molecular species. The tunneling energy scale t_{σ} is set by the recoil energy, and so will be similar for other alkali dimers. The dipole–dipole energy scale $U_{\sigma\sigma'}$ is fixed by the permanent dipole moment, and so will change with the molecular species. For example, LiCs have a dipole moment roughly ten times larger than that of KRb, and so $U_{\sigma\sigma'}$ will be of order 25 kHz. The scale of the ac term is determined by the power of the microwave field E_{ac}, which is readily tunable. The range of energies we have quoted represents the most interesting regime where the basic assumptions of our derivation hold.

In the following sections we will justify the HMHH and list the essential assumptions underlying its derivation, but we first pause to note some of its unusual properties.

Table 4.1 Table of energy scales of the hyperfine molecular Hubbard Hamiltonian

Term	Energy scale
Δ_σ	~1–100 kHz (depends on static field strengths)
t_σ	~1 kHz
$U_{\sigma\sigma'}$	~250 Hz
$d_{\sigma\sigma'}E_{\mathrm{ac}}$	~1–50 kHz

From top to bottom: energy Δ_σ of internal state $|\sigma\rangle$, relative to the ground state; tunneling t_σ; dipole–dipole interaction $U_{\sigma\sigma'}$; transition dipole moment $d_{\sigma\sigma'}$ due to the ac electric drive E_{ac}

4.2.1 Quantum Dephasing

The first property, which we call *quantum dephasing*, was investigated previously for a molecular Hubbard Hamiltonian involving only rotational degrees of freedom [15]. The effect, which is purely many-body in nature, may be summarized in this context as the destruction of coherent Rabi flopping due to the population of many spatial degrees of freedom in a many-body system driven at a single-molecule resonance. This effect is also of interest in the more general context of oscillations in a many-body system that are damped by some intrinsic mechanism following a quench [17, 18].

Dephasing is strongest when the Rabi frequency is on the order of the tunneling energies and the difference in tunneling energies for the two internal modes is also comparable to these two scales. For a system with two single-particle levels 0 and 1 and tunneling energies t_0 and t_1, respectively, this gives the condition $\Omega \sim t_0 \sim t_1 \sim |t_0 - t_1|$, which can be achieved with the HMHH for reasonable parameter values. The Rabi oscillations between the two internal states connected by the single molecule resonance damp out exponentially in time with an emergent timescale τ which can be measured experimentally, see Fig. 4.1. Dephasing can be observed in the structure factors

$$S_\pi^{\sigma\sigma'} = \frac{1}{L}\sum_{i,j=1}^{L}(-1)^{i-j}\langle\hat{n}_{i\sigma}\hat{n}_{j\sigma'}\rangle\,, \tag{4.2}$$

where L is the number of lattice sites; $S_\pi^{\sigma\sigma'}$ can be measured in scattering experiments [19].

4.2.2 Internal State Dependence of Phase Diagram

The dependence of the tunneling energy t_σ on the internal state σ makes the borders of the phase diagram shift strongly (e.g. by a factor of 2). This dependence is shown explicitly in Fig. 4.2. Thus, by preparing a collection of molecules in multiple

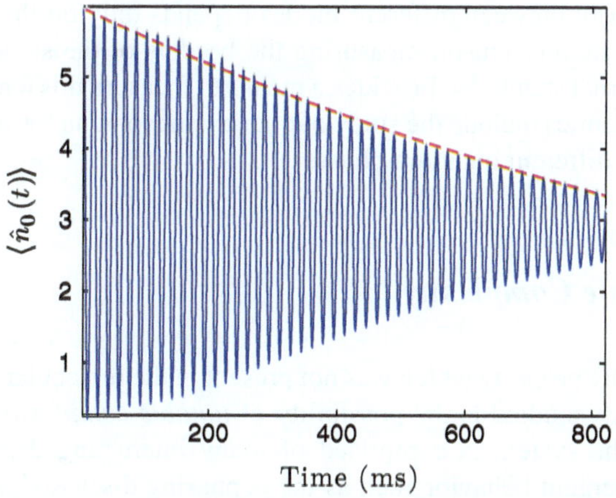

Fig. 4.1 Quantum dephasing in the HMHH. The *plot* shows the behavior of the total number in state 0: $\langle \hat{n}_0 \rangle \equiv \langle \sum_i \hat{n}_{i0} \rangle$ when the system evolves under the Hamiltonian (4.1). Quantum dephasing produces an emergent exponential envelope on the Rabi oscillation pattern between states 0 and 1. Only the number of state 0 is shown for clarity. The *dashed red curve* is an exponential envelope fit to $N \exp(-t/\tau)$ with $\tau = 1,441.17$ ms. The nonexponential behavior near $t = 200$ is due to the finite size of the lattice

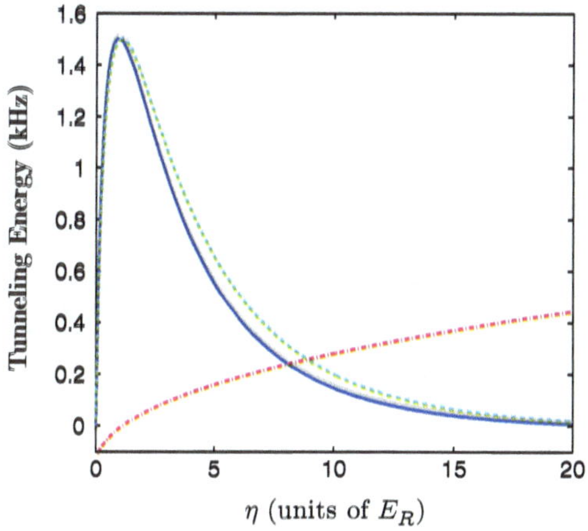

Fig. 4.2 Tunneling matrix elements in a dc electric field. Tunneling energies (in kHz) of the $N = 0$ (*solid blue line*) and $N = 1$ (*dashed green line*) rotational states and their difference divided by their arithmetic mean, $2(t_1 - t_0)/(t_1 + t_0)$, (*dash-dotted red line*) for KRb in a field of 10 kV/cm as a function of the effective isotropic lattice height $\eta \equiv \bar{\alpha} \left|\mathbf{E}_{\text{opt}}\right|^2$ (in recoil energy units). The values of the polarizability tensor are taken from [20]

internal states one can study interactions of many-body systems in different quantum phases and possibly far from equilibrium. Possibilities for quantum statics include studying the properties of phase equilibria as a function of population imbalance and effective mass (as determined by the tunneling energy) [21]. Also, as the difference in tunneling energy between different modes depends only on the elements of the molecular polarizability tensor, measuring the borders of the static phase diagram for different internal states also provides a means to measure this tensor. Possibilities for quantum dynamics include the study of quench phenomena for interacting many-body systems in different quantum phases.

4.2.3 Tunable Complexity

A final noteworthy property which was not present in the molecular Hubbard Hamiltonians previously studied is the possibility of *tunable complexity*. By complexity we mean that the system is comprised of many interacting degrees of freedom and displays emergent behavior such as the dephasing discussed above. Tunability refers to the fact that we may alter the *number* of internal degrees of freedom that are accessed dynamically as well as the *timescale* of their relative interactions. The key point for tunability is that the electric and magnetic fields affect different degrees of freedom: the electric dipole moment and nuclear spins, respectively. We illustrate this concept, and the corresponding geometries and polarizations needed for experiments, in Fig. 4.3.

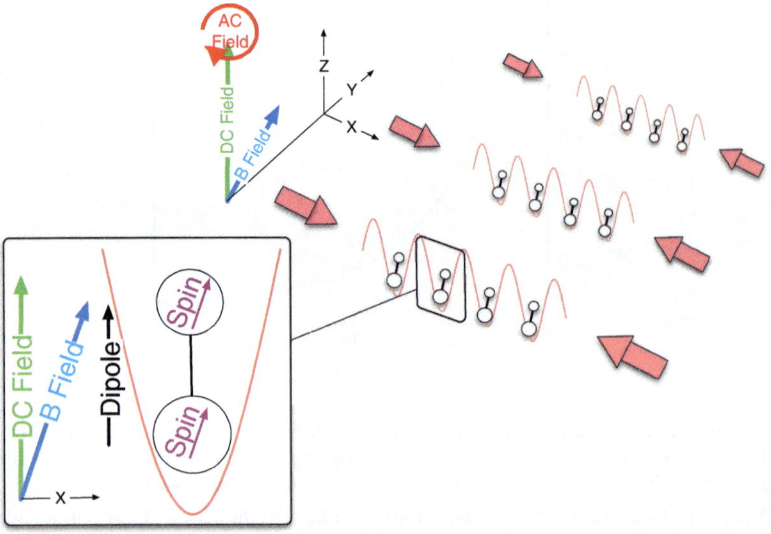

Fig. 4.3 Geometry of the HMHH. Counter-propagating laser beams along the y and z directions create an array of 1D tubes, and an additional pair of laser beams along x creates a lattice potential. A strong dc field orients the dipoles along the direction perpendicular to motion, and a magnetic field orients the nuclear spins. An ac field of circular space-fixed polarization drives transitions between internal levels

In slightly more detail, tunability is achieved as follows. In the presence of an electric field aligned along the z direction, dipole moments are induced between states having the same nuclear spin projection along the field. The introduction of a strong magnetic field defines an effective axis of quantization for the nuclear spins while leaving the rotational structure unchanged because of the strong nuclear Zeeman effect, the weak rotational Zeeman effect, and the presence of only weak (quadrupole) coupling between the rotational and nuclear spin degrees of freedom. In the presence of a strong magnetic field that is not collinear with the electric field it is therefore possible to induce dipole moments between states with different hyperfine quantum numbers.

Thus, by changing the relative angle between the electric and magnetic fields one can control the number of states accessible from a particular state. The power of the applied ac field determines the interaction scale and the Rabi frequency of these dipole couplings, and the strength of the magnetic field determines the energetic splittings between states, in turn determining the relative rates of internal state population. The HMHH may therefore be used as a quantum simulator of a *quantum complex system* where the number and timescale of the internal components may be dynamically altered. Precise measures of complexity and simulations displaying characteristic behavior in various regimes will be discussed in future work [22].

4.3 Derivation of the HMHH

We consider the experimental setup shown schematically in Fig. 4.3. Counter-propagating laser beams along the y and z directions create a series of 1D optical lattice "tubes." The intensity of these beams is such that the tubes are isolated from one another, and the lattice spacing is chosen (e.g., by crossed beams) such that the dipole–dipole interaction along y and z is negligible on experimental timescales. An additional pair of beams creates a lattice potential along the x-direction. The experimental techniques required to create this setup have been well established for ultracold atoms [23–25]. In addition to the lattice potential there is a uniform dc electric field along the z direction, a uniform magnetic field which lies in the xz plane, and an ac microwave field propagating in z which is assumed to have circular polarization $q = 1$ in the space-fixed spherical basis.

In the lattice is an ultracold quantum degenerate gas of $^1\Sigma$ heteronuclear molecules characterized by permanent electric dipole moment d, rotational constant B_N, rotational angular momentum $\mathbf{N}$,[1] and nuclear spins $\mathbf{I}_1$ and $\mathbf{I}_2$. Both nuclear spins are taken to be greater than one-half, so that both nuclei have nonzero electric quadrupole moments. In second quantization the full low-energy Hamiltonian for this setup is

[1] We reserve $\mathbf{J}$ for future studies involving nonzero orbital or electronic spin angular momentum.

$$\hat{H} = \int d\mathbf{r}\, \hat{\psi}^{\dagger}(\mathbf{r}) \left[\hat{H}_{\text{in}} + \hat{H}_{\text{F}} + \hat{H}_{\text{ac}} + \hat{H}_{\text{kin}} + \hat{H}_{\text{opt}} \right] \hat{\psi}(\mathbf{r})$$
$$+ \frac{1}{2} \int d\mathbf{r} d\mathbf{r}'\, \hat{\psi}^{\dagger}(\mathbf{r})\, \hat{\psi}^{\dagger}(\mathbf{r}')\, \hat{H}_{\text{DD}}(|\mathbf{r}-\mathbf{r}'|)\, \hat{\psi}(\mathbf{r}')\, \hat{\psi}(\mathbf{r}), \tag{4.3}$$

where

$$\hat{H}_{\text{in}} = \hat{H}_{\text{rot}} + \hat{H}_{\text{scal}} + \hat{H}_{\text{tens}} + \hat{H}_{r-s} + \hat{H}_{\text{quad}} \tag{4.4}$$
$$= B_N \hat{\mathbf{N}}^2 + c_4 \hat{\mathbf{I}}_1 \cdot \hat{\mathbf{I}}_2 + c_3 \hat{\mathbf{I}}_1 \cdot \tilde{T} \cdot \hat{\mathbf{I}}_2 + \sum_{i=1}^{2} c_i \hat{\mathbf{N}} \cdot \hat{\mathbf{I}}_i$$
$$+ \sum_{i=1}^{2} \hat{\hat{\mathbf{V}}}_i \cdot \hat{\hat{\mathbf{Q}}}_i\,,$$

$$\hat{H}_{\text{F}} = -\, g_r \mu_N \hat{\mathbf{N}} \cdot \mathbf{B} - \sum_{i=1}^{2} g_i \mu_N \,(1-\sigma_i)\, \hat{\mathbf{I}}_i \cdot \mathbf{B} \tag{4.5}$$
$$- \mathbf{E}_{\text{dc}} \cdot \hat{\mathbf{d}}\,,$$

$$\hat{H}_{\text{ac}} = -\, \mathbf{E}_{\text{ac}} \cdot \hat{\mathbf{d}}\,, \tag{4.6}$$

$$\hat{H}_{\text{kin}} = \frac{\hat{\mathbf{p}}^2}{2m}\,, \tag{4.7}$$

$$\hat{H}_{\text{opt}} = -\, \mathbf{E}^{\star}_{\text{opt}} \cdot \hat{\tilde{\alpha}}\left(\omega_{\text{opt}}\right) \cdot \mathbf{E}_{\text{opt}}\,, \tag{4.8}$$

$$\hat{H}_{\text{DD}}(R) = \frac{\hat{\mathbf{d}}_1 \cdot \hat{\mathbf{d}}_2 - 3\left(\hat{\mathbf{d}}_1 \cdot \mathbf{e}_R\right)\left(\mathbf{e}_R \cdot \hat{\mathbf{d}}_2\right)}{R^3}\,. \tag{4.9}$$

The first line of Eq. (4.3) is comprised of single-molecule terms. In order, these are $\hat{H}_{\text{in}}$, the Hamiltonian governing the internal rotational and nuclear spin degrees of freedom; $\hat{H}_{\text{F}}$, the interaction of the molecule with externally applied dc electric and magnetic fields; $\hat{H}_{\text{ac}}$, the interaction of the molecule with an ac microwave field; $\hat{H}_{\text{kin}}$, the kinetic energy of the molecule; and $\hat{H}_{\text{opt}}$, the interaction of the molecule with the optical lattice potential. The second line of Eq. (4.3) is the two-molecule resonant dipole–dipole force. The main assumptions underlying this Hamiltonian and our subsequent analysis are the following.

First, the individual molecules are assumed to be in their electronic and vibrational ground states, and it is assumed that none of these degrees of freedom can be excited at the large intermolecular separations and low temperatures/relative energies that we consider.

Second, the characteristic trapping potential length is chosen large enough compared to the internuclear axis to assume spherical symmetry, i.e. a locally constant potential.

Third, we consider only the lowest two rotational levels. All ac fields will be sufficiently weak to allow this assumption. We also work in the rotating wave approximation, which requires that the detuning be small compared to the driving frequency.

Fourth, we consider all molecules to be in the lowest Bloch band. The ac Rabi frequencies are chosen to be small ($\sim$1–50 kHz) in comparison with the lattice bandwidth ($\sim 10E_R \sim$100 kHz) to ensure this assumption.

Fifth, we work in the "hard-core" limit where at most one molecule is allowed per site. This is enforced by strongly repulsive dipole–dipole interactions on-site, caused by our z-alignment of the electric field, as sketched in Fig. 4.3. We consider the lattice spacing large enough to include only nearest-neighbor dipole–dipole interactions. We neglect the effects of chemical reactions or hyperfine changing collisions which occur at very short range.

Sixth, we neglect dipole–dipole interactions between molecules in different 1D "tubes." For a consistent level of approximation this requires the tubes to be separated by twice the lattice spacing. This can be achieved in principle using crossed beams to create larger lattice spacings.

Seventh, we consider only pairwise interactions of the molecules, neglecting three and higher-body interactions. This is valid for KRb because the permanent dipole moment $d = 0.566\,\mathrm{D}$ is rather small. For molecules such as LiCs with larger permanent dipole moments, the three-body interaction can play a significant role [26].

To derive a Hamiltonian of Hubbard type from Eq. (4.3) we follow the standard prescription [13] of expanding the field operators of our second-quantized Hamiltonian in a Wannier basis of single-molecule states centered at a particular discrete position $\mathbf{r}_i$:

$$\hat{\psi} = \sum_i \sum_\sigma \hat{a}_{i\sigma} w_\sigma \left(\mathbf{r} - \mathbf{r}_i\right) , \tag{4.10}$$

where i is a site index and σ an index denoting the internal state of the molecule. The Wannier basis we use is the basis which diagonalizes the internal plus static field Hamiltonians $\hat{H}_{\mathrm{in}} + \hat{H}_{\mathrm{F}}$ and in which all states with $N = 1$ rotate with frequency ω, where ω is the frequency of the applied ac electric field. With the field operator written in this manner, we find the Hubbard parameters

$$t_\sigma \equiv -\int d\mathbf{r}\, w_\sigma^\star \left(\mathbf{r} - \mathbf{r}_i\right) \left[\hat{H}_{\mathrm{kin}} + \hat{H}_{\mathrm{opt}}\right] w_\sigma \left(\mathbf{r} - \mathbf{r}_{i+1}\right) , \tag{4.11}$$

$$\Delta_\sigma \equiv \int d\mathbf{r}\, w_\sigma^\star \left(\mathbf{r} - \mathbf{r}_i\right) \left[\hat{H}_{\mathrm{in}} + \hat{H}_{\mathrm{F}}\right] w_\sigma \left(\mathbf{r} - \mathbf{r}_i\right) , \tag{4.12}$$

and

$$-d_{\sigma\sigma'} E_{\mathrm{ac}} \equiv \int d\mathbf{r}\, w_\sigma^\star \left(\mathbf{r} - \mathbf{r}_i\right) \hat{H}_{\mathrm{ac}} w_{\sigma'} \left(\mathbf{r} - \mathbf{r}_i\right) , \tag{4.13}$$

$$U_{\sigma\sigma'} \equiv \int d\mathbf{r} d\mathbf{r}' \, w_{\sigma}^{\star}(\mathbf{r}-\mathbf{r}_i)\, w_{\sigma'}^{\star}(\mathbf{r}'-\mathbf{r}_{i+1}) \times H_{\mathbf{DD}}(\mathbf{r}-\mathbf{r}')\, w_{\sigma}(\mathbf{r}-\mathbf{r}_i)\, w_{\sigma'}(\mathbf{r}'-\mathbf{r}_{i+1}) \,. \tag{4.14}$$

The detunings Δ_σ are determined by the single-molecule spectra, which are well known [9, 27]. In the interest of the present article's completeness, we have included appendices reviewing the basic results and explaining them in the context of the present problem. In the following sections we discuss the remaining Hubbard parameters.

4.3.1 Tunneling Energies

A key component of the realization of many-body Hamiltonians using ultracold molecules is the presence of a far off-resonant optical lattice which confines the molecules in a reduced geometry. The Hamiltonian of this interaction is

$$\hat{H}_{\text{opt}} = -\mathbf{E}_{\text{opt}}^{\star}(\mathbf{r}, \omega_{\text{opt}}) \cdot \hat{\tilde{\alpha}}(\omega_{\text{opt}}) \cdot \mathbf{E}_{\text{opt}}(\mathbf{r}, \omega_{\text{opt}}) \,, \tag{4.15}$$

where $\mathbf{E}_{\text{opt}}(\mathbf{r}, \omega_{\text{opt}})$ is the optical lattice field and $\hat{\tilde{\alpha}}(\omega_{\text{opt}})$ is the polarizability tensor operator of the molecule, evaluated at the optical lattice frequency ω_{opt}. In our notation, the circumflex accent (the "hat") denotes an operator, the tilde denotes a rank 2 tensor, and boldface denotes a rank 1 tensor, or vector. This optical potential couples to the electronic degrees of freedom and is detuned from resonance by an amount several orders of magnitude larger than any hyperfine splittings. Thus dependence on the hyperfine quantum numbers in negligible. For tight optical traps, the optical trap potential at each well is close to that of a harmonic trap plus a small state-dependent tensor shift of the trap frequency affecting levels with $N > 0$ due to the polarizability anisotropy [15].

When the optical potential is combined with the kinetic portion of the Hamiltonian and evaluated in the Wannier basis one obtains the tunneling energies. As the tunneling energies are independent of the hyperfine quantum numbers, we can use results obtained in the case of only rotational degrees of freedom, derived in our earlier work [15]. Then the tunneling energies in the eigenbasis of $\hat{H}_{\text{rot}}$, $|NM_N\rangle$, are given by

$$\frac{\tilde{t}_{NM_N}}{E_R} = A\left(\frac{V_{NM_N}}{E_R}\right)^B \exp\left(-C\sqrt{\frac{V_{NM_N}}{E_R}}\right), \tag{4.16}$$

where $A = 1.397$, $B = 1.051$, and $C = 2.121$ are fit parameters [28], E_R the recoil energy, and

$$V_{NM_N} = \left|\mathbf{E}_{\mathrm{opt}}\right|^2 \left[\bar{\alpha} + \frac{2\Delta\alpha}{3} \frac{N(N+1) - 3M_N^2}{(2N-1)(2N+3)}\right] \tag{4.17}$$

is the effective lattice height for the $|NM_N\rangle$ level. Here $\bar{\alpha}$ is the average polarizability and $\Delta\alpha$ the polarizability anisotropy.

In the presence of a dc field the rotational levels become mixed, leading to new effective tunneling energies which we denote as t_{NM_N}, with N and M_N the corresponding zero field values. This hybridization of rotational levels in principle also allows tunneling events which change the rotational state of the molecule, but we can ignore such events because the rotational level separation is much larger than the tunneling energies. The effective tunneling for the $N = 0$ and $N = 1$, $M_N = \pm 1$ levels is shown in Fig. 4.2. The scale is set by the recoil energy, which is $2\pi \times 1.44$ kHz for KRb in a 1,054-nm optical lattice.

4.3.2 *Two-Molecule Interactions*

Heteronuclear $^1\Sigma$ molecules possess permanent dipole moments, and thus interact via a dipole–dipole interaction

$$\hat{H}_{\mathrm{DD}}(\mathbf{R}) = \frac{\hat{\mathbf{d}}_1 \cdot \hat{\mathbf{d}}_2 - 3\left(\hat{\mathbf{d}}_1 \cdot \mathbf{e}_R\right)\left(\mathbf{e}_R \cdot \hat{\mathbf{d}}_2\right)}{R^3}, \tag{4.18}$$

where $\mathbf{R} \equiv \mathbf{r}_2 - \mathbf{r}_1$, $\mathbf{e}_R$ is a unit vector in the direction of $\mathbf{R}$, and $\hat{\mathbf{d}}_i$ is the vector dipole operator of the ith molecule. In the absence of external fields, this interaction is off-resonant, leading to a van der Waals interaction $\hat{H}_{\mathrm{DD}}(\mathbf{R}) \sim R^{-6}$, but in the presence of electric fields resonant dipoles are induced and the interaction displays a resonant R^{-3} behavior in addition to the R^{-6} behavior.

The anisotropic nature of the dipole–dipole force has been experimentally shown to dominate the rethermalization behavior of a molecular gas via inelastic collisions [29]. This is because a "head-to-tail" arrangement of molecules leads to an attractive potential, whereas "side-to-side" interactions are repulsive. To ensure the stability of an ultracold molecular ensemble and to prevent losses from inelastic collisions it is crucial therefore not only to orient the dipoles using a dc field, but also to confine the molecules in a reduced geometry. A thorough discussion of the nature of the two-molecule spectra for $^1\Sigma$ molecules without hyperfine structure and its implications for stability in two dimensions is presented in [14]. Diagonalization of the full two-molecule Hamiltonian is impractical when hyperfine structure is included due to the very large matrices that result. Instead, we argue based on comparisons of length and energy scales that the hyperfine structure is negligible during the collisional processes which occur in our proposed setup.

Our reduced geometry is imposed by the optical lattice described earlier. Namely, we consider the case where the molecules are confined to move only along the x direction and a dc field polarized along the z direction orients the dipoles such that all collisions are side-to-side and repulsive. The dipole–dipole interaction in this geometry reduces to

$$\hat{H}_{\mathrm{DD}} = \frac{1}{R^3}\Big[\hat{d}_0 \otimes \hat{d}_0 + \frac{1}{2}\left(\hat{d}_{-1} \otimes \hat{d}_1 + \hat{d}_1 \otimes \hat{d}_{-1}\right) - 3\left(\hat{d}_{-1} \otimes \hat{d}_{-1} + \hat{d}_1 \otimes \hat{d}_1\right)\Big], \tag{4.19}$$

where $\hat{d}_q \equiv \hat{\mathbf{d}} \cdot \mathbf{e}_q$ is the component of the dipole operator along the q direction in the space-fixed spherical basis. For z-polarized electric field, the only diagonal components are those involving d_0. The components of the interaction involving $d_{\pm 1}$ couple states with $\Delta M_N = \pm 1$ that are separated in energy by an amount of order the rotational constant for the dc fields we consider (see Fig. 4.4). Contributions from these components are suppressed at distances greater than $r_B \equiv \left(d^2/B\right)^{1/3}$, of order a few nanometers. Thus, at the nearest-neighbor distance in a 1,054-nm optical lattice we consider only the diagonal elements of the dipole–dipole interaction. This restriction gives rise to the two-body term

$$\hat{H}_{\mathrm{DD}} = \frac{1}{2}\sum_{\sigma\sigma'} U_{\sigma\sigma'} \sum_{\langle i,j\rangle} \hat{n}_{i\sigma}\hat{n}_{j\sigma'}, \tag{4.20}$$

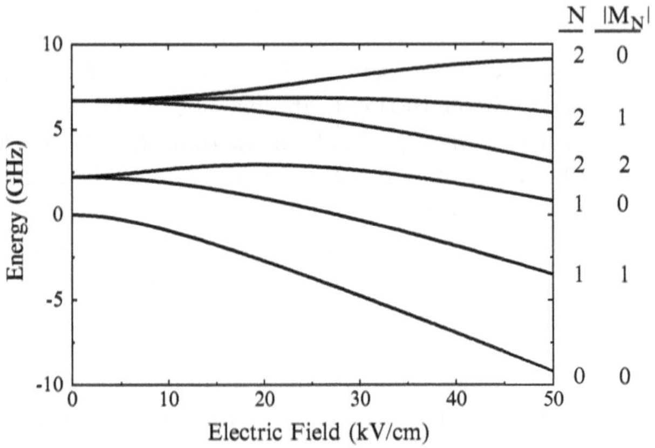

Fig. 4.4 Gigahertz scale view of the Stark effect for KRb. Introduction of a dc field breaks the degeneracy between all states with the same N but different $|M_N|$. The large electric dipole moment causes GHz scale energy shifts which completely obscure the hyperfine splittings on the scale of this plot. Because the dipole moment is the same for any isotope of KRb, the Stark effect on this scale is the same for all isotopes

where

$$U_{\sigma\sigma'} = \frac{d_\sigma d_{\sigma'}}{(\lambda/2)^3}. \tag{4.21}$$

In Eq. (4.21) d_σ is the resonant dipole moment of state $|\sigma\rangle$ and λ is the wavelength of the optical lattice. We assume that the long-range repulsive diagonal d_0 portion of the dipole–dipole interaction is strong enough to prevent both the occupation of any one lattice site by more than one molecule and access to the region where hyperfine-changing collisions involving the $d_{\pm 1}$ dipole moments occur.

4.3.3 Interactions with Static External Fields

The spectral properties of $^1\Sigma$ molecules in collinear dc electric and magnetic fields have been elucidated elsewhere in the literature [9, 10, 27], and the basic results of the analysis are given in Appendix 2 for the reader's convenience. In this section, we focus on the properties of such molecules in *noncollinear* fields, in particular on the dipole moments.

The behavior of the molecular dipole moments is controlled by an external dc electric field which mixes rotational levels of opposite parity and thus orients the molecule. However, a dc field does not couple to the nuclear spins. So for a z-polarized field the selection rules $\Delta M_1 = 0$, $\Delta M_2 = 0$ are enforced, where M_1 and M_2 are the nuclear spin projections along the field direction. In contrast, a magnetic field couples strongly to the nuclear spins but only weakly to the rotational angular momentum due to the relative sizes of the g factors [27]. The magnetic field Hamiltonian thus has eigenstates which are energetically distinct nuclear spin states with a quantization axis given by the field direction. It is in this sense that we say the magnetic field defines an effective axis of quantization for the nuclear spins. Thus, in the absence of internal couplings of the rotational and hyperfine degrees of freedom they may be manipulated independently: the rotational angular momentum with an electric field and the nuclear spin angular momenta with a magnetic field.

The presence of nuclear quadrupole couplings in alkali dimer molecules couples states with the same total angular momentum projection M_F but different rotational and nuclear spin projections. For example, in KRb, the interaction couples $|\sigma\rangle = |N = 1, M_N = 0, M_{\mathrm{K}}, M_{\mathrm{Rb}} \pm 1\rangle$ to $|\sigma'\rangle = |N = 1, M_N = \pm 1, M_{\mathrm{K}}, M_{\mathrm{Rb}}\rangle$ with the latter accounting for ~10 % of the state in the absence of fields.[2] Clearly, since the $N = 0$ state has only one projection $M_N = 0$, the nuclear quadrupole interaction leaves this level unaffected. In a dc electric field where the rotational levels become mixed, the states correlating with the $N = 0$ levels and $N = 1$ levels

[2]The interaction also couples the $M_{\mathrm{K}} \pm 1$ states, but the coupling constant $(eqQ)_{\mathrm{K}}$ is significantly smaller than $(eqQ)_{\mathrm{Rb}}$ and so the mixing is negligible in comparison [11].

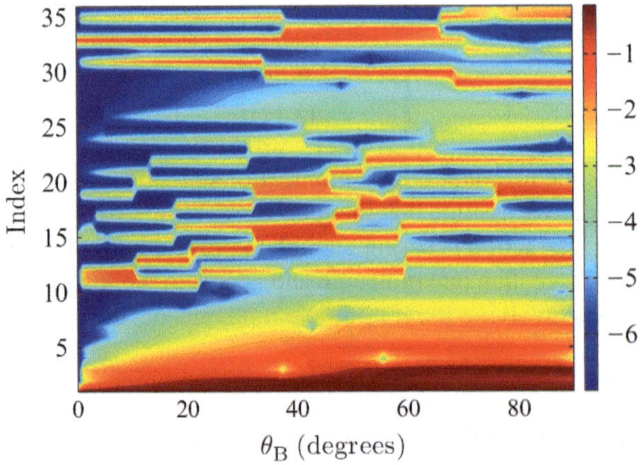

Fig. 4.5 Distribution of dipolar character. The *colorbar* shows the logarithm of the transition dipole moment with the ground state, $\log\langle \text{g.s.}|\hat{d}_1|i\rangle$, as a function of the angle between the magnetic and electric fields θ_B and the state index (ordered by energy). Changing the angle between the electric and magnetic fields breaks the nuclear spin projection selection rule and allows for transition dipole moments between many states. Only dipole moments greater than 10^{-7} are displayed

both display quadrupole effects, but these effects are still not identical. In strong fields the Zeeman effect dominates over the quadrupole coupling, allowing control over the nuclear spins that displays a weak dependence on the rotational level.

Thus, a strong magnetic field defines an effective axis of quantization for the nuclear spins, resulting in nuclear spin states which are superpositions of states in the basis with the axis of quantization along the electric field axis. This implies that by changing the angle of the magnetic field with respect to the electric field, it is possible to change the number of states which are coupled by transition dipole moments. This is illustrated in Fig. 4.5, which shows the logarithm of the transition dipole moment with the ground state as a function of the angle between the dc and magnetic fields θ_B and a state index (ordered by energy). The lowest state index denotes the lowest energy state in the $N = 1$ manifold. When the fields are collinear, one state dominates the dipole spectrum. As the angle changes the dipolar character becomes spread over many states. These transition dipole moments allow the states to couple in an ac microwave field and generate complex dynamics.

4.3.4 Interaction with an ac Microwave Field

The introduction of an ac microwave field contributes to the Hamiltonian in a similar way to a dc field. In addition, the inherent time dependence allows for circular and linear polarization as well as the possibility of driving transitions between internal states. In the absence of hyperfine structure, an ac field of spherical polarization q couples the $|N = 0, M_N = 0\rangle$ and $|N = 1, M_N = q\rangle$ levels, leading to an effective two-level system in the Floquet picture [14]. In the presence of hyperfine structure, states with different total angular momentum projections M_F in the $N \geq 1$ manifolds become mixed due to the electric quadrupole interaction. Thus no rigorous selection rules can be established. This complicates the issue of addressing single hyperfine states using microwave fields, but it also allows the hyperfine state to be changed using microwave fields. Addressing a single hyperfine state can be achieved by the application of a strong magnetic field such as those used in the STIRAP procedure, which defines the projections sufficiently to suppress transitions to non-target hyperfine states [9]. In the presence of an electric field, this last comment holds only in the case where the two fields are collinear. When the fields are not collinear many states can be accessed from any one state via a microwave transition due to the behavior of the transition dipole moments in crossed fields, as was described in Sect. 4.3.3.

We choose the polarization of the ac field to be purely circular, $q_{\text{ac}} = 1$. A component along $q = 0$ would lead to rapid oscillation of the eigenenergies because the d_0 moments induced by the electric field couple to the ac field, and this complicates the analysis. Furthermore, we consider Rabi frequencies which are much less than the bandwidth of the optical lattice so that our approximation of being in the lowest Bloch band remains valid and we are also justified in using a rotating wave approximation. The above considerations together with the single-molecule ac Hamiltonian

$$\hat{H}_{\text{ac}} = -\hat{\mathbf{d}} \cdot \mathbf{E}_{\text{ac}} = -\hat{d}_q E_{\text{ac}} e^{-i\omega t} + \text{h.c.} \tag{4.22}$$

lead directly to the second quantized Hamiltonian

$$\hat{H}_{\text{ac}} = -\frac{1}{2} \sum_{\sigma\sigma'} d_{\sigma\sigma'} E_{\text{ac}} \sum_i \left[\hat{a}^\dagger_{i\sigma} \hat{a}_{i\sigma'} e^{i\omega t} + \text{h.c.} \right]. \tag{4.23}$$

In Eq. (4.23) the label σ refers to the eigenstate $|\sigma\rangle$ of the internal plus static field Hamiltonian $\hat{H}_{\text{in}} + \hat{H}_{\text{F}}$, $d_{\sigma\sigma'} \equiv \langle \sigma | d_1 | \sigma' \rangle$, and E_σ is the energy of state $|\sigma\rangle$.

Assembling all the many-body terms expressed in this basis, we obtain the time-dependent Hamiltonian

$$\begin{aligned} \hat{H} = & \sum_\sigma E_\sigma \sum_i \hat{n}_{i\sigma} - \sum_\sigma t_\sigma \sum_{\langle i,j \rangle} \left[\hat{a}^\dagger_{i\sigma} \hat{a}_{j\sigma} + \text{h.c.} \right] \\ & + \frac{1}{2} \sum_{\sigma,\sigma'} U_{\sigma\sigma'} \sum_{\langle i,j \rangle} \hat{n}_{i\sigma} \hat{n}_{j\sigma'} \\ & - \frac{1}{2} \sum_{\sigma\sigma'} d_{\sigma\sigma'} E_{\text{ac}} \sum_i \left[\hat{a}^\dagger_{i\sigma} \hat{a}_{i\sigma'} e^{i\omega t} + \text{h.c.} \right]. \end{aligned} \tag{4.24}$$

If we change to a basis where all single-molecule states with $N = 1$ rotate with frequency ω we have, finally:

$$\begin{aligned}\hat{H} &= \sum_{\sigma} \Delta_{\sigma} \sum_{i} \hat{n}_{i\sigma} - \sum_{\sigma} t_{\sigma} \sum_{\langle i,j \rangle} \left[\hat{a}^{\dagger}_{i\sigma} \hat{a}_{j\sigma} + \text{h.c.} \right] \\ &+ \frac{1}{2} \sum_{\sigma,\sigma'} U_{\sigma\sigma'} \sum_{\langle i,j \rangle} \hat{n}_{i\sigma} \hat{n}_{j\sigma'} \\ &- \frac{1}{2} \sum_{\sigma\sigma'} d_{\sigma\sigma'} E_{\text{ac}} \sum_{i} \left[\hat{a}^{\dagger}_{i\sigma} \hat{a}_{i\sigma'} + \text{h.c.} \right], \end{aligned} \quad (4.25)$$

where $\Delta_{\sigma} = E_{\sigma}$ for states with $N = 0$ and $E_{\sigma} - \omega$ for states with $N = 1$.

4.4 Conclusions

We have presented and derived the HMHH. The HMHH is a lattice Hamiltonian describing the effective low-energy physics of an ultracold gas of heteronuclear alkali dimer molecules with hyperfine structure loaded into a 1D optical lattice and interacting with external dc electric, ac microwave, and static magnetic fields. By tuning the angle between the electric and magnetic fields and the strength of the magnetic and ac fields it is possible to change the number and timescale of internal states contributing to many-body dynamics. The Hamiltonian also displays emergent quantum dephasing, and has a phase diagram which depends strongly on the initial state. These features make the HMHH an ideal candidate for a model quantum complex system.

Future work will involve time-evolving block decimation simulations of the HMHH similar to past studies of molecular Hubbard Hamiltonians [15]. In particular, we will discuss measures of complexity and how they relate to experimentally measurable quantities. Future work on the Hamiltonian itself will include realistic models of molecule loss due to inelastic and chemical processes. Such dissipative processes are key to dissipative quantum phase transitions, which is a major area of interest in quantum many-body theory [30–32].

We acknowledge useful discussions with Immanuel Bloch, John Bohn, Silke Ospelkaus, Luis Santos, and Peter Zoller. This work was supported by the National Science Foundation under Grant PHY-0903457.

Appendix 1: The Internal Hamiltonian

A $^1\Sigma$ molecule in its electronic and vibrational ground states has three angular momentum degrees of freedom: the rotational angular momentum **N** and the nuclear spins $\mathbf{I}_1$ and $\mathbf{I}_2$. In this work we shall use the coupling schemes $|(I_1 I_2)\, INFM_F\rangle$ and $|I_1 M_1 I_2 M_2 N M_N\rangle$, which we refer to as the coupled and uncoupled bases, respectively. Explicit expressions for all single-molecule matrix elements in both bases are provided in Appendix 3. The relevant Hamiltonian for the internal degrees of freedom $\hat{H}_{\text{in}}$ may be written as a sum of rotational and hyperfine terms as

$$\hat{H}_{\text{in}} = \hat{H}_{\text{rot}} + \hat{H}_{\text{hf}}, \tag{4.26}$$

where

$$\hat{H}_{\text{rot}} = B_N \mathbf{N}^2, \tag{4.27}$$

$$\hat{H}_{\text{hf}} = \sum_{i=1}^{2} c_i \mathbf{N} \cdot \mathbf{I}_i + c_3 \mathbf{I}_1 \cdot \tilde{T} \cdot \mathbf{I}_2 + c_4 \mathbf{I}_1 \cdot \mathbf{I}_2 + \sum_{i=1}^{2} \mathbf{V}_i \cdot \mathbf{Q}_i\,. \tag{4.28}$$

The rotational term Eq. (4.27) corresponds to the Hamiltonian of a rigid spherical rotor with $(2N+1)$-fold degenerate eigenstates $|NM_N\rangle$, M_N being the projection of $\mathbf{N}$ on a space-fixed quantization axis [33]. The eigenenergies are given by $E_{NM_N} = B_N N\,(N+1)$, where B_N is the rotational constant of the molecule (we use the notation B_N instead of the more common B to avoid confusion with the magnetic field magnitude B). In the case of ^{40}K^{87}Rb, B_N=1.114 GHz [27]. The rotational level splitting defines the largest intrinsic energy scale for $^1\Sigma$ molecules.

The first term of the hyperfine Hamiltonian, $\sum_{i=1}^{2} c_i \mathbf{N} \cdot \mathbf{I}_i$ represents the interaction of the nuclear spins with the magnetic field created by the rotation of the molecule, and is governed by two coupling constants c_{K} and c_{Rb} related to the nuclear shielding tensor. For ^{40}K^{87}Rb, these have been determined from density functional calculations to be ∼20 Hz and ∼100 Hz, respectively [27]. Because of the smallness of these constants and the fact that this term does not couple states with different N, this term plays a very small role in the spectra.

The two nuclear spins have nuclear magnetic moments which interact via a resonant dipole–dipole interaction

$$\hat{H}_{\text{hf-dd}} = g_H^2 \mu_N^2 \,(\mu_0/4\pi) \left[\frac{\mathbf{I}_1 \cdot \mathbf{I}_2}{R^3} - \frac{3\,(\mathbf{I}_1 \cdot \mathbf{R})\,(\mathbf{R} \cdot \mathbf{I}_2)}{R^5}\right], \tag{4.29}$$

where g_H is the proton g factor and $\mathbf{R}$ the vector joining the two nuclei [33]. This may be written as the contraction of two rank-2 spherical tensors as

$$\hat{H}_{\text{hf-dd}} = -g_H^2 \mu_N^2 \,(\mu_0/4\pi)\, \langle R^{-3} \rangle \sqrt{6}\, (\mathbf{C})^{(2)} \cdot (\mathbf{T}\,(\mathbf{I}_1, \mathbf{I}_2))^{(2)}\,, \tag{4.30}$$

where $(\mathbf{C})^{(2)}$ is an unnormalized spherical harmonic in the relative degrees of freedom. The nuclear spins can also interact indirectly through the electron spins, and do so even for $^1\Sigma$ configurations [33]. This indirect interaction is represented by a tensor $\tilde{J}$ which may be decomposed into its isotropic part J_{iso} and its anisotropy

$\Delta J = J_{\parallel} - J_{\perp}$. The combination of direct and indirect nuclear spin-nuclear spin interaction may thus be written as the sum of a scalar interaction and a tensor interaction as

$$\hat{H}_{\text{hf-dd}} + \hat{H}_{\text{indirect}} = c_4 \mathbf{I}_1 \cdot \mathbf{I}_2 + c_3 \mathbf{I}_1 \cdot \tilde{T} \cdot \mathbf{I}_2, \tag{4.31}$$

where $c_4 \equiv J_{\text{iso}}$, $c_3 \equiv g_H^2 \mu_N^2 (\mu_0/4\pi) \langle R^{-3} \rangle - \Delta J/3$, and the tensor $\tilde{T}$ contains the angular dependence of the tensor interaction. c_3 is of order 10 Hz for the various isotopes of KRb, and so plays a very small role in the spectra. c_4 splits the various levels according to their total nuclear spin I as

$$\begin{aligned} &\langle (I_1 I_2)\, INFM_F | c_4 \mathbf{I}_1 \cdot \mathbf{I}_2 | (I_1 I_2)\, I'N'F'M_F' \rangle \\ &= \delta_{I,I'} \delta_{N,N'} \delta_{F,F'} \delta_{M_F,M_F'} \\ &\quad \times \frac{c_4}{2} \left[I(I+1) - I_1(I_1+1) - I_2(I_2+1) \right]. \end{aligned} \tag{4.32}$$

c_4 is of order 100 Hz–10 kHz for isotopes of KRb, and so is the dominant hyperfine contribution for $N = 0$ in the absence of external fields, see Fig. 4.6. Note that c_4 may be either positive or negative. For ^{40}K^{87}Rb, $c_4 = -20.304$ kHz [27], and so the lowest energy states for $N = 0$ in zero field are the highest nuclear spin states $I = 11/2$.

The final term in the hyperfine Hamiltonian is the interaction of the quadrupole moment of the nuclei with the gradient of the electric field produced by the

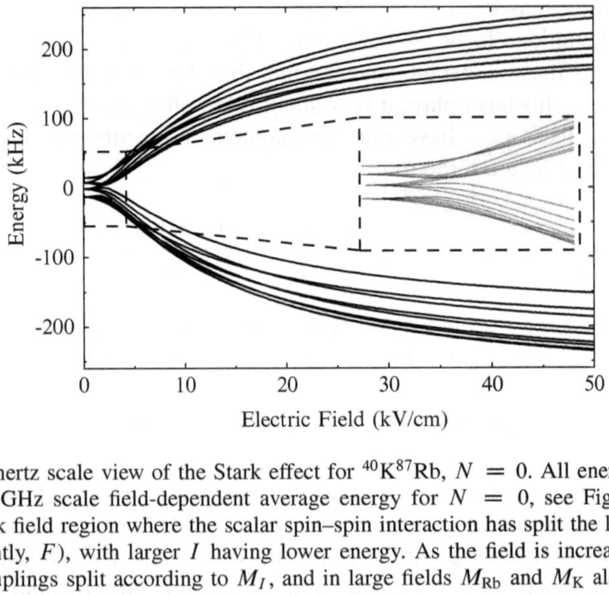

Fig. 4.6 Kilohertz scale view of the Stark effect for ^{40}K^{87}Rb, $N = 0$. All energies are shown relative to the GHz scale field-dependent average energy for $N = 0$, see Fig. 4.4. The *inset* shows the weak field region where the scalar spin–spin interaction has split the levels according to I (equivalently, F), with larger I having lower energy. As the field is increased the nuclear quadrupole couplings split according to M_I, and in large fields M_{Rb} and M_{K} also become well defined. See text for details

electrons. We may represent this interaction by the sum $\sum_{i=1}^{2} \mathbf{V}_i \cdot \mathbf{Q}_i$ where $\mathbf{V}_i$ is a second rank spherical tensor describing the electric field gradient at the ith nucleus and $\mathbf{Q}_i$ is a second rank spherical tensor describing the nuclear quadrupole of the ith nucleus. The pertinent coupling constants $(eqQ)_i$ which arise in the matrix elements of this Hamiltonian are of order 100–1,000 kHz, making it the largest term in the hyperfine Hamiltonian. The quadrupole term doesn't affect the $N = 0$ level, however, and so the scalar spin–spin coupling dominates there. In a strong dc field the rotational levels become deeply mixed and the nuclear quadrupole thus becomes the dominant hyperfine contribution for all states.

Appendix 2: Interactions with Static External Fields

Polar molecules such as heteronuclear dimers can couple to external fields either through their permanent electric dipole moment, through magnetic moments generated from their rotation or nuclear spin, or through their polarizability tensor. The Hamiltonian representing interaction of the molecule with a static dc electric field $\mathbf{E}_{\text{dc}}$ and a static magnetic field $\mathbf{B}$ may be written

$$\hat{H}_{\text{F}} = -\mathbf{d} \cdot \mathbf{E}_{\text{dc}} - g_r \mu_N \mathbf{N} \cdot \mathbf{B} - \sum_{i=1}^{2} g_i \mu_N \mathbf{I}_i \cdot \mathbf{B} (1 - \sigma_i) . \tag{4.33}$$

For $^1\Sigma$ molecules the permanent dipole moment $\mathbf{d}$ lies along the internuclear axis which defines the $p = 0$ axis in a spherical coordinate system rotating with the molecule. Because this basis leads to anomalous commutation relations $[J_i, J_k] = -i\hbar\epsilon_{ijk} J_k$ [34] we find it convenient to transform to the space-fixed frame where the angular momentum operators satisfy the normal commutation relations $[J_i, J_k] = i\hbar\epsilon_{ijk} J_k$, giving $\mathbf{d}\cdot\mathbf{e}_q \equiv d_q = dC_q^{(1)}(\theta, \phi)$, where $\mathbf{e}_q$ is a unit vector along the space fixed spherical q direction and $C_q^{(1)}(\theta, \phi)$ is an unnormalized spherical harmonic whose arguments θ and ϕ are the polar and azimuthal angles of the internuclear axis in the space fixed frame. Taking matrix elements of d_q in our two basis sets yields

$$\begin{aligned} &\langle I_1 M_1 I_2 M_2 N M_N | d_q | I_1 M_1' I_2 M_2' N' M_N' \rangle \\ &= \delta_{M_1, M_1'} \delta_{M_2, M_2'} d \sqrt{(2N+1)(2N'+1)} (-1)^{M_N} \\ &\times \begin{pmatrix} N & 1 & N' \\ 0 & 0 & 0 \end{pmatrix} \begin{pmatrix} N & 1 & N' \\ -M_N & q & M_N' \end{pmatrix} , \end{aligned} \tag{4.34}$$

$$
\begin{aligned}
&\langle (I_1 I_2)\, INFM_F | d_q | (I_1 I_2)\, I'N'F'M_F' \rangle \\
&= \delta_{I,I'} d\, (-1)^{2F-M_F+I+N'+N+1} \begin{pmatrix} N & 1 & N' \\ 0 & 0 & 0 \end{pmatrix} \\
&\times \sqrt{(2N+1)(2N'+1)(2F+1)(2F'+1)} \\
&\times \begin{pmatrix} N & 1 & N' \\ -M_N & q & M_N' \end{pmatrix} \begin{Bmatrix} N & F & I \\ F' & N' & 1 \end{Bmatrix},
\end{aligned} \tag{4.35}
$$

where $\begin{pmatrix} j_1 & j_2 & j_3 \\ m_1 & m_2 & m_3 \end{pmatrix}$ is a Wigner 3-j coefficient and $\begin{Bmatrix} j_1 & j_2 & j_3 \\ j_4 & j_5 & j_6 \end{Bmatrix}$ is a Wigner 6-j coefficient [35]. We see that the rotational eigenstates have no net dipole moment, but that the dipole operator couples the state $|N, F, M_F\rangle$ with the states $|N \pm 1, F \pm 1, M_F + q\rangle$. The introduction of a dc electric field $\mathbf{E}_{\mathrm{dc}}$ with Hamiltonian $-\mathbf{d} \cdot \mathbf{E}_{\mathrm{dc}}$ couples these levels and induces dipole moments, breaking the rotational symmetry and removing the $(2N+1)$-fold degeneracy. Typical molecular dipole moments are measured in Debye (D), where 1D=503.4 MHz/(kV/cm), and so the dc field becomes the dominant contribution to the Hamiltonian for modest fields of a few kV/cm. The permanent dipole moment of KRb has been experimentally determined to be 0.566 D [3].

On the scale of the rotational constant, the effect of a dc field on the single-molecule energy spectrum is as in Fig. 4.4. It is quadratic for field energies small compared to the rotational energy but becomes linear in stronger fields because the field strongly mixes states of opposite parity [36]. We consider the quantization axis to lie along the field direction, and so states with the same value of $|M_N|$ remain degenerate. A universal plot for all $^1\Sigma$ molecules results on this scale if the energy and field strength dE_{dc} are both scaled to the rotational constant.

The average orientation of the molecule with the electric field can be obtained with the Feynman–Hellman theorem as

$$
\langle \cos\theta \rangle = -\frac{\partial E}{\partial (dE_{\mathrm{dc}})}, \tag{4.36}
$$

where E is the energy eigenvalue. The energy eigenvalue is dominated by the GHz scale structure, thus the degree of alignment with the field is essentially independent of the hyperfine structure. From the degree of orientation we can also determine the effective space-fixed dipole moment as $d\langle\cos\theta\rangle$. Figure 4.7 shows the behavior of the induced dipoles as the field strength is increased. For all field strengths the $N=0$ and $N=1$, $M_N = \pm 1$ states align with the field and so have a positive induced dipole moment. In contrast, the $N=1$, $M_N=0$ state antialigns with the field for weak fields and aligns with the field for stronger fields.

The magnitude of the field energy completely obscures the hyperfine splittings, and so to see the effects of hyperfine structure we subtract from each state with a given N the field-dependent average energy of all hyperfine states with the

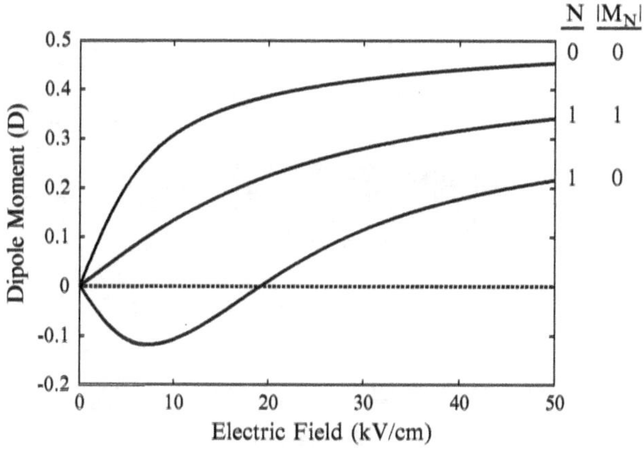

Fig. 4.7 Induced dipoles for KRb in an electric field. The $N = 0$ and $N = 1$, $M_N = \pm 1$ levels orient along the field, giving rise to positive dipole moments. The $N = 1$, $M_N = 0$ state antialigns with the field for small fields, but aligns in stronger fields. All resonant dipole moments approach the "permanent" value 0.566 D as the field strength increases

same N. For $N = 0$ the results are shown in Fig. 4.6. For low fields the hyperfine splittings are dominated by the scalar spin–spin coupling and are of order c_4, a few kHz. As the field is increased the various hyperfine states split according to $|M_I|$. For large fields M_1 and M_2 also become well defined, which occurs because the energetic differences between states with $\Delta M_N = \pm 1$ become larger than the quadrupole coupling constants [see Eq. (4.45)]. Pairs of M_1 and M_2 which have the same $|M_1 + M_2|$ are degenerate, and the state with $|M_1 + M_2| = 0$ is degenerate due to reflection symmetry in the plane of the electric field vector.

Because of the signs of the quadrupole couplings for $^{40}K^{87}Rb$, the lowest energy states are those with M_{Rb} the largest and M_K the smallest. Because the kHz scale Stark effect depends on several molecular parameters it cannot be put into a universal form for all $^1\Sigma$ molecules like the GHz scale Stark effect. However, the qualitative structure will be similar for all $^1\Sigma$ molecules with nuclear quadrupole couplings; key differences being the number of nondegenerate levels and the energetic ordering of the magnetic quantum numbers [27]. The hyperfine Stark effect for $N = 1$ and other molecular species as well as the effects of electric fields on microwave spectra may be found in [10].

Magnetic fields couple to the magnetic moments generated by the rotation of the molecule and by the nuclear spins. The former interaction is given by $-g_r \mu_N \mathbf{N} \cdot \mathbf{B}$, where g_r is the rotational g factor of the molecule and μ_N is the nuclear magneton $e\hbar/2m_p$=762.259 Hz/G [37]. The latter interaction is given by $-\sum_{i=1}^{2} g_i \mu_N \mathbf{I}_i \cdot \mathbf{B}\,(1 - \sigma_i)$, where g_1 and g_2 are the g-factors of nucleus 1 and 2, respectively, and σ_i is the isotropic part of the nuclear shielding tensor for nucleus i. The rotational contribution is typically much smaller than the contributions from the nuclei, due

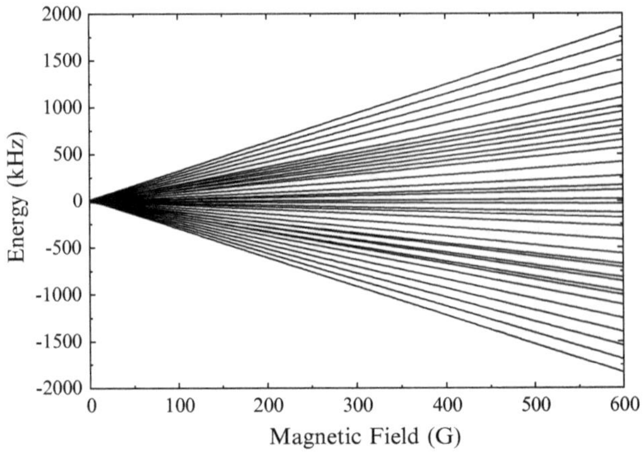

Fig. 4.8 Zeeman effect for $^{40}K^{87}Rb$, N=0. The magnetic field splits the hyperfine levels according to their projections M_K and M_{Rb} with splittings between adjacent levels of order kHz for the experimentally relevant range $B \sim 550$ G. The lowest (highest) energy state corresponds to $m_F = -4 + 3/2 = -5/2$ (5/2). The zero field splitting is set by c_4 and is not visible on the scale of this plot

to smaller g-factors and the fact that the isotropic parts of the nuclear shielding tensors are typically only a few parts per thousand. For example, in $^{40}K^{87}Rb$ $g_r = 0.0140$, $g_K = -0.324$, $g_{Rb} = 1.834$, $\sigma_K = 1{,}321$ ppm, and $\sigma_{Rb} = 3{,}469$ ppm [27]. We neglect diamagnetic contributions to the Zeeman effect, as these contributions are small for the fields we consider.

Typical experimental magnetic fields are ~550 G because of the Feshbach association stage of the STIRAP procedure [2]. In Fig. 4.8 we show the Zeeman effect for the $N = 0$ level of $^{40}K^{87}Rb$ for fields up to this range. We see that the magnetic field splits the spectrum according to the nuclear spin projections M_K and M_{Rb}, with larger (smaller) M_K (M_{Rb}) having lower energy due to the signs of the g factors for $^{40}K^{87}Rb$. Because the nuclear quadrupole interaction doesn't affect the $N = 0$ level, the zero field splittings are determined by the small scalar spin–spin coupling parameter c_4. The Zeeman term dominates over the scalar spin–spin coupling at very low fields and so the effects of the scalar spin–spin coupling are not discernible on the scale of this plot. Additionally, the Zeeman contribution at these fields is larger than the hyperfine Stark splittings from the largest electric fields accessible in current experiments, see Fig. 4.6.

The spectrum for the $N = 1$ level of $^{40}K^{87}Rb$ is shown in Fig. 4.9. It is greatly complicated by the fact that there are three times as many states as the $N = 0$ case (corresponding to the allowed M_N). Also, the nuclear quadrupole interaction affects the $N = 1$ level, causing the large zero field splittings. These larger zero field splittings delay the separation of the levels into well-defined M_1 and M_2, and also causes a complicated series of avoided crossings between states with the same M_F.

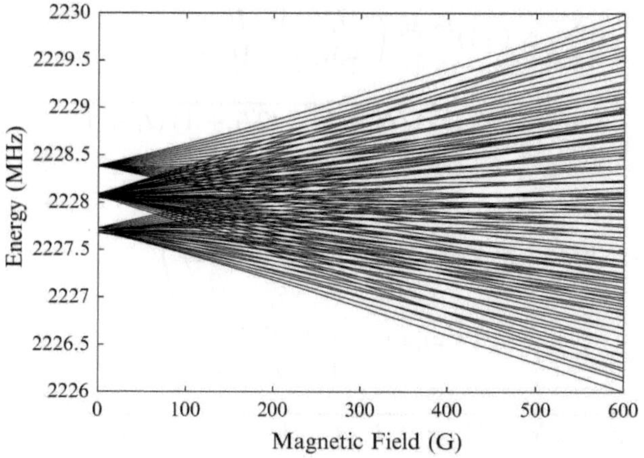

Fig. 4.9 Zeeman effect for ^{40}K^{87}Rb, N=1. The zero field splitting is caused mainly by the nuclear quadrupole interaction and separates the levels into groups of well-defined F. The much larger zero field splitting causes avoided crossings between states with the same M_F to occur at much higher fields than in the $N = 0$ case

Appendix 3: Explicit Values for the Single-Molecule Matrix Elements

Here we present the matrix elements of the single-molecule terms of the Hamiltonian (4.26) in the coupled and uncoupled basis sets. We adopt the conventions of Zare [35].

The matrix elements of the rotational Hamiltonian are given by

$$\langle I_1 M_1 I_2 M_2 N M_N | B_N \mathbf{N}^2 | I_1 M_1' I_2 M_2' N' M_N' \rangle$$
$$= \delta_{M_1,M_1'} \delta_{M_2,M_2'} \delta_{N,N'} \delta_{M_N,M_N'} B_N N (N+1) , \tag{4.37}$$

$$\langle (I_1 I_2) I N F M_F | B_N \mathbf{N}^2 | (I_1 I_2) I' N' F' M_F' \rangle$$
$$= \delta_{I,I'} \delta_{N,N'} \delta_{F,F'} \delta_{M_F,M_F'} B_N N (N+1) . \tag{4.38}$$

The matrix elements of the rotation–spin Hamiltonian are given by

$$\left\langle I_1 M_1 I_2 M_2 N M_N \left| \sum_{i=1}^{2} c_i \mathbf{N} \cdot \mathbf{I}_i \right| I_1 M_1' I_2 M_2' N' M_N' \right\rangle$$
$$= \delta_{N,N'} \sum_q (-1)^{q+N-M_N} \begin{pmatrix} N & 1 & N \\ -M_N & q & M_N' \end{pmatrix}$$

$$\times \sum_i c_i (-1)^{I_i - M_i} \begin{pmatrix} I_i & 1 & I_i \\ -M_i & -q & M_i' \end{pmatrix}$$
$$\times \sqrt{N(2N+1)(N+1)I_i(2I_i+1)(I_i+1)}\,, \tag{4.39}$$

$$\left\langle (I_1 I_2) INFM_F \left| \sum_{i=1}^{2} c_i \mathbf{N}\cdot\mathbf{I}_i \right| (I_1 I_2) I' N' F' M_F' \right\rangle$$
$$= \delta_{N,N'} (-1)^{I+N+F+I_1+I_2+1} \begin{Bmatrix} I & N & F \\ N & I' & 1 \end{Bmatrix}$$
$$\times \sqrt{N(2N+1)(N+1)(I+1)(2I+1)(2I'+1)}$$
$$\times \left[\delta_{M_2,M_2'} (-1)^{I'} c_1 \begin{Bmatrix} I_1 & I & I_2 \\ I' & I_1 & 1 \end{Bmatrix} \sqrt{I_1(2I_1+1)(I_1+1)} \right.$$
$$\left. + \delta_{M_1,M_1'} (-1)^{I} c_2 \begin{Bmatrix} I_2 & I & I_1 \\ I' & I_2 & 1 \end{Bmatrix} \sqrt{I_2(2I_2+1)(I_2+1)} \right]. \tag{4.40}$$

The matrix elements of the scalar spin–spin coupling are

$$\langle I_1 M_1 I_2 M_2 N M_N | c_4 \mathbf{I}_1 \cdot \mathbf{I}_2 | I_1 M_1' I_2 M_2' N' M_N' \rangle$$
$$= \delta_{N,N'} \delta_{F,F'} \delta_{M_F,M_F'} c_4 (-1)^{I_1 - M_1 + I_2 - M_2}$$
$$\times \sqrt{(2I_1+1)I_1(I_1+1)(2I_2+1)I_2(I_2+1)}$$
$$\times \sum_q (-1)^q \begin{pmatrix} I_1 & 1 & I_1 \\ -M_1 & q & M_1' \end{pmatrix} \begin{pmatrix} I_2 & 1 & I_2 \\ -M_2 & -q & M_2' \end{pmatrix}, \tag{4.41}$$

$$\langle (I_1 I_2) INFM_F | c_4 \mathbf{I}_1 \cdot \mathbf{I}_2 | (I_1 I_2) I' N' F' M_F' \rangle$$
$$= \delta_{I,I'} \delta_{N,N'} \delta_{F,F'} \delta_{M_F,M_F'}$$
$$\times \frac{c_4}{2} \left[I(I+1) - I_1(I_1+1) - I_2(I_2+1) \right]. \tag{4.42}$$

The matrix elements of the tensor spin–spin coupling are

$$\langle I_1 M_1 I_2 M_2 N M_N | c_3 \mathbf{I}_1 \cdot \tilde{T} \cdot \mathbf{I}_2 | I_1 M_1' I_2 M_2' N' M_N' \rangle$$
$$= -c_3 \sqrt{6} \begin{pmatrix} N & 2 & N' \\ 0 & 0 & 0 \end{pmatrix} \sqrt{(2N+1)(2N'+1)}$$

$$\times \sqrt{I_1 I_2 (2I_1+1)(2I_2+1)(I_1+1)(I_2+1)}$$
$$\times \sum_q (-1)^{q-M_N+I_1-M_1+I_2-M_2} \begin{pmatrix} N & 2 & N' \\ -M_N & q & M'_N \end{pmatrix}$$
$$\times \sum_m \langle 1, m; 1, -q-m | 2, -q \rangle \begin{pmatrix} I_1 & 1 & I_1 \\ -M_1 & m & M'_1 \end{pmatrix}$$
$$\times \begin{pmatrix} I_2 & 1 & I_2 \\ -M_2 & -q-m & M'_2 \end{pmatrix}, \tag{4.43}$$

$$\langle (I_1 I_2)\, INFM_F | c_3 \mathbf{I}_1 \cdot \tilde{T} \cdot \mathbf{I}_2 | (I_1 I_2)\, I'N'F'M'_F \rangle$$
$$= -c_3 \delta_{F,F'} \delta_{M_F,M'_F} (-1)^{I'+F} \begin{Bmatrix} I & N & F \\ N' & I' & 2 \end{Bmatrix}$$
$$\times \sqrt{(2N+1)(2N'+1)} \begin{pmatrix} N & 2 & N' \\ 0 & 0 & 0 \end{pmatrix} \begin{Bmatrix} I_1 & I_1 & 1 \\ I_2 & I_2 & 1 \\ I & I' & 2 \end{Bmatrix}$$
$$\times \sqrt{30 (2I+1)(2I'+1) I_1 I_2}$$
$$\times \sqrt{(I_1+1)(I_2+1)(2I_1+1)(2I_2+1)}. \tag{4.44}$$

The matrix elements of the nuclear quadrupole Hamiltonian are given by

$$\left\langle I_1 M_1 I_2 M_2 N M_N \left| \sum_{i=1}^{2} \mathbf{V}_i \cdot \mathbf{Q}_i \right| I_1 M'_1 I_2 M'_2 N' M'_N \right\rangle$$
$$= \sum_{i=1}^{2} \frac{(eqQ)_i}{4} \sum_q (-1)^{q-M_N+I_i-M_i} \sqrt{(2N+1)(2N'+1)}$$
$$\times \begin{pmatrix} N & 2 & N' \\ -M_N & q & M'_N \end{pmatrix} \begin{pmatrix} I_i & 2 & I_i \\ -M_i & -q & M'_i \end{pmatrix}$$
$$\times \begin{pmatrix} N & 2 & N' \\ 0 & 0 & 0 \end{pmatrix} \begin{pmatrix} I_i & 2 & I_i \\ -I_i & 0 & I_i \end{pmatrix}^{-1}, \tag{4.45}$$

$$\left\langle (I_1 I_2)\, INFM_F \left| \sum_{i=1}^{2} \mathbf{V}_i \cdot \mathbf{Q}_i \right| (I_1 I_2)\, I'N'F'M'_F \right\rangle$$
$$= \delta_{F,F'} \delta_{M_F,M'_F} \frac{1}{4} (-1)^{I'+F+I_1+I_2}$$

$$
\begin{aligned}
&\times \sqrt{(2N+1)(2N'+1)(2I'+1)(2I+1)} \\
&\times \begin{pmatrix} N & 2 & N' \\ 0 & 0 & 0 \end{pmatrix} \begin{Bmatrix} I & N & F \\ N' & I' & 2 \end{Bmatrix} \\
&\times \left[\delta_{I_2, I_2'} (eqQ)_1 (-1)^{I'} \begin{Bmatrix} I_1 & I & I_2 \\ I' & I_1 & 2 \end{Bmatrix} \begin{pmatrix} I_1 & 2 & I_1 \\ -I_1 & 0 & I_1 \end{pmatrix}^{-1} \right. \\
&\left. + \delta_{I_1, I_1'} (eqQ)_2 (-1)^{I} \begin{Bmatrix} I_2 & I & I_1 \\ I' & I_2 & 2 \end{Bmatrix} \begin{pmatrix} I_2 & 2 & I_2 \\ -I_2 & 0 & I_2 \end{pmatrix}^{-1} \right].
\end{aligned}
\tag{4.46}
$$

References

1. Carr, L.D., Demille, D., Krems, R.V., Ye, J.: Cold and ultracold molecules: Science, technology, and applications. New J. Phys. **11**, 055049 (2009)
2. Ni, K.K., Ospelkaus, S., de Miranda, M.H.G. Peér, A., Neyenhuis, B., Zirbel, J.J., Kotochigova, S., Julienne, P.S., Jin, D.S., Ye, J.: A high phase-space-density gas of polar molecules. Science **322**, 231–235 (2008)
3. Ospelkaus, S., Ni, K.K., de Miranda, M.H.G., Pe'er, A., Nyenhuis, B., Wang, D., Kotochigova, S., Julienne, P.S., Jin, D.S., Ye, J.: Ultracold polar molecules near quantum degeneracy. Faraday Discuss. **142**, 351–359 (2009)
4. Deiglmayr, J., Grochola, A., Repp, M., Mörtlbauer, K., Glück, C., Lange, J., Dulieu, O., Wester, R., Weidemüller, M.: Formation of ultracold polar molecules in the rovibrational ground state. Phys. Rev. Lett. **101**(13), 133004 (2008). doi:10.1103/PhysRevLett.101.133004
5. Danzl, J.G., Mark, M.J., Haller, E., Gustavsson, M., Hart, R., Aldegunde, J., Hutson, J.M., Nägerl, H.C.: An ultracold high-density sample of rovibronic ground-state molecules in an optical lattice. Nat. Phys. **6**, 265–270 (2010)
6. Winkler, K., Lang, F., Thalhammer, G., Straten, P.V.D., Grimm, R., Hecker Denschlag, J.: Coherent optical transfer of feshbach molecules to a lower vibrational state. Phys. Rev. Lett. **98**(4), 043201 (2007). doi:10.1103/PhysRevLett.98.043201
7. Pilch, K., Lange, A.D., Prantner, A., Kerner, G., Ferlaino, F., Nägerl, H.-C., Grimm, R.: Observation of interspecies Feshbach resonances in an ultracold Rb-Cs mixture. Phys. Rev. A **79**(4), 042718 (2009). doi:10.1103/PhysRevA.79.042718.
8. Voigt, A.-C., Taglieber, M., Costa, L., Aoki, T., Wieser, W., Hänsch, T.W., Dieckmann, K.: Ultracold heteronuclear Fermi-Fermi molecules. Phys. Rev. Lett. **102**(2), 020405 (2009). doi:10.1103/PhysRevLett.102.020405
9. Aldegunde, J., Ran, H., Hutson, J.M.: Manipulating ultracold polar molecules with microwave radiation: The influence of hyperfine structure. Phys. Rev. A **80**, 043410 (2009)
10. Ran, H., Aldegunde, J., Hutson, J.M.: Hyperfine structure in the microwave spectra of ultracold polar molecules. New J. Phys. **12**(4), 043015 (2010). http://stacks.iop.org/1367-2630/12/i=4/a=043015
11. Ospelkaus, S., Ni, K.K., Quemener, G., Neyenhuis, B., Wang, D., de Miranda, M.H.G., Bohn, J.L., Ye, J., Jin, D.S.: Controlling the hyperfine state of rovibronic ground-state polar molecules. Phys. Rev. Lett. **104**, 030402 (2010)
12. Feynman, R.P.: Simulating physics with computers. Int. J. Theor. Phys. **21**, 467 (1982)
13. Lewenstein, M.S.A., Ahufinger, V., Damski, B., Sen De, A., Sen, U.: Ultracold atomic gases in optical lattices: Mimicking condensed matter physics and beyond. Adv. Phys. **56**, 243–379 (2007)

14. Micheli, A., Pupillo, G., Büchler, H.P., Zoller, P.: Cold polar molecules in two-dimensional traps: Tailoring interactions with external fields for novel quantum phases. Phys. Rev. A **76**, 043604 (2007)
15. Wall, M.L., Carr, L.D.: Emergent timescales in entangled quantum dynamics of ultracold molecules in optical lattices. New J. Phys. **11**, 055027 (2009)
16. Brennen, G.K., Micheli, A., Zoller, P.: Designing spin-1 lattice models using polar molecules. New J. Phys. **9**, 138 (2007)
17. Huber, S.D., Altman, E.: Universal dephasing of many-body rabi oscillations of atoms in one-dimensional traps. Phys. Rev. Lett. **103**, 160402 (2009)
18. Barmettler, P., Punk, M., Gritsev, V., Demler, E., Altman, E.: Quantum quenches in the anisotropic spin-1/2 heisenberg chain: Different approaches to many-body dynamics far from equilibrium. New J. Phys. **12**(5), 055017 (2010). http://stacks.iop.org/1367-2630/12/i=5/a=055017
19. Ashcroft, N., Mermin, D.: Solid State Physics. Saunders College Publishing, Orlando (1976)
20. Deiglmayr, J., Aymar, M., Wester, R., Weidemüller, M., Dulieu, O.: Calculations of static dipole polarizabilities of alkali dimers: Prospects for alignment of ultracold molecules. J. Chem. Phys. **129**, 064309 (2008)
21. Iskin, M., Sa de Melo, C.A.R.: Mixtures of ultracold fermions with unequal masses. Phys. Rev. A. **76**, 013601 (2007)
22. Wall, M.L., Carr, L.D.: (2010, in preparation)
23. Paredes, B., Widera, A., Murg, V., Mandel, O., Folling, S., Cirac, I., Shlyapnikov, G.V., Hansch, T.W.: Tonks-Girardeau gas of ultracold atoms in an optical lattice. Nature **429**, 277–281 (2005)
24. Tolra, B.L., O'Hara, K.M., Huckans, J.H., Phillips, W.D., Rolston, S.L., Porto, J.V.: Observation of reduced three-body recombination in a correlated 1D degenerate bose gas. Phys. Rev. Lett. **92**, 190401–1–4 (2004)
25. Wenger, T., Kinoshita, T., Weiss, D.S.: A quantum Newton's cradle. Nature **440**, 900–903 (2006)
26. Büchler, H.P., Micheli, A., Zoller, P.: Three-body interactions with cold polar molecules. Nat. Phys. **3**, 726–731 (2007)
27. Aldegunde, J., Rivington, B.A., Zuchowski, P.S., Hutson, J.M.: Hyperfine energy levels of alkali-metal dimers: Ground-state polar molecules in electric and magnetic fields. Phys. Rev. A **78**, 033434 (2008)
28. Rey, A.M.: Ultracold Bosonic Atoms in Optical Lattices. Ph.D. thesis, University of Maryland (2004)
29. Ni, K.K., Ospelkaus, S., Wang, D., Quemener, G., Neyenhuis, B., de Miranda, M.H.G., Bohn, J.L., Ye, J., Jin, D.S.: Dipolar collisions of polar molecules in the quantum regime. Nat. Lond. **464**, 1324 (2010)
30. Chakravarty, S.: Quantum phase transition, dissipation, and measurement. e-print http://arxiv.org/abs/0909.2316 (2010)
31. Le Hur, K.: Quantum phase transitions in spin-boson systems: Dissipation and light phenomena. e-print http://arxiv.org/abs/0909.4822 (2010)
32. Werner, P., Troyer, M.: Cluster monte carlo algorithms for dissipative quantum phase transitions. In: Carr, L.D. (ed.) Understanding Quantum Phase Transitions. CRC Press, Boca Raton (2010)
33. Brown, J., Carrington, A.: Rotational Spectroscopy of Diatomic Molecules. Cambridge University Press, Cambridge (2003)
34. Van Vleck, J.H.: The coupling of angular momentum vectors in molecules. Rev. Mod. Phys. **23**, 213–227 (1951)
35. Zare, R.: Angular Momentum: Understanding Spatial Aspects in Chemistry and Physics. Wiley, New York (1988)
36. Townes, R.N., Schawlow, A.L.: Microwave Spectroscopy. Dover, New York (1975)
37. Mohr, P.J., Taylor, B.N.: CODATA recommended values of the fundamental physical constants: 1998. Rev. Mod. Phys. **72**, 351 (2000)

Part III
The Fermi Resonance Hamiltonian

Chapter 5
Microscopic Model for Feshbach Interacting Fermions in an Optical Lattice with Arbitrary Scattering Length and Resonance Width

Abstract We numerically study the problem of two fermions in a three dimensional optical lattice interacting via a zero-range Feshbach resonance, and display the dispersions of the bound states as a two-particle band structure with unique features compared to typical single-particle band structures. We show that the exact two-particle solutions of a projected Hamiltonian may be used to define an effective two-channel, few-band model for the low energy, low density physics of many fermions at arbitrary s-wave scattering length. Our method applies to resonances of any width, and can be adapted to multichannel situations or higher-ℓ pairing. In strong contrast to usual Hubbard physics, we find that pair hopping is significantly altered by strong interactions and the presence of the lattice, and the lattice induces multiple molecular bound states.

The crossover of a system of attractive two-component fermions from a condensate of loosely bound Cooper pairs to a condensate of tightly bound bosonic molecules has a long history [1], and appears in many contexts, including high-temperature superconductivity [2] and ultracold atoms [3]. Furthermore, near the crossover such a system enters the *unitary regime* where the scattering length is larger than any other length scale in the problem. The physics of this regime is relevant to many different fields, bringing together quantum chromodynamics, holographic duality, and ultracold quantum gases [4]. Theoretical study of the unitary regime is generally difficult due to the absence of any small parameter.

Theoretical analysis becomes even more difficult in a lattice, as the center of mass, relative, and internal degrees of freedom become coupled, leading to composite particles whose properties depend on their center of mass motion [5]. Furthermore, strong interactions require the inclusion of a large number of Bloch bands for an accurate description, and this cannot be handled efficiently by modern analytical or numerical many-body techniques. In addition to general theoretical interest in how fermions pair to form bosons in a discrete lattice setting, the study of pairing in lattices is of significant practical importance. For example, an accurate,

---

Published previously as *Microscopic Model for Feshbach Interacting Fermions in an Optical Lattice with Arbitrary Scattering Length and Resonance Width*, M.L. Wall and L.D. Carr, Phys. Rev. Lett. **109**, 055302 (2012).

M.L. Wall, *Quantum Many-Body Physics of Ultracold Molecules in Optical Lattices*, Springer Theses, DOI 10.1007/978-3-319-14252-4_5

systematically correctable, and computationally feasible many-body Hamiltonian is necessary for calibrating fermionic quantum simulators as has been done in the bosonic case [6].

In this Letter, we describe a general method to derive an effective few-band low-energy Hamiltonian for Feshbach interacting fermions in a lattice from the numerical solution of the two-body problem. We call this Hamiltonian the *Fermi Resonance Hamiltonian* (FRH). This method applies to Feshbach resonances of any width and for arbitrary scattering length, and all parameters appearing in the effective model can be computed microscopically from the properties of the two-body solution. The difference between the bare model and the FRH is sketched in Fig. 5.1.

The simplest approach to describing Feshbach interacting fermions is to replace the interaction with a pseudo-potential chosen to reproduce the correct scattering length. When restricted to a single Bloch band, this leads to the popular Hubbard model [7] which has been shown to break down for scattering lengths which are far from being resonant, even when the parameters appearing in the model are determined self-consistently from few-body physics [8]. Our work instead defines a "dressed" closed channel whose properties are chosen to reproduce both the scattering and bound states correctly. In contrast to past two-channel approaches [9], we construct the dressed fields using the full lattice solution and not an approximation where the center of mass and relative coordinates separate, such as the harmonic oscillator potential. The use of any separable approximation leads to qualitative errors, such as the lack of tunneling along non-principal axes, and quantitative errors, such as the underestimation of principal axis tunneling matrix elements, often by an order of magnitude.

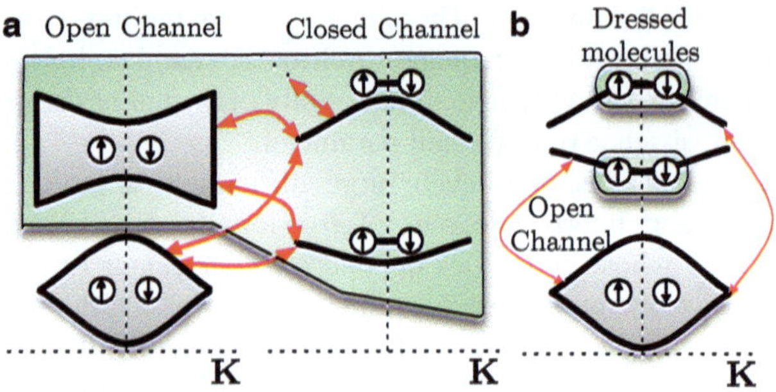

Fig. 5.1 Schematic of the FRH transformation. (**a**) In a broad Feshbach resonance, all two-particle scattering continua (*gray shading*) are strongly coupled to bare molecular bands (*solid lines*). Thus all scattering continua are virtually strongly coupled. (**b**) By correctly dressing the molecular bands, one obtains a single scattering continuum (*gray*) plus well-separated dressed molecular bands (*green*), with much simpler couplings. This is our efficient, numerically tractable, FRH

5.1 Exact Solution for Two-Particles

The basic concept of a two-channel model is for an *open* channel to describe scattering between two atoms and a separate *closed* channel to describe bound pairs. While each channel represents a single scattering or bound state in the continuum, in the lattice it also acquires a band index. Because of an inter-channel coupling, the actual molecule is a superposition of bands from both channels.

To treat this problem, we begin with the nonlinear eigenvalue problem developed by Büchler [8] for $E_{\mathbf{K}}$, the bound state energy at total quasimomentum $\mathbf{K}$, and $R_{\mathbf{s}}^{\mathbf{K}}$, the coefficients of the closed channel portion of the wavefunction. As shown in [8], for the bound states of two fermions in an optical lattice interacting via a zero-range Feshbach resonance in a two-channel model:

$$\left[E_{\mathbf{K}} - \nu - E_{\mathbf{sK}}^{M}\right] R_{\mathbf{s}}^{\mathbf{K}} = \frac{g^2}{a^3}\sum_{\mathbf{t}}\chi_{\mathbf{st}}^{\mathbf{K}}\left(E_{\mathbf{K}}\right) R_{\mathbf{t}}^{\mathbf{K}}\,, \tag{5.1}$$

$$\chi_{\mathbf{st}}^{\mathbf{K}}\left(E_{\mathbf{K}}\right) \equiv \int \frac{d\mathbf{q}}{v_0}\sum_{\mathbf{mn}}\frac{h_{\mathbf{sK}}^{\mathbf{nm}}(\mathbf{q})h_{\mathbf{tK}}^{\mathbf{nm}\star}(\mathbf{q})}{E_{\mathbf{K}}-E_{\mathbf{nm}}^{\mathbf{K}}(\mathbf{q})+i\eta} - \bar{\chi}_{\mathbf{st}}^{\mathbf{K}}\,, \tag{5.2}$$

$$\bar{\chi}_{\mathbf{st}}^{\mathbf{K}} \equiv -\int \frac{d\mathbf{q}}{v_0}\sum_{\mathbf{mn}}\bar{h}_{\mathbf{sK}}^{\mathbf{nm}}\left(\mathbf{q}\right)\bar{h}_{\mathbf{tK}}^{\mathbf{nm}\star}\left(\mathbf{q}\right)/\bar{E}_{\mathbf{nm}}^{\mathbf{K}}\left(\mathbf{q}\right)\,. \tag{5.3}$$

Here ν is the renormalized detuning between the open and closed channels, g is the inter-channel coupling, a is the lattice spacing, v_0 is the volume of the Brillouin zone (BZ), and the bars in Eqs. (5.2) and (5.3) denote quantities computed in the absence of an optical lattice. We assume that spin–spin interactions which change the orbital angular momentum are negligible so that the scattering is purely s-wave. The optical lattice is assumed to be simple cubic with lattice spacing a and potential $V\left(\mathbf{x}\right) = V\sum_{j\in\{x,y,z\}}\sin^2\left(\pi j/a\right)$. The overlaps of the dimensionless coupling between the open and closed channels are

$$\frac{h_{\mathbf{sK}}^{\mathbf{nm}}\left(\mathbf{q}\right)}{\sqrt{N^3a^3}} = \int d\mathbf{x}d\mathbf{y}\left[\psi_{\mathbf{nq}}\left(\mathbf{x}\right)\psi_{\mathbf{m},\mathbf{K}-\mathbf{q}}\left(\mathbf{y}\right)\right]^{\star}\alpha\left(\mathbf{r}\right)\phi_{\mathbf{sK}}\left(\mathbf{R}\right)\,,$$

where N^3 is the number of unit cells, the $\psi_{\mathbf{nq}}\left(\mathbf{x}\right)$ are Bloch functions with energies $E_{\mathbf{nq}}$ for particles with mass m spanning the open channel, and $\phi_{\mathbf{sK}}\left(\mathbf{z}\right)$ are Bloch functions with energies $E_{\mathbf{sK}}^{M}$ for particles with mass $2m$ in a lattice potential $2V$ spanning the closed channel. We have also defined relative $\mathbf{r} \equiv \mathbf{x} - \mathbf{y}$ and center of mass $2\mathbf{R} = \mathbf{x} + \mathbf{y}$ coordinates, and $\alpha\left(\mathbf{r}\right)$ is a regularization of the inter-channel coupling. The sum of the noninteracting energies of the open channel is denoted $E_{\mathbf{nm}}^{\mathbf{K}}\left(\mathbf{q}\right) = E_{\mathbf{nq}} + E_{\mathbf{m},\mathbf{K}-\mathbf{q}}$ and the zero of energy is $E_{\mathbf{11}}^{\mathbf{0}}\left(\mathbf{0}\right)$. Here and throughout the rest of this work $\mathbf{n}$ and $\mathbf{m}$ are band indices for the open channel, $\mathbf{s}$ and $\mathbf{t}$ are band indices for the closed channel, $\mathbf{q}$ is a single-particle quasimomentum, and $\mathbf{K}$ is the total quasimomentum.

While Eqs. (5.1)–(5.3) apply to resonances of any width, we focus on the experimentally relevant limit of a broad resonance. Narrow resonances are treated in the supplementary material. A broad resonance in the few-body sense is the limit of effective range much smaller than the lattice spacing, $r_B \ll a$, and so we can take the limits $g/E_R a^{3/2} = 4\sqrt{a/r_B\pi^3} \to \infty$, $\nu/E_R \to \infty$, $a_s/a = -\pi g^2/8a^3 E_R \nu =$ const., to obtain $(8a_s E_R/\pi a)\chi^{\mathbf{K}}(E_{\mathbf{K}})\,\mathbf{R}^{\mathbf{K}} - \mathbf{R}^{\mathbf{K}} = 0$, where $E_R = \hbar^2\pi^2/2ma^2$ is the recoil energy and a_s is the s-wave scattering length. How can we then obtain the dispersion relation $E_{\mathbf{K}}$ for fixed a_s, ν, etc.? First, fix the energy eigenvalue $E_{\mathbf{K}}$ and solve the resulting *linear* eigenproblem for $1/a_s$. This provides exact eigentuples $\left(E_{\mathbf{K}}, a_s, \mathbf{R}^{\mathbf{K}}\right)$ of the nonlinear eigenproblem, though it may not be the a_s we seek. Second, fix a_s and use the exact tuple nearest this value as initialization for a Newton–Raphson iteration [10]. This two-stage approach converges to a relative accuracy of 0.01 % in a few tens of iterations [11].

Because the eigenequation (5.1) is invariant under translation by any Bravais lattice vector, its eigenvalues can be classified according to the total quasimomentum and shown as a two-particle band structure. A complete classification of the solutions is given in the supplemental material. In Fig. 5.2 we show only the energy of the low-energy bound states with completely even parity under inversion as a function of $\mathbf{K}$ for several a_s/a in a lattice with $V/E_R = 12$. We see the appearance of several bound states for a fixed s-wave scattering length, in contrast to the continuum. These additional bound states arise from the coupling of quasimomenta modulo a reciprocal lattice vector induced by the reduced translational symmetry. One salient feature is the emergence of universality, which is the independence of the dispersion from the sign of a_s when $|a_s/a|$ becomes large. For non-resonant and negative a_s/a, picturing the bound states as Fermi pairs with twice the mass and twice the polarizability captures the relative spacings between energy levels quite accurately, but generally *overestimates* the effective mass of the bound states. This effective mass difference is an indication of the coupling between the center of mass and relative motion which leads to important properties of the FRH.

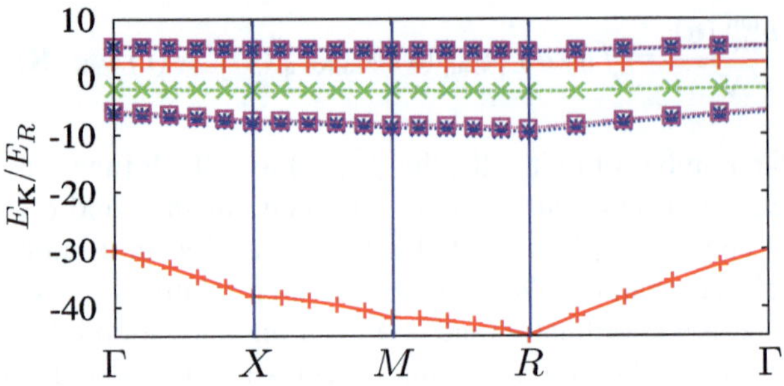

Fig. 5.2 Exact two-particle band structures for various a_s in a strong optical lattice. The bound state energies for $a_s/a = -5$ (*purple boxes*), -0.1 (*red pluses*), 0.1 (*green crosses*), and 5 (*blue asterisks*) as a function of the total quasimomentum $\mathbf{K}$ along a path connecting the high-symmetry points in the irreducible BZ $\Gamma = (0,0,0)$, $X = (-\pi/a, 0, 0)$, $M = (-\pi/a, -\pi/a, 0)$, $R = (-\pi/a, -\pi/a, -\pi/a)$ for a lattice with $V/E_R = 12$. The near-resonant points $a_s/a = \pm 5$ lie nearly on *top* of one another, demonstrating universality

5.2 Fermi Resonance Hamiltonian

A promising route to describing Feshbach interacting ultracold gases is by a lattice projection of a two-channel model in which the closed channel appears explicitly in the Hamiltonian. However, for a typical broad resonance such models require a large number of both open and closed channel bands to solve accurately, and so cannot be treated efficiently. Because the modern context of this problem involves extremely low temperatures and densities, we can look for an effective model valid in these limits which still reproduces the correct physics. This is done by replacing the model containing couplings between all open channel bands with all closed channel bands with a model describing an effective resonance between the lowest open channel band with a suitable set of effective closed channel bands whose properties are set by the two-body solution for low densities. This process is displayed schematically in Fig. 5.1. The purpose of this section is to derive such an effective Hamiltonian using our two-particle theory.

We begin by separating our two-particle Hamiltonian using projectors $\mathbb{L}$ into the lowest open channel band and $\mathbb{D} = 1 - \mathbb{L}$ into all excited open channel bands and all closed channel bands. A similar approach was taken in [12] for the 1D case. An analysis analogous to that leading to Eqs. (5.1) and (5.2) gives a nonlinear eigenequation for the closed channel components of $\mathbb{D}|\psi\rangle$ as

$$\left[E_{\mathbf{K}} - \nu - E^{M}_{\mathbf{sK}}\right] R^{\mathbf{K}}_{\mathbf{s}} = \frac{g^2}{a^3} \sum\nolimits_{\mathbf{t}} \tilde{\chi}^{\mathbf{K}}_{\mathbf{st}} \left(E_{\mathbf{K}}\right) R^{\mathbf{K}}_{\mathbf{t}}\,, \tag{5.4}$$

$$\tilde{\chi}^{\mathbf{K}}_{\mathbf{st}} \left(E_{\mathbf{K}}\right) \equiv \sum\nolimits'_{\mathbf{mn};\mathbf{q}} \frac{h^{\mathbf{nm}}_{\mathbf{sK}}(\mathbf{q}) h^{\mathbf{nm}\star}_{\mathbf{tK}}(\mathbf{q})}{E_{\mathbf{K}} - E^{\mathbf{K}}_{\mathbf{nm}}(\mathbf{q}) + i\eta} - \bar{\chi}^{\mathbf{K}}_{\mathbf{st}}\,, \tag{5.5}$$

where the prime on the sum indicates $(\mathbf{m},\mathbf{n}) \neq (\mathbf{1},\mathbf{1})$. Here $\tilde{\chi}$ differs from χ in Eqs. (5.1) and (5.2) in that the summation excludes the lowest band. We emphasize that the renormalization $\bar{\chi}$ includes all bands and so the detuning and scattering length used in this projected model are those of the full (non-projected) and properly renormalized two-body problem. We call the eigenstates of this projected system *dressed molecules*. Here we label distinct eigenstates of Eq. (5.4) by the parameter α. These solutions share many features of the full solution presented above. However, the divergence of the s-wave scattering length for the lowest energy completely even parity state occurs near $E_{\mathbf{K}} = 0$, indicating that scattering resonances in the lowest open channel band are generated by coupling to this state.

We now assume that, at low temperatures and low densities, two-particles which are separated by a distance large compared to the effective range of the potential will remain in the lowest band to minimize their energy. When two-particles come together they interact strongly and populate many of the excited open channel bands as well as the closed channel bands. Because it is rare for more than two-particles to come together, the particular populations of the excited states are fixed by the two-particle solution. The dressed molecules encapsulate the short distance, high energy physics and couple it to the long wavelength, low energy physics of the lowest band fermions through the Feshbach coupling. The point of connection between the few- and many-body physics is the two-particle scattering length (equivalently g and ν for narrow resonances), which appears as a parameter in Eq. (5.4) defining the dressed molecules.

The FRH is

$$\begin{aligned}\hat{H}_{\mathrm{eff}} = &-t_f\sum_{\sigma\in\{\uparrow,\downarrow\}}\sum_{\langle i,j\rangle}\hat{a}^{\dagger}_{i\sigma}\hat{a}_{j\sigma} + E_0\sum_{\sigma\in\{\uparrow,\downarrow\}}\sum_i\hat{n}^{(f)}_{i\sigma}\\ &-\sum_{\alpha\in\mathcal{M}}\sum_{i,j}t^{\alpha}_{i,j}\hat{d}^{\dagger}_{i,\alpha}\hat{d}_{j,\alpha} + \sum_{\alpha\in\mathcal{M}}\bar{\nu}_{\alpha}\sum_i\hat{n}^{(b)}_{i\alpha}\\ &+\sum_{\alpha\in\mathcal{M}}\sum_{ijk}g^{\alpha}_{i-k,k-j}\left[\hat{d}^{\dagger}_{i,\alpha}\hat{a}_{j,\uparrow}\hat{a}_{k,\downarrow} + \mathrm{h.c.}\right],\end{aligned} \tag{5.6}$$

where $\hat{a}^{\dagger}_{i\sigma}$ creates a particle with spin σ in the lowest open channel band Wannier state centered at lattice site i, $w_i(\mathbf{x})$; $\hat{d}^{\dagger}_{i,\alpha}$ creates a particle in the αth dressed molecule Wannier state centered at site i, $\mathcal{W}_{i,\alpha}(\mathbf{x},\mathbf{y})$; $\hat{n}^{(f)}_{i\sigma}$ is the number operator for fermions in the lowest Bloch band; and $\hat{n}^{(b)}_{i\alpha}$ is the number operator for the αth dressed molecule state. The set of dressed molecules $\mathcal{M}$ which are included dynamically can be determined on energetic and symmetry grounds from the two-particle solution. At low energies, only the completely even parity dressed molecule in the lowest sheet is relevant to the set $\mathcal{M}$, as all others either have vanishing on-site couplings from parity considerations or are very far off-resonance. In order, the terms in Eq. (5.6) represent tunneling of atoms in the lowest Bloch band between neighboring lattice sites i and j; the energy $E_0 = \sum_{\mathbf{q}} E_{1,\mathbf{q}}/N^3$ of a fermion in the lowest band with respect to the zero of energy; tunneling of the dressed molecular center of mass between two lattice sites i and j, not necessarily nearest-neighbors; detunings of the dressed molecules from the lowest band two-particle scattering continuum; and resonant coupling between the lowest band fermions at sites j and k in different internal states and a dressed molecule at site i. The FRH is a two-channel resonance model, between unpaired fermions in the lowest band, and dressed molecules nearby in energy.

We now describe how to calculate the Hubbard parameters appearing in Eq. (5.6). The first term is well known from single-band Hubbard models [13] and we do not discuss it here. Due to the fact that the solutions of the projected nonlinear eigenequation (5.4) are also eigenstates of the total quasimomentum, the second and third terms may be written as $\bar{\nu}_{\alpha} = \int d\mathbf{K}E^{\alpha}_{\mathbf{K}}/v_0$ and $t^{\alpha}_{i,j} = -\int d\mathbf{K}e^{i\mathbf{K}\cdot(\mathbf{R}_i-\mathbf{R}_j)}E^{\alpha}_{\mathbf{K}}/v_0$. Because the band structure is not separable, $E_{\mathbf{K}} \neq \sum_{l=\{x,y,z\}} E_{K_l}$, dressed molecules can tunnel along directions which are not the principal axes of the lattice. This is in stark contrast to single-particle tunneling in Bravais lattices which always occurs along the principal axes. Thus *diagonal hopping* is a key feature neglected in previous approaches. In Fig. 5.3a we show that diagonal hopping is often of the same order of magnitude as the tunneling of open channel fermions in the lowest band. The signs and magnitudes of the tunnelings and particularly the dressed-molecule atom couplings are crucially affected by the parities of the dressed molecular Wannier functions. We stress that

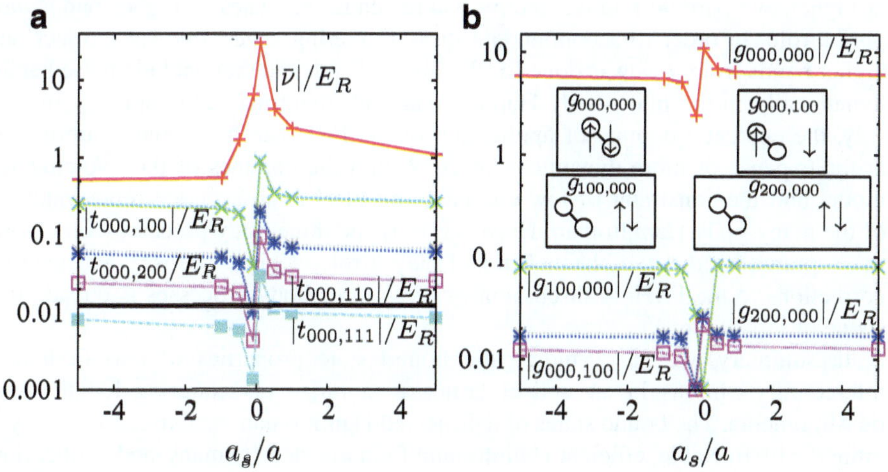

Fig. 5.3 Hubbard parameters for the FRH. (**a**) The detunings and tunnelings of the completely even parity dressed molecule in the lowest sheet as a function of a_s/a. The detuning is negative for $a_s < 0$ and positive otherwise. The *solid black horizontal line* is the tunneling of a single open channel fermion in the lowest band. The nearest-neighbor dressed molecular tunneling is nearly two orders of magnitude larger than the open channel tunneling near resonance. (**b**) The effective atom-dressed molecule couplings of the completely even parity dressed molecule in the lowest sheet as a function of a_s/a. Schematics of the spatial dependence of the various coupling processes are shown in the *boxes*

only a full lattice solution can reproduce these important properties of the Hubbard parameters; the frequently used harmonic oscillator approximation will fail even qualitatively to do so.

The remaining Hubbard parameter is the dressed molecule-atom coupling, which becomes in the limit of a broad resonance $g/E_R a^{3/2} \to \infty$

$$g^{\alpha}_{i-k,k-j} = \sum_{\mathbf{s}} \int \frac{d\mathbf{K}}{v_0} R^{\mathbf{K}}_{\alpha\mathbf{s}} g_{\alpha\mathbf{K}} \int \frac{d\mathbf{q}}{v_0} e^{i(\mathbf{K}\cdot\mathbf{R}_{ik}+\mathbf{q}\cdot\mathbf{R}_{kj})} h^{\mathbf{11}}_{\mathbf{sK}}(\mathbf{q}), \tag{5.7}$$

where the renormalized coupling is $g_{\alpha\mathbf{K}} = E_R/\sqrt{-\mathbf{R}^{\mathbf{K}}_{\alpha} \cdot (\partial\tilde{\chi}_{\mathbf{K}}/\partial E^{\alpha}_{\mathbf{K}}) \cdot \mathbf{R}^{\mathbf{K}}_{\alpha}}$ and $\mathbf{R}_{ij} = \mathbf{R}_i - \mathbf{R}_j$. We emphasize that $g^{\alpha}_{j,k}$ has only implicit dependence on the divergent parameter $g/E_R a^{3/2}$ through $g_{\alpha\mathbf{K}}$ and so remains finite, see Fig. 5.3b. As $g_{\alpha\mathbf{K}} \ll g/a^{3/2}$, the transformation to the FRH has the effect of narrowing the resonance. In Fig. 5.3b we also see that the on-site coupling $g_{000,000}$ is the dominant energy scale of the problem for large a_s/a, and that off-site couplings can also be large compared to other Hubbard parameters such as the open channel tunneling. Atoms which do not lie along a principal axis can pair to form a molecule, but this effect is much weaker than diagonal tunneling for the completely even parity dressed molecule.

In the derivation of the FRH we use only the bound states of the projected problem and neglect scattering states in higher bands. This captures the scattering states in the lowest band and nearby bound states, but will fail to capture the physics at higher two-particle relative energy where scattering states in higher bands can play a role. In order to accommodate these scattering states, one can project out higher bands from χ as was done for the lowest band, and then include these bands dynamically in the many-body Hamiltonian with renormalized couplings. In this way, the energetic domain of application of the FRH can be extended arbitrarily at the expense of more dynamical fields. Within the confines of the two-channel model and the constraint of low energies, the FRH is an accurate representation of the many-body Hamiltonian. However, intrinsic three-body processes which are not captured by the two-channel model play a role at higher density and lead to corrections to the FRH. A discussion of these three-body processes is outside the scope of this paper.

In summary, we have studied the bound state properties of two Feshbach interacting fermions in an optical lattice at a range of scattering lengths and quasimomenta. The bound states of a projected Hamiltonian were used to identify a numerically tractable, efficient Hamiltonian for a low density many-body collection of lattice fermions at arbitrary scattering length and low energies, the FRH. Our results provide the appropriate starting point for future investigations of strongly interacting lattice fermions.

We acknowledge useful discussions with J.L. Bohn, H.P.Büchler, C.W. Clark, D.E. Schirmer, D.M. Wood, and Zhigang Wu. We also thank H.P. Büchler for providing computer code for comparison. This work was supported by the Alexander von Humboldt Foundation, AFOSR, NSF, and GECO.

5.3 Supplemental Material: Derivation of the Nonlinear Eigenequation

Here we review the derivation of Eq. (5.2) of the main text for the bound states of two fermions in an optical lattice interacting via a zero-range Feshbach resonance. All quantities have their same meaning as in the main text. The starting point of our analysis is a two-channel model with the open channel spanned by states of two fermions in different internal states with equal mass m and the closed channel spanned by molecular states with twice the fermionic mass and twice the fermionic polarizability. We describe their interaction via an inter-channel coupling g which couples the pair of open channel fermions to a closed channel molecule at the center of mass and a detuning ν between the two channels. This gives rise to the scattering amplitude

$$f(\mathbf{k}) = -\frac{1}{1/a_s + ik + r_b k^2}, \tag{5.8}$$

with s-wave scattering length $a_s = -2\mu g^2/4\pi\hbar^2\nu$ and effective range $r_B = \pi\hbar^4/\mu^2 g^2$. Here μ is the reduced mass and $\mathbf{k}$ the incident wavevector.

Denoting the wave function of the two fermions in the open channel as $\psi(\mathbf{x},\mathbf{y})$ and the wave function of the closed channel molecules as $\phi(\mathbf{z})$, the two-channel Schrödinger equation in position representation is

$$[E - \hat{H}_0(\mathbf{x}) - \hat{H}_0(\mathbf{y})]\psi(\mathbf{x},\mathbf{y}) = g\int d\mathbf{z}\alpha(\mathbf{r})\phi(\mathbf{z})\delta(\mathbf{z}-\mathbf{R})\,,$$
$$[E - \nu_0 - \hat{H}_0^M(\mathbf{z})]\phi(\mathbf{z}) = g\int d\mathbf{x}d\mathbf{y}\alpha(\mathbf{r})\psi(\mathbf{x},\mathbf{y})\delta(\mathbf{z}-\mathbf{R})\,.$$

In this expression $\hat{H}_0(\mathbf{x}) = -\frac{\hbar^2}{2m}\nabla_\mathbf{x}^2 + V(\mathbf{x})$ is the single particle Hamiltonian for the open channel and $\hat{H}_0^M(\mathbf{z}) = -\frac{\hbar^2}{4m}\nabla_\mathbf{z}^2 + 2V(\mathbf{z})$ is the single particle Hamiltonian for the closed channel. The subscript 0 in ν_0 denotes that this is a bare detuning entering the microscopic theory which is related to the physically observable detuning ν in the limit as the regularization cutoff $\Lambda \to \infty$. Additionally, we note that the Feshbach regularization $\alpha(\mathbf{r}) \to \delta(\mathbf{r})$ in the limit $\Lambda \to \infty$, where $\delta(\mathbf{r})$ is the Dirac delta function.

The open channel solution with total quasimomentum $\mathbf{K}$ may be parameterized as

$$\psi_\mathbf{K}(\mathbf{x},\mathbf{y}) = \frac{1}{\sqrt{N^3}}\sum_{\mathbf{nm}}\sum_{\mathbf{q}}\varphi_{\mathbf{nm}}^{\mathbf{Kq}}\psi_{\mathbf{n},\mathbf{q}}(\mathbf{x})\psi_{\mathbf{m},\mathbf{K}-\mathbf{q}}(\mathbf{y})\,, \tag{5.9}$$

where N^3 is the total number of unit cells in a 3D lattice with periodic boundary conditions and $\psi_{\mathbf{nq}}(\mathbf{x})$ is a Bloch eigenfunction of the single-particle Hamiltonian. As in the main text, quantities denoted in bold represent three-component vectors, e.g. $\mathbf{n} = (n_x, n_y, n_z)$. Similarly, we parameterize the closed channel wave function as a sum over Bloch states computed for twice the mass and twice the polarizability $\phi_{\mathbf{sK}}(\mathbf{z})$ as

$$\phi_\mathbf{K}(\mathbf{z}) = \sum_{\mathbf{s}} R_\mathbf{s}^\mathbf{K}\phi_{\mathbf{sK}}(\mathbf{z})\,. \tag{5.10}$$

Inserting these expansions into the two-channel Schrödinger equation yields

$$\left[E_\mathbf{K} - E_{\mathbf{nm}}^\mathbf{K}(\mathbf{q})\right]\varphi_{\mathbf{nm}}^{\mathbf{Kq}} = \frac{g}{\sqrt{a^3}}\sum_{\mathbf{s}} h_{\mathbf{sK}}^{\mathbf{nm}}(\mathbf{q})R_\mathbf{s}^\mathbf{K}\,,$$
$$\left[E_\mathbf{K} - \nu_0 - E_{\mathbf{sK}}^M\right]R_\mathbf{s}^\mathbf{K} = \frac{g}{\sqrt{a^3}}\sum_{\mathbf{nm}}\int\frac{d\mathbf{q}}{v_0}h_{\mathbf{sK}}^{\mathbf{nm}\star}(\mathbf{q})\varphi_{\mathbf{nm}}^{\mathbf{Kq}}\,.$$

Formally solving the first of the two equations with a Green's function and inserting into the second equation gives

$$\left[E_\mathbf{K} - \nu_0 - E_{\mathbf{sK}}^M\right]R_\mathbf{s}^\mathbf{K} = \frac{g^2}{a^3}\int\frac{d\mathbf{q}}{v_0}\sum_{\mathbf{mnt}}\frac{h_{\mathbf{sK}}^{\mathbf{nm}}(\mathbf{q})h_{\mathbf{tK}}^{\mathbf{nm}\star}(\mathbf{q})}{E_\mathbf{K}-E_{\mathbf{nm}}^\mathbf{K}(\mathbf{q})+i\eta}R_\mathbf{t}^\mathbf{K}\,,$$

where η is a positive infinitesimal. This expression diverges in the limit $\Lambda \to \infty$, as is well known for two-channel theories involving a pointlike boson [14]. We remove this divergence through renormalization, replacing the bare detuning ν_0 with the physical detuning ν by subtracting the infinite constant $\tilde{\chi}^{\mathbf{K}}$, yielding Eq. (5.2) of the main text. The divergent parts of Eq. (5.3) in the main text cancel and we may safely take the limit $\Lambda \to \infty$.

Following [15], we use the regularization

$$\alpha(\mathbf{r}) = \int_{v(\Lambda)} \frac{d\mathbf{k}}{(2\pi)^3} e^{i\mathbf{k}\cdot\mathbf{r}}, \tag{5.11}$$

where the cubical volume $v(\Lambda) = v_0\Lambda^3$ is centered around $\mathbf{k} = 0$ with v_0 the volume of the BZ. We also define a shell summation over bands with shell parameter S, $\sum_{\mathbf{nm};S}$, as the summation over all band indices $\mathbf{n}$ and $\mathbf{m}$ less than or equal to S with at least one of the band indices being S. The correct limiting procedure to obtain $\chi^{\mathbf{K}}_{\mathbf{st}}(\mathbf{K})$ in the limit of an infinite summation over bands and vanishing short-distance cutoff is

$$\lim_{\Lambda\to\infty} \left[\lim_{S\to\infty} \chi^{\mathbf{K}}_{\mathbf{st}}(E_{\mathbf{K}}) \right]. \tag{5.12}$$

The Λ limit is taken using the asymptotic scaling relation

$$\left[\chi^{\mathbf{K}}_{\mathbf{st}}(E_{\mathbf{K}})\right](\Lambda) = a_{\mathbf{st}}/\Lambda + \chi^{\mathbf{K}}_{\mathbf{st}}(E_{\mathbf{K}}). \tag{5.13}$$

On the right-hand side, $a_{\mathbf{st}}$ the slope defining the scaling with Λ and $\chi^{\mathbf{K}}_{\mathbf{st}}(E_{\mathbf{K}})$ is the value as $\Lambda \to \infty$.

One may be concerned that the scaling relation Eq. (5.13) only holds for $\chi^{\mathbf{K}}$ of the full model and not for $\tilde{\chi}^{\mathbf{K}}$ in the projected model. To show that this is not the case, we note that in the limit of an infinite number of unit cells $N \to \infty$ the overlaps $h^{\mathbf{nm}}_{\mathbf{sK}}(q)$ may be written as products of 1D overlaps $h^{nm}_{sK}(q)$ of the form

$$\begin{aligned} h^{nm}_{sK}(q) = \lim_{\ell\to\infty} \sum_{r,r'=-\ell}^{\ell} & c^{r}_{nq} c^{r'}_{mK-q} c^{r+r'+f/2\pi}_{M;sK} \\ & \times \operatorname{rect}\left(\frac{2q - K - f + 2\pi(r - r')}{2\pi\Lambda}\right), \end{aligned} \tag{5.14}$$

where f is an integer multiple of 2π which shifts $K-q$ into the BZ, $\operatorname{rect}(x)$ denotes the rectangle function, and the vectors $\mathbf{c}_{nq}$ denote the Fourier expansion of the open channel Bloch functions as

$$\psi_{nq}(x) = e^{iqx} \lim_{\ell\to\infty} \sum_{r=-\ell}^{\ell} c^{r}_{nq} e^{-2\pi irx/a} / \sqrt{Na}. \tag{5.15}$$

ℓ represents a finite Fourier cutoff used in numerics. Similarly, $\mathbf{c}_{M;sK}$ denote the Fourier coefficients of the closed channel Bloch functions. $\tilde{\chi}^{\mathbf{K}}$ differs from $\chi^{\mathbf{K}}$ in the exclusion of all terms with $n = m = 1$. However, provided that Λ is large enough to capture the support of the vectors $\mathbf{c}_{nq}$ with $n = 1$, Eq. (5.14) demonstrates that these terms are no longer functions of Λ. Thus, the scaling relation Eq. (5.13) also holds for $\tilde{\chi}^{\mathbf{K}}$. Similar arguments show that the same scaling holds for $\tilde{\chi}^{\mathbf{K}}$ when any finite number of open channel bands have been projected out.

5.4 Supplemental Material: Classification of the Two-Particle Bound States

For the simple cubic lattice we consider the Hamiltonian is invariant under reflection in any Cartesian direction: $H(\theta_{\mathcal{R}}\mathbf{x}) = H(\mathbf{x}') = H(\mathbf{x})$ where $\mathbf{x}' \equiv \theta_{\mathcal{R}}\mathbf{x}$ is related to $\mathbf{x}$ by changing the sign of all coordinates in some set $\mathcal{R}$: $x_j \to -x_j$, $j \in \mathcal{R}$. Because the generators of reflection and translation do not commute we cannot find simultaneous eigenfunctions except at high-symmetry points of the BZ. However, the fact that the Hamiltonian commutes with both operators implies that parity transformations yield relationships between degenerate sets of Bloch functions. In particular, for the given lattice potential, the invariance under the reflection symmetry generated by $\theta_{\mathcal{R}}$ implies that the Bloch functions transform as

$$\psi_{\mathbf{nq}}(\mathbf{x}') = \prod_{j\in\mathcal{R}} (-1)^{n_j+1} \psi_{\mathbf{n},\mathbf{q}'}(\mathbf{x}) , \tag{5.16}$$

where we begin indexing the bands from 1. We can thus characterize the bands according to whether they are even or odd under inversions by the triple $\mathbf{p} = (p_x, p_y, p_z)$, where $p_\nu \equiv (-1)^{n_\nu+1}$. This inversion relationship implies that the inter-channel overlaps transform as

$$h^{\mathbf{nm}}_{\mathbf{sK}'}(\mathbf{q}') = \prod_{j\in\mathcal{R}} (-1)^{n_j+m_j+s_j+1} h^{\mathbf{nm}}_{\mathbf{sK}}(\mathbf{q}) , \tag{5.17}$$

and $\chi^{\mathbf{K}}$ transforms as

$$\chi^{\mathbf{K}'}_{\mathbf{st}}(E_{\mathbf{K}}) = \prod_{j\in\mathcal{R}} (-1)^{s_j+t_j} \chi^{\mathbf{K}}_{\mathbf{st}}(E_{\mathbf{K}}) . \tag{5.18}$$

It can be proven that this transformation leaves the eigenvalues invariant, but the eigenvectors $\mathbf{R}^{\mathbf{K}}_\alpha$ transform according to

$$R^{\mathbf{K}'}_{\mathbf{s}\alpha} = \prod_{j\in\mathcal{R}} (-1)^{s_j+1} R^{\mathbf{K}}_{\mathbf{s}\alpha} . \tag{5.19}$$

For a total quasimomentum $\mathbf{K}$ all of whose components consist of either 0 or $-\pi/a$, this implies that only molecular bands which transform identically under *complete* inversions mix. Hence, at these exceptional points of the BZ, we can unambiguously determine the parity of the two-particle state $\psi_{\mathbf{K}\alpha}(\mathbf{x},\mathbf{y})$ by the components of its associated eigenvector $\mathbf{R}^{\mathbf{K}}_{\alpha}$. The parity is then chosen to depend only on the eigenstate index α by requiring that $\mathbf{R}^{\mathbf{K}}_{\alpha}$ is a smooth function of $\mathbf{K}$. This construction follows that of the 1D case studied by Kohn [16], which leads to maximally localized Wannier functions.

With this construction, there is still an undefined global phase under inversion that we can fix in the following way. The complete two-particle bound state solution is

$$\Psi_{\mathbf{K}\alpha}(\mathbf{x},\mathbf{y}) = \frac{1}{\mathcal{N}_{\mathbf{K}\alpha}}\left[\sum_{\mathbf{s}} R^{\mathbf{K}}_{\mathbf{s}\alpha}\phi_{\mathbf{sK}}(\mathbf{x})\tilde{r}(\mathbf{x}-\mathbf{y}) + \frac{g}{\sqrt{N^3a^3}}\sum_{\mathbf{nms;q}}\frac{R^{\mathbf{K}}_{\mathbf{s}\alpha}h^{\mathbf{nm}}_{\mathbf{sK}}(\mathbf{q})\psi_{\mathbf{nq}}(\mathbf{x})\psi_{\mathbf{mK-q}}(\mathbf{y})}{E^{\alpha}_{\mathbf{K}} - E^{\mathbf{K}}_{\mathbf{nm}}(\mathbf{q})}\right], \tag{5.20}$$

where $\mathcal{N}_{\mathbf{K}\alpha}$ is a normalizing factor and $\tilde{r}(\mathbf{x}-\mathbf{y})$ denotes a relative wavefunction for the closed channel which has characteristic width a/Λ. As the theory has already been regularized, we may take $\Lambda \to \infty$ with the understanding that this relative wavefunction has a probability density of 1, and forces the closed channel to contribute only at the center of mass. Because of the partitioning of Hilbert space into open and closed channels, the normalization coefficient is

$$\mathcal{N}^2_{\mathbf{K}\alpha} = 1 - \left(\frac{g}{E_R a^{3/2}}\right)^2 \mathbf{R}^{\mathbf{K}}_{\alpha}\cdot\chi'(E^{\alpha}_{\mathbf{K}}/E_R)\cdot\mathbf{R}^{\mathbf{K}}_{\alpha}\,. \tag{5.21}$$

Here $\chi'(E)$ is the derivative of χ with respect to E. Using the transformation properties under $\theta_{\mathcal{R}}$, we find

$$\Psi_{\mathbf{K}'\alpha}(\mathbf{x},\mathbf{y}) = P_{\alpha}\Psi_{\mathbf{K}\alpha}(\mathbf{x}',\mathbf{y}')\,. \tag{5.22}$$

Accordingly, we set $P_{\alpha} = \prod_{j\in\mathcal{R}} p_{\nu}$. This implies that the dressed molecular Wannier functions transform as $\mathcal{W}_{i\alpha}(\mathbf{x}',\mathbf{y}') = P_{\alpha}\mathcal{W}_{i\alpha}(\mathbf{x},\mathbf{y})$.

In Fig. 5.4 we display the bound state energies at $\mathbf{K} = 0$ for a lattice of depth $V/E_R = 12$ classified according to their parity. The red points correspond to $\mathbf{p} = (1,1,1)$, the green points to $\mathbf{p} = (1,1,-1)$ et cyc, the blue points to $\mathbf{p} = (1,-1,-1)$ et cyc, and the magenta points to $\mathbf{p} = (-1,-1,-1)$. In contrast to the continuum where there exists at most one bound state for fixed scattering length a_s, there is the possibility of several bound states for fixed a_s in the lattice due to the reduced translational symmetry. Thus, the parity and the quasimomentum are

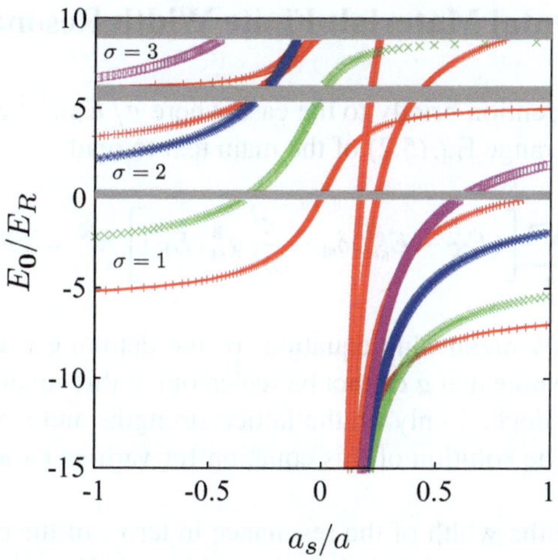

Fig. 5.4 Classification of two-particle bound states. The bound state energies at $\mathbf{K} = 0$ for a lattice of depth $V/E_R = 12$ are classified according to their parity and sheet indices. *Red* corresponds to $\mathbf{p} = (1, 1, 1)$, *green* to $\mathbf{p} = (1, 1, -1)$, *blue* to $\mathbf{p} = (1, -1, -1)$, and *magenta* to $\mathbf{p} = (-1, -1, -1)$. In addition, higher molecular bands for the $\mathbf{p} = (1, 1, 1)$ level are shown. These give rise to weak scattering resonances with the lowest open channel band for $a_s/a > 0$ and avoided crossings in the higher sheets for $a_s/a < 0$

not sufficient to completely describe the states. For a fixed s-wave scattering length a_s, we provide two other indices which we call the *sheet index* $\sigma \in \{1, 2, \ldots, \infty\}$ and the *molecular band index* $s \in \{1, 2, \ldots, \infty\}$. The sheet index labels the open channel two-particle scattering bands, with the convention that the first sheet lies below the first band, the second sheet lies between the first and second bands, etc. as indicated in the figure. The open channel scattering bands are denoted by solid grey stripes. The molecular band index labels eigenstates which have the same parity and sheet indices but differ in energy. The number of molecular bands obtained is restricted by the number of closed channel bands used to construct $\chi^{\mathbf{K}}$. Let us define m to be the maximum value of the closed channel band index along any Cartesian direction. In Fig. 5.4, the solid red line corresponds to the completely even parity state computed with $m = 2$ and the red points correspond to the complete even parity states computed with $m = 3$. The choice $m = 2$ captures the physics well near the lowest open channel scattering band. For $a_s/a > 0$, the higher molecular bands cross the lowest open channel scattering continuum at narrow ranges of a_s/a, leading to weak scattering resonances. For $a_s/a < 0$, the higher molecular bands are present at extended ranges of a_s/a, and avoided crossings between these molecular bands can lead to differences with lower m computations, see, e.g., the third sheet near $a_s/a = -0.2$.

5.5 Supplemental Material: Finite Width Resonances

We now turn our attention briefly to the case where $g/E_R a^{3/2}$ and ν/E_R are finite. In this case we rearrange Eq. (5.2) of the main text to read

$$\sum_{\mathbf{t}} \left[(E_{\mathbf{K}} - E^{M}_{\mathbf{sK}})\delta_{\mathbf{st}} - \frac{g^2}{a^3}\chi^{\mathbf{K}}_{\mathbf{st}}(E_{\mathbf{K}}) \right] R^{\mathbf{K}}_{\mathbf{t}} = \nu R^{\mathbf{K}}_{\mathbf{s}} \tag{5.23}$$

which is an ordinary eigenvalue equation for the detuning ν when $E_{\mathbf{K}}$ and g are treated as fixed. We note that g cannot be scaled out of this equation as the molecular band energies $E^{M}_{\mathbf{sK}}$ depend only on the lattice strengths and masses and not on the resonance width. The solution of this equation for various r_B and $\mathbf{K} = 0$ is shown in Fig. 5.5.

We characterize the width of the resonance in terms of the experimentally measurable effective range r_B which defines the width as $g/E_R a^{3/2} = \sqrt{16a/\pi^3 r_B}$. For narrow resonances with large r_B only the lowest resonance can be seen, and a_s/a is greater than 1, corresponding to strong interactions, only in a very narrow energy range. As the resonance becomes broader the energy range over which the system is strongly interacting widens, and we begin to see resonant behavior near higher scattering continua. Additionally, the positions of the narrow resonances are shifted downwards in energy with respect to the broad resonances, eventually becoming

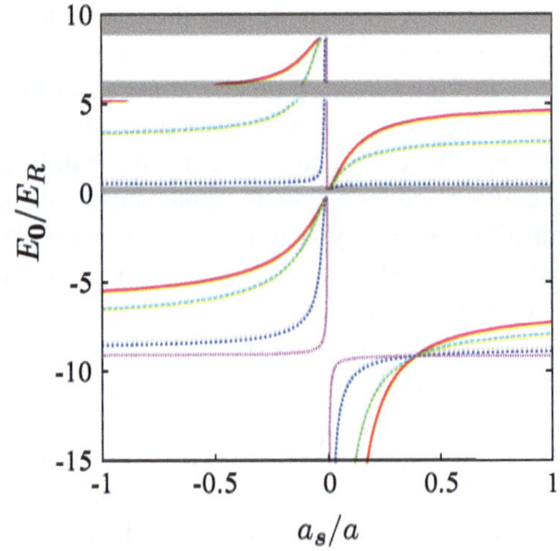

Fig. 5.5 Bound state energies for finite width resonances. Shown are the lowest energy bound state energies for $\mathbf{p} = (1, 1, 1)$ at $\mathbf{K} = 0$ in a strong optical lattice with $V/E_R = 12$. The *red solid line* is $r_B/a = 0.01$, the *green dashed line* is $r_B/a = 0.1$, the *blue dotted line* is $r_B/a = 1$, and the *magenta short-dashed line* is $r_B/a = 10$. For narrow resonances (large r_B) the divergence of a_s is sharply pronounced around a narrow energy range, and is shifted downwards from the broad resonance value, compare Fig. 5.4

the free molecular band energies. We note that the broadest resonance shown is in fact narrower than typical broad resonances found in the experimentally relevant ultracold atomic systems, but differs from the infinitely broad resonance results by at most a few percent. This justifies our use of the infinitely broad resonance limit in the main text.

References

1. Eagles, D.M.: Phys. Rev. **186**, 456 (1969); Chen, Q., Stajic, J., Tan, S., Levin, K.: Phys. Rep. **412**, 1 (2005)
2. Ranninger, J., Robin, J.M.: Physica C **253**, 279 (1995)
3. Kohl, M., et al.: Phys. Rev. Lett. **94**, 080403 (2005)
4. Adams, A., Carr, L.D., Schaefer, T., Steinberg, P., Thomas, J.E.: New J. Phys. **14**, 115009 (2012)
5. Cafolla, A.A., Schnatterly, S.E., Tarrio, C.: Phys. Rev. Lett. **55**, 2818 (1985)
6. Trotzky, S., et al.: Nat. Phys. **6**, 998 (2010)
7. Esslinger, T.: Ann. Rev. Condens. Matter Phys. **1**, 129 (2010)
8. Büchler, H.P.: Phys. Rev. Lett. **104**, 090402 (2010); Büchler, H.P.: Phys. Rev. Lett. **108**, 069903 (2012)
9. Carr, L.D., Holland, M.J.: Phys. Rev. A **72**, 033622(R) (2005); Dickerscheid, D.B.M., Khawaja, U.A., van Oosten, D., Stoof, H.T.C.: Phys. Rev. A **71**, 043604 (2005); Diener, R.B., Ho, T.-L.: Phys. Rev. Lett. **96**, 010402 (2006); Duan, L.-M.: Phys. Rev. Lett. **95**, 243202 (2005); Duan, L.-M.: EPL **81**, 20001 (2008); Cui, X., Wang, Y., Zhou, F.: Phys. Rev. Lett. **104**, 153201 (2010)
10. Lancaster, P.: Lambda-Matrices and Vibrating Systems. Pergamon Press, Oxford (1966)
11. Wall, M.L., Carr, L.D.: Phys. Rev. A **87**, 033601 (2013)
12. von Stecher, J., Gurarie, V., Radzihovsky, L., Rey, A.M.: Phys. Rev. Lett. **106**, 235301 (2011)
13. Jaksch, D., et al.: Phys. Rev. Lett. **81**, 3108 (1998)
14. Kokkelmans, S.J.J.M.F., Milstein, J.N., Chiofalo, M.L., Walser, R., Holland, M.J.: Phys. Rev. A **65**, 053617 (2002)
15. Büchler, H.P.: Phys. Rev. Lett. **108**, 069903 (2012)
16. Kohn, W.: Phys. Rev. **115**, 809 (1959)

Part IV
Matrix Product States

Chapter 6
Matrix Product States: Foundations

The bulk of the numerical results for many-body systems contained in this thesis are obtained by variational algorithms on a class of states known as matrix product states (MPSs). The theory of MPSs as a variational ansatz for eigenstates and dynamics of general finite-sized one-dimensional (1D) systems is expounded at length in Chap. 7. In addition, the definitive review of MPSs at the time of the writing of this thesis is [1]. In this chapter, we aim instead to give intuitive notions of what MPSs are and why they are useful as variational ansätze.

6.1 Bird's Eye View of MPSs

The quantum many-body problem is in principle completely solved given the microscopic degrees of freedom and their interactions, as the relevant Schrödinger equation is known once the Hamiltonian has been specified. However, in practice, the Hilbert space of a typical many-body ensemble grows exponentially with the number of constituents of the system, and so an algorithm which explicitly forms a matrix representation of the Hamiltonian and diagonalizes it to find eigenstates is limited to very small systems. Even when all symmetries of the Hamiltonian have been carefully accounted for and sparse diagonalization routines such as the Lanczos [2] or Davidson [3] algorithms are used to find only extremal eigenstates, cutting-edge *exact diagonalization* is limited to approximately 40 two-component spins or a 20-site fermionic Hubbard model on the square lattice at half filling.

The full Hilbert space of a many-body system is in fact too big for physical discussions, as can be shown by the following argument [4]. Consider a time-dependent Hamiltonian acting on N particles

$$\hat{H}(t) = \sum_{X \in \{1,2,\dots,N\}} \hat{H}_X(t)\,, \tag{6.1}$$

M.L. Wall, *Quantum Many-Body Physics of Ultracold Molecules in Optical Lattices*, Springer Theses, DOI 10.1007/978-3-319-14252-4_6

where X labels subsets of the N particles. We restrict each term in the summation to have bounded norm $\|\hat{H}_X(t)\| \leq E$ and to act on no more than k particles at a time, where k is independent of the number of particles. An operator acting on k particles simultaneously is said to be *k-local*. Note that we make no assumptions about the range of the Hamiltonian, and so our arguments apply to the long-ranged Hamiltonians relevant for molecules. We now wish to expand the full time-ordered propagator of this Hamiltonian in terms of a series of at most k-local unitary operators. This can be accomplished by using the time-ordered Trotter expansion for two non-commuting terms $\hat{H}_1(t)$ and $\hat{H}_2(t)$ [5]

$$\hat{U}(t, t+\delta t) = \mathcal{T}\exp\left[-i\int_t^{t+\delta t} dt' \left(\hat{H}_1(t') + \hat{H}_2(t')\right)\right], \tag{6.2}$$

$$\begin{aligned} &\approx \hat{U}_{\text{Trotter}}(t, t+\delta t) \\ &= \mathcal{T}\exp\left[-i\int_t^{t+\delta t} dt'\, \hat{H}_1(t')\right]\mathcal{T}\exp\left[-i\int_t^{t+\delta t} dt'\, \hat{H}_2(t')\right], \end{aligned} \tag{6.3}$$

which is accurate in the operator norm as

$$\|\hat{U}(t, t+\delta t) - \hat{U}_{\text{Trotter}}(t, t+\delta t)\| \leq c_{\text{Trotter}}(\delta t)^2, \tag{6.4}$$

$$c_{\text{Trotter}} = \frac{1}{(\delta t)^2}\int_t^{t+\delta t} dt' \int_t^{t'} dt'' \left\| \left[\hat{H}_1(t''), \hat{H}_2(t')\right] \right\|. \tag{6.5}$$

Here $\mathcal{T}\exp[\bullet]$ is the time-ordered exponential [6]. We note that these bounds do not depend on the smoothness of the Hamiltonian, and so are also valid for non-analytic time dependence. Iterating this expansion for two operators $\log_2(L)$ times, where $L \in \text{poly}(N)$ is the number of k-particle terms in Eq. (6.1), we may write our full propagator as a product of time-ordered k-local unitaries with the total error bounded by

$$\frac{1}{2}\left|\max_X \sup_{0\leq t'\leq t} \|\hat{H}_X(t')\|\right|^2 L^2(\delta t)^2. \tag{6.6}$$

The notation $L \in \text{poly}(N)$ denotes that L is a polynomial function of N. This error may be made as small as desired by taking $(\delta t)^{-1} \in \text{poly}(N)$. The number of k-body unitaries which is required to bound this error by ϵ is inversely proportional to ϵ and polynomial in both t, the final time desired, and L. One can replace these polynomially many k-body unitaries with a discrete set of fixed one and two-body operators which is also polynomial in L due to the Solovay–Kitaev theorem [7–9].

From the above, we conclude that any state which can be produced from a given state $|0\rangle$ under evolution by an arbitrary time-dependent Hamiltonian can be well approximated by the state $|0\rangle$ acted on by a polynomial-size quantum circuit with a fixed set of discrete operations. It is known [8] that such circuits can only reach an exponentially small subset of states, and so we come to the conclusion that physically relevant states, those that can be reached from a reference state to a desired accuracy from any time-dependent Hamiltonian in a time polynomial in the size of the system, form an exponentially small subset of the complete Hilbert space. We note that this general argument is independent of dimension, and can be extended from pure states to open quantum systems described by a k-local time-dependent Liouvillian [10]. Hence, the argument also applies at finite temperature. The emphasis of the many-body problem now shifts from, "How do we deal with the largeness of Hilbert space?" to, "How do we parameterize the set of states relevant to our particular many-body model?"

The answer to this latter question can be provided assuming that our Hamiltonian has finite-range interactions and a gap to excitations[1] and we consider our system at zero temperature. The ground states of such systems obey an *area law* [11–15], which is to say that the von Neumann entropy of entanglement of the ground state in a region A,

$$S_{\mathrm{vN}}(A) = -\mathrm{Tr}\left[\hat{\rho}_A \log \hat{\rho}_A\right], \tag{6.7}$$

where $\hat{\rho}_A$ is the reduced density operator describing region A, scales only with the border of region A and not with its volume. For 1D systems, the presence of an area law implies that the entanglement between any two complementary subregions is independent of their size. This is in contrast to a random state in Hilbert space, for which the entropy of entanglement is extensive. The existence of area laws has been proven rigorously in 1D for spin-1/2 systems obeying the above constraints by Hastings [16], and has been observed to hold for the vast majority of other known systems satisfying our hypotheses regardless of dimension.[2] At finite temperature a condition similar to the area law can be made rigorous in any dimension as the *mutual information* between a region A and its complement $\bar{A}$ in a system at thermal equilibrium,

$$M\left(A;\bar{A}\right) = S_{\mathrm{vN}}(A) + S_{\mathrm{vN}}\left(\bar{A}\right) - S_{\mathrm{vN}}\left(A+\bar{A}\right), \tag{6.8}$$

[1]By which we mean a *spectral gap* in the thermodynamic limit, $E_1 - E_0 > 0$ where E_1 is the energy of the first excited state and E_0 is the energy of the ground state. Finite-sized systems whose infinite counterparts are gapless typically have a gap which vanishes as an inverse polynomial in the system volume.

[2]For a counterexample in a non-translationally invariant 1D chain, see [17]. Here, a variant of the Dasgupta–Ma–Fisher renormalization group procedure for random spin chains is used to explicitly construct a spin chain satisfying a volume law. Additionally, as this system is not translationally invariant, the volume law depends crucially on how the system is divided. In fact, there exist divisions for which the entropy of entanglement is identically zero.

satisfies an area law whose bound is inversely proportional to the temperature [18]. Here the density matrix used to obtain $S_{\mathrm{vN}}(X)$, $\hat{\rho}_X$, is the reduced density matrix describing region X in a system at thermal equilibrium.

A related concept in physics which gives an enlightening alternative perspective on area laws for entanglement entropy is the notion of area laws for black holes [19]. Here, it is the thermodynamic entropy of a black hole that scales with the area enclosing the black hole rather than its volume [20]. As the entropy of any volume of spacetime is bounded by the entropy of a black hole that fits inside of that volume [21], we have that entropy in gravity is always sub-extensive. This fact coupled with *holographic dualities* [19] mapping the properties of $D+1$-dimensional bulk spacetime to a D-dimensional quantum field theory on its boundary, which are being proved on a case-by-case basis, provide strong evidence that area laws are ubiquitous in strongly correlated quantum systems. A similar holographic principle applies for MPSs and their generalizations, as expectation values of operators may be determined in terms of the dynamics of a dissipative boundary theory [22, 23].

A notable class of exceptions to our hypotheses are systems at the critical point of a quantum phase transition in which the gap vanishes. For systems of free fermions in D dimensions, the corrections to the area law are logarithmic in the linear dimension of region A, L_A, [24, 25]

$$S_{\mathrm{vN}}(A) \sim L_A^{D-1} \log L_A, \tag{6.9}$$

while area laws in bosonic systems in dimensions $D \geq 2$ appear to be insensitive to criticality [11, 26–28]. For 1D systems whose universal critical theory is a conformal field theory [29] the logarithmic corrections to the area law are made precise by the *Calabrese–Cardy formula* [12, 14, 30]

$$S_{\mathrm{vN}}(A) \sim \frac{c}{6} \log L_A. \tag{6.10}$$

Here c is the *central charge* of the conformal field theory [29] which, in an intuitive picture, counts the number of universal bosonic degrees of freedom.[3] For example, the central charge of a free bosonic system is 1 and a free fermionic system has $c = 1/2$, corresponding to "half a boson." The above results suggest that a reasonable class of states to be used for variational studies of 1D systems should satisfy an area law, and possibly be able to handle weak logarithmic violations efficiently.

A construction of a 1D state which obeys an area law is to take a chain of sites and replace each site with a pair of χ-level systems which are maximally entangled with their neighbors [32]. This construction is shown in Fig. 6.1. If we are now to isolate any region A such as that enclosed by the red box in Fig. 6.1b, the entanglement of this region with the remainder of the system is generated only by the entanglement

[3] Rigorously, the central charge is determined by the coefficient of the anomalous term in the commutator of the energy-momentum tensor at two different positions [31]. Hence, c describes the behavior of a conformally invariant system when a macroscopic length scale is introduced.

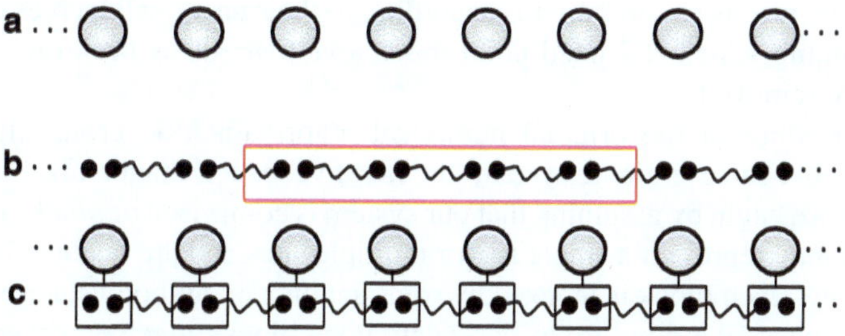

Fig. 6.1 Schematic of the MPS construction. (**a**) A 1D system comprised of local Hilbert spaces (*spheres*) arranged in a regular pattern. (**b**) Each Hilbert space is replaced by two fictitious maximally entangled χ-level systems, where the *wiggly lines* indicate which pairs are maximally entangled. Partitioning the system as with the *red box*, entanglement of the *boxed region* with its complement occurs only at the boundary. (**c**) The maximally entangled fictitious systems are projected to produce an entangled state in the physical degrees of freedom satisfying an area law

between the pairs of χ-level systems at the boundaries. This boundary entanglement scales as $\log \chi$. These χ-level subsystems thus provide a fictitious set of states which can be manipulated such as to produce the desired entanglement structure in the physical degrees of freedom. We manipulate these fictitious systems by projecting from this fictitious Hilbert space onto the physical Hilbert space via a set of isometric tensors. These tensors form the parameters of an MPS. Finally, we note that the relationship $S_{\mathrm{vN}} \sim \log \chi$ implies that χ, the Hilbert space dimension of the fictitious system, is exponential in the entanglement entropy S_{vN}.[4] For systems with weak logarithmic violations of the area law obeying Eq. (6.10), this gives that χ is polynomially related to the system size, with polynomial degree depending on the central charge [33]. Hence, any algorithm which is polynomial in χ represents a method which is at most polynomial in the system size, an exponential improvement over exact diagonalization![5]

An enlightening alternative view of MPSs is that they are states which are the result of a real-space *renormalization group* (RG) iteration [35]. The essential idea of the renormalization group is to coarse-grain a system by integrating out the irrelevant degrees of freedom. As coarse graining is iteratively applied, only the degrees of freedom which describe the macroscopic behavior of the system emerge, and we can characterize these macroscopic degrees of freedom by the *universality classes* which represent fixed points of the RG iteration. Models whose microscopic details differ greatly may fall into the same universality class, and hence have the same macroscopic behavior. The renormalization group provides

[4]That is to say, the χ that is required to accurately reflect the entanglement structure grows exponentially with the von Neumann entropy.

[5]Strictly speaking, this is true only for fixed error. If we require that the error be bounded by an inverse polynomial in the system size, finding a ground state with the given representation is still very difficult [34].

a quantitative means to sort systems according to their universal degrees of freedom via determining which RG fixed point they move towards as irrelevant degrees of freedom are removed.

The procedure of the original numerical renormalization group algorithm of Wilson is to coarse-grain a system by iterative exact diagonalization. In one dimension, we begin by assuming that our system is comprised of *blocks* $B^{(0)}$, where each block may represent a single site or multiple sites and are taken to be identical for simplicity. Each block is indexed by some set of χ many-body states which span its Hilbert space. We now group two contiguous blocks together, diagonalize the Hamiltonian of this two-block system, and keep only the χ states which are lowest in energy. The transformation to the new configuration where two of the blocks $B^{(0)}$ are combined into a single *superblock* $B^{(1)} = B^{(0)} \otimes B^{(0)}$ may be represented by an isometric matrix A[6] mapping from $B^{(0)} \otimes B^{(0)} \rightarrow B^{(1)}$. We now combine $B^{(1)}$ and one of its neighboring blocks $B^{(0)}$ together, diagonalize the Hamiltonian in this space, and keep only the low energy states again, resulting in an isometric tensor mapping from $B^{(1)} \otimes B^{(0)} \rightarrow B^{(2)}$. This outlines the general structure of the iteration.

Let us now assume that out initial blocks are the two leftmost sites of a 1D lattice, and denote the isometric tensors constructed at iteration n as $A^{[n]i_n}_{\alpha\beta}|\alpha\rangle|i\rangle\langle\beta|$, where $|\alpha\rangle$ are the states spanning $B^{(n-1)}$, $|i\rangle$ are the states spanning $B^{(0)}$, and $|\beta\rangle$ are the states spanning $B^{(n)}$. Furthermore, let us take open boundaries on a chain of L sites, which amounts to the left basis of $A^{[1]i_1}$ and the right basis of $A^{[L]i_L}$ being one-dimensional.[7] Then, the many-body state resulting at iteration L may be written as

$$|\psi\rangle = \sum_{i_1 \dots i_L} A^{[1]i_1} \dots A^{[L]i_L} |i_1 \dots i_L\rangle \,, \tag{6.11}$$

which, as we shall see, is the form of an MPS.

This procedure produced excellent results for the Kondo problem [35], but fails miserably for the toy model of a single particle on a 1D chain! The reasons for this failure were investigated by White [36], and led him to the *density-matrix renormalization group* (DMRG) algorithm, in which the relevant states which are kept after diagonalization of the superblock Hamiltonian are not those with the lowest energy, but those which have the largest eigenvalues in the block reduced density matrix. Examining Eq. (6.7), we can see that this corresponds intuitively to the states which are most strongly entangled to the "environment" formed by all

[6]By isometric, we mean that this matrix has orthonormal rows. If it were square, it would be unitary, but we are transforming from a χ^2 dimensional space to a χ dimensional space, and so only the rows are orthonormal.

[7]That is, at the first iteration we map from the product of the vacuum and a block $B^{(0)}$ to a new set of states which is of course identical to the block $B^{(0)}$. A similar comment applies to the last iteration.

previous iterations.[8] The more precise statement we can make is that by keeping the states which have largest eigenvalues in the reduced density matrix we minimize the 2-norm distance between the system in the truncated Hilbert space and the true ground state with the given environment, which is borne out in the *Schmidt decomposition* [8, 38]

$$|\psi\rangle = \sum_{\alpha} \lambda_{\alpha}|A_{\alpha}\rangle|B_{\alpha}\rangle. \tag{6.12}$$

The Schmidt decomposition represents a general quantum state $|\psi\rangle$ in terms of orthonormal bases $|A\rangle$ and $|B\rangle$ for two complementary subsystems and the eigenvalues of the reduced density matrix $\hat{\rho}_A$, $\{\lambda_{\alpha}^2\}$. With this change to the original numerical renormalization group procedure, DMRG not only overcomes the failure for the single-particle case, but turns this real-space renormalization group procedure into a method of unparalleled power for strongly-correlated 1D systems.

After the development of DMRG, it was realized that DMRG can be formulated as a variational method in terms of MPSs [39, 40]. While this does not lead to any significant numerical improvement in the algorithm to find the ground state of 1D Hamiltonians [1], other variational algorithms, for example to find excited states or perform generic time evolution, benefit greatly from this observation. The reason is that in DMRG all states involved in a calculation are represented as a single MPS, which requires that they share common bases in the matrix product. Hence, representing states together as an MPS often requires vastly more resources than representing each state separately as an MPS for a fixed error. These issues are discussed at length in Chap. 7. Hence, in this thesis, we focus on the explicit formulation of variational algorithms within the class of MPSs rather than the implicit MPS representation used in DMRG.

6.2 Definitions

We now move to explicit definitions of the mathematical objects which are used when discussing MPSs. We define a *tensor* as a map from a product of Hilbert spaces to the complex numbers

$$T : \mathbb{H}_1 \otimes \mathbb{H}_2 \otimes \cdots \otimes \mathbb{H}_r \to \mathbb{C}. \tag{6.13}$$

[8]Note that, strictly speaking, DMRG does not maximize entanglement due to a renormalization of the density-matrix spectra induced by truncation [37]. However, for most physical systems this intuition does not cause any pitfalls.

Here r is the *rank*[9] of the tensor. If we evaluate the elements of the tensor T in a fixed basis $\{|i_j\rangle\}$ for each Hilbert space $\mathbb{H}_j$, then equivalent information is carried in the multidimensional array $T_{i_1\dots i_r}$. We will also refer to this multidimensional array as a tensor. The information carried in a tensor does not change if we change the order in which its indices appear. We will call such a generalized transposition a *permutation* of the tensor. As an example, the permutations of the rank-3 tensor T are

$$T_{ijk} = \left[T'\right]_{kij} = \left[T''\right]_{jki} = \left[T'''\right]_{jik} = \left[T''''\right]_{kji} = \left[T'''''\right]_{ikj} . \tag{6.14}$$

Here, the primes indicate that the tensor differs from its unprimed counterpart only by a permutation of indices. Similarly, by combining two such indices together using the Kronecker product we can define an equivalent tensor of lower rank, a process we call *index fusion*. We denote the Kronecker product of two indices a and b using parentheses as (ab), and a representation is provided by

$$(ab) = (a-1)\, d_b + b, \tag{6.15}$$

where d_b is the dimension of $\mathbb{H}_b$ and a and b are both indexed starting from 1. An example of fusion is

$$T_{ijk} = \left[T'\right]_{i(jk)} . \tag{6.16}$$

Here, T is a rank-3 tensor of dimension $d_i \times d_j \times d_k$ and T' is a matrix of dimension $d_i \times d_j d_k$. The inverse operation of fusion, which involves creating a tensor of higher rank by splitting a composite index, we refer to as *index splitting*.

Just as permutations generalize the notion of matrix transposition, *tensor contraction* generalizes the notion of matrix multiplication. In a contraction of two tensors A and B some set of indices $\mathbf{c}_A$ and $\mathbf{c}_B$ which describe a common Hilbert space are summed, and the resulting tensor C consists of products of the elements of A and B as

$$C_{\bar{\mathbf{c}}_A\bar{\mathbf{c}}_B} = \sum_{\mathbf{c}} A_{\bar{\mathbf{c}}_A\mathbf{c}} B_{\mathbf{c}\bar{\mathbf{c}}_B} . \tag{6.17}$$

Here $\bar{\mathbf{c}}_B$ denotes the indices of A which are not contracted and likewise for $\bar{\mathbf{c}}_B$. The rank of C is $r_A + r_B - 2n_c$, where n_c is the number of indices contracted (i.e., the number of indices in $\mathbf{c}$) and r_A and r_B are the ranks of A and B, respectively. In writing expression Eq. (6.17) we have permuted all of the indices $\mathbf{c}_A$ to be contracted

[9]This should not be confused with the rank of a matrix, which is the number of nonzero singular values. We shall avoid confusion in this text by referring to the number of nonzero singular values as the *matrix rank* or rank of a matrix and referring to the definition given here as the *tensor rank*, the rank of a tensor, or a rank-r tensor.

to the furthest right position in A and the indices $\mathbf{c}_B$ to the furthest leftmost position in B for notational simplicity. If we were also to fuse the elements in $\bar{\mathbf{c}}_A$, $\mathbf{c}$, and $\bar{\mathbf{c}}_B$ together, then we would recognize Eq. (6.17) as a matrix–matrix multiplication. This is a valuable insight, as matrix–matrix multiplication routines such as DGEMM[10] in BLAS[11] [41] have been highly optimized and this lends efficiency to tensor contraction algorithms.

At this stage, it is advantageous to develop a graphical notation for tensors and their operations [42]. A tensor is represented graphically by a point with lines extending upwards from it. The number of lines is equal to the rank of the tensor. The order of the indices from left to right is the same as the ordering of lines from left to right. A contraction of two tensors is represented by a line connecting two points. Finally, the complex conjugate of a tensor is denoted by a point with lines extending downwards. Some basic tensor operations are shown in graphical notation in Fig. 6.2.

Following a similar line of reasoning as for contractions above, we may also *decompose* tensors into contractions of tensors using permutation, fusion, and any of the well-known matrix decompositions such as the singular value decomposition (SVD) or the QR decomposition. For example, a rank-3 tensor T can be factorized as

$$T_{ijk} = \sum_l U_{(ij)l} S_l V_{lk}, \tag{6.18}$$

where U and V are unitary and S is a positive semidefinite real vector. Such decompositions are of great use in enforcing canonical forms on MPSs, see Sect. 6.3.

A *tensor network* is now defined as a set of tensors whose indices are connected in a network pattern, see Fig. 6.3. Let us consider that some set of the network's indices $\mathbf{c}$ are contracted over, and the complement $\bar{\mathbf{c}}$ remain uncontracted. Then, this network is a decomposition of some tensor $T_{\bar{\mathbf{c}}}$. The basic idea of tensor network

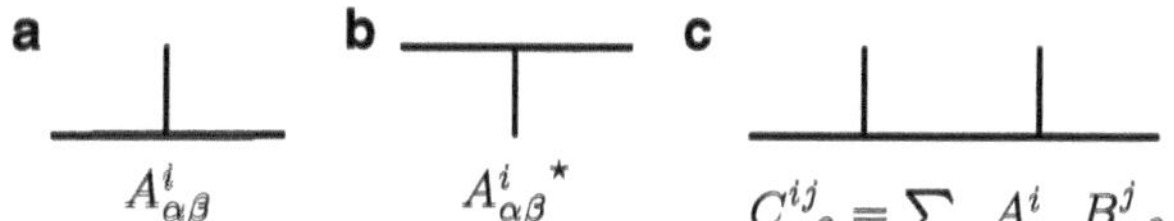

Fig. 6.2 Examples of basic tensor operations in diagrammatic notation. (**a**) A rank-3 tensor. (**b**) The conjugate of a rank-3 tensor. (**c**) The contraction of two rank-3 tensors over a single index produces a rank-4 tensor

[10] Double precision General Matrix Multiply.

[11] Basic Linear Algebra Subprograms, a large collection of numerical routines which were designed to take advantage of the cache structure of modern computers. Using BLAS routines versus naive loops for contractions leads to speedups often of a factor of 4 or more, even when using aggressively optimizing compilers.

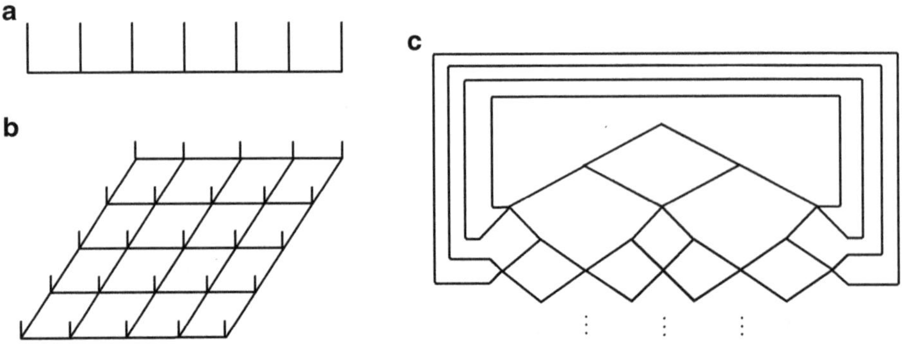

Fig. 6.3 Examples of tensor networks. (**a**) An MPS with seven sites and open boundary conditions (OBC). (**b**) A square lattice PEPS with 25 sites and OBC. (**c**) Two-levels of a 1D MERA with periodic boundary conditions. This network consists of alternating rows of rank-3 and rank-4 tensors

algorithms utilizing MPSs and their higher-dimensional generalizations such as projected entangled-pair states (PEPS) [43, 44] and the multiscale entanglement renormalization algorithm (MERA) [45, 46] are to represent the high-rank tensor $c_{i_1 \ldots i_L}$ encoding a many-body wavefunction in a Fock basis,

$$|\psi\rangle = \sum_{i_1 \ldots i_L} c_{i_1 \ldots i_L} |i_1 \ldots i_L\rangle, \tag{6.19}$$

as a tensor network with tensors of small rank. We set the convention in the remainder of this chapter that indices which are contracted over in the tensor network decomposition will be denoted by Greek indices, and indices which are left uncontracted will be denoted by Roman indices. The former type of index will be referred to as a bond index, and the latter as a physical index.

In particular, an MPS imposes a one-dimensional topology on the tensor network such that all the tensors appearing in the decomposition are rank-3. The resulting decomposition has the structure shown in Fig. 6.3a. Explicitly, an MPS may be written in the form

$$|\psi_{\mathrm{MPS}}\rangle = \sum_{i_1,\ldots i_L=1}^{d} \mathrm{Tr}\left(A^{[1]i_1} \ldots A^{[L]i_L}\right) |i_1 \ldots i_L\rangle. \tag{6.20}$$

Here, $i_1 \ldots i_L$ label the L distinct sites, each of which contains a d-dimensional Hilbert space. We will call d the *local dimension*. The superscript index in brackets $[j]$ denotes that this is the tensor of the jth site, as these tensors are not all the same in general. Finally, the trace effectively sums over the first and last dimensions of $A^{[1]i_1}$ and $A^{[L]i_L}$ concurrently, and is necessary only for periodic boundary conditions (PBC) where these dimensions are greater than 1. Obscured within the matrix product of Eq. (6.20) is the size of the matrix $A^{[j]i_j}$ formed from the tensor $A^{[j]}$ with its physical index held constant. We will refer to the left and right dimensions of this matrix as χ_j and χ_{j+1}, and the maximum value of χ_j for any tensor, the *bond dimension*, will be denoted as χ. The bond dimension is the parameter which determines the efficiency of an MPS simulation, and also its dominant computational scaling. From the relation $S_{\mathrm{vN}} \leq \log \chi$, we also have that χ represents an entanglement cutoff for MPSs. The scaling of variational MPS algorithms is discussed in Chap. 7.

In what follows, we consider MPSs with open boundary conditions (OBC) unless explicitly indicated otherwise. Details on algorithms for PBC are provided in Chap. 9. The reasoning for using OBC is twofold. First, finding the normalized eigenvector corresponding to the minimum eigenvalue of a Hermitian operator $\hat{Q}$ may be expressed as the minimization of the Rayleigh quotient [2]

$$\min_{|\psi\rangle} \langle\psi|\hat{Q}|\psi\rangle/\langle\psi|\psi\rangle. \tag{6.21}$$

In variational MPS algorithms this quotient is minimized locally by holding all tensors except for some subnetwork $\mathcal{A}$ fixed and minimizing the Rayleigh quotient with respect to the parameters of $\mathcal{A}$. The minimization problem for the subnetwork $\mathcal{A}$ becomes a generalized eigenvalue problem

$$\mathcal{Q}_{\mathrm{eff}}\mathcal{A} = \lambda\mathcal{N}\mathcal{A}, \tag{6.22}$$

where $\mathcal{Q}_{\mathrm{eff}}$ is the action of the operator $\hat{Q}$ on the subnetwork to be optimized with the rest of the network held fixed, and $\mathcal{N}$ is the action of the identity on the subnetwork to be optimized with the rest of the network held fixed. There exists a canonical form for MPSs with OBC such that $\mathcal{N}$ is the identity, and the generalized eigenvalue problem becomes an eigenvalue problem. For PBC this operator cannot generally be made the identity, and may be singular or poorly conditioned, leading to numerical instabilities and a reduction in accuracy.

The second reason for avoiding PBC is deeper. When we divide a system with PBC into two contiguous regions, one of length L and one of length ℓ, generally the two ends of subsystem L which surround that of subsystem ℓ are correlated. This is to be contrasted with OBC, where an arbitrary partition creates an environment for ℓ which is uncorrelated. This idea is demonstrated in Fig. 6.4. A consequence of this is that the entanglement entropy of a critical system with PBC grows twice as fast as one with OBC, that is

$$S_{\mathrm{PBC}}(A) \sim \frac{c}{3}\log L_A. \tag{6.23}$$

This should be compared with Eq. (6.10) for OBC. Hence, the corresponding bond dimension for PBC grows as the square of the bond dimension for OBC. This is sometimes referred to as $\mathcal{O}\left(\chi^6\right)$ scaling, as introducing a fictitious OBC system with

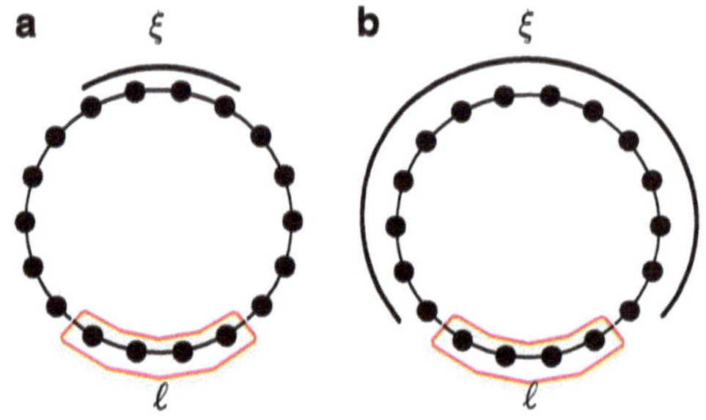

Fig. 6.4 Correlation structure of PBC states. (**a**) For a state with a correlation length $\xi \ll L$, the two ends of subregion L are weakly correlated and the entanglement of L with ℓ is constant, as in the OBC case. (**b**) As the correlation length grows comparable to L, the two ends of subregion L are strongly correlated and give rise to twice as much entanglement between L and ℓ as in the case of OBC

a long-range interaction between the first and last sites will require a bond dimension χ_{PBC} that scales as the square of the bond dimension χ_{OBC} in the absence of the long-range interaction to represent the system with the same level of accuracy. That is, the system with the long-range interaction will require $\mathcal{O}\left(\chi_{\text{PBC}}^3\right) = \mathcal{O}\left(\chi_{\text{OBC}}^6\right)$ time for a fixed accuracy. This scaling can be brought down to $\mathcal{O}\left(\chi^5\right)$ by introducing an MPS which has the correct entanglement structure [32],[12] and recent efforts [47, 48] have shown how to reduce this scaling to $\mathcal{O}\left(p\chi^3\right)$, where p is the number of relevant correlation lengths of the transfer operator, see Sect. 6.5.

6.3 Canonical Forms for MPSs

The matrix product structure of an MPS implies that we can insert any invertible matrix X and its inverse between any two matrices appearing in Eq. (6.20) without affecting the state $|\psi_{\text{MPS}}\rangle$. Hence, the MPS decomposition is highly non-unique. We shall now describe canonical forms for MPSs which remove this non-uniqueness.

The first canonical form requires that a given tensor A satisfies

$$\sum_i A^{i\dagger} A^i = I. \tag{6.24}$$

We say that a tensor satisfying Eq. (6.24) is in *left-canonical* form, or it is left-canonical for succinctness. We can enforce left-canonical form on a particular tensor via fusing the leftmost two indices and performing a SVD

[12] That is, the algorithm scales as $\mathcal{O}\left(\chi^5\right)$ but $\chi_{\text{PBC}} = \chi_{\text{OBC}}$ for a fixed error in the improved ansatz.

$$A_{(\alpha i)\beta} = A^{[\ell]i}_{\alpha\beta}, \tag{6.25}$$

$$\sum_{\gamma} U_{(\alpha i)\gamma} S_{\gamma} V^{\dagger}_{\gamma\beta} = A_{(\alpha i)\beta}. \tag{6.26}$$

Because the matrix U returned from the SVD is unitary, Eq. (6.24) is satisfied by making the replacement

$$A^{[\ell]i}_{\alpha\gamma} = U_{(\alpha i)\gamma}. \tag{6.27}$$

In order for the total state to remain unchanged, we must contract the remaining information returned from the SVD into the neighboring tensor as

$$A^{[\ell+1]i}_{\gamma\eta} = \sum_{\beta} S_{\gamma} V^{\dagger}_{\gamma\beta} A^{[\ell+1]i}_{\beta\eta}. \tag{6.28}$$

Hence, the procedure of bringing a tensor into canonical form affects two tensors at a time. By carrying out this recursion all the way from the left (open) boundary to the right boundary, we end up with an MPS in which each tensor satisfies Eq. (6.24) and a 1×1 matrix whose trace is the norm of the MPS. Such an MPS is said to be in left-canonical form.

One can imagine performing the recursion Eqs. (6.25)–(6.28) in the opposite direction, instead replacing A with the unitary matrix V as

$$A_{\alpha(i\beta)} = A^{[\ell]i}_{\alpha\beta}, \tag{6.29}$$

$$\sum_{\gamma} U_{\alpha\gamma} S_{\gamma} V^{\dagger}_{\gamma(i\beta)} = A_{\alpha(i\beta)}, \tag{6.30}$$

$$A^{[\ell]i}_{\gamma(i\beta)} = V^{\dagger}_{\gamma(i\beta)}, \tag{6.31}$$

$$A^{[\ell-1]i}_{\eta\gamma} = \sum_{\alpha} A^{[\ell-1]i}_{\eta\alpha} U_{\alpha\gamma} S_{\gamma}. \tag{6.32}$$

The resulting tensor $A = A^{[\ell]}$ satisfies the *right-canonical* condition

$$\sum_{i} A^{i} A^{i\dagger} = I. \tag{6.33}$$

As with the left-canonical form, one can perform the recursions Eqs. (6.29)–(6.32) from the right boundary to the left boundary to obtain a representation in which all tensors satisfy Eq. (6.33) together with a 1×1 matrix whose trace is the norm of the MPS. In the remainder of this chapter, we will reserve the notation A for a left-canonical tensor and B for a right-canonical tensor. The left-canonical and right-canonical MPS forms thus read

$$|\psi_{\text{left-canonical}}\rangle = \sum_{i_1,\ldots i_L=1}^{d} \mathrm{Tr}\left(A^{[1]i_1} \ldots A^{[L]i_L}\right)|i_1 \ldots i_L\rangle, \tag{6.34}$$

$$|\psi_{\text{right-canonical}}\rangle = \sum_{i_1,\ldots i_L=1}^{d} \mathrm{Tr}\left(B^{[1]i_1} \ldots B^{[L]i_L}\right)|i_1 \ldots i_L\rangle. \tag{6.35}$$

The consequences of left- and right-canonical form in graphical notation are provided as part of the publication in Chap. 7.

We now consider taking a state $|\psi\rangle$ and performing Eqs. (6.25)–(6.28) to put all tensors $A^{[j]}$ from $j = 1, \ldots, k$ into left-canonical form and performing Eqs. (6.29)–(6.32) such that all tensors $B^{[j]}$ from $j = k+1, \ldots, L$ are right-canonical. In order for the state to be consistent, we must insert a matrix M between $A^{[k]}$ and $B^{[k+1]}$ which represents $S^{[k]}V^{[k]\dagger}$ from the right-moving contraction Eq. (6.28) multiplied by $U^{[k+1]}S^{[k+1]}$ from the left-moving contraction Eq. (6.32). That is, the state is

$$|\psi\rangle = \sum_{i_1,\ldots i_L=1}^{d} \mathrm{Tr}\left(A^{[1]i_1} \ldots A^{[k]i_k} M B^{[k+1]i_{k+1}} \ldots B^{[L]i_L}\right)|i_1 \ldots i_L\rangle, \tag{6.36}$$

where $M = S^{[k]}V^{[k]\dagger}U^{[k+1]}S^{[k+1]}$. Performing a SVD $M = U\Lambda V$ and absorbing U into $A^{[k]}$ and V into $B^{[k+1]}$,[13] we may write this as

$$|\psi\rangle = \sum_{i_1,\ldots i_L=1}^{d} \mathrm{Tr}\left(A^{[1]i_1} \ldots A^{[k]i_k} \Lambda B^{[k+1]i_{k+1}} \ldots B^{[L]i_L}\right)|i_1 \ldots i_L\rangle, \tag{6.37}$$

where now Λ is a diagonal matrix. This form of the state corresponds identically with the Schmidt decomposition mentioned in Sect. 6.1. As the matrix Λ lies at the boundary between the left-orthogonal and right-orthogonal parts of the state, we will call it the *orthogonality center*. To be more precise, we call it the *bond-centered orthogonality center* as it resides on the bond between two-sites. The state Eq. (6.37) is said to be in *bond-centered mixed canonical form*. We may absorb Λ into $A^{[k]}$ to give a tensor which is no longer left-canonical, but does carry all the information that was contained in the orthogonality center. We will denote this tensor with a tilde to indicate its lack of left-canonical form:

$$|\psi\rangle = \sum_{i_1,\ldots i_L=1}^{d} \mathrm{Tr}\left(A^{[1]i_1} \ldots \tilde{A}^{[k]i_k} B^{[k+1]i_{k+1}} \ldots B^{[L]i_L}\right)|i_1 \ldots i_L\rangle. \tag{6.38}$$

[13] Note that this does not affect the canonical form of these two tensors.

We now say that $\tilde{A}^{[k]}$ is the orthogonality center of the MPS, more precisely the *site-centered orthogonality center*, and we call this particular canonical form *site-centered mixed canonical form*. When there is no chance of confusion, we will refer to both Eqs. (6.37) and (6.38) as mixed canonical form. The most important feature of the orthogonality center is that it expresses the wavefunction in terms of a tensor whose indices run over orthonormal bases.

The final canonical form we consider takes the mixed canonical form to its logical conclusion by ensuring that a bipartite splitting at any bond within the lattice results in the Schmidt decomposition, and we shall call it the *Vidal canonical form* [49, 50]:

$$|\psi_{\mathrm{VCF}}\rangle = \sum_{\alpha_1\ldots\alpha_{L+1}} \sum_{i_1\ldots i_L} \lambda^{[1]}_{\alpha_1}\Gamma^{[1]i_1}_{\alpha_1\alpha_2}\lambda^{[2]}_{\alpha_2}\ldots\lambda^{[L]}_{\alpha_L}\Gamma^{[L]i_L}_{\alpha_L\alpha_{L+1}}\lambda^{[L+1]}_{\alpha_{L+1}}. \tag{6.39}$$

The Γ tensors and λ tensors are chosen such that a bipartite splitting of our system at the bond between sites l and $l+1$ is exactly the Schmidt decomposition

$$|\psi\rangle = \sum_{\alpha_l=1}^{\chi_S} \lambda^{[l+1]}_{\alpha_l}|\phi^{[1\ldots l]}_{\alpha_l}\rangle|\phi^{[l+1\ldots n]}_{\alpha_l}\rangle, \tag{6.40}$$

with the Schmidt vectors

$$|\phi^{[1\ldots l]}_{\alpha_l}\rangle = \sum_{\alpha_0,\ldots,\alpha_{l-1}}^{\chi_S} \lambda^{[1]}_{\alpha_0}\Gamma^{[1]i_1}_{\alpha_0\alpha_1}\lambda^{[2]}_{\alpha_1}\Gamma^{[2]i_2}_{\alpha_1\alpha_2}\lambda^{[3]}_{\alpha_2}\Gamma^{[3]i_3}_{\alpha_2\alpha_3}\ldots\Gamma^{[l]i_l}_{\alpha_{l-1}\alpha_l}|i_1\rangle\ldots|i_l\rangle \tag{6.41}$$

and

$$|\phi^{[l+1\ldots L]}_{\alpha_l}\rangle = \sum_{\alpha_{l+1},\ldots,\alpha_L}^{\chi_S} \Gamma^{[l+1]i_{l+1}}_{\alpha_l\alpha_{l+1}}\lambda^{[l+1]}_{\alpha_{l+2}}\Gamma^{[l+2]i_{l+2}}_{\alpha_{l+1}\alpha_{l+2}}\ldots\Gamma^{[L]i_L}_{\alpha_{L-1}\alpha_L}\lambda^{[L]}_{\alpha_L}|i_{l+1}\rangle\ldots|i_L\rangle, \tag{6.42}$$

and the Schmidt coefficients $\lambda^{[l+1]}_{\alpha_l}$. We can translate between the left, right, and Vidal canonical forms by using the translation formulae

$$A^{[j]i_j}_{\alpha\beta} = \lambda^{[j]}_{\alpha}\Gamma^{[j]i_j}_{\alpha\beta}, \tag{6.43}$$

$$B^{[j]i_j}_{\alpha\beta} = \Gamma^{[j]i_j}_{\alpha\beta}\lambda^{[j+1]}_{\beta}. \tag{6.44}$$

These will be useful in Chap. 8, when we discuss translationally invariant MPSs.

As an example of why the various MPS canonical forms are useful, we now consider finding the expectation value of an observable $\hat{O}_k$ which acts only on a single lattice site k. Assuming that site k is the orthogonality center, this expectation value is

$$\langle\psi|\hat{O}_k|\psi\rangle = \sum_{i_1\dots i_k\dots i_L}\sum_{i_k'} O_k^{i_k i_k'}\mathrm{Tr}\left(A^{[1]i_1}\dots\tilde{A}^{[j]i_k'}\dots B^{[L]i_L}B^{[L]i_L\dagger}\dots\tilde{A}^{[k]i_k\dagger}\dots A^{[1]i_1\dagger}\right),$$
$$= \sum_{i_1\dots i_k\dots i_L}\sum_{i_k'} O_k^{i_k i_k'}\mathrm{Tr}\left(A^{[1]i_1\dagger}A^{[1]i_1}\dots\tilde{A}^{[k]i_k'}\dots B^{[L]i_L}B^{[L]i_L\dagger}\dots\tilde{A}^{[k]i_k\dagger}\dots\right),$$
$$= \sum_{i_2\dots i_k\dots i_{L-1}}\sum_{i_k'} O_k^{i_k i_k'}$$
$$\mathrm{Tr}\left(A^{[2]i_2\dagger}A^{[2]i_2}\dots\tilde{A}^{[k]i_k'}\dots B^{[L-1]i_{L-1}}B^{[L-1]i_{L-1}\dagger}\dots\tilde{A}^{[k]i_k\dagger}\dots\right),$$
$$= \sum_{i_k i_k'} O_k^{i_k i_k'}\mathrm{Tr}\left(\tilde{A}^{[k]i_k'}\tilde{A}^{[k]i_k\dagger}\right). \qquad (6.45)$$

Here, the second line used the cyclic nature of the trace and the third and fourth lines used the left- and right-canonical conditions Eqs. (6.24) and (6.33) for the tensors to the left and right of k, respectively. Thus, mixed canonical form projects many operations involving the entire wavefunction into operations involving only the orthogonality center.

6.4 Examples of MPSs

Several important and well-known states can be cast exactly as MPSs with constant bond dimension. The purpose of this section is to provide some examples of states with exact MPS representations.

6.4.1 *Product State*

The simplest MPS is a product state $|k_1k_2\dots k_L\rangle$, which is a product of 1×1 matrices $A^{[j]i_j} = \left(\delta_{i_j k_j}\right)$. For the Vidal canonical form, all of the λ tensors are (1) and the Γ tensors are represented by the A tensors here. In this case, $\chi = 1$ for any subsystem. This is the only class of states in which all of the canonical forms above have identical MPS representations.

6.4.2 *GHZ State*

A nontrivial MPS with bond dimension two is the (unnormalized) Greenberger–Horne–Zeilinger (GHZ) state

$$|\mathrm{GHZ}\rangle = |00\dots0\rangle + |11\dots1\rangle. \qquad (6.46)$$

This state represents a realization of Schrödinger's famous cat paradox [51] in which a quantum system exists in two very different macroscopic states simultaneously. Also, the GHZ state is closely related to *NOON states* in which N particles exist in a superposition of all particles in state a and all particles in state b simultaneously. The GHZ state garnered interest because of its very strong non-classical correlations [52]. The state is given in left- or right-canonical form by the matrices

$$A^i = \begin{pmatrix} \delta_{i,0} & 0 \\ 0 & \delta_{i,1} \end{pmatrix}. \tag{6.47}$$

Here we have neglected the boundary conditions. For PBC all matrices are equivalent, and for OBC the first and last sites are 1×2 and 2×1 matrices whose elements are the diagonal elements of A^i, respectively. The normalized GHZ state can be represented in mixed canonical form by introducing the orthogonality center

$$A^i = \begin{pmatrix} \delta_{i,0}/\sqrt{2} & 0 \\ 0 & \delta_{i,1}/\sqrt{2} \end{pmatrix}. \tag{6.48}$$

In the Vidal canonical form, the normalized GHZ state takes the form

$$\lambda^{[j]} = \begin{pmatrix} \frac{1}{\sqrt{2}} \\ \frac{1}{\sqrt{2}} \end{pmatrix}, \Gamma^{[j]i} = \begin{pmatrix} \delta_{i,0}\sqrt{2} & 0 \\ 0 & \delta_{i,1}\sqrt{2} \end{pmatrix}, \tag{6.49}$$

for $2 \leq j \leq L-1$ together with the boundaries

$$\lambda^{[1]} = \lambda^{[L+1]} = (1), \; \Gamma^{[1]i} = \begin{pmatrix} \delta_{i,0} & \delta_{i,1} \end{pmatrix}, \; \Gamma^{[L]i} = \begin{pmatrix} \delta_{i,0} \\ \delta_{i,1} \end{pmatrix}. \tag{6.50}$$

As this is a state with $\chi = 2$, it has minimally nontrivial spatial entanglement.

6.4.3 *W State*

The W state is the equal superposition of all translates of $|10\ldots0\rangle$,

$$|W\rangle = \frac{1}{\sqrt{L}} \sum_{i=1}^{L} |0\rangle^{\otimes i-1} |1\rangle |0\rangle^{\otimes L-i}, \tag{6.51}$$

and can also be represented as an MPS with bond dimension 2. If we interpret $|0\rangle$ as being spin down and $|1\rangle$ as being spin up, the W state with $L = 2$ constituents is the spin triplet. If $|0\rangle$ and $|1\rangle$ are interpreted as lattice sites containing 0 and 1 particles, respectively, the W state represents the state of a single particle on a ring with quasimomentum $q = 0$. The Vidal canonical form is

$$\lambda^{[j]} = \begin{pmatrix} \sqrt{\frac{L-j+1}{L}} \\ \sqrt{\frac{j-1}{L}} \end{pmatrix}, \Gamma^{[j]i} = \begin{pmatrix} \delta_{i,0}\sqrt{\frac{L}{L-j+1}} & \delta_{i,1}\sqrt{\frac{L}{j(L-j+1)}} \\ \delta_{i,1}\sqrt{\frac{L}{j}} & 0 \end{pmatrix}, \tag{6.52}$$

for $2 \leq j \leq L-1$ together with the same boundary tensors as the GHZ state. In the MPS representation we can choose

$$A^{[1]i} = \begin{pmatrix} \delta_{i,0}/\sqrt{L} \\ \delta_{i,1}/\sqrt{L} \end{pmatrix}, \; A^{[j]i} = \begin{pmatrix} \delta_{i,0} & 0 \\ \delta_{i,1} & \delta_{i,0} \end{pmatrix}, \; A^{[L]i} = \begin{pmatrix} \delta_{i,0} & \delta_{i,1} \end{pmatrix}, \tag{6.53}$$

which amounts to mixed canonical form with the first site being the orthogonality center.

6.4.4 AKLT State

Affleck, Kennedy, Lieb, and Tasaki (AKLT) considered the following Hamiltonian [53]

$$\hat{H}_{\text{AKLT}} = \sum_i \hat{\mathbf{S}}_i \cdot \hat{\mathbf{S}}_{i+1} + \frac{1}{3}\left(\hat{\mathbf{S}}_i \cdot \hat{\mathbf{S}}_{i+1}\right)^2. \tag{6.54}$$

Here, $\hat{\mathbf{S}}_i$ is the vector of three-component spin operators at site i. We will call a three-component quantum system, isomorphic to the internal space of a particle with a spin of 1, a *qutrit*. Similarly, a two-component quantum system will be referred to as a *qubit*. By defining a projector onto the Hilbert space of total spin equal to two as

$$\hat{P} = \frac{1}{6 \cdot 4}\left(\hat{\mathbf{S}}^2_{\text{total}} - 2\right)\hat{\mathbf{S}}^2_{\text{total}}, \tag{6.55}$$

and setting $\hat{\mathbf{S}}_{\text{total}} = \hat{\mathbf{S}}_i + \hat{\mathbf{S}}_{i+1}$, we find that the projector of the spin on bond i onto the spin-2 subspace is

$$\hat{P}_i = \frac{1}{6 \cdot 4}\left(\hat{\mathbf{S}}^2_i + \hat{\mathbf{S}}^2_{i+1} + 2\hat{\mathbf{S}}_i \cdot \hat{\mathbf{S}}_{i+1} - 2\right)\left(\hat{\mathbf{S}}^2_i + \hat{\mathbf{S}}^2_{i+1} + 2\hat{\mathbf{S}}_i \cdot \hat{\mathbf{S}}_{i+1}\right) \tag{6.56}$$

$$= \frac{1}{2}\left[\frac{2}{3} + \hat{\mathbf{S}}_i \cdot \hat{\mathbf{S}}_{i+1} + \frac{1}{3}\left(\hat{\mathbf{S}}_i \cdot \hat{\mathbf{S}}_{i+1}\right)^2\right]. \tag{6.57}$$

Thus,

$$\hat{H}_{\text{AKLT}} = \sum_i \left(2\hat{P}_i - \frac{2}{3}\hat{I}\right). \tag{6.58}$$

Let us now consider each qutrit to be comprised of two qubits with internal states $\{|\uparrow\rangle, |\downarrow\rangle\}$. For this to be consistent, we must require these two qubits to be completely symmetrized so as to lie in the subspace with total spin equal to 1. Now, consider the many-body state where adjacent pairs of these qubits not forming a qutrit are joined in a singlet state such that the bonds have spin zero. The projector $\hat{P}_i$ acting on this state gives zero, and from $\langle \hat{P}_i \rangle \geq 0$ we have that the given construction produces a ground state of the AKLT Hamiltonian. It can also be shown that this state is unique [53]. We will call the state formed from the above construction the *AKLT state*.

We can write the AKLT state as an MPS by considering the chain of qubits which has length $2L$, with L being the length of the original qutrit chain. We define the qubit sites a_i and b_i, $i = 1, \ldots L$, such that the bonds $a_i b_i$ are connected in a symmetric fashion and $b_i a_{i+1}$ are connected in an antisymmetric fashion. We will also consider periodic boundaries so as to easily facilitate taking the limit of an infinite chain. A matrix encapsulating the state of the singlet bonds, $(|\uparrow\downarrow\rangle - |\downarrow\uparrow\rangle)/\sqrt{2}$, is

$$S = \begin{pmatrix} 0 & \frac{1}{\sqrt{2}} \\ -\frac{1}{\sqrt{2}} & 0 \end{pmatrix}. \tag{6.59}$$

Similarly, matrices representing the three triplet bonds $|1\rangle = |\uparrow\uparrow\rangle$, $|0\rangle = (|\uparrow\downarrow\rangle + |\downarrow\uparrow\rangle)/\sqrt{2}$, and $|-1\rangle = |\downarrow\downarrow\rangle$, are

$$T^i = \begin{pmatrix} \delta_{i,1} & \frac{1}{\sqrt{2}}\delta_{i,0} \\ \frac{1}{\sqrt{2}}\delta_{i,0} & \delta_{i,-1} \end{pmatrix}. \tag{6.60}$$

Hence, the state of the qutrit chain may be parameterized as

$$|\psi\rangle = \sum_{i_1 a_1 b_1 \ldots i_L a_L b_L} T^{i_1}_{a_1 b_1} S_{b_1 a_2} T^{i_2}_{a_2 b_2} S_{b_2 a_3} \ldots S_{b_{L-1} a_L} T^{i_L}_{a_L b_L} S_{b_L a_1} |i_1 \ldots i_L\rangle, \tag{6.61}$$

$$= \sum_{i_1 \ldots i_L} \mathrm{Tr}\left(A^{[1]i} \ldots A^{[L]i_L}\right) |i_1 \ldots i_L\rangle, \tag{6.62}$$

where

$$A^i = T^i S = \begin{pmatrix} -\frac{1}{2}\delta_{i,0} & \frac{1}{\sqrt{2}}\delta_{i,1} \\ -\frac{1}{\sqrt{2}}\delta_{i,-1} & \frac{1}{2}\delta_{i,0} \end{pmatrix}. \tag{6.63}$$

Noting that $\sum_i A^{i\dagger} A^i = \frac{3}{4} I$, we can normalize the state in the thermodynamic limit[14] by scaling by $2/\sqrt{3}$ to obtain

$$A^i = \begin{pmatrix} -\frac{1}{\sqrt{3}}\delta_{i,0} & \sqrt{\frac{2}{3}}\delta_{i,1} \\ -\sqrt{\frac{2}{3}}\delta_{i,-1} & \frac{1}{\sqrt{3}}\delta_{i,0} \end{pmatrix}. \tag{6.64}$$

We close this section by expressly pointing out the similarity between the AKLT construction and the general construction of MPSs as projections from maximally entangled pairs of χ-level subsystems onto physical states mentioned in Sect. 6.1. In fact, the similarity extends from the state construction to the Hamiltonian, as every MPS is the ground state of a *parent Hamiltonian* built of projectors which is gapped, is *frustration-free* in the sense that each term in the Hamiltonian minimizes the energy locally, is k-local with $k \sim 2\log\chi/\log d$, and allows for a detailed analysis of the ground state degeneracy [54, 55].

6.5 Correlations Within MPSs and the Transfer Operator

Let us now turn to the structure of correlations within MPSs. By a correlation, we mean a two-point correlation function, for example the density–density correlation function $\langle \hat{n}_p \hat{n}_q \rangle$. Assuming that the state is in mixed canonical form with the orthogonality center k satisfying $p \leq k \leq q$, the correlation becomes a finite tensor network contraction between sites p and q. Written out explicitly, we have

$$\langle \hat{O}_p \hat{O}_q \rangle \tag{6.65}$$

$$= \mathrm{Tr}\left(\sum_{i_p i'_p} \hat{O}_{i_p i'_p} A^{[p] i_p \star} \otimes A^{[p] i'_p} \prod_{j=p+1}^{q-1} \left(\sum_{i_j} A^{[j] i_j \star} \otimes A^{[j] i_j} \right) \right.$$
$$\left. \sum_{i_q i'_q} \hat{O}_{i_q i'_q} A^{[q] i_q \star} \otimes A^{[q] i'_q} \right),$$

$$= \mathrm{Tr}\left(E^{[p]}_{\hat{O}} \prod_{j=p+1}^{q-1} E^{[j]}_{\hat{I}} E^{[q]}_{\hat{O}} \right), \tag{6.66}$$

[14] See Eq. (6.69) for the condition that a left-canonical MPS on an infinite lattice is normalized.

where we have defined the transfer operator

$$E_{\hat{O}} = \sum_{ii'} \hat{O}_{ii'} A^{i*} \otimes A^{i'}. \tag{6.67}$$

We will also use the alternate notation

$$E_{\hat{O}}[M] = \sum_{ii'} A^{i\dagger} \hat{O}_{ii'} M A^{i'}, \tag{6.68}$$

which expresses the action of the transfer operator on a matrix M [1]. The difference in notation is demonstrated by comparing the expectation Eqs. (6.65)–(6.45). The transfer operator $E_{\hat{O}}$ either acts upon length χ^2 vectors or takes $\chi \times \chi$ matrices to $\chi \times \chi$ matrices,[15] depending on which interpretation we use.

While the transfer operator is not generally symmetric, we can still venture to find its eigenvalues and (left and right) eigenvectors/eigenmatrices. It can be shown that the transfer operator of a normalized state has a spectral radius of 1 [1]. For matrices which are left-canonical, we have that

$$E_{\hat{I}}\left[\hat{I}\right] = \sum_{i} A^{i\dagger} A^{i} = \hat{I}, \tag{6.69}$$

and so the left eigenmatrix of $E_{\hat{I}}$ with eigenvalue 1 is the identity matrix. Similarly, an appropriately defined transfer operator of right-canonical MPS matrices has the identity matrix as a right eigenmatrix with eigenvalue 1. For simplicity, let us now consider a state which is translationally invariant such that all matrices appearing in the contraction Eq. (6.65) are identical, and let $E_{\hat{I}}$ have a non-degenerate maximal eigenvalue 1. Expanding the product of transfer operators $\prod_{j=p+1}^{q-1} E_{\hat{I}}^{[j]}$ in terms of its eigenspectrum, we have that

$$\langle \hat{O}_p \hat{O}_q \rangle = \sum_{k} \langle 1 | E_{\hat{O}}^{[p]} | k \rangle \lambda_k^{q-p-1} \langle k | E_{\hat{O}}^{[q]} | 1 \rangle, \tag{6.70}$$

where $|k\rangle$ and $\langle k|$ are the right and left eigenvectors corresponding to eigenvalue λ_k. There are now two possibilities. The first is that this correlation function is long-ranged, which occurs when $\langle 1 | E_{\hat{O}}^{[p]} | 1 \rangle \langle 1 | E_{\hat{O}}^{[q]} | 1 \rangle$ is nonzero, and the second is a superposition of exponential decays with decay lengths $\xi_k = -1/\log \lambda_k$. This may be written compactly as

[15] Strictly speaking, the transfer operator using the MPS tensors at site j take $\chi_j \times \chi_j$ matrices to $\chi_{j-1} \times \chi_{j-1}$ matrices with the given order of operations, but the great numerical use is in infinite systems where χ is uniform across all bonds.

$$\frac{\langle\psi|\hat{O}^p\hat{O}^q|\psi\rangle}{\langle\psi|\psi\rangle} = \langle 1|E_{\hat{O}}^{[p]}|1\rangle\langle 1|E_{\hat{O}}^{[q]}|1\rangle + \sum_{k=2}^{\chi^2}\langle 1|E_{\hat{O}}^{[p]}|k\rangle\langle k|E_{\hat{O}}^{[q]}|1\rangle e^{-|q-p-1|/\xi_k}. \tag{6.71}$$

To illustrate these ideas, we can use the AKLT state Eq. (6.64) from above. The transfer operator is

$$E_{\hat{I}} = \sum_i A^i \otimes A^i = \begin{pmatrix} \frac{1}{3} & 0 & 0 & \frac{2}{3} \\ 0 & -\frac{1}{3} & 0 & 0 \\ 0 & 0 & -\frac{1}{3} & 0 \\ \frac{2}{3} & 0 & 0 & \frac{1}{3} \end{pmatrix}, \tag{6.72}$$

which has eigenvalues 1 and $-1/3$, the latter being triply degenerate. The dominant eigenmatrix is $\hat{I}$, as discussed for left-canonical matrices above. The other eigenspace is spanned by $\hat{\sigma}_z$, $\hat{\sigma}_+$, and $\hat{\sigma}_-$, the spin-1/2 Pauli matrices. Using the transfer operators

$$E_{\hat{S}_z} = \begin{pmatrix} 0 & 0 & 0 & \frac{2}{3} \\ 0 & 0 & 0 & 0 \\ 0 & 0 & 0 & 0 \\ -\frac{2}{3} & 0 & 0 & 0 \end{pmatrix}, \tag{6.73}$$

$$E_{\exp(i\pi\hat{S}_z)} = \begin{pmatrix} \frac{1}{3} & 0 & 0 & -\frac{2}{3} \\ 0 & -\frac{1}{3} & 0 & 0 \\ 0 & 0 & -\frac{1}{3} & 0 \\ -\frac{2}{3} & 0 & 0 & \frac{1}{3} \end{pmatrix}, \tag{6.74}$$

we find exponential decay of antiferromagnetic correlations, $\langle\hat{S}_z^i\hat{S}_z^j\rangle \sim (-1/3)^{i-j}$, but long-range order in the string order parameter $\langle\hat{S}_z^i\prod_{k=i+1}^{j-1}\exp\left(i\pi\hat{S}_z^k\right)\hat{S}_z^j\rangle = -4/9$. Here string order refers to the fact that, although true long-range antiferromagnetic order of the classical Néel type $|\ldots 1,-1,1,-1\ldots\rangle$ is absent, any site with $S^z = \pm 1$ is followed by a site with the opposite spin projection $S^z = \mp 1$, and these two-sites are connected by a string of $S^z = 0$ sites which can have arbitrarily long length.

Given that an MPS has correlations which decay exponentially asymptotically, how can they accurately represent a critical state which displays power-law decay of some correlator? The answer is that the general correlation structure of MPSs is a sum of exponentials, and so on length scales short compared to the dominant correlation length this sum can approximate an algebraic decay. This idea is demonstrated in Fig. 6.5, where approximations to the function $1/r^3$ are provided by fitting a sum of exponentials to this function. The accuracy of the fit extends to longer distances as more exponentials are included. The minimization procedure used to fit the exponentials is precisely that used to define matrix product operators with long-range interactions, see Sect. 7.2.2.

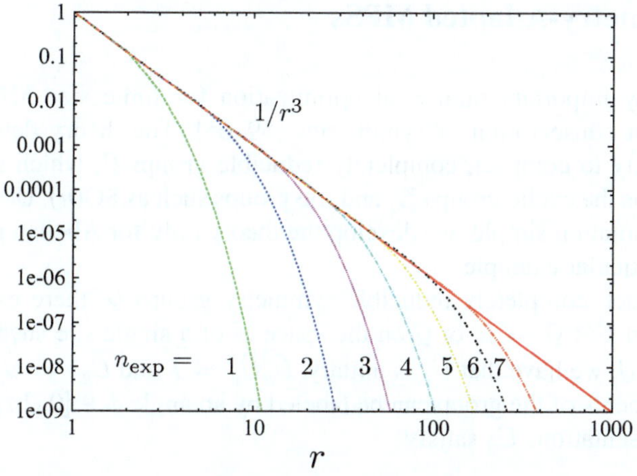

Fig. 6.5 Approximating algebraic decay by sums of exponentials. The *solid red line* represents the function $1/r^3$, and the *other lines* represent approximations to $1/r^3$ obtained by a least squares minimization of a sum of exponentials with the indicated number of terms

Given that the accuracy of an MPS representation of a state monotonically increases with the bond dimension and the correlation length of a critical state diverges, we find that the correlation length must increase with χ. In fact, χ obeys a scaling relationship with the correlation length [56]

$$\chi \sim \xi^{\kappa}, \tag{6.75}$$

where

$$\kappa = \frac{6}{\sqrt{12c} + c}. \tag{6.76}$$

Here c is the central charge of the conformal field theory describing criticality, see Eq. (6.10) and the surrounding discussion. Results such as Eq. (6.75) fall under the heading of *finite-entanglement scaling*, as contrasted with finite-size scaling [57]. A nice feature of finite-entanglement scaling is that scaling relationships depend only on truly universal quantities such as the central charge rather than scaling dimensions as in finite-size scaling. Provided we are in the finite-entanglement scaling regime rather than the finite-size scaling regime [58], for example because we have taken the size of the system much larger than the correlation length,[16] finite-entanglement scaling provides us with a powerful means of extracting universal quantities for relatively small amounts of computational effort, see Sect. 8.4.

[16]For critical systems, the diverging of the correlation length requires us to consider systems with an infinite number of sites. For an MPS algorithm which operates in this limit, see Chap. 8.

6.6 Symmetry-Adapted MPSs

A particularly important numerical optimization for finite size MPS algorithms is the explicit conservation of symmetries [59–65]. The theory developed below extends readily to compact, completely reducible groups $\mathcal{G}$, which includes finite groups such as the cyclic groups $\mathbb{Z}_q$ and Lie groups such as SO(n), U(n), and SU(n). To keep the notation simple, we develop the theory only for Abelian groups, taking U(1) as a particular example.

For compact, completely reducible symmetry groups $\mathcal{G}$, there exists a unitary representation $\hat{U} : \mathcal{G} \to \mathbb{H}$ of $\mathcal{G}$ on the space $\mathbb{H}$ of a single site such that for each element $g \in \mathcal{G}$, we have that $\hat{U}_g$ is unitary, $\hat{U}_g\hat{U}_g^{\dagger} = I$ and $\hat{U}_{gg'} = \hat{U}_g\hat{U}_{g'}$ [66]. For U(1) the elements of the group can be labeled by an angle $\phi \in [0, 2\pi)$ such that the unitary representations $\hat{U}_{\phi}$ satisfy

$$\hat{U}_{\phi}^{\dagger}\hat{U}_{\phi} = \hat{U}_{\phi}\hat{U}_{\phi}^{\dagger} = \hat{I}, \tag{6.77}$$

$$\hat{U}_{\phi_1}\hat{U}_{\phi_2} = \hat{U}_{\phi_2}\hat{U}_{\phi_1} = \hat{U}_{\phi_3}, \tag{6.78}$$

where $\phi_3 = \phi_1 + \phi_2 \bmod 2\pi$. $\mathbb{H}$ now decomposes into possibly degenerate one-dimensional[17] irreducible representations (*irreps*) of $\mathcal{G}$ as

$$\mathbb{H}_j = \oplus_q \mathbb{H}_q, \tag{6.79}$$

where the d_q-dimensional subspaces $\mathbb{H}_q$ are labeled by an integer q which we call the *charge* of the irrep. Here d_q, the *degeneracy dimension*, denotes the number of copies of the irrep of charge q which are present in $\mathbb{H}$. Because the irreps of Abelian groups are one-dimensional, these are in fact copies and not higher-dimensional irreps. We will refer to the spaces $\mathbb{H}_q$ as *degeneracy spaces*. We can construct the unitary representations of U(1) appearing in Eq. (6.77) by using a local Hermitian operator $\hat{Q}$ whose expectation over all sites gives the total conserved charge as

$$\hat{U}_{\phi} = \exp\left(-i\hat{Q}\phi\right). \tag{6.80}$$

The eigenstates of $\hat{Q}$ with eigenvalue q form a basis for $\mathbb{H}_q$. We will index the states in the degeneracy space $\mathbb{H}_q$ by the *degeneracy index* t_q which runs from $1, \ldots, d_q$ such that

$$\hat{Q}|qt_q\rangle = q|qt_q\rangle, \tag{6.81}$$

$$\hat{U}_{\phi}|qt_q\rangle = e^{-iq\phi}|qt_q\rangle, \tag{6.82}$$

$$\langle qt_q|q't'_q\rangle = \delta_{q,q'}\delta_{t_q t'_q}. \tag{6.83}$$

[17] These irreps are only one-dimensional in the Abelian case. In the non-Abelian case it is still true that the space decomposes into degeneracy spaces and irreps; however, the irreps can have dimensions larger than one.

Hence, any state $|\psi\rangle$ which is an eigenstate of $\hat{Q}$ transforms symmetrically as

$$\hat{U}_\phi|\psi\rangle = \exp(-iq\phi)\,|\psi\rangle, \tag{6.84}$$

and so can be expanded in the basis states $|qt_q\rangle$ which span the degeneracy space $\mathbb{H}_q$:

$$|\psi\rangle = \sum_{t_q}\langle qt_q|\psi\rangle|qt_q\rangle. \tag{6.85}$$

Similarly, a linear operator $\hat{T}$ which is covariant in the sense that

$$\hat{U}_\phi\hat{T}\hat{U}_\phi^\dagger = e^{-i\Delta q\phi}\hat{T} \tag{6.86}$$

decomposes as

$$\hat{T} = \oplus_q\hat{T}_{q,q-\Delta q}, \tag{6.87}$$

where the $\hat{T}_{q,q-\Delta q}$ are $d_q \times d_{q-\Delta q}$ operators mapping $\mathbb{H}_{q-\Delta q} \to \mathbb{H}_q$. We will denote the matrix elements of $T_{q,q'}$ in the basis $|qt_q\rangle\langle q't_q'|$ as $\left[T_{qq'}\right]_{t_qt_q'}$. A linear operator which is covariant with $\Delta q = 0$ is said to be invariant, and commutes with the group operation. This is Schur's lemma [67]. Symmetric states and covariant operators are strongly constrained by the fact that they act only in the smaller subspaces $\mathbb{H}_q$ rather than the entire space $\mathbb{H}$.

To make these ideas more concrete, let us consider some explicit examples. First, let us consider a single lattice site which can accommodate up to two bosons. The charge in this case is the number of bosons, and so the operator $\hat{Q}$ is $\hat{n}$, the boson number operator. In the basis $\{|0\rangle, |1\rangle, |2\rangle\}$ where $|N\rangle$ is the Fock state with N bosons, this operator is

$$\hat{n} = \begin{pmatrix} 0 & 0 & 0 \\ 0 & 1 & 0 \\ 0 & 0 & 2 \end{pmatrix}. \tag{6.88}$$

The Fock states $|0\rangle$, $|1\rangle$, and $|2\rangle$ span the spaces $\mathbb{H}_0$, $\mathbb{H}_1$, and $\mathbb{H}_2$ with charges 0, 1, and 2, respectively. Hence, all degeneracy spaces are one-dimensional, $d_q = 1\ \forall q$. The operator $\hat{n}$ is invariant, as it does not connect states with different charges. It may be written in the basis $|qt_q\rangle \in \{|01\rangle, |11\rangle, |21\rangle\}$ as

$$\hat{n} = \hat{n}_{00} \oplus \hat{n}_{11} \oplus \hat{n}_{22} = \begin{pmatrix} (0) & 0 & 0 \\ 0 & (1) & 0 \\ 0 & 0 & (2) \end{pmatrix}. \tag{6.89}$$

The fact that this operator is invariant has constrained that only the elements in the blocks denoted by parentheses may be nonzero. As the blocks $\hat{n}_{qq}$ are 1×1 matrices, the operator $\hat{n}$ is in fact specified by three numbers. An example of an operator which is covariant but not invariant is the bosonic destruction operator

$$\hat{b} = \begin{pmatrix} 0 & 1 & 0 \\ 0 & 0 & \sqrt{2} \\ 0 & 0 & 0 \end{pmatrix}. \tag{6.90}$$

This operator reduces the particle number by 1, and so $\Delta q = -1$. It may be written in the basis $|qt_q\rangle \in \{|01\rangle, |11\rangle, |21\rangle\}$ as

$$\hat{b} = \hat{b}_{01} \oplus \hat{b}_{12} = \begin{pmatrix} 0 & (1) & 0 \\ 0 & 0 & \left(\sqrt{2}\right) \\ 0 & 0 & 0 \end{pmatrix}. \tag{6.91}$$

As a second example, consider a single lattice site in a system comprised of spin-1/2 fermions. The charge is now the total number of fermions. In the basis $\{|00\rangle, |1-\frac{1}{2}\rangle, |1\frac{1}{2}\rangle, |20\rangle\}$, where $|NS_z\rangle$ is the state with N particles and a total spin projection of S_z along the z direction, the total fermion number operator is

$$\hat{n} = \begin{pmatrix} 0 & 0 & 0 & 0 \\ 0 & 1 & 0 & 0 \\ 0 & 0 & 1 & 0 \\ 0 & 0 & 0 & 2 \end{pmatrix}. \tag{6.92}$$

Hence, the degeneracy space $\mathbb{H}_1$ spanned by $|1 \pm \frac{1}{2}\rangle$ is two-dimensional due to the spin degree of freedom. In the basis $|qt_q\rangle \in \{|01\rangle, |11\rangle, |12\rangle, |21\rangle\}$ we may write this invariant operator as

$$\hat{n} = \hat{n}_{00} \oplus \hat{n}_{11} \oplus \hat{n}_{22} = \begin{pmatrix} (0) & 0\;0 & 0 \\ 0 & \begin{pmatrix} 1 & 0 \\ 0 & 1 \end{pmatrix} & 0 \\ 0 & 0\;0 & (2) \end{pmatrix}. \tag{6.93}$$

It is specified by the 1×1 matrices $\hat{n}_{00}$ and $\hat{n}_{22}$ and the 2×2 matrix $\hat{n}_{11}$. An example of an invariant operator with off-diagonal matrix elements in the degeneracy space is $\hat{S}^x = \hat{a}_\uparrow^\dagger \hat{a}_\downarrow + \hat{a}_\downarrow^\dagger \hat{a}_\uparrow^\dagger$, where $a_\uparrow$ destroys a fermion with $S_z = \frac{1}{2}$ and $a_\downarrow$ destroys a fermion with $S_z = -\frac{1}{2}$. This operator may be written in invariant form as

$$\hat{S}^x = \begin{pmatrix} (0) & 0\;0 & 0 \\ 0 & \begin{pmatrix} 0 & 1 \\ 1 & 0 \end{pmatrix} & 0 \\ 0 & 0\;0 & (0) \end{pmatrix}. \tag{6.94}$$

These examples illustrate the general utility of symmetry conservation in tensor network algorithms. Conservation of a symmetry breaks a covariant operator into two pieces. The first piece is the structure of nonzero elements of the operator, specified by a direct sum of operators appropriately transforming between degeneracy spaces. This piece is completely determined by the symmetry. The remaining piece, which distinguishes a particular operator from all others which transform under the group operation in the same way, is the matrix elements of the operators acting on the degeneracy spaces.

Let us now consider the product of two spaces $\mathbb{H}^A$ and $\mathbb{H}^B$ which admit representations of U(1). The action of U(1) on the coupled system $\mathbb{H}^{AB} = \mathbb{H}^A \otimes \mathbb{H}^B$ is generated by the total charge operator $\hat{Q}^{AB} = \hat{Q}^A \otimes \hat{1}^B + \hat{1}^A \otimes \hat{Q}^B$, and so the space $\mathbb{H}^{AB}$ decomposes into irreps with total charge q_{AB} and degeneracy $d_{q_{AB}}$. This coupled basis carries the same information as the decoupled basis indexed by the charges of the subsystems q_A and q_B, and so a one-to-one correspondence must exist between the two. We define

$$C^{q_{AB}t_{AB}}_{q_At_A;q_Bt_B} \equiv \langle q_{AB}t_{AB}|q_At_Aq_Bt_B\rangle \tag{6.95}$$

as the elements of the unitary matrix which encapsulate this one-to-one correspondence. For the Abelian case at hand, each element of this transformation is either zero or one, but for non-Abelian groups these elements represent the corresponding Clebsch–Gordan coefficients.[18]

As an example of the transformation to the coupled system, consider two lattice sites which can contain up to two bosons. Using Eq. (6.89), we can write the total charge operator in the uncoupled basis $|q_1t_{q_1}q_2t_{q_2}\rangle = \{|0101\rangle, |0111\rangle, |0121\rangle, |1101\rangle, |1111\rangle, |1121\rangle, |2101\rangle, |2111\rangle, |2121\rangle\}$ as

$$\hat{n} = \hat{n}_1 \otimes \hat{I} + \hat{I} \otimes \hat{n}_2 = \begin{pmatrix} 0&0&0&0&0&0&0&0&0 \\ 0&1&0&0&0&0&0&0&0 \\ 0&0&2&0&0&0&0&0&0 \\ 0&0&0&1&0&0&0&0&0 \\ 0&0&0&0&2&0&0&0&0 \\ 0&0&0&0&0&3&0&0&0 \\ 0&0&0&0&0&0&2&0&0 \\ 0&0&0&0&0&0&0&3&0 \\ 0&0&0&0&0&0&0&0&4 \end{pmatrix}. \tag{6.96}$$

Ordering the vectors in terms of their total charges, we can identify the invariant form of the total charge operator

[18] In the physics literature, the Clebsch–Gordan coefficients typically refer to this unitary transformation for the case of SU(2). Here we use it in the more general mathematical sense as the unitary matrix connecting the tensor product of the representation spaces of two irreps of a group to a direct sum of irreducible representation spaces [68].

$$\hat{n} = \hat{n}_{00} \oplus \hat{n}_{11} \oplus \hat{n}_{22} \oplus \hat{n}_{33} \oplus \hat{n}_{44} \tag{6.97}$$

$$= \begin{pmatrix} (0) & 0\,0 & 0\,0\,0 & 0\,0 & 0 \\ 0 & \begin{pmatrix} 1 & 0 \\ 0 & 1 \end{pmatrix} & \begin{matrix} 0\,0\,0 \\ 0\,0\,0 \end{matrix} & \begin{matrix} 0\,0 \\ 0\,0 \end{matrix} & \begin{matrix} 0 \\ 0 \end{matrix} \\ 0 & 0\,0 & \begin{pmatrix} 2 & 0 & 0 \\ 0 & 2 & 0 \\ 0 & 0 & 2 \end{pmatrix} & \begin{matrix} 0\,0 \\ 0\,0 \\ 0\,0 \end{matrix} & \begin{matrix} 0 \\ 0 \\ 0 \end{matrix} \\ 0 & \begin{matrix} 0\,0 \\ 0\,0 \end{matrix} & \begin{matrix} 0\,0\,0 \\ 0\,0\,0 \end{matrix} & \begin{pmatrix} 3 & 0 \\ 0 & 3 \end{pmatrix} & \begin{matrix} 0 \\ 0 \end{matrix} \\ 0 & 0\,0 & 0\,0\,0 & 0\,0 & (4) \end{pmatrix}, \tag{6.98}$$

and the corresponding nonzero Clebsch–Gordan coefficients $C^{01}_{01;01} = C^{11}_{01;11} = C^{12}_{11;01} = C^{21}_{01;21} = C^{22}_{11;11} = C^{23}_{21;01} = C^{31}_{11;21} = C^{32}_{21;11} = C^{41}_{21;21} = 1$. This construction also demonstrates that even if the irreps of a single lattice site are non-degenerate, a *combinatoric degeneracy* in the space of fixed total charge arises when multiple lattice sites are considered.

If we now consider the general case of an L-fold tensor product of a space admitting a representation of U(1), then the total charge operator is

$$\hat{Q} = \sum_{i=1}^{L} \hat{Q}_i, \tag{6.99}$$

where $\hat{Q}_i$ is the local charge operator at site i. The total charge operator generates the unitary transformations

$$\hat{U}_{\phi;L} = \exp\left(-i\hat{Q}\phi\right) = \left(\hat{U}_\phi\right)^{\otimes L}. \tag{6.100}$$

The total tensor product space thus decomposes into spaces of fixed total charge. From Schur's lemma [67], an invariant Hamiltonian has no matrix elements between states that have differing total charge, and so we can diagonalize the Hamiltonian in subspaces of fixed total charge. Thus, all eigenstates of an invariant Hamiltonian can be chosen to be symmetric in the sense of Eq. (6.84). This makes rigorous the intuitive notion that the eigenvectors of a Hamiltonian which conserves the number of particles can be chosen to have a definite number of particles.[19]

We can generalize the commutation relation Eq. (6.86) for linear operators to tensors by introducing the notion of *incoming* and *outgoing* indices which encapsulate the overall charge flow described by the tensor. The distinction is

[19]Here we say only that the eigenvectors can be chosen in this way to account for possible energetic degeneracies of two states with differing total particle number. In the presence of such a degeneracy, any linear combination of the degenerate states is also an eigenstate.

that incoming indices transform as $\hat{U}_\phi$ under group action while outgoing indices transform as $\hat{U}_\phi^\dagger$. Intuitively, incoming indices denote charge flowing into the tensor and outgoing indices describe charge flow out from the tensor. With this definition, we have that a tensor T is invariant under the group action if

$$\sum_{\mathrm{I}'\in I}\left[\hat{U}_\phi\otimes\cdots\otimes\hat{U}_\phi\right]_{\mathrm{II}'}\sum_{\mathrm{J}'\in O}\left[\hat{U}_\phi^\dagger\otimes\cdots\otimes\hat{U}_\phi^\dagger\right]_{\mathrm{JJ}'}T_{\mathrm{I}'\mathrm{J}'}=T_{\mathrm{IJ}}. \tag{6.101}$$

Here I denotes all incoming indices and O all outgoing indices, we have permuted the indices of T such that all incoming indices lie to the left and all outgoing indices to the right, and the notation $[A\otimes B]_{ij}$ denotes the ijth component of the tensor product. Now, we consider decomposing each on-site state $|i_j\rangle$ into its local degeneracy spaces spanned by $|q_j t_{q_j}\rangle$ using Eq. (6.84). This implies that the tensor decomposes as

$$T_{i_1\dots i_L}|i_1\dots i_L\rangle=\left[T_{q_1\dots q_L}\right]_{t_{q_1}\dots t_{q_L}}|q_1t_{q_1}\dots q_Lt_{q_L}\rangle, \tag{6.102}$$

and the condition that the tensor be invariant becomes that the total incoming and outgoing charge be the same, where the incoming and outgoing charges are defined as $Q_{\text{incoming}}=\sum_{i\in I}q_i$ and $Q_{\text{outgoing}}=\sum_{i\in O}q_i$, respectively. This implies that any invariant tensor takes the form

$$T_{i_1\dots i_L}=\left[T_{q_1\dots q_L}\right]_{t_{q_1}\dots t_{q_L}}\delta_{Q_{\text{incoming}}Q_{\text{outgoing}}}. \tag{6.103}$$

This canonical form is the key result of our analysis. Using the symmetry, we have broken each invariant tensor into a part determined by the symmetry (the delta function) and a part which is not (indexed by the t_j). This canonical form may be viewed as a conservation of charge by invariant tensors, as the amount of charge flowing in is equal to the amount of charge flowing out.

For MPS tensors $A^i_{\alpha\beta}$, we define α and i to be incoming and β to be outgoing such that q_γ with a Greek index denotes the total charge to the left of a given bond and q_i with a Roman index denotes the charge of the particular on-site irrep. The tensors A then become arrays of tensors $\left[A^{q_i}_{q_\alpha q_\beta}\right]^{t_{q_i}}_{t_{q_\alpha}t_{q_\beta}}$. This convention of incoming and outgoing indices can be displayed in a modified tensor network diagram as in Fig. 6.6, compare Fig. 6.3c.[20] The arrows indicate the flow of charge through an

Fig. 6.6 Structure of charge flow through an MPS in diagrammatic notation

[20]This construction can also be compared with circuit diagrams demonstrating conservation of charge as specified by Kirchoff's laws.

MPS tensor, with the local states $|q_j t_{q_j}\rangle$ acting as sources of charge. We shall refer to the canonical form of a tensor which is covariant under the action of U(1) as a *q-tensor*. All of the results discussed above extend immediately to products of Abelian groups by considering the charge q to be a vector $\mathbf{q}$ whose entries are the charges of the individual Abelian symmetries under consideration. Multiple symmetries are relevant in models whose constituents are particles with spin, as the total number and the total magnetization are often independently conserved, for example.

To perform index fusion on a q-tensor we use Eq. (6.95) in the form

$$\left[T'_{q_{(\alpha i)}q_\beta}\right]_{t_{q_{(\alpha i)}}t_{q_\beta}} = \sum_{q_\alpha q_i t_{q_\alpha} t_{q_i}} C^{q_{(\alpha i)}t_{q_{(\alpha i)}}}_{q_\alpha t_{q_\alpha};q_i t_{q_i}} \left[T^{q_i}_{q_\alpha q_\beta}\right]^{t_{q_i}}_{t_{q_\alpha}t_{q_\beta}}, \tag{6.104}$$

where $q_{(\alpha i)}$ is the charge of $|\alpha\rangle|i\rangle$. The transformation tensor C does not change the total charge, and so it is an invariant tensor. Taking α and i to be incoming indices and (αi) to be an outgoing index hence requires that $q_{\alpha i} = q_\alpha * q_i$, where $*$ is the group operation and we can characterize the degeneracy index of the composite system as the Kronecker product of the degeneracy indices from the constituent systems, $t_{q_{(\alpha i)}} = \left(t_{q_\alpha} t_{q_i}\right)$. That is, the symmetry determines which irreps can combine and then the degeneracy indices are combined under ordinary index fusion.[21] Writing this out explicitly, we have

$$\left[T'_{q_\alpha * q_i, q_\beta}\right]_{\left(t_{q_\alpha}t_{q_i}\right)t_{q_\beta}} = \left[T^{q_i}_{q_\alpha q_\beta}\right]^{t_{q_i}}_{t_{q_\alpha}t_{q_\beta}}. \tag{6.105}$$

Splitting is performed by reading this expression in reverse. Using these fusion and splitting rules we can also apply the matrix decompositions discussed earlier to develop canonical forms for q-tensors.

It is remarkable that the only place where the specific form of the Abelian group arises is in the charge fusion rule $q_{\alpha i} = q_\alpha * q_i$. For U(1) $*$ is ordinary addition while for $\mathbb{Z}_p$ $*$ is addition mod p. In general, we can use the Cayley table [69] of the group under consideration to construct the fusion rule.

The contraction of two q-tensors is restricted by the fact that incoming indices in one covariant tensor can only be contracted with outgoing indices of another covariant tensor in order for the result to also be a covariant tensor. Furthermore, as can be gleaned from the fusion rule above, the charges of each index which are contracted have to agree between the two tensors in order to yield a nonzero result. Hence, the procedure to contract the sets of indices A and B of two q-tensors U and V together is to find all matching charges within the two sets q_A and q_B, and

[21] This is the most significant difference between the Abelian and non-Abelian cases. In the latter, the allowed irreps are enumerated by the Clebsch–Gordan series, and the elements of the unitary matrix relating $t_{q_{(\alpha i)}}$ to t_{q_α} and t_{q_i} are nontrivial.

then contract the degeneracy tensors using ordinary tensor contraction, Eq. (6.17), resulting in

$$\left[T_{q_{\bar{A}}q_{\bar{B}}}\right]_{t_{\bar{A}}t_{\bar{B}}} = \sum_{q}\sum_{t_q}\left[U_{q_{\bar{A}}q}\right]_{t_{\bar{A}}t_q}\left[V_{qq_{\bar{B}}}\right]_{t_q t_{\bar{B}}}. \tag{6.106}$$

Hence, essentially all expressions using MPSs can be translated directly for their (Abelian) symmetry-adapted counterparts by replacing summations over indices with simultaneous summations over the charges and degeneracy dimensions.

When written naively, Eq. (6.106) implies an algorithm which scales as $\mathcal{O}(N_U N_V)$, where N_T is the number of irreps in T. This scaling can be improved by sorting the quantum numbers and using a binary search to find matches. We note that, because the charges can be put into one-to-one correspondence with some subset of the integers, we can always define the vacuum charge to be zero for our finite Hilbert spaces. Rather than working directly with the charges themselves, it is useful to define a *hash function* [70] which takes unique arrays of nonnegative integers to unique values.[22] That is, the function is *injective*. A hash function satisfying this criterion, which has a number of other useful properties for our purposes, is the square-root of primes hash

$$H(\mathbf{q}) = \sum_i q_i\sqrt{p_i}. \tag{6.107}$$

Here p_i is the ith prime, and the injectivity of H follows from the fact that the numbers $\sqrt{p_i}$ are incommensurable. A nice feature of $H(\mathbf{q})$ is that it is linear in the elements of $\mathbf{q}$, and so given the values of $\sqrt{p_i}$ the hash function can be obtained for arbitrary combinations of charges very quickly. In practice one never needs more than a few primes, and these may be generated once using a sieve and stored for later use. With hashing, the procedure for contracting two q-tensors becomes

1. Hash U and V according to A and B in $\mathcal{O}(N_U + N_V)$ time.
2. Sort the smaller of the two sets A and B according to its hashes in $\mathcal{O}(N_M \log N_M)$ time, where $N_M = \min(N_U, N_V)$.
3. Loop through the elements of the larger list and perform a binary search to find matching hashes in the smaller list in $\mathcal{O}(N_X \log N_M)$ time, where $N_X = \max(N_U, N_V)$.
4. Contract all matching hashes using the contraction algorithm for ordinary tensors on the degeneracy spaces.

Using the square-root of primes hash makes the process of contracting tensors efficient even when permutations, large numbers of indices, or multiple Abelian symmetries are involved.

[22] This is in fact the definition of a perfect hash function, which is harder to find than a hash function in general but can be explicitly found for our purposes.

As an example of a symmetric MPS, let us again consider the W state from Sect. 6.4.3

$$|W\rangle = \frac{1}{\sqrt{L}} \sum_{i=1}^{L} |0\rangle^{\otimes i-1} |1\rangle |0\rangle^{\otimes L-i}. \tag{6.108}$$

If we interpret $|0\rangle$ as a state with no particles and $|1\rangle$ as a state with 1 particle, then the W state is an eigenstate of the total particle number operator with eigenvalue 1. We will use charge interchangeably with the number of particles in what follows. We can write the W state as

$$\begin{aligned} |W\rangle =& \frac{1}{\sqrt{L}} \left[|0\rangle^{\otimes j-1} |1\rangle |0\rangle^{\otimes L-j} + |W; j-1\rangle |0\rangle |0\rangle^{\otimes L-j} \right. \\ & \left. + |0\rangle^{\otimes j-1} |0\rangle |W; L-j\rangle \right], \end{aligned} \tag{6.109}$$

where

$$|W; \ell\rangle = \sum_{i=1}^{\ell} |0\rangle^{\otimes i-1} |1\rangle |0\rangle^{\otimes \ell-i} \tag{6.110}$$

is an unnormalized W state on ℓ sites. Hence, if site j is in state $|1\rangle$, then it follows that the state to the left of site j is $|0\rangle^{\otimes j-1}$ and the state to the right of site j is $|0\rangle^{\otimes L-j}$. The state $|1\rangle$ is the unique on-site state with charge $q = 1$, and so $|1\rangle = |11\rangle$ in the $|q_j t_{q_j}\rangle$ basis. Furthermore, $|00\ldots0\rangle$ is the unique state with charge 0, and so $|00\ldots0\rangle = |01\rangle$ in the $|q_\alpha t_{q_\alpha}\rangle$ basis. This implies that the contribution to the MPS at site j with quantum numbers $(q_\alpha q_i q_\beta) = (010)$ is

$$\left[A_{00}^{[j]1} \right]_{11}^{1} = 1. \tag{6.111}$$

If site j is in state $|0\rangle$, then either the state to the left is $|W; j-1\rangle$ and the state to the right is $|00\ldots0\rangle$ or the state to the left is $|00\ldots0\rangle$ and the state to the right is $|W; \ell - j\rangle$. Because the W state on any number of sites has charge 1 and the W state is the only state with charge 1 relevant to any subsystem, we have the elements

$$\left[A_{01}^{[j]0} \right]_{11}^{1} = 1, \quad \left[A_{10}^{[j]0} \right]_{11}^{1} = 1. \tag{6.112}$$

Hence, the complete MPS q-tensor at site j is the direct sum of the elements

$$\left[A_{01}^{[j]0} \right]_{11}^{1} = 1, \quad \left[A_{10}^{[j]0} \right]_{11}^{1} = 1, \quad \left[A_{00}^{[j]1} \right]_{11}^{1} = 1, \tag{6.113}$$

compare Eq. (6.53). Both the MPS tensor Eq. (6.53) and the MPS q-tensor Eq. (6.113) have three independent elements. The difference is that Eq. (6.113) writes these three elements as a direct sum with the topology of index contraction determined by the charges of the subsystems. The utility of the q-tensor decomposition is that contractions over tensors become direct sums of contractions over tensors with smaller linear dimensions, see Eq. (6.106). Because of the polynomial scaling of contractions with the bond dimension, this leads to a significant speedup when symmetries are explicitly utilized. In the present case the reduction of the bond dimension is complete, as the tensors in Eq. (6.113) are all $1 \times 1 \times 1$ dimensional.

The present chapter gives an intuition for why MPSs are useful as variational ansätze for strongly correlated 1D systems, demonstrates how to extract physical information from MPSs, and explains how to express MPSs in terms of irreducible tensors such that Abelian symmetries are explicitly conserved. In Chap. 7, we discuss how to generate MPS representations of eigenstates of 1D systems through variational means. Additionally, it is shown how to time-evolve the MPS form of a wavefunction in a generic way. These algorithms all apply to lattice systems of finite extent. In Chap. 8 we present an algorithm for variationally finding the ground state of a homogenous 1D system which has infinite extent. This is done by introducing an MPS decomposition of a unit cell which, when infinitely repeated, generates the full state. This unit cell is then optimized using variations of the finite-size algorithms. Finally, in Chap. 9, we discuss how algorithms for time evolution of MPSs may be adapted to extract finite-temperature properties of 1D systems. This chapter also discusses the form of MPSs and the structure of their algorithms for systems with PBC.

References

1. Schollwöck, U.: The density-matrix renormalization group in the age of matrix product states. Ann. Phys. **326**(1), 96–192 (2011). doi:10.1016/j.aop.2010.09.012. http://www.sciencedirect.com/science/article/pii/S0003491610001752 (January 2011 Special Issue) [ISSN 0003-4916]
2. Golub, G.H., Van Loan, C.F.: Matrix Computations, 3rd edn. Johns Hopkins Studies in Mathematical Sciences. The Johns Hopkins University Press, Baltimore (1996)
3. Davidson, E.R.: The iterative calculation of a few of the lowest eigenvalues and corresponding eigenvectors of large real-symmetric matrices. J. Comput. Phys. **17**(1), 87–94 (1975). doi:10.1016/0021-9991(75)90065-0. http://www.sciencedirect.com/science/article/pii/0021999175900650 [ISSN 0021-9991]
4. Poulin, D., Qarry, A., Somma, R., Verstraete, F.: Quantum simulation of time-dependent hamiltonians and the convenient illusion of hilbert space. Phys. Rev. Lett. **106**, 170501 (2011). doi:10.1103/PhysRevLett.106.170501. http://link.aps.org/doi/10.1103/PhysRevLett.106.170501
5. Huyghebaert, J., De Raedt, H.: Product formula methods for time-dependent Schrodinger problems. J. Phys. A **23**(24), 5777 (1990). http://stacks.iop.org/0305-4470/23/i=24/a=019
6. Fetter, A.L., Walecka, J.D.: Quantum Theory of Many-Particle Systems. McGraw-Hill, New York (1971)
7. Kitaev, A., Shen, A.H., Vyalyi, M.N.: Classical and Quantum Information. The American Mathematical Society, Providence (2002)

8. Nielsen, M., Chuang, I.: Quantum Computation and Quantum Information. Cambridge University Press, Cambridge (2000)
9. Dawson, C.M., Nielsen, M.A.: The Solovay-Kitaev algorithm. Quantum Inf. Comput. **6**(1), 81–95 (2006)
10. Kliesch, M., Barthel, T., Gogolin, C., Kastoryano, M., Eisert, J.: Dissipative quantum church-turing theorem. Phys. Rev. Lett. **107**, 120501 (2011). doi:10.1103/PhysRevLett.107.120501. http://link.aps.org/doi/10.1103/PhysRevLett.107.120501
11. Srednicki, M.: Entropy and area. Phys. Rev. Lett. **71**, 666–669 (1993). doi:10.1103/PhysRevLett.71.666. http://link.aps.org/doi/10.1103/PhysRevLett.71.666
12. Holzhey, C., Larsen, F., Wilczek, F.: Geometric and renormalized entropy in conformal field theory. Nucl. Phys. B **424**(3), 443–467 (1994). doi:10.1016/0550-3213(94)90402-2. ISSN 0550-3213. http://www.sciencedirect.com/science/article/pii/0550321394904022
13. Vidal, G., Latorre, J.I., Rico, E., Kitaev, A.: Entanglement in quantum critical phenomena. Phys. Rev. Lett. **90**, 227902 (2003). doi:10.1103/PhysRevLett.90.227902. http://link.aps.org/doi/10.1103/PhysRevLett.90.227902
14. Calabrese, P., Cardy, J.: Entanglement entropy and quantum field theory. J. Stat. Mech. **2004**(06), P06002 (2004). http://stacks.iop.org/1742-5468/2004/i=06/a=P06002
15. Eisert, J., Cramer, M., Plenio, M.B.: Colloquium: area laws for the entanglement entropy. Rev. Mod. Phys. **82**, 277–306 (2010). doi:10.1103/RevModPhys.82.277. http://link.aps.org/doi/10.1103/RevModPhys.82.277
16. Hastings, M.B.: An area law for one-dimensional quantum systems. J. Stat. Mech. **2007**(10), P08024 (2007)
17. Vitagliano, G., Riera, A., Latorre, J.I.: Volume-law scaling for the entanglement entropy in spin-1/2 chains. New J. Phys. **12**(11), 113049 (2010). http://stacks.iop.org/1367-2630/12/i=11/a=113049
18. Wolf, M.M., Verstraete, F., Hastings, M.B., Cirac, J.I.: Area laws in quantum systems: Mutual information and correlations. Phys. Rev. Lett. **100**, 070502 (2008). doi:10.1103/PhysRevLett.100.070502. http://link.aps.org/doi/10.1103/PhysRevLett.100.070502
19. Adams, A., Carr, L.D., Schaefer, T., Steinberg, P., Thomas, J.E.: Strongly correlated quantum fluids: Ultracold quantum gases, quantum chromodynamic plasmas, and holographic duality. New J. Phys. arXiv:1205.5180v1 (2012)
20. Bekenstein, J.D.: Black holes and entropy. Phys. Rev. D **7**, 2333–2346 (1973). doi:10.1103/PhysRevD.7.2333. http://link.aps.org/doi/10.1103/PhysRevD.7.2333
21. Bousso, R.: The holographic principle. Rev. Mod. Phys. **74**, 825–874 (2002). doi:10.1103/RevModPhys.74.825. http://link.aps.org/doi/10.1103/RevModPhys.74.825
22. Osborne, T.J., Eisert, J., Verstraete, F.: Holographic quantum states. Phys. Rev. Lett. **105**, 260401 (2010). doi:10.1103/PhysRevLett.105.260401. http://link.aps.org/doi/10.1103/PhysRevLett.105.260401
23. Cirac, J.I., Poilblanc, D., Schuch, N., Verstraete, F.: Entanglement spectrum and boundary theories with projected entangled-pair states. Phys. Rev. B **83**, 245134 (2011). doi:10.1103/PhysRevB.83.245134. http://link.aps.org/doi/10.1103/PhysRevB.83.245134
24. Wolf, M.M.: Violation of the entropic area law for fermions. Phys. Rev. Lett. **96**, 010404 (2006). doi:10.1103/PhysRevLett.96.010404. http://link.aps.org/doi/10.1103/PhysRevLett.96.010404
25. Gioev, D., Klich, I.: Entanglement entropy of fermions in any dimension and the widom conjecture. Phys. Rev. Lett. **96**, 100503 (2006). doi:10.1103/PhysRevLett.96.100503. http://link.aps.org/doi/10.1103/PhysRevLett.96.100503
26. Plenio, M.B., Eisert, J., Dreißig, J., Cramer, M.: Entropy, entanglement, and area: Analytical results for harmonic lattice systems. Phys. Rev. Lett. **94**, 060503 (2005). doi:10.1103/PhysRevLett.94.060503. http://link.aps.org/doi/10.1103/PhysRevLett.94.060503
27. Cramer, M., Eisert, J., Plenio, M.B., Dreißig, J.: Entanglement-area law for general bosonic harmonic lattice systems. Phys. Rev. A **73**, 012309 (2006). doi:10.1103/PhysRevA.73.012309. http://link.aps.org/doi/10.1103/PhysRevA.73.012309

28. Cramer, M., Eisert, J.: Correlations, spectral gap and entanglement in harmonic quantum systems on generic lattices. New J. Phys. **8**(5), 71 (2006). http://stacks.iop.org/1367-2630/8/i=5/a=071
29. Cardy, J.: Scaling and Renormalization in Statistical Physics. Cambridge University Press, Cambridge (1996)
30. Calabrese, P., Cardy, J.: Entanglement entropy and conformal field theory. J. Phys. A Math. Theor. **42**(50), 504005 (2009). http://stacks.iop.org/1751-8121/42/i=50/a=504005
31. Di Francesco, P., Mathieu, P., Sénéchal, D.: Conformal Field Theory. Springer, New York (1997)
32. Verstraete, F., Porras, D., Cirac, J.I.: Density matrix renormalization group and periodic boundary conditions: A quantum information perspective. Phys. Rev. Lett. **93**(22), 227205 (2004). doi:10.1103/PhysRevLett.93.227205
33. Verstraete, F., Cirac, J.I.: Matrix product states represent ground states faithfully. Phys. Rev. B **73**, 094423 (2006). http://link.aps.org/abstract/PRB/v73/e094423
34. Schuch, N., Cirac, I., Verstraete, F.: Computational difficulty of finding matrix product ground states. Phys. Rev. Lett. **100**, 250501 (2008). doi:10.1103/PhysRevLett.100.250501. http://link.aps.org/doi/10.1103/PhysRevLett.100.250501
35. Wilson, K.G.: The renormalization group: Critical phenomena and the kondo problem. Rev. Mod. Phys. **47**, 773–840 (1975)
36. White, S.R.: Density matrix formulation for quantum renormalization groups. Phys. Rev. Lett. **69**(19), 2863–2866 (1992). doi:10.1103/PhysRevLett.69.2863
37. Schollwock, U.: The density-matrix renormalization group. Rev. Mod. Phys. **77**, 259 (2005)
38. Schmidt, E.: Zur Theorie der linearen und nichtlinearen Integralgleichungen. Math. Ann. **63**, 433–476 (1907). doi:10.1007/BF01449770. ISSN 0025-5831. http://dx.doi.org/10.1007/BF01449770.
39. Östlund, S., Rommer, S.: Thermodynamic limit of density matrix renormalization. Phys. Rev. Lett. **75**(19), 3537–3540 (1995). doi:10.1103/PhysRevLett.75.3537
40. Rommer, S., Östlund, S.: Class of ansatz wave functions for one-dimensional spin systems and their relation to the density matrix renormalization group. Phys. Rev. B **55**(4), 2164–2181 (1997). doi:10.1103/PhysRevB.55.2164
41. Basic Linear Algebra Subprograms: http://www.netlib.org/blas
42. Shi, Y.-Y., Duan, L.-M., Vidal, G.: Classical simulation of quantum many-body systems with a tree tensor network. Phys. Rev. A **74**(2), 022320 (2006). doi:10.1103/PhysRevA.74.022320. http://link.aps.org/abstract/PRA/v74/e022320
43. Verstraete, F., Cirac, J.I.: Renormalization algorithms for quantum many-body systems in two and higher dimensions. arXiv:cond-mat/0407066v1 (2004)
44. Verstraete, F., Murg, V., Cirac, J.I.: Matrix product states, projected entangled pair states, and variational renormalization group methods for quantum spin systems. Adv. Phys. **57**(2), 143–224 (2008). doi:10.1080/14789940801912366. http://www.tandfonline.com/doi/abs/10.1080/14789940801912366
45. Vidal, G.: Entanglement renormalization. Phys. Rev. Lett. **99**, 220405 (2007)
46. Evenbly, G., Vidal, G.: Entanglement renormalization in two spatial dimensions. Phys. Rev. Lett. **102**(18), 180406 (2009). doi:10.1103/PhysRevLett.102.180406
47. Pippan, P., White, S.R., Evertz, H.G.: Efficient matrix-product state method for periodic boundary conditions. Phys. Rev. B **81**(8), 081103 (2010). doi:10.1103/PhysRevB.81.081103
48. Pirvu, B., Verstraete, F., Vidal, G.: Exploiting translational invariance in matrix product state simulations of spin chains with periodic boundary conditions. Phys. Rev. B **83**, 125104 (2011). doi:10.1103/PhysRevB.83.125104. http://link.aps.org/doi/10.1103/PhysRevB.83.125104
49. Vidal, G.: Efficient classical simulation of slightly entangled quantum computations. Phys. Rev. Lett. **91**, 147902–1–4 (2003)
50. Vidal, G.: Efficient simulation of one-dimensional quantum many-body systems. Phys. Rev. Lett. **93**, 040502–1–4 (2004)
51. Wheeler, J.A., Zurek, W.H. (eds.): Quantum Theory and Measurement. Princeton University Press, Princeton, NJ (1983)

52. Bouwmeester, D., Pan, J.-W., Daniell, M., Weinfurter, H., Zeilinger, A.: Observation of three-photon greenberger-horne-zeilinger entanglement. Phys. Rev. Lett. **82**, 1345–1349 (1999). doi:10.1103/PhysRevLett.82.1345. http://link.aps.org/doi/10.1103/PhysRevLett.82.1345
53. Affleck, I., Kennedy, T., Lieb, E.H., Tasaki, H.: Rigorous results on valence-bond ground states in antiferromagnets. Phys. Rev. Lett. **59**, 799–802 (1987). doi:10.1103/PhysRevLett.59.799. http://link.aps.org/doi/10.1103/PhysRevLett.59.799
54. Hastings, M.B.: Solving gapped hamiltonians locally. Phys. Rev. B **73**, 085115 (2006). doi:10.1103/PhysRevB.73.085115. http://link.aps.org/doi/10.1103/PhysRevB.73.085115
55. Perez-Garcia, D., Verstraete, F., Wolf, M.M., Cirac, J.I.: Matrix product state representations. Quantum Inf. Comput. **7**(401), (2007)
56. Pollmann, F., Mukerjee, S., Turner, A.M., Moore, J.E.: Theory of finite-entanglement scaling at one-dimensional quantum critical points. Phys. Rev. Lett. **102**, 255701 (2009). doi:10.1103/PhysRevLett.102.255701. http://link.aps.org/doi/10.1103/PhysRevLett.102.255701
57. Barber, M.N.: In: Domb, C., Lebowitz, J.L. (eds) Phase Transitions and Critical Phenomena, vol. 8. Academic, New York (1983)
58. Pirvu, B., Vidal, G., Verstraete, F., Tagliacozzo, L.: Matrix product states for critical spin chains: Finite size scaling versus finite entanglement scaling (2012). http://arxiv.org/abs/1204.3934
59. Daley, A.J.: Manipulation and simulation of cold Atoms in optical lattices. Ph.D. Thesis, Leopold-Franzens-Universität Innsbruck (2005)
60. McCulloch, I.P., Gulácsi, M.: The non-Abelian density matrix renormalization group algorithm. EPL (Europhys. Lett.) **57**(6), 852 (2002). http://stacks.iop.org/0295-5075/57/i=6/a=852
61. McCulloch, I.P.: From density-matrix renormalization group to matrix product states. J. Stat. Mech. **2007**(10), P10014 (2007). http://stacks.iop.org/1742-5468/2007/i=10/a=P10014
62. Singh, S., Zhou, H.-Q., Vidal, G.; Simulation of one-dimensional quantum systems with a global SU(2) symmetry. New J. Phys. **12**(3), 033029 (2010). http://stacks.iop.org/1367-2630/12/i=3/a=033029
63. Singh, S., Pfeifer, R.N.C., Vidal, G.: Tensor network states and algorithms in the presence of a global U(1) symmetry. Phys. Rev. B **83**, 115125 (2011). doi:10.1103/PhysRevB.83.115125. http://link.aps.org/doi/10.1103/PhysRevB.83.115125
64. Singh, S., Pfeifer, R.N.C., Vidal, G.: Tensor network decompositions in the presence of a global symmetry. Phys. Rev. A **82**, 050301 (2010). doi:10.1103/PhysRevA.82.050301. http://link.aps.org/doi/10.1103/PhysRevA.82.050301
65. Bauer, B., Corboz, P., Orús, R., Troyer, M.: Implementing global Abelian symmetries in projected entangled-pair state algorithms. Phys. Rev. B **83**, 125106 (2011). doi:10.1103/PhysRevB.83.125106. http://link.aps.org/doi/10.1103/PhysRevB.83.125106
66. Wigner, E.P.: Group Theory and Its Application to the Quantum Mechanics of Atomic Spectra. Academic, New York (1959)
67. Hamermesh, M.: Group Theory and Its Application to Physical Problems. Dover, New York (1989)
68. Alex, A., Kalus, M., Huckleberry, A., von Delft, J.: A numerical algorithm for the explicit calculation of SU(N) and SL(N, $\mathbb{C}$) clebschgordan coefficients. J. Math. Phys. **52**, 023507 (2011)
69. Keown, R.M.: An Introduction to Group Representation Theory. Academic, New York (1975)
70. Knuth, D.E.: The Art of Computer Programming. Addison-Wesley, Reading (1973)

Chapter 7
Out-of-Equilibrium Dynamics with Matrix Product States

Abstract Theoretical understanding of strongly correlated systems in one spatial dimension (1D) has been greatly advanced by the density-matrix renormalization group (DMRG) algorithm, which is a variational approach using a class of entanglement-restricted states called matrix product states (MPSs). However, DMRG suffers from inherent accuracy restrictions when multiple states are involved due to multi-state targeting and also the approximate representation of the Hamiltonian and other operators. By formulating the variational approach of DMRG explicitly for MPSs one can avoid errors inherent in the multi-state targeting approach. Furthermore, by using the matrix product operator (MPO) formalism, one can exactly represent the Hamiltonian and other operators relevant for the calculation. The MPO approach allows 1D Hamiltonians to be templated using a small set of finite state automaton (FSA) rules without reference to the particular microscopic degrees of freedom. We present two algorithms which take advantage of these properties: eMPS to find excited states of 1D Hamiltonians and tMPS for the time evolution of a generic time-dependent 1D Hamiltonian. We properly account for time-ordering of the propagator such that the error does not depend on the rate of change of the Hamiltonian. Our algorithms use only the MPO form of the Hamiltonian, and so are applicable to microscopic degrees of freedom of any variety, and do not require Hamiltonian-specialized implementation. We benchmark our algorithms with a case study of the Ising model, where the critical point is located using entanglement measures. We then study the dynamics of this model under a time-dependent quench of the transverse field through the critical point. Finally, we present studies of a dipolar, or long-range Ising model, again using entanglement measures to find the critical point and study the dynamics of a time-dependent quench through the critical point.

Published previously as *Out-of-Equilibrium Dynamics with Matrix Product States*, M.L. Wall and L.D. Carr, New J. Phys. **14**, 125015 (2012).

M.L. Wall, *Quantum Many-Body Physics of Ultracold Molecules in Optical Lattices*, Springer Theses, DOI 10.1007/978-3-319-14252-4_7

7.1 Introduction

The great success of experimental ultracold atomic physics has made the study of strongly correlated one-dimensional (1D) quantum systems a major avenue of current physics research. Examples of novel condensed matter physics realized with 1D atomic systems include the role of integrability in thermalization [1] and static [2] and dynamic [3] quantum simulators of Hubbard models. Furthermore, as ultracold molecules approach quantum degeneracy [4–6], lattice models with complex internal degrees of freedom, and long-range interactions become relevant [7–9]. As more and more complex models become amenable to study, the need for numerical methods which can adapt to different degrees of freedom, different Hamiltonians, and different dynamical processes thus becomes essential.

In addition to practical interest in understanding and benchmarking atomic and molecular quantum simulators, the ability to simulate the dynamics of 1D systems also provides insight into fundamental questions such as the universality of dynamics approach quantum critical points and the effects of integrability on the thermalization process [10]. The natural setting for studying dynamics near critical points is a quantum quench where one of the parameters of the Hamiltonian is driven through a quantum critical point following a time-dependent protocol, for example

$$g\left(t\right)=g_0+\frac{v\left(t-t_0\right)^r}{r!}\theta\left(t-t_0\right), \tag{7.1}$$

with $\theta\left(t\right)$ the step function. Such quenches pose a special difficulty for numerical studies as by definition they involve evolution with a time-dependent Hamiltonian which does not commute with itself at different times. The propagator is then generally a time-ordered exponential whose precise form may be difficult to ascertain. Standard methods such as the Suzuki–Trotter expansion which ignore the time dependence of the Hamiltonian require [11] that the infinitesimal time step used be much less than the fluctuation time-scale of $H\left(t\right)$ to be valid, $\delta t \ll |\partial H/\partial t|^{-1}$. This can cause simulations with rapid quench rates to become numerically very costly, and invalidates the approach altogether for non-analytic time dependence.

Currently the only unbiased method available for the dynamics of quantum systems is exact diagonalization (ED). By unbiased, we refer to the fact that the other methods available for dynamics are generally variational, and so have a bias towards a particular ansatz. ED is limited in an essential way by the exponential growth of the size of the Hilbert space with the physical size of the system. The current state of the art is ~40 spins for spin-$1/2$ models and 20 sites for a fermionic Hubbard model at half filling. These sizes are often too small for accurate finite-size scaling. An extremely powerful method for the low-energy properties of 1D systems is White's density matrix renormalization group (DMRG) algorithm, which uses a variational ansatz based on a class of states known as matrix product states (MPSs). MPSs will be reviewed in Sect. 7.2.1. DMRG uses an iterative procedure to develop a set of reduced bases that the full many-body problem is projected into, and then variationally minimizes an energy functional in this reduced space, enlarging it if necessary. DMRG uses an implicit MPS representation, which is to say that the state is not stored explicitly. This also means that the Hamiltonian and other operators in the calculation are stored in an approximate way, as they are represented within the reduced basis describing the variational state. This does not cause problems in practice when a single state is sought using the DMRG process. In fact, one can show that the algorithm to variationally find the ground state is identical when phrased in the implicit formulation of DMRG and when using an explicit MPS representation for the variational state, other than the representation of the Hamiltonian [12]. However, because of the exact representation of operators independent of the state, MPSs can put rigorous bounds on distances from exact quantum states such as eigenstates by considering quantities like the variance $\langle\psi|\left(\hat{H}-E\right)^2|\psi\rangle$ with $\hat{H}$ the Hamiltonian operator and E the energy expectation of the MPS $|\psi\rangle$. In contrast, DMRG can only return the distance of the variational state from an eigenstate of the approximate operator $\hat{H}$ in the given variational basis, and is unable to determine how well the given variational basis represents the true operator. A particularly clear indication of the failures this can cause is given in [13] where time evolution of a particular initial state in DMRG fails because the Hamiltonian has no nonzero matrix elements in the initial DMRG basis.

The situation becomes much different when multiple states are sought using the DMRG procedure. In this case the reduced density matrix used to determine the optimal reduced bases for the algorithm is a convex sum of the reduced density matrices for the desired states. This is called *multi-state targeting*. In contrast, an explicit MPS representation stores each of the desired states separately as an MPS. In multi-state targeting, none of the states can be represented with the same accuracy available if DMRG targeted that state alone. The MPS representation also deals automatically with the fact that each state has its own optimal bases for representation, whereas in DMRG these bases are all tied together by the multi-state targeting. In this work we present two algorithms which take advantage of MPSs' ability to deal with multiple states, eMPS to find excited states of 1D Hamiltonians and tMPS to simulate the dynamics of a generic time-dependent Hamiltonian. In the first algorithm a projector orthogonal to a set of lower-lying eigenstates is constructed from their MPS representations and used to orthogonalize a variational state against this set. In the second algorithm Krylov vectors in a Lanczos approximation to the matrix exponential are stored separately as MPSs and combined in an optimal way only at the end of the calculation. While Krylov-based MPS approaches have been used [13, 14] to study time-dependent processes, the errors in these approaches were set by time derivative of the Hamiltonian. In contrast, by taking explicit account of the time ordering of the propagator, the errors in our approach are set only by commutators of the Hamiltonian at different times, and hence allow for larger time steps. Because of the explicit MPS representation, we are able to put bounds on the errors in each step of the calculations.

Finally, MPSs have a natural operator-valued extension known as matrix product operators (MPOs) which allow for the exact representation of all operators used in the calculation. We present a general framework for constructing MPOs from a set of rules which is independent of the nature of the microscopic degrees of freedom. This allows for the templating of 1D Hamiltonians for general purpose software. In addition, the ability to perform arithmetic operations on MPOs exactly enables us to perform time-evolution using our tMPS algorithm with knowledge only of the MPO form of the Hamiltonian and the time-dependent functional form of its parameters. To emphasize the general nature of our algorithms, we include a generic simulation protocol for the out-of-equilibrium dynamics of strongly correlated 1D systems using the algorithms presented in this paper.

The remainder of this paper is organized as follows. In Sect. 7.2 we review the theory of MPSs, MPOs, and their canonical forms. In addition to providing a canonical form for operators within the matrix product formalism, we define finite state automaton (FSA) rules for MPOs and demonstrate how 1D Hamiltonians can be constructed from a small set of such rules. In Sect. 7.3 we review the algorithm for finding ground states using MPSs as variational ansätze, and in Sect. 7.4 we present the eMPS algorithm which extends the ground state search to general excited states. Section 7.5 discusses how to extract observable quantities from MPSs. In Sect. 7.6 we discuss methods for time evolution with MPS. In particular, we provide the tMPS algorithm to time evolve an MPS using only the MPO representation of a Hamiltonian and the functional form of its time-dependent parameters. We contrast our approach with other Krylov subspace approaches and identify the possible sources of error. In Sect. 7.10 we present two case studies. The first is of the Ising model in a transverse field, where we study both the statics and the dynamics of a linear quench of the transverse field through the quantum critical point. The second is of a dipolar, or long-range Ising model in a transverse field, where we also determine the critical point from the statics and study a linear quench of the transverse field. Finally, in Sect. 7.11, we conclude. Details concerning numerically exact solutions for the Ising model which are used to benchmark our algorithms are given in Appendix.

7.2 Brief Review of Matrix Product Formalism

7.2.1 Matrix Product States

The Hilbert space of a quantum mechanical many-body system is exponentially large in the physical size of the system, for example the number of unit cells in a lattice or the number of particles. Stated another way, a state picked at random from the Hilbert space of a quantum many-body system will have entanglement (as quantified by the Schmidt measure [15]) which grows exponentially with the system size. Here and throughout this paper, when we say entanglement we mean the spatial entanglement between two spatially separated parts of the system which are comprised of well-defined local Hilbert spaces. In contrast to this random state, it has been shown that the class of states which are physically relevant in the sense that they can be prepared from some reference state by generic time evolution in polynomial time [11] or are useful for quantum computation [16] is much smaller than the full Hilbert space. In 1D, the physically relevant class of states appears to be those which have entanglement which is either constant or polynomially growing as a function of system size. A convenient representation of states with entanglement restricted in this manner is known as MPSs [12, 17, 18].

We consider our physical system to be comprised of a 1D lattice of *sites*, where each site i is a d-dimensional Hilbert space $\mathbb{H}_i$ spanned by the vectors $\{|i\rangle\}$. We will refer to d as the *local dimension*, and take all sites to be identical for simplicity. We define an MPS on a lattice with L sites as

$$|\psi_{\mathrm{MPS}}\rangle = \sum_{i_1,\ldots,i_L=1}^{d} \mathrm{Tr}\left(A^{[1]i_1}\ldots A^{[L]i_L}\right)|i_1\ldots i_L\rangle, \tag{7.2}$$

where the object $A^{[k]i_k}$ is a $\chi_k \times \chi_{k+1}$ matrix (with $\chi_1 = \chi_{L+1}$) and Tr denotes the matrix trace. We will refer to the maximum linear dimension of any of the matrices $A^{[k]i_k}$, $\max_k \chi_k$, as the *bond dimension* of the MPS, and denote this quantity by χ. χ may be used as a measure of the entanglement of the state [12]. In this work we will focus on systems with open boundary conditions (OBC). MPS algorithms can also be devised for systems with periodic boundary conditions, as discussed in [7, 19–21], but these algorithms have worse scaling and are generally less numerically stable than their OBC counterparts. For OBC, $\chi_1 = \chi_{L+1} = 1$, and arguments using the Schmidt decomposition demonstrate that $\chi_k \leq \min(d^{k-1}, d^{L-k})$ [22].

MPSs have been used for many years to represent exact ground states of parent Hamiltonians [23] which are formed from projectors onto local high-symmetry subspaces [24, 25]. However, it was not until the great success of the DMRG algorithm pioneered by White [26] that MPSs became valuable as variational *ansätze* in their own right [19]. Why are MPSs useful as variational *ansätze*? It has been shown [27] that the ground states of gapped 1D systems have bipartite entanglement which does not depend on the system size. Such states can be represented exactly as MPSs with a fixed bond dimension [28]. This is an example of an *area law* [29]; the entanglement between two disjoint subsystems depends only on the boundary of the two regions and not on their volume. For systems near a quantum critical point which is described by a conformal field theory (CFT), this area law is subject to weak logarithmic violations such that the entropy of entanglement between two subsystems of size L is given by the Calabrese–Cardy formula [30, 31]

$$S_L \sim a + \frac{c}{6}\log L, \tag{7.3}$$

where a is a constant and c is the central charge of the underlying CFT. Here $\sim$ denotes scaling equivalence in the bulk of an infinite system. In finite systems there are often oscillating boundary and finite size contributions [32, 33]. Hence, the bond dimension of an MPS describing a conformally invariant critical system is given as

$\chi_L \sim \exp S_L \sim e^a L^{c/6}$, which grows only polynomially in the system size. Typical values of c range from 1/2 for the Ising model [34] and 1 for the Bose–Hubbard model [35] to 2 for more exotic phases like the gapless Mott insulator of the JK model [36]. We note that, strictly speaking, finding an MPS which approximates the ground state of an arbitrary 1D Hamiltonian to an accuracy which is an inverse polynomial in the system size is still NP-complete [37], but practical experience demonstrates that this method is extremely useful and robust for physical systems of interest.

We adopt the following conventions for the representation of tensors: We use Roman indices for indices which correspond to physical states and Greek indices for indices which are summed over in the matrix-product ansatz. Explicitly writing out Eq. (7.2) with these indices, we have

$$|\psi_{\mathrm{MPS}}\rangle = \sum_{\alpha_1 \ldots \alpha_L} \sum_{i_1, \ldots i_L = 1}^{d} A^{[1] i_1}_{\alpha_1 \alpha_2} \ldots A^{[L] i_L}_{\alpha_L \alpha_1} |i_1 \ldots i_L\rangle. \tag{7.4}$$

A superscript index in square brackets [] denotes the lattice site that the physical indices of the tensor describe. A superscript index in curly braces {} denotes association with a particular many-body state. For example, the MPS tensors at site j of the MPSs $|\phi_k\rangle$, $k = 1, \ldots, n$ would be denoted $A^{[j]\{k\}i_j}_{\alpha\beta}$. Finally, indices which appear together in parentheses, e.g. $(\alpha\beta)$, represent a composite index which runs over the Cartesian product of the indices in the parentheses. As an example, if $\alpha = 1, \ldots, \chi_\alpha$ and $\beta = 1, \ldots \chi_\beta$, $(\alpha\beta) = (\alpha - 1)\chi_\beta + \beta$. To lighten the notation, we will leave off indices when they are unnecessary.

A particularly useful means to visualize MPSs and manipulations with them is provided by *tensor network* diagrams like those shown in Fig. 7.1 [38]. Here a rank-k tensor is represented by a circle with k lines extending from it. Each line represents one of the indices of the tensor. Whenever a line connects two tensors, that index is summed over, and disconnected lines represent free indices. Hence, an MPS can be represented as a chain of rank-3 tensors as in Fig. 7.1c. Note that the first and last MPS tensors are rank two because we have assumed OBC and so $\chi_1 = \chi_{L+1} = 1$.

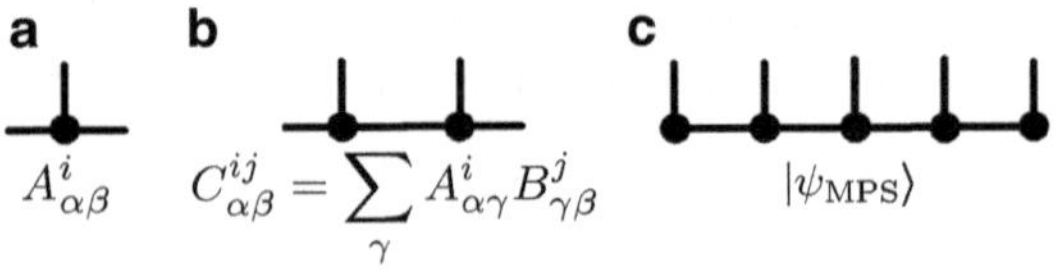

Fig. 7.1 Diagrammatic notation for tensor networks. (**a**) A rank-three tensor is represented as a *circle* with three lines extending from it. (**b**) Contraction of two rank-three tensors is accomplished by connecting the contracted index, and produces a tensor of rank four. (**c**) An MPS on five sites with open boundary conditions is represented as a contraction over rank-three tensors with two rank-two boundary tensors

We note that the MPS definition Eq. (7.2) does not uniquely specify the tensors A. That is, we can insert an invertible matrix X and its inverse X^{-1} between any two MPS matrices without altering the physical content of the state:

$$|\psi_{\mathrm{MPS}}\rangle = \sum_{i_1,\dots i_L=1}^{d} \mathrm{Tr}\left(A^{[1]i_1}\dots A^{[L]i_L}\right)|i_1\dots i_L\rangle, \tag{7.5}$$

$$\tilde{A}^{[j]i_j} = A^{[j]i_j}X,\quad \tilde{A}^{[j+1]i_{j+1}} = X^{-1}A^{[j+1]i_{j+1}}, \tag{7.6}$$

$$|\tilde{\psi}_{\mathrm{MPS}}\rangle = \sum_{i_1,\dots i_L=1}^{d} \mathrm{Tr}\left(A^{[1]i_1}\dots \tilde{A}^{[j]i_j}\tilde{A}^{[j+1]i_{j+1}}\dots A^{[L]i_L}\right)|i_1\dots i_L\rangle = |\psi_{\mathrm{MPS}}\rangle. \tag{7.7}$$

This is referred to as *gauge freedom* in the literature [22]. For OBC, we can specify the state uniquely[1] by choosing a site k, which we call the *orthogonality center* of the MPS, and requiring that all sites i to the left and right of k satisfy the left

$$\sum_i A^{i\dagger}A^i = I \tag{7.8}$$

and right

$$\sum_i A^iA^{i\dagger} = I \tag{7.9}$$

gauge conditions, respectively. In these expressions, I is the appropriately dimensioned identity matrix. These conditions are shown in graphical notation in Fig. 7.2a, b, respectively.[2] Graphically it is clear that the norm squared of the state is

$$\langle\psi|\psi\rangle = \sum_i \mathrm{Tr}\left(A^{[k]i\dagger}A^{[k]i}\right), \tag{7.10}$$

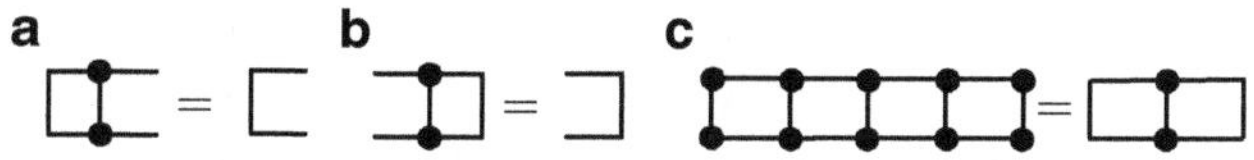

Fig. 7.2 Gauge conditions for MPSs. (**a**) *Left gauge* condition Eq. (7.8). (**b**) *Right gauge* condition Eq. (7.9). (**c**) Using (**a**) and (**b**), the norm squared of an MPS reduces to the trace of its orthogonality center squared, here chosen to be the third site

[1] The state is unique up to possible degeneracies in the Schmidt decomposition.

[2] Here we also establish the graphical convention that downwards pointing lines correspond to Hermitian conjugates of tensors.

as shown in Fig. 7.2c, and so this site carries all information about the norm of the state. This particular canonical form for an MPS is called *mixed canonical form* [12]. The mixed canonical form is crucial for improving the speed and numerical stability of variational algorithms with MPSs.

We can impose the left gauge conditions via the following recursion:

$$\tilde{A}_{(\alpha i)\beta} = A^{[\ell]i}_{\alpha\beta}, \tag{7.11}$$

$$\sum_{\gamma} U_{(\alpha i)\gamma} \Sigma_{\gamma} V^{\dagger}_{\gamma\beta} = \tilde{A}_{(\alpha i)\beta}, \tag{7.12}$$

$$A^{[\ell]i}_{\alpha\gamma} = U_{(\alpha i)\gamma}, \tag{7.13}$$

$$A^{[\ell+1]i}_{\gamma\eta} = \sum_{\beta} \Sigma_{\gamma} V^{\dagger}_{\gamma\beta} A^{[\ell+1]i}_{\beta\eta}, \tag{7.14}$$

where Eq. (7.12) represents the singular value decomposition (SVD) of $\tilde{A}$ with Σ the vector of singular values and U and V unitary matrices.[3] Because U returned from the SVD is unitary, Eq. (7.8) is satisfied by construction. Similarly, the recursion for the right gauge conditions is

$$\tilde{A}_{\alpha(i\beta)} = A^{[\ell]i}_{\alpha\beta}, \tag{7.15}$$

$$\sum_{\gamma} U_{\alpha\gamma} \Sigma_{\gamma} V^{\dagger}_{\gamma(i\beta)} = \tilde{A}_{\alpha(i\beta)}, \tag{7.16}$$

$$A^{[\ell]i}_{\gamma(i\beta)} = V^{\dagger}_{\gamma(i\beta)}, \tag{7.17}$$

$$A^{[\ell-1]i}_{\eta\gamma} = \sum_{\alpha} A^{[\ell-1]i}_{\eta\alpha} U_{\alpha\gamma} \Sigma_{\gamma}. \tag{7.18}$$

To put a general state into mixed canonical form with orthogonality center k we begin at site 1 and iterate Eqs. (7.11)–(7.14) until we reach site k, then start at site L and iterate Eqs. (7.15)–(7.18) until we again reach site k.

Finally, we note that the set of all MPSs with a fixed bond dimension χ is not a vector space, as the sum of two MPSs with bond dimensions χ_A and χ_B has a bond dimension χ which is bounded by the sum of the two bond dimensions $\chi \leq \chi_A + \chi_B$. This can be seen from considering the sum of the two states $|0\dots 0\rangle$ and $|1\dots 1\rangle$, with MPS representations

[3]Note that any matrix decomposition of $\tilde{A}$ which returns a unitary matrix as part of the decomposition will suffice in place of the SVD. In particular, the QR decomposition [39] is particularly efficient when the rank of $\tilde{A}$ is not required.

$$|0\ldots0\rangle = \sum_{i_1\ldots i_L} A^{i_1}\ldots A^{i_L}|i_1\ldots i_L\rangle, \quad A^i = (\delta_{i,0}), \tag{7.19}$$

$$|1\ldots1\rangle = \sum_{i_1\ldots i_L} B^{i_1}\ldots B^{i_L}|i_1\ldots i_L\rangle, \quad B^i = (\delta_{i,1}). \tag{7.20}$$

The matrices A^i and B^i have a bond dimension of 1, as these are product states. Their sum is

$$|0\ldots0\rangle + |1\ldots1\rangle = \sum_{i_1\ldots i_L} C^{[1]i_1}C^{[2]i_2}\ldots C^{[L-1]i_{L-1}}C^{[L]i_L}|i_1\ldots i_L\rangle, \tag{7.21}$$

$$C^{[1]i} = \begin{pmatrix}\delta_{i0} & \delta_{i1}\end{pmatrix};$$
$$C^{[j]i} = \begin{pmatrix}\delta_{i0} & 0\\ 0 & \delta_{i1}\end{pmatrix}, 2 \le j \le L-1; \quad C^{[L]i} = \begin{pmatrix}\delta_{i0}\\ \delta_{i1}\end{pmatrix}; \tag{7.22}$$

which has a bond dimension of 2.

7.2.2 *Matrix Product Operators*

The natural operator generalization of MPSs is a MPO, defined as

$$\hat{O}_{\mathrm{MPO}} = \sum_{i_1,\ldots,i_L=1}^{d}\ \sum_{i'_1,\ldots,i'_L=1}^{d} \mathrm{Tr}\left(W^{[1]i_1i'_1}\ldots W^{[L]i_Li'_L}\right)|i_1\ldots i_L\rangle\langle i'_1\ldots i'_L|. \tag{7.23}$$

Here $W^{[k]ii'}$ is a $\chi^O_k \times \chi^O_{k+1}$ dimensional matrix, and we will again refer to the maximum value of χ^O as the bond dimension of the operator. Note that this bond dimension χ^O need not be the same as the bond dimension χ appearing in the MPS representation, Eq. (7.2). That an MPO takes MPSs to MPSs can be seen clearly from the graphical representation of Fig. 7.3. We also see from this representation

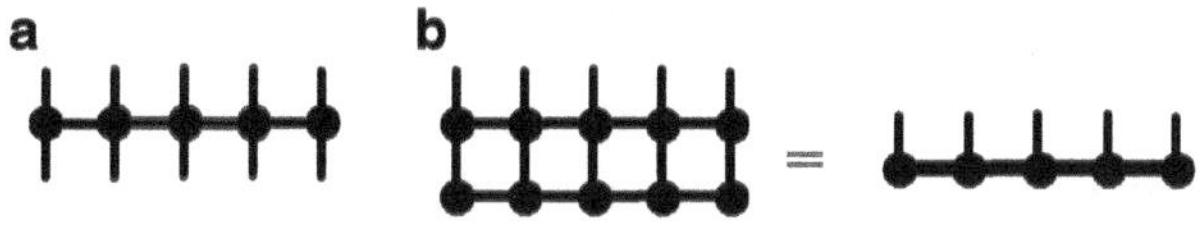

Fig. 7.3 MPO in diagrammatic notation. (**a**) A MPO consists of a contraction of rank-four tensors. We adopt the graphical convention that the line below the horizontal corresponds to i' and the line above the horizontal to i for an operator $O_{ii'}$ which takes the local state from $|i'\rangle$ to $|i\rangle$. (**b**) The product of an MPO and an MPS produces another MPS whose bond dimension is the product of the bond dimensions of the original MPS and the MPO, here denoted by *thicker lines*

that the bond dimension of the MPS representing the product of an MPO and an MPS is the product of the bond dimensions of the MPO and the MPS. Because the states $|i_1 \dots i_L\rangle$ are tensor products, we can also use the notation

$$\hat{O}_{\mathrm{MPO}} = \sum_{i_1,\dots i_L=1}^{d} \sum_{i'_1,\dots i'_L=1}^{d} \mathrm{Tr}\left(\mathcal{W}^{[1]i_1 i'_1} \dots \mathcal{W}^{[L]i_L i'_L}\right), \tag{7.24}$$

where each one of the objects $\mathcal{W}^{[j]i_j i'_j} \equiv W^{[j]i_j i'_j}|i_j\rangle\langle i'_j|$. That is to say, we can consider the matrices which appear in the matrix-product ansatz of an MPO to be operator-valued.

It is tempting to look for canonical forms for MPOs just as we did for MPSs, but the relevant norm for MPOs is the Frobenius norm $\langle \hat{O}_1, \hat{O}_2\rangle = \mathrm{Tr}(\hat{O}_1^{\dagger}\hat{O}_2)$ which scales exponentially in the local dimension with the number of lattice sites. Thus, for a typical many-body system with an exponentially large Hilbert space, the elements of the orthogonality center can differ in magnitude greatly, causing a catastrophic loss of precision during orthogonalization. However, most physically relevant MPOs such as one-dimensional Hamiltonians can be written down exactly in terms of an MPO canonical form which is analogous to an LU decomposition [40]. To discuss this canonical form, it is useful to recast an MPO as an FSA [41, 42].

To recast an MPO as a FSA, we first enumerate all of the physical operators we use to define our local Hilbert space, $\mathcal{O} = \{\hat{O}^{\alpha}\}$, where α labels distinct operators. We will call this set our *operator alphabet* (OA). As examples, the OA for the Ising model would be $\left\{\hat{I}, \hat{\sigma}^x, \hat{\sigma}^z\right\}$ with $\hat{I}$ the identity operator and $\hat{\sigma}^{\nu}$ the Pauli operator along the νth Cartesian direction. The OA for the Bose–Hubbard model [43, 44] would be $\left\{\hat{I}, \hat{b}^{\dagger}, \hat{b}, \hat{n}\right\}$, where $\hat{b}$ is a bosonic destruction operator and $\hat{n} = \hat{b}^{\dagger}\hat{b}$ the number operator. The particular matrix representation of the OA fixes the local basis states $\{|i\rangle\}$.

Using the OA, we now introduce a set of FSA *rules* $\{\mathcal{R}_p(\{\hat{O}^{p_1}, \dots, \hat{O}^{p_n}\}, \{h^p\}, w^p)\}$ which generate strings of the n operators $\{\hat{O}^{p_1}, \dots, \hat{O}^{p_n}\} \in \mathcal{O}$ weighted by scalar w^p and variables $\{h^p\}$. We will call the variables $\{h^p\}$ *Hamiltonian parameters* and the scalars w^p *weights*. Each rule consists of three operator-valued matrices in the case where the $\{h^p\}$ do not depend on position. The first matrix is the rightmost matrix in the MPO representation, $\mathcal{W}^{[L]}$, and represents the initial configuration. Next, we have the MPO matrix of the bulk of the chain $\mathcal{W}^{[j]}$, $2 \le j \le L-1$, which takes an input vector of operators on k sites and produces a vector of operators on $k+1$ sites according to some deterministic pattern. Finally, we have the leftmost matrix in the MPO representation, $\mathcal{W}^{[1]}$, which extends the operators according to the patten of $\mathcal{W}^{[j]}$ and returns a 1×1 operator-valued matrix. The trace of this matrix as in Eq. (7.23) is the desired Hamiltonian term. The generalization to position-dependent Hamiltonian parameters requires $L-2$ matrices in place of the bulk matrix $\mathcal{W}^{[j]}$, $2 \le j \le L-1$, but the only modification is that h^p becomes $h^p(j)$.

As a concrete example, consider the Ising model

$$\hat{H} = -J \sum_{\langle i,j \rangle} \hat{\sigma}_i^z \hat{\sigma}_j^z - h \sum_i \hat{\sigma}_i^x . \tag{7.25}$$

The Hamiltonian consists of two rules. The first is a site rule $\mathcal{R}_{\text{site}}(\hat{\sigma}^x, h, -1)$ which generates the string $-h\sum_i \hat{\sigma}_i^x$. The three matrices which provide this rule are

$$\mathcal{W}_{\text{site}}^{[1]} = \begin{pmatrix} -h\hat{\sigma}^x & \hat{I} \end{pmatrix}, \quad \mathcal{W}_{\text{site}}^{[2\le j\le L-1]} = \begin{pmatrix} \hat{I} & 0 \\ -h\hat{\sigma}^x & \hat{I} \end{pmatrix},$$
$$\mathcal{W}_{\text{site}}^{[L]} = \begin{pmatrix} \hat{I} \\ -h\hat{\sigma}^x \end{pmatrix}. \tag{7.26}$$

As can be verified,

$$\prod_{j=L-1}^{k} \mathcal{W}_{\text{site}}^{[j]} \mathcal{W}_{\text{site}}^{[L]} = \begin{pmatrix} \hat{I} \dots \hat{I} \\ -h\sum_{i=k}^{L} \hat{\sigma}_i^x \end{pmatrix}, \tag{7.27}$$

and so this rule produces the desired operator. Similarly, there is a bond rule $\mathcal{R}_{\text{bond}}(\{\hat{\sigma}^z, \hat{\sigma}^z\}, J, -1)$ given by

$$\mathcal{W}_{\text{bond}}^{[1]} = \begin{pmatrix} 0 & -J\hat{\sigma}^z & \hat{I} \end{pmatrix}, \quad \mathcal{W}_{\text{bond}}^{[2\le j\le L-1]} = \begin{pmatrix} \hat{I} & 0 & 0 \\ \hat{\sigma}^z & 0 & 0 \\ 0 & -J\hat{\sigma}^z & \hat{I} \end{pmatrix},$$
$$\mathcal{W}_{\text{bond}}^{[L]} = \begin{pmatrix} \hat{I} \\ \hat{\sigma}^z \\ 0 \end{pmatrix}, \tag{7.28}$$

which produces $-J\sum_{\langle i,j \rangle} \hat{\sigma}_i^z \hat{\sigma}_j^z$, with $\langle i, j \rangle$ denoting a sum over nearest-neighbors i and j. The full Hamiltonian is given by the direct sum of the matrices. Collecting rows of the direct sum which are exactly parallel, we have the MPO representation of the full operator

$$\mathcal{W}_{\text{Ising}}^{[1]} = \begin{pmatrix} -h\hat{\sigma}^x & -J\hat{\sigma}^z & \hat{I} \end{pmatrix}, \quad \mathcal{W}_{\text{Ising}}^{[2\le j\le L-1]} = \begin{pmatrix} \hat{I} & 0 & 0 \\ \hat{\sigma}^z & 0 & 0 \\ -h\hat{\sigma}^x & -J\hat{\sigma}^z & \hat{I} \end{pmatrix},$$
$$\mathcal{W}_{\text{Ising}}^{[L]} = \begin{pmatrix} \hat{I} \\ \hat{\sigma}^z \\ -h\hat{\sigma}^x \end{pmatrix}. \tag{7.29}$$

This construction can be readily extended to general sums of nearest-neighbor and on-site interactions with Hamiltonian

$$\hat{H} = -\sum_{\alpha=1}^{n_b} J_\alpha \sum_{\langle i,j \rangle} \hat{O}_i^{b_{1:\alpha}} \hat{O}_j^{b_{2:\alpha}} - \sum_{\beta=1}^{n_s} h_\beta \sum_i \hat{O}_i^{s_\beta} \tag{7.30}$$

by summing up the individual rules:

$$\begin{aligned}
\mathcal{W}^{[1]} &= \left(-\sum_{\beta=1}^{n_s} h_\beta \hat{O}^{s_\beta} \; -J_1 \hat{O}^{b_{1:1}} \; \dots \; -J_{n_b} \hat{O}^{b_{1:n_b}} \; \hat{I} \right) \\
\mathcal{W}^{[2\leq j\leq L-1]} &= \begin{pmatrix} \hat{I} & 0 & \dots & 0 & 0 \\ \hat{O}^{b_{2:1}} & 0 & \dots & 0 & 0 \\ \vdots & \vdots & \ddots & \vdots & \vdots \\ \hat{O}^{b_{2:n_b}} & 0 & \dots & 0 & 0 \\ -\sum_{\beta=1}^{n_s} h_\beta \hat{O}^{s_\beta} & -J_1 \hat{O}^{b_{1:1}} & \dots & -J_{n_b} \hat{O}^{b_{1:n_b}} & \hat{I} \end{pmatrix}, \\
\mathcal{W}^{[L]} &= \begin{pmatrix} \hat{I} \\ \hat{O}^{b_{2:1}} \\ \vdots \\ \hat{O}^{b_{2:n_b}} \\ -\sum_{\beta=1}^{n_s} h_\beta \hat{O}^{s_\beta} \end{pmatrix}.
\end{aligned} \tag{7.31}$$

The bond dimension of the MPO representation of this Hamiltonian is $2 + n_B$. MPO representations are not restricted to nearest-neighbor terms. Exponentially decaying terms of the form[4]

$$\hat{H} = -J \sum_{i<j} e^{-\lambda(j-i-1)} \hat{O}_i^1 \hat{O}_j^2 \tag{7.32}$$

can also be accommodated with the rule $\mathcal{R}_{\exp}(\{\hat{O}^1, \hat{O}^2\}, \{J, \lambda\}, -1)$

$$\begin{aligned}
\mathcal{W}_{\exp}^{[1]} &= \left(0 \; -J\hat{O}^1 \; \hat{I} \right), \quad \mathcal{W}_{\exp}^{[2\leq j\leq L-1]} = \begin{pmatrix} \hat{I} & 0 & 0 \\ \hat{O}^2 & e^{-\lambda}\hat{I} & 0 \\ 0 & -J\hat{O}^1 & \hat{I} \end{pmatrix}, \\
\mathcal{W}_{\exp}^{[L]} &= \begin{pmatrix} \hat{I} \\ \hat{O}^2 \\ 0 \end{pmatrix}.
\end{aligned} \tag{7.33}$$

[4]Note in this form that the nearest-neighbor coupling is J and only longer-ranged couplings contain λ.

We can also produce a general monotonically decaying term

$$\hat{H} = J\sum_{i<j} f(j-i)\,\hat{O}_i^1\hat{O}_j^2 \tag{7.34}$$

by approximating the term as a sum of exponentials to a desired distance r_{cutoff}. That is, we minimize the functional

$$F(\mathbf{a},\mathbf{b}) = \sum_{i=1}^{n_{\exp}}\sum_{x=1}^{r_{\mathrm{cutoff}}} \left|f(x) - a_i b_i^{x-1}\right|^2 \tag{7.35}$$

with respect to the exponential weights a_i and decay parameters b_i, where $\mathbf{a}$ and $\mathbf{b}$ are the elements a_i and b_i, respectively, arranged as vectors and $n_{\exp}$ is a convergence parameter controlling the number of exponentials used in the expansion [45–47]. Note that the fit is only guaranteed to be accurate to r_{cutoff} while the term has infinite range. This does not cause difficulties in practice for decaying functions, as the resulting fit is also guaranteed to be monotonically decaying. The decaying function rule is then just a sum of these $n_{\exp}$ exponential rules, $\mathcal{R}_{\mathrm{df}}(\{\hat{O}^1,\hat{O}^2\},\{J,f(x),n_{exp},r_{\mathrm{cutoff}}\},1) = \sum_{i=1}^{n_{\exp}}\mathcal{R}_{\exp}(\{\hat{O}^1,\hat{O}^2\},\{Ja_i,-\log b_i\},1)$. In contrast to the other rules presented above this rule is not exact, but the number of exponentials can be increased to any desired accuracy.[5] This procedure is surprisingly accurate even for small numbers of exponentials; 5 exponentials suffice to accurately represent a $1/r^3$ interaction out to distances of 1,000 sites to an error of 10^{-9} [47]. While the bond dimension increases linearly with the number of exponentials, the number of nonzero terms in the MPO also grows only linearly with the number of exponentials, and so operations with the MPO scale well as $n_{\exp}$ increases. However, the eigenstate of a Hamiltonian with larger $n_{\exp}$ may be more highly entangled than with smaller $n_{\exp}$, leading to an increase in χ and longer runtimes. That is to say, the relationship between the bond dimension of an MPO and the bond dimension of an MPS representing an eigenstate of the MPO is difficult to predict.

While the pure functional interaction Eq. (7.34) is appealing from a theoretical point of view, in practical applications infinite-range interactions do not represent a consistent level of approximation. That is to say, at some distance interactions are screened, where the screening length is set by energetics or lifetime constraints in the case of ultracold atoms or molecules [48]. Hence, we provide the finite-ranged rule $\mathcal{R}_{\mathrm{FiniteFunction}}(\{\hat{O}^1,\hat{O}^2\},\{h,f,r_{\mathrm{cutoff}}\},w)$ which generates the Hamiltonian

$$\hat{H} = wh\sum_i\sum_{j=i+1}^{i+r_{\mathrm{cutoff}}} f(j-i)\,\hat{O}_i^1\hat{O}_j^2 \tag{7.36}$$

[5] By accuracy we mean that the functional Eq. (7.35) is smaller than a given tolerance.

and is given by

$$
\begin{aligned}
\mathcal{W}^{[1]} &= \left(0\;0\;\dots\;0\;\dots\;0\;whf\left(1\right)\hat{O}^1\;\hat{I}\right)\\
\mathcal{W}^{[2\le j\le L-1]} &= \begin{pmatrix}
\hat{I} & 0 & \dots & 0 & \dots & 0 & 0 & 0\\
\hat{O}^2 & 0 & \dots & 0 & \dots & 0 & 0 & 0\\
\hat{O}^2 & \frac{f\left(r_{\mathrm{cutoff}}\right)}{f\left(r_{\mathrm{cutoff}}-1\right)}\hat{I} & \dots & 0 & \dots & 0 & 0 & 0\\
\vdots & \vdots & \ddots & \vdots & \ddots & \vdots & \vdots & \vdots\\
\hat{O}^2 & 0 & \dots & \frac{f\left(k\right)}{f\left(k-1\right)} & \dots & 0 & 0 & 0\\
\vdots & \vdots & \ddots & \vdots & \ddots & \vdots & \vdots & \vdots\\
\hat{O}^2 & 0 & \dots & 0 & \dots & \frac{f\left(2\right)}{f\left(1\right)}\hat{I} & 0 & 0\\
0 & 0 & \dots & 0 & \dots & 0 & whf\left(1\right)\hat{O}^1 & \hat{I}
\end{pmatrix},\\
\mathcal{W}^{[L]} &= \begin{pmatrix}
\hat{I}\\ \hat{O}^2\\ \hat{O}^2\\ \vdots\\ \hat{O}^2\\ \vdots\\ \hat{O}^2\\ 0
\end{pmatrix}.
\end{aligned}
\tag{7.37}
$$

For a vector f which contains zero elements a small modification must be made, but the bond dimension remains the same.

A small set of rules like those presented above allow us to template operators for MPS simulations. That is, given a method to produce a Hamiltonian from a set of basic rules, we can generate a wide variety of different Hamiltonians by specifying different OAs, weights, and Hamiltonian parameters as inputs to a program. The resulting MPOs are lower triangular and usually very sparse, and so sparse matrix structures can be used for efficiency. Beyond the flexibility that MPOs provide to MPS algorithms, they also allow for arithmetic operations such as addition and multiplication to be performed exactly, albeit at the expense of a growing bond dimension. The sum of two operators expressed as MPOs has MPO matrices which are the direct sums of the constituent matrices, and the product of two MPOs has MPO matrices which are the direct product of the constituent MPO matrices. Both of these operations preserve the sparse lower triangular structure of the MPOs.

7.3 Variational Ground State Search

We now turn to using MPSs as variational ansätze for the eigenstates of a Hamiltonian expressed as an MPO. The ground state is found by minimizing the functional

$$\mathcal{E}\left[|\psi\rangle\right] = \langle\psi|\hat{H}|\psi\rangle - E\langle\psi|\psi\rangle \tag{7.38}$$

with respect to the parameters of $|\psi\rangle$, where E is a Lagrange multiplier enforcing normalization. The general minimization of this functional is an NP-hard problem, so we instead adopt a local search heuristic that has proven to work well in practice. Let us consider fixing all parameters in the MPS except for a contiguous block of s MPS tensors $A^{[j]}\dots A^{[j+s-1]}$. The single-site ($s = 1$) and two-site ($s = 2$) algorithms are the most commonly used variants. We then find the extremum as

$$\frac{\partial}{\partial A^{[j]\star}\dots A^{[j+s-1]\star}}\left(\langle\psi|\hat{H}|\psi\rangle - E\langle\psi|\psi\rangle\right) = 0 \tag{7.39}$$

which corresponds to the diagrammatic equation shown in Fig. 7.4. Here, the partial derivative with respect to a tensor is defined to be a tensor whose elements are the partial derivatives with respect to the elements of the tensor. If we assume that the block of tensors to be optimized contains the orthogonality center, then the rightmost diagram reduces to the block of tensors being optimized and the leftmost diagram is the action of the *effective Hamiltonian* on this same block. Thus, minimization consists of finding the eigenvector corresponding to the smallest eigenvalue of the effective Hamiltonian eigenvalue problem

$$\hat{H}_{\text{eff}}^{[j]}A^{[j]}\dots A^{[j+s-1]} = EA^{[j]}\dots A^{[j+s-1]}, \tag{7.40}$$

where the effective Hamiltonian is

$$\hat{H}_{\text{eff}\,\alpha\beta\alpha'\beta'}^{[j]i_j\dots i_{j+s-1}i'_j\dots i'_{j+s-1}} = \sum_{\kappa\kappa'}\sum_{\gamma_1\dots\gamma_s} L^{[j]}_{\kappa\alpha\alpha'} W^{[j]i_j i'_j}_{\kappa\gamma_1}\dots W^{[j+s-1]i_{j+s-1}i'_{j+s-1}}_{\gamma_s\kappa'} R^{[j+s]}_{\kappa'\beta'\beta}, \tag{7.41}$$

L and R are the partial overlaps of the Hamiltonian MPO with the state as in Fig. 7.4, and $W^{[j]}$ is the MPO tensor at site j of the Hamiltonian. $\hat{H}_{\text{eff}}^{[j]}$ represents the Hamiltonian for the variational degrees of freedom in the block to be optimized with the rest of the state held fixed. This justifies our use of E as the eigenvalue, as E obtained from the solution of this equation is the current best estimate for the energy. We can view Eq. (7.40) as a linear eigenvalue problem by combining indices as

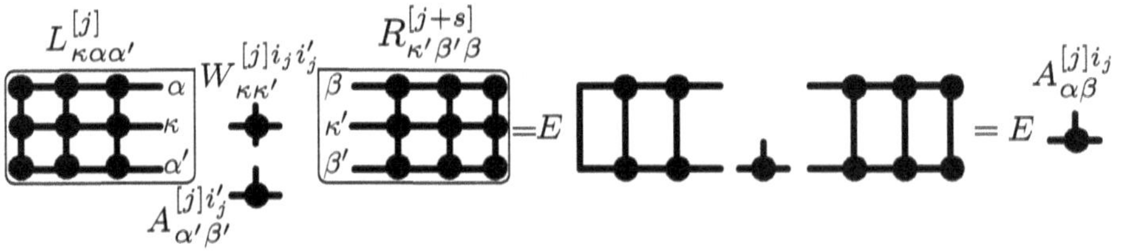

Fig. 7.4 Variational ground state search in diagrammatic notation. Here we display the single site ($s = 1$) effective Hamiltonian eigenvalue problem for simplicity. The contractions between the block of tensors to be optimized, the MPO, and the rest of the diagram have been omitted to accentuate the structure of the effective Hamiltonian. The rightmost equality follows from assuming that the variational site is the orthogonality center

$$\sum_{\alpha' i'_j \ldots i'_{j+s-1}\beta'} \hat{H}_{\text{eff}\,(\alpha i_j \ldots i_{j+s-1}\beta)(\alpha' i'_j \ldots i'_{j+s-1}\beta')} \left[A^{[j]} \ldots A^{[j+s-1]}\right]_{(\alpha' i'_j \ldots i'_{j+s-1}\beta')}$$
$$= E\left[A^{[j]} \ldots A^{[j+s-1]}\right]_{(\alpha i_j \ldots i_{j+s-1}\beta)}. \quad (7.42)$$

The linear dimension of this matrix representation of $\hat{H}^{[j]}_{\text{eff}}$ is $\chi_j d^s \chi_{j+s}$, and so a solution of this problem with dense methods would require $\mathcal{O}(\chi_j^3 d^{3s} \chi_{j+s-1}^3)$ basic operations, leading to a very slow algorithm of order $\mathcal{O}(\chi^6)$. In contrast, by taking advantage of the separable form of the effective Hamiltonian Eq. (7.41) multiplication of our block of tensors by the effective Hamiltonian can be done in $\mathcal{O}(\chi^3)$ time [12]. Thus, sparse eigensolvers such as the Lanczos [39] or Davidson [49] algorithms, which require only matrix–vector multiplies, should be employed to solve this eigenvalue problem.

The general algorithm for ground state search is thus as follows. We begin with an initial state with orthogonality center at site k. We choose a block of tensors containing k and optimize them by solving the effective Hamiltonian eigenvalue problem. We then shift the orthogonality center and the block of tensors one site to the right and again optimize. We continue shifting to the right until we reach the right boundary. We then reverse direction, shifting the orthogonality center and the block of tensors to be optimized to the left and solving the effective Hamiltonian eigenvalue problem until we reach the left boundary, at which point we reverse direction again. A single iteration of this optimization cycle which affects each tensor twice is called a *sweep*, and sweeping is continued until convergence. In addition, using the MPO form of the Hamiltonian, it is possible to develop a caching algorithm for the overlaps L and R such that the solution of this problem requires $\mathcal{O}(L)$ scaling in the number of lattice sites [42].[6] To do so, we begin the iteration

[6] This scaling does not account for possible L dependence of the bond dimension χ such as exists for conformal critical points.

with a guess for the ground state $|\psi\rangle$ assumed to have orthogonality center k. We then use the left recursion

$$L^{[1]}_{\kappa\alpha\alpha'} = \delta_{\alpha,1}\delta_{\alpha',1}\delta_{\kappa,1}, \tag{7.43}$$

$$L^{[\ell+1]}_{\kappa\alpha\alpha'} = \left[\sum_{\gamma' i'}\left[\sum_{i\kappa'}\left[\sum_{\gamma} A^{[\ell]i\star}_{\gamma\alpha} L^{[\ell]}_{\kappa'\gamma\gamma'}\right] W^{[\ell]ii'}_{\kappa'\kappa}\right] A^{[\ell]i'}_{\gamma'\alpha'}\right], \tag{7.44}$$

to generate the L overlaps up to k and the right recursion

$$R^{[L+1]}_{\kappa\beta'\beta} = \delta_{\beta,1}\delta_{\beta',1}\delta_{\kappa,1}, \tag{7.45}$$

$$R^{[\ell]}_{\kappa\beta'\beta} = \left[\sum_{\gamma' i'} A^{[\ell]i'}_{\beta'\gamma'}\left[\sum_{\kappa' i} W^{[\ell]ii'}_{\kappa\kappa'}\left[\sum_{\gamma} R^{[\ell+1]}_{\kappa'\gamma'\gamma} A^{[\ell]i\star}_{\beta\gamma}\right]\right]\right], \tag{7.46}$$

to generate the R overlaps down to $k+s$. Here the square braces indicate the order in which the contraction should be performed to achieve ideal scaling. Once the eigenvalue problem has been solved and the orthogonality center shifted, we use the recurrence Eq. (7.44) to update the overlaps when we are sweeping to the right and the recurrence Eq. (7.46) to update the overlaps when we are sweeping to the left.

Convergence is achieved when the variance

$$\Delta \equiv \left\langle\left(\widehat{H^2} - E^2\right)\right\rangle, \tag{7.47}$$

with E the energy eigenvalue, drops below a user-specified tolerance ϵ. Given the MPO form of the Hamiltonian, the variance operator $\hat{\Delta} \equiv \widehat{H^2} - E^2$ can be constructed by constructing an MPO whose matrices $\tilde{W}^{[i]}$ consist of the direct product of the corresponding matrices from $\hat{H}$, $\tilde{W}^{[i]} = W^{[i]} \otimes W^{[i]}$, and then subtracting $-\hat{I}E^2/L$ from the lower leftmost element of each $\tilde{W}^{[i]}$, where $\hat{I}$ is the identity operator. This representation is exact, in contrast to DMRG-based approaches where the basis of the Hamiltonian is tied together with the basis of the state. The variance is a much better measure of convergence of the *state* than the so-called discarded weight which is used to measure convergence of the two-site DMRG algorithm. This is because it is a property of the actual MPS state and not of the eigenvalue. As a note of caution, the variance only guarantees that the state found is *an* eigenstate to the given tolerance, it does not specify that it is the ground state. This has not proven to cause problems in practice for non-disordered systems.

In summary, the complete algorithm for variational ground state search is:

1. *Input*: Input a Hamiltonian $\hat{H}$ in MPO form, an initial guess $|\psi\rangle$ for the ground state in MPS form, and a tolerance ϵ for the variance.
2. *Initialization*: Construct the LR overlaps using the recursions Eqs. (7.44) and (7.46).

3. *Sweeping*:
   (a) Solve the effective Hamiltonian eigenvalue problem Eq. (7.40) and replace the variational block of tensors with the eigenvector corresponding to the lowest eigenvalue.
   (b) Shift the orthogonality center to the right and update the *LR* overlaps using the recursion Eq. (7.44).
   (c) Continue to iterate (a) and (b) (*right sweeping*) until the right boundary is reached.
   (d) Solve the effective Hamiltonian eigenvalue problem Eq. (7.40) and replace the variational block of tensors with the eigenvector corresponding to the lowest eigenvalue.
   (e) Shift the orthogonality center to the left and update the *LR* overlaps using the recursion Eq. (7.46).
   (f) Continue to iterate (d) and (e) (*left sweeping*) until the left boundary is reached.
   (g) Iterate (a) and (b) until k is reached.

4. *Check convergence*: Using the most recent estimate of the energy eigenvalue $\tilde{E}$ from the last effective Hamiltonian solution, construct the variance operator $\hat{\Delta}$ and find the variance. If $\Delta < \epsilon$, exit, otherwise return to (a).

7.4 Variational Excited State Search

We now turn to finding excited states. We find the nth excited state by minimizing the functional

$$\mathcal{E}\left[|\psi\rangle\right] = \langle\psi|\hat{H}|\psi\rangle - E\langle\psi|\psi\rangle - \sum_{k=0}^{n-1}\mu_k\langle\psi|\phi_k\rangle \tag{7.48}$$

where $\{|\phi_k\rangle\}$ are the $n-1$ lower-lying eigenstates of $\hat{H}$ and the $\{\mu_k\}$ are Lagrange multipliers enforcing the orthogonality constraints $\langle\psi|\phi_k\rangle = 0$. Again fixing a block of s tensors, the minimization of this functional with respect to this block is given by the *projected* effective Hamiltonian eigenvalue problem

$$\hat{P}^{[j]\dagger}\hat{H}^{[j]}_{\text{eff}}\hat{P}^{[j]}A^{[j]}\dots A^{[j+s-1]} = EA^{[j]}\dots A^{[j+s-1]}, \tag{7.49}$$

where $\hat{P}^{[j]}$ is a projector into the space orthogonal to the $\{|\phi_k\rangle\}$. Given the states $\{|\phi_k\rangle\}$ as MPSs, we construct these projectors as follows. The diagram corresponding to the overlap $\langle\psi|\phi_k\rangle$ is shown in Fig. 7.5a, with the bold lines and black circles corresponding to $|\phi_k\rangle$ and the thin lines and red circles to $\langle\psi|$. This is a *linear* form in all of the MPS tensors of $\langle\psi|$, and so the condition that $|\psi\rangle$

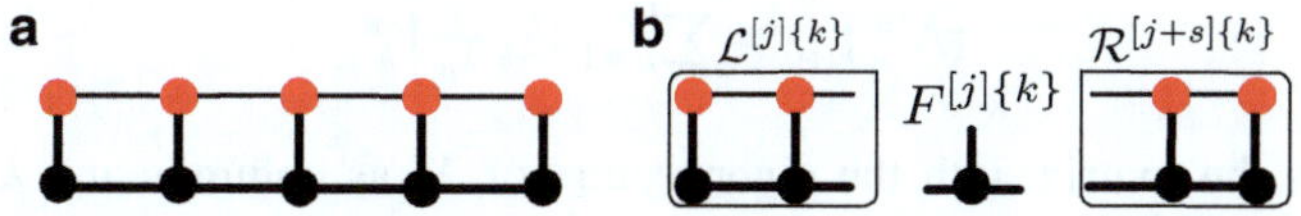

Fig. 7.5 Linear forms enforcing orthogonality. (**a**) The overlap $\langle\psi|\phi_k\rangle$ in diagrammatic notation. The *red circles* and *thin lines* correspond to $\langle\psi|$ and the *black circles* and *thick lines* to $|\phi_k\rangle$. (**b**) The linear form $F^{[j]\{k\}}$ in diagrammatic notation for the single-site case. As before, we leave the bottom tensor uncontracted to accentuate the definitions of the $\mathcal{LR}$ overlaps

be orthogonal to this state for the given block of tensors with all others held fixed may thus be stated as

$$\langle\psi|\phi_k\rangle = 0 \Rightarrow \left(\frac{\partial}{\partial A^{[j]\star}\dots A^{[j+s-1]\star}}\langle\psi|\phi_k\rangle\right) A^{[j]\star}\dots A^{[j+s-1]\star} = 0 \quad (7.50)$$

$$\equiv F^{[j]\{k\}} A^{[j]\star}\dots A^{[j+s-1]\star} = 0. \quad (7.51)$$

Here the linear form enforcing orthogonality $F^{[j]\{k\}}$ is shown diagrammatically in Fig. 7.5b. This object is a rank-$(s+2)$ tensor whose first and last dimensions are respectively χ_j, the left bond dimension of $A^{[j]}$, and χ_{j+s}, the right bond dimension of $A^{[j+s-1]}$. All other dimensions are d. Hence, $F^{[j]\{k\}}$ has the same shape as $A^{[j]}\dots A^{[j+s-1]}$, as implied by Eq. (7.50). We can construct the projector $\hat{P}^{[j]}$ as

$$\hat{P}^{[j]} = \hat{1} - \sum_{kk'} F^{[j]\{k\}} \left(\mathcal{N}^{-1}\right)_{kk'} F^{[j]\{k'\}\dagger}, \quad (7.52)$$

where $(\mathcal{N}^{-1})_{kk'}$ is the kk'th element of the inverse of the Gram matrix

$$\mathcal{N}_{kk'} = \mathrm{Tr}\left(F^{[j]\{k\}\dagger} F^{[j]\{k'\}}\right). \quad (7.53)$$

The projector $\hat{P}^{[j]}$ is a rank-$(2s+4)$ tensor which takes a tensor with the same shape as $A^{[j]}\dots A^{[j+s-1]}$ into another tensor of the same shape. The Gram matrix inverse is important to ensure that the projector is idempotent. As before, direct construction of the projected effective Hamiltonian leads to an algorithm which scales poorly as $\mathcal{O}(\chi^6)$. Hence, it is important to use sparse methods which require only the application of $\hat{P}$ and $\hat{H}$ on some block of tensors $A^{[j]}\dots A^{[j+s-1]}$. Direct application of $\hat{P}^{[j]}$ as written requires $\mathcal{O}(\chi^4)$ operations and also scales quadratically in the number of eigenstates desired N_E due to the double sum in Eq. (7.52). To find a total of N_E eigenstates by this method thus requires $\mathcal{O}(N_E^3\chi^4)$ operations, which is unacceptably slow. A simple idea to reduce this scaling would be to find the eigenvectors of the inverse Gram matrix and re-express the projectors $F^{[j]\{k\}}$ in terms of them, rendering the double sum a single sum. However, while the Gram matrix $\mathcal{N}$ is Hermitian and positive semidefinite it may also be badly conditioned and singular. A numerically stable alternative to this idea is to construct the Moore–Penrose pseudoinverse [39] of the Gram matrix

$$\left(\mathcal{N}^{+}\right)_{kk'}=\sum_{\mu=1}^{n_p}V_{k\mu}\frac{1}{\lambda_\mu}V^{\star}_{k'\mu},\tag{7.54}$$

where V is the matrix with the eigenvectors of $\mathcal{N}$ as columns and λ are the n_p eigenvalues of $\mathcal{N}$ which are greater than $n\lambda_{\max}\epsilon$, where n is the linear dimension of $\mathcal{N}$, $\lambda_{\max}$ its largest eigenvalue, and ϵ the machine precision. We use this pseudoinverse to transform to a new set of linear forms

$$G^{[J]\{\mu\}}=\frac{1}{\sqrt{\lambda_\mu}}\sum_k V_{k\mu}F^{[J]\{k\}}\tag{7.55}$$

such that

$$\hat{P}^{[J]}=\hat{1}-\sum_\mu G^{[J]\{\mu\}}G^{[J]\{\mu\}\dagger}.\tag{7.56}$$

Often, the dimension of the set $\left\{G^{[J]\{\mu\}}\right\}$ is much smaller than N_E. The diagonalization of the Gram matrix requires $\mathcal{O}(N_E^3)$ operations, independent of χ, and its construction and the construction of G in Eq. (7.55) both require $\mathcal{O}(\chi^2)$ operations. The operation of $\hat{P}^{[J]}$ on the variational block of tensors is now

$$\begin{aligned}\hat{P}^{[J]}A^{[J]}\dots A^{[J+s-1]}&=A^{[J]}\dots A^{[J+s-1]}\\&-\sum_\mu \mathrm{Tr}\left(G^{[J]\{\mu\}\dagger}A^{[J]}\dots A^{[J+s-1]}\right)G^{[J]\{\mu\}}\end{aligned}\tag{7.57}$$

which is linear in N_E and scales only as $\mathcal{O}(\chi^2)$. Thus, the dominant scaling for typical parameters $\chi\gg N_E$ is still the $\mathcal{O}(\chi^3)$ scaling of the effective Hamiltonian multiply, and the algorithm to find N_E excited states scales as $\mathcal{O}(N_E\chi^3)$. A sweeping approach is used as in the ground state search algorithm, and the iteration is stopped when the variance drops below a user-specified tolerance. As before, the variance does not guarantee that the state found is the next lowest-lying eigenstate, but this does not usually cause problems in practice.

As with the *LR* overlaps used in the variational ground state search, one can also cache the overlaps $\mathcal{LR}$ used to construct the linear forms F using the recursions

$$\mathcal{L}^{[1]\{k\}}_{\alpha\alpha'}=\delta_{\alpha,1}\delta_{\alpha',1},\tag{7.58}$$

$$\mathcal{L}^{[\ell+1]\{k\}}=\sum_i A^{[\ell]i\dagger}\mathcal{L}^{[\ell]k}B^{[\ell]\{k\}i}\tag{7.59}$$

and

$$\mathcal{R}^{[L+1]k}_{\beta'\beta}=\delta_{\beta,1}\delta_{\beta',1}\delta_{k,1},\tag{7.60}$$

$$\mathcal{R}^{[\ell]\{k\}}=\sum_i B^{[\ell]\{k\}i}\mathcal{R}^{[\ell+1]\{k\}}A^{[\ell]i\dagger},\tag{7.61}$$

where $B^{[\ell]\{k\}}$ is the MPS tensor of $|\phi_k\rangle$ at site ℓ. The linear forms are constructed using these overlaps as

$$F_{\alpha\beta}^{[j]\{k\}i_j \dots i_{j+s-1}} = \sum_{\gamma_1 \dots \gamma_{s+1}} \mathcal{L}_{\alpha\gamma_1}^{[j]\{k\}i_j} B_{\gamma_1\gamma_2}^{[j]\{k\}i_j} \dots B_{\gamma_s\gamma_{s+1}}^{[j+s-1]\{k\}i_{j+s-1}} \mathcal{R}_{\gamma_{s+1}\beta}^{[j+s]\{k\}}, \tag{7.62}$$

see Fig. 7.5b.

The variational ground state algorithm presented above is essentially equivalent to standard DMRG, aside from the calculation of the variance [12]. When finding excited states, however, DMRG-based approaches target multiple excited states in a single MPS, which causes the bond dimensions to grow and the quality of each individual eigenstate to degrade. Furthermore, as the ground state and all excited states are solved together in that approach, the sparse eigensolver must be able to converge interior eigenvalues, which is known to be troublesome [39, 40]. We call the present algorithm, which is a sparse and numerically stable variant of that proposed in [50] for PBC, *eMPS* to accentuate the difference.

In our experience, there are two main limitations of eMPS. The first is that it is difficult to construct good variational guesses for the excited states in contrast to the ground state where the infinite size MPS algorithm [26, 51] is applicable. Here, the usefulness of the variance becomes readily apparent, as the discarded weight can be 10^{-12} or less while the variance is of order 10^{-2} in early sweeps. The second is that the area law considerations which make MPS algorithms so practical for ground states do not in general apply to bulk eigenstates, and so the bond dimension required to accurately represent a general eigenstate may be exponential in the system size, rendering eMPS inapplicable.

The ability to find excited states is useful in many contexts. It provides access to the dynamical gap for determining the location of second order quantum phase transitions [52] and Kibble–Zurek scalings, even when the gap is not between different symmetry sectors. It can help in understanding the structure of CFTs by providing access to the primary scaling fields [53]. Excited states yield the structure function and other dynamic response functions of low-lying excitations. Such response functions are of great use for comparing to experimental measurements. For the case of periodic boundaries, MPS variants of the single-mode approximation [50, 54–56] together with a modified version of eMPS suitable for periodic boundary conditions could be used to variationally search for excited states of a definite momentum. Finally, by considering more complex functionals such as $\langle\psi|(\hat{H}-\epsilon)^2|\psi\rangle - \lambda\langle\psi|\psi\rangle$ for minimization, one can determine level spacing statistics in a desired energy range for systems much larger than are amenable to exact diagonalization. Such studies are immensely useful in discussions of integrability and quantum chaos, as well as investigations of the thermalization hypothesis [57–59].

7.5 Calculation of Observables

We now turn to how we can extract information from a state expressed as an MPS. We do so by the expectation values of Hermitian operators, or *observables*. We will demonstrate how to compute observables of three different types: local observables, two-point correlation functions, and general MPOs.

We define a local observable as an operator which acts only on the Hilbert space of a single site: $\hat{O}^{[k]} = \sum_{i,i'} O^{[k]}_{ii'}|i\rangle\langle i'|$. If this site corresponds to the orthogonality center of $|\psi\rangle$, then the expectation value reduces to

$$\langle\psi|\hat{O}^{[k]}|\psi\rangle = \sum_{i_k i_k'} \mathrm{Tr}\left(A^{[k]i_k\dagger} O^{[k]}_{i_k i_k'} A^{[k]i_k'}\right). \tag{7.63}$$

The overall scaling for fixed site index k is $\mathcal{O}\left(\chi^2 d^2\right)$.

A two-point correlation function is an expectation value of the form $\langle\hat{O}^{[q]\dagger}\hat{O}^{[r]}\rangle$ where we take $q < r$ without loss of generality. If the orthogonality center of the MPS, k, lies within the range $q \le k \le r$, then we can evaluate the expectation value using only the tensors in this range. The most efficient way to proceed is first to form the matrix

$$R^{[r]} = \sum_{i_r i_r'} A^{[r]i_r'} O^{[r]}_{i_r i_r'} A^{[r]i_r\dagger}, \tag{7.64}$$

recursively generate

$$R^{[r-\ell]} = \sum_i A^{[r-\ell]i} R^{[r-\ell+1]} A^{[r-\ell]i\dagger} \tag{7.65}$$

for $\ell = 1, \ldots, r-q+1$, and then evaluate

$$\langle\hat{O}^{[q]\dagger}\hat{O}^{[r]}\rangle = \sum_{i_q, i_q'} \mathrm{Tr}\left(O^{[q]}_{i_q i_q'} A^{[q]i_q'} R^{[q+1]} A^{[q]i_q\dagger}\right). \tag{7.66}$$

For fixed q and r, the algorithm scales as $\mathcal{O}\left(\chi^3 d + \chi^2 d^2\right)$.

To compute the expectation of a general many-body observable $\hat{O}$ expressed as an MPO we start from the right (left) boundary and follow the recursion Eq. (7.44) (Eq. (7.46)) all the way to the opposite boundary, at which point the remaining $1\times1\times1$ tensor is the expectation value. The overall scaling is $\mathcal{O}\left(\chi^3 d\chi_O + \chi^2 d^2 M\left(\chi_O\right)\right)$ where χ_O is the bond dimension of the MPO and $M\left(\chi_O\right)$ is the number of nonzero elements in the MPO, which usually scales as χ_O.

Entanglement measures such as the bond entropy

$$S_j \equiv -\mathrm{Tr}\hat{\rho}_j \log\hat{\rho}_j, \tag{7.67}$$

$$\hat{\rho}_j \equiv \mathrm{Tr}_{i<j+1}|\psi\rangle\langle\psi|, \tag{7.68}$$

can be calculated from the singular values $\boldsymbol{\Sigma}$ of the MPS tensor $A^{[j]i}_{\alpha\beta}$ as

$$S_j = -\sum_{\gamma} \Sigma_{\gamma}^2 \log \Sigma_{\gamma}^2 \tag{7.69}$$

when this tensor is the orthogonality center. These singular values are computed automatically as part of the algorithm to shift the orthogonality center, see Sect. 7.2.1.

7.6 Time Evolution with MPSs

We now turn our attention to a variational solution of the time-dependent Schrödinger equation

$$i\hbar\frac{\partial}{\partial t}|\psi(t)\rangle = \hat{H}(t)|\psi(t)\rangle \tag{7.70}$$

using MPSs. The general strategy is to find some representation of the propagator over some time interval $[t, t+\delta t]$, $\hat{U}(t, t+\delta t)$, and variationally optimize the functional

$$\left||\psi(t+\delta t)\rangle - \hat{U}(t, t+\delta t)|\psi(t)\rangle\right|^2 \tag{7.71}$$

with respect to the MPS tensors of $|\psi(t+\delta t)\rangle$. Several complications arise in this case which were not present in the earlier algorithms. The first practical consideration is that the MPO form of the propagator may be difficult and very expensive to calculate. The second difficulty is more physical; the time-dependent state following a global quench of a Hamiltonian parameter has entanglement which generally grows linearly in time [60]. This causes the bond dimension χ to grow exponentially in time, and so there is some finite time where an MPS simulation will exhaust the available computational resources. However, many important questions regarding non-equilibrium dynamics can still be answered by considering moderately sized systems and short times. In addition, consideration of a situation in which the Hamiltonian changes only locally can greatly increase the accessible system sizes and simulation times [61–63].

The most common approach to time evolution for MPSs is to use the Suzuki–Trotter expansion

$$\exp\left[-i\delta t\left(\sum_{n=1}^{L-1}\hat{H}_n\right)\right] = \prod_{n=1}^{L-1}\exp\left(-\frac{i\delta t}{2}\hat{H}_n\right) \times \prod_{n=L-1}^{1}\exp\left(-\frac{i\delta t}{2}\hat{H}_n\right) + \mathcal{O}\left(\delta t^3\right) \tag{7.72}$$

or its higher order variants to construct a series of two-site propagators which can be constructed and applied easily. This is the basis of the equivalent [12] TEBD [64] and tDMRG [65, 66] algorithms. Here $\hat{H}_n$ is the nearest-neighbor bond Hamiltonian acting on-sites n and $n+1$. This approach is no longer viable when the Hamiltonian has longer-ranged terms, and attempts to accommodate such longer-ranged terms often exhibit poor scaling [67, 68] and require Hamiltonian-specialized implementation, resulting in inefficient, sometimes prohibitively inefficient code. Krylov-based time evolution, which will form the basis for our approach, has been considered in both DMRG [69] and MPS [13, 70] variants for the time-independent case. We note that the latter approach has been used [13, 14] to study time-dependent systems, but this necessitated very small time steps set by the rate of change of the Hamiltonian in order to provide accurate results. Our approach generalizes the latter method to the time-dependent case where the error is independent of the rate of change of the Hamiltonian and demonstrates how the algorithm can be formulated entirely in terms of FSA rules for MPOs.

Finally, we would like to mention one other recent development for time evolution of MPSs using MPOs, which is the transverse folding method proposed in [71, 72]. In these works, it is shown that contracting the tensor network which represents some time-evolved expectation value along the spatial direction rather than the time direction can lead to greater efficiency and hence extrapolations of results to longer times. However, this method requires that the propagator of the Hamiltonian have a known MPO representation in order to contract the network. The representation of the propagator as an MPO is possible for nearest-neighbor Hamiltonians using the Suzuki–Trotter expansion and exact MPO representations of the propagator are known in some specific cases such as the Ising model [45], which was the focus of the above works. At present, we know of no method for representing a unitary approximation to the propagator of a general Hamiltonian as an MPO. Our approach does not require the MPO form of the propagator. Rather, the action of the propagator on a given state represented as an MPS is determined using only the MPO form of the Hamiltonian. Hence, our method applies to any system whose Hamiltonian is represented as an MPO, including long-range interacting systems.

7.7 Commutator-Free Magnus Expansions

The propagator of a general time-dependent Hamiltonian which does not commute with itself at different times is given as a time-ordered exponential

$$\hat{U}\left(t,t+\delta t\right) \equiv \mathcal{T}\left[\exp\left(-i\int_t^{t+\delta t} dt'\hat{H}\left(t'\right)\right)\right] \tag{7.73}$$

whose most well-known form is the Dyson series

$$\hat{U}(t, t+\delta t) = \hat{1} - i\int_t^{t+\delta t} dt_1 \hat{H}(t_1) - \int_t^{t+\delta t} dt_1 \int_t^{t+\delta t} dt_2 \hat{H}(t_1)\hat{H}(t_2) + \ldots \tag{7.74}$$

This formulation of the propagator is not convenient numerically, as the Dyson series is an asymptotic series and so it can be difficult to determine an appropriate criteria for termination of the series. Furthermore, keeping only a finite number of terms in the Dyson series does not preserve the Lie group structure of the propagator; that is, the finite approximation is not unitary. An alternative approach which produces unitary approximations to the propagator was given by Magnus [73] who used the ansatz

$$\hat{U}(0, t) = \exp\left(\hat{\Omega}(t)\right) \tag{7.75}$$

to define the *Magnus series*

$$\hat{\Omega}(t) = \sum_{n=1}^{\infty} \hat{\Omega}_n(t), \tag{7.76}$$

where the nth term is of order t^n in the sense that its power series in t starts with t^n. The term $\hat{\Omega}_n(t)$ involves n nested integrations over $n-1$ nested commutators of $\hat{H}(t)$ at different times. Explicitly, the first few terms are:

$$\hat{\Omega}_1(t) = -i\int_0^t dt_1 \hat{H}(t_1), \tag{7.77}$$

$$\hat{\Omega}_2(t) = -\frac{1}{2}\int_0^t dt_1 \int_0^{t_1} dt_2 \left[\hat{H}(t_1), \hat{H}(t_2)\right], \tag{7.78}$$

$$\hat{\Omega}_3(t) = i\frac{1}{6}\int_0^t dt_1 \int_0^{t_1} dt_2 \int_0^{t_2} dt_3 \Big(\left[\hat{H}(t_1), \left[\hat{H}(t_2), \hat{H}(t_3)\right]\right] + \left[\left[\hat{H}(t_1), \hat{H}(t_2)\right], \hat{H}(t_3)\right]\Big). \tag{7.79}$$

While approximations obtained from truncating the series yield unitary propagators, these expressions are still formidable numerically, involving nested commutators and multidimensional integrals. The commutators pose a special difficulty for MPOs, as exact multiplication of MPOs involves multiplication of the bond dimensions of the MPOs and hence the algorithm scales exponentially in the number of terms kept in the series. Optimization algorithms which attempt to variationally shrink the bond dimension of an MPO sum or product such as those proposed in [47] may also be used, but these become numerically unstable for large systems, and when MPOs are subtracted as in commutators large cancellations can cause these algorithms to become stuck far from the variational optimum.

Hence, rather than work directly with the Magnus series, Eq. (7.76), we start from ansätze of the form

$$\hat{U}\left(t,t+\delta t\right)=\prod_{i=1}^{s}\exp\left(-i\delta t\hat{\Omega}_i\right) \tag{7.80}$$

where each one of the $\hat{\Omega}_i$ is a linear combination of $\hat{H}$ at different times in the interval $[t, t+\delta t]$, and require that our ansatz matches the Magnus expansion (equivalently, the full propagator) up to order δt^{N+1}. We will call such an ansatz a commutator-free Magnus expansion (CFME) [74, 75]. This ansatz has a number of features which make it desirable for our purposes. It is exactly unitary and so the norm is conserved. Also, provided that we consider the case where only the Hamiltonian parameters change in time and the operators are time-independent, the sums of the Hamiltonian at different times can be represented exactly as an MPO using the rules of Sect. 7.2.2. Thus, the need for complex operations with MPOs vanishes. Finally, because the ansatz takes into account the time dependence of the Hamiltonian explicitly, the time step is not necessarily fixed by the rate of variation of the Hamiltonian, allowing for more coarse stepping in time with fixed error.

Following [76], the procedure for generating an Nth-order CFME is to expand the function $H(t)$ in terms of (shifted) Legendre polynomials P_n,

$$\hat{H}_n=(2n-1)\,\delta t\int_0^1 dx\hat{H}\left(t+x\delta t\right)P_{n-1}\left(x\right). \tag{7.81}$$

The orthogonality properties of the Legendre polynomials allow the nested integration to be done exactly, leaving a series of nested commutators of the $\hat{H}_n$. Working in a Hall basis [77], this series of commutators is matched with the original Magnus expansion to yield the order conditions $f_{i,n}$ such that

$$\hat{\Omega}_i=\sum_{n=1}^{N}f_{i,n}\hat{H}_n. \tag{7.82}$$

We note that these order conditions are independent of $\hat{H}(t)$ by construction, and so are set by the choice of CFME alone. As we only require the result to be valid to order δt^{N+1}, the integration required for the coefficients $\hat{H}_n$ may be performed using Gauss–Legendre quadrature of order $N/2+1$. The end result of the analysis is that an Nth order expansion with s exponentials may be written as

$$\hat{U}\left(t+\delta t,t\right)=e^{-i\delta t\hat{\Omega}_1}\dots e^{-i\delta t\hat{\Omega}_s} \tag{7.83}$$

$$\hat{\Omega}_i=\sum_{m=1}^{N/2+1}g_{i,m}\hat{H}\left(t+x_m\delta t\right) \tag{7.84}$$

$$g_{i,m}=w_m\sum_{n=1}^{N/2+1}\left(2n-1\right)P_{n-1}\left(x_m\right)f_{i,n}, \tag{7.85}$$

where x_m and w_m are the points and weights for Gauss–Legendre quadrature [78]. In this work we use a fourth order expansion with three exponentials ($N = 4$, $s = 3$). The optimal order conditions for this case, obtained in [76], are

$$f_{1,1} = 11/40, \quad f_{1,2} = 20/87, \quad f_{1,3} = 7/50, \quad f_{2,1} = 9/20, \quad f_{2,3} = -7/25, \tag{7.86}$$

with $f_{s-i+1,n} = (-1)^{n+1} f_{i,n}$. Order conditions for higher order expansions may also be found in [76].

We now consider that our time-dependent Hamiltonian MPO is constructed from a set of FSA rules $\left\{\mathcal{R}_p\left(\left\{\hat{O}^{p_1},\dots,\hat{O}^{p_n}\right\},\{h^p(t)\},w^p\right)\right\}$ in which the OA and the weights are chosen without loss of generality to be time-independent. In this case, the expansion Eq. (7.81) is applied individually to each Hamiltonian parameter $h^p(t)$, resulting in the parameters $\{h_n^p\}$. Now, because of the canonical decomposition of Sect. 7.2.2, the MPO forms of $\hat{\Omega}_i$ from Eq. (7.84) at time t can be constructed exactly using the FSA rule set $\left\{\mathcal{R}_p\left(\left\{\hat{O}^{p_1},\dots,\hat{O}^{p_n}\right\},\left\{\sum_{m=1}^{N/2+1} g_{i,m}h^p(t+x_m\delta t)\right\},w^p\right)\right\}$. We note that each one of these operators has the same bond dimensions as the original Hamiltonian, and the updates of operators $\hat{\Omega}_i$ at each time step can be done in $\mathcal{O}(L\chi_O)$ time, which is essentially negligible.

The fact that we can construct the $\hat{\Omega}_i$ from the same FSA rules as the Hamiltonian implies that our CFME ansatz is equivalent to evolving our system according to piecewise constant Hamiltonians of the same form but with differing Hamiltonian parameters. Additionally, as also occurs in high-order Suzuki–Trotter expansions, evolution backwards in time may occur. Finally, we note that even terms in the Hamiltonian whose parameters do not vary in time have their magnitude altered by Eq. (7.84), as $\sum_m g_{i,m} \neq 1$ in general.

7.8 Krylov Subspace Propagation

Using the CFME Eq. (7.83) we never need to explicitly form an MPO representation of the propagator provided we can find an MPS representation of the exponential of an MPO applied to an MPS. We find such a representation from minimizing functionals of the form

$$\left| |\phi\rangle - \exp\left(-i\delta t\hat{\Omega}\right)|\psi\rangle \right|^2, \tag{7.87}$$

where, importantly, $\hat{\Omega}$ has a known MPO representation. We do so by forming a Krylov subspace approximation to the exponential [79] in which the Krylov vectors are represented as MPSs. Specifically, we do so via the Lanczos algorithm for the matrix exponential, which can be stated as

1. *Input*: Input $\hat{\Omega}$ in MPO form, $|\psi\rangle$ in MPS form, and a tolerance ϵ for truncating the recursion.
2. *Initialize*: Set $\beta_0 = \langle\psi|\psi\rangle = 1$ and $|r\rangle = |\psi\rangle$.
3. *Iterate*: For $j = 1, 2, \ldots$ until convergence
   (a) $|v_j\rangle = |r\rangle/\beta_{j-1}$.
   (b) $|r\rangle = \hat{\Omega}|v_j\rangle$.
   (c) $\alpha_j = \langle v_j|r\rangle$.
   (d) Orthogonalize $|r\rangle$ against $|v_j\rangle$ and $|v_{j-1}\rangle$.
   (e) Re-orthogonalize $|r\rangle$ against all v_k, $k \leq j$ if necessary.
   (f) $\beta_j = \langle r|r\rangle$
   (g) Form the matrix exponential of $T^{(j)}$, $U^{(j)}$, and obtain $c^{(j)} = U^{(j)}_{1:j,1}$.
   (h) Test for convergence.
4. *Finalize:* Set $|\phi\rangle = \sum_{i=1}^{j} c_i^{(j)}|v_i\rangle$

Here $T^{(j)}$ is the symmetric tridiagonal matrix with the α_i, $1 \leq i \leq j$ on the diagonal and β_i, $1 \leq i \leq j-1$ on the super-diagonal. It is important to use a matrix exponentiation method which produces a unitary matrix to machine precision in order to not lose the Lie group structure. Because of the small linear dimensions of the matrix $T^{(j)}$, exponentiation by direct diagonalization is practical. An a posteriori estimate for convergence of the Lanczos recursion is that $\left|2\beta_{j-1}c_j^{(j)}\right| < \epsilon$, where ϵ is the tolerance [79]. This can be compared with residual estimates in the ordinary Lanczos algorithm for finding eigenvalues. A rigorous bound on the approximation $\left||\psi\rangle_{\text{krylov}} - |\psi\rangle\right| \leq 12\exp\left\{-\frac{(\rho\delta t)^2}{16n}\right\}\left(\frac{e\rho\delta t}{4n}\right)^n$ can be established [80] when $n \geq \rho\delta t/2$ with n the number of Lanczos vectors and $\rho = |E_{\text{max}} - E_{\text{min}}|$ the spectral width of $\hat{\Omega}$. This estimate shows that for typical tolerances $\epsilon = 10^{-6}$ to 10^{-10}, 6–20 Lanczos vectors suffice.

As stated before, MPSs do not form a vector space and so the multiplication by $\hat{\Omega}$, the orthogonalization, and the final summation cannot be done exactly while keeping the bond dimension of our MPS fixed. However, just as with the eigenstate search, we can devise variational algorithms for these three operations which are iteratively performed until a desired tolerance is reached and use this tolerance to bound the bond dimension of our time-evolved MPS. We begin by briefly reviewing the standard algorithm [12] for finding the optimal MPS $|\phi\rangle$ representing a sum of MPSs $\sum_{k=1}^{N} c_k|\psi_k\rangle$ to a given tolerance, as the other algorithms are similar but more complex. In this case we have a set of $\mathcal{LR}$ overlaps defined between our variational state $\langle\phi|$ and the states $|\psi_k\rangle$ as in Eqs. (7.59) and (7.61). We now sweep through the lattice and make the replacement

$$A^{[j]}\ldots A^{[j+s-1]} = \sum_k c_k F^{[j]\{k\}}, \tag{7.88}$$

where the $F^{[j]\{k\}}$ are formed as in Eq. (7.62), see also Fig. 7.5b. The orthogonality center of $|\phi\rangle$ is then shifted, the $\mathcal{LR}$ overlaps updated, and sweeping continued. Convergence can be monitored via

$$\left| |\phi\rangle - \sum_k c_k |\psi_k\rangle \right|^2 = \mathrm{Tr}\left[\left(A^{[j]}\dots A^{[j+s-1]}\right)^{\dagger} A^{[j]}\dots A^{[j+s-1]}\right] + 1 \quad (7.89)$$

$$-2\sum_k \mathrm{Re}\,\mathrm{Tr}\left[c_k\left(A^{[j]}\dots A^{[j+s-1]}\right)^{\dagger} F^{[j]\{k\}}\right] < \epsilon ,$$

with Re denoting the real part. Because we do not have to solve an eigenvalue problem at each iteration, this algorithm is often much less costly than the iterative eigenstate search. Also, when we have that the coefficient vector c and all of the $\{|\psi_k\rangle\}$ have length 1, we can normalize the state $|\phi\rangle$ at the end of the calculation if required.

The algorithm to variationally fit an MPS to $\hat{\Omega}|\psi\rangle$ is similar. In this case we have a set of LR overlaps defined via the recursions

$$L^{[1]}_{\kappa\alpha\alpha'} = \delta_{\alpha,1}\delta_{\alpha',1}\delta_{\kappa,1}, \quad (7.90)$$

$$L^{[\ell+1]}_{\kappa\alpha\alpha'} = \left[\sum_{\gamma' i'}\left[\sum_{i\kappa'}\left[\sum_{\gamma} A^{[\ell]i\star}_{\gamma\alpha} L^{[\ell]}_{\kappa'\gamma\gamma'}\right] W^{[\ell]ii'}_{\kappa'\kappa}\right] B^{[\ell]i'}_{\gamma'\alpha'}\right], \quad (7.91)$$

$$R^{[L+1]}_{\kappa\beta'\beta} = \delta_{\beta,1}\delta_{\beta',1}\delta_{\kappa,1}, \quad (7.92)$$

$$R^{[\ell]}_{\kappa\beta'\beta} = \left[\sum_{\gamma' i'} B^{[\ell]i'}_{\beta'\gamma'}\left[\sum_{\kappa' i} W^{[\ell]ii'}_{\kappa\kappa'}\left[\sum_{\gamma} R^{[\ell+1]}_{\kappa'\gamma'\gamma} A^{[\ell]i\star}_{\beta\gamma}\right]\right]\right], \quad (7.93)$$

where the MPS tensors of $|\phi\rangle$ are denoted by A and those of $|\psi\rangle$ denoted by B. We now sweep through the lattice and make the replacement

$$A^{[j]}\dots A^{[j+s-1]} = \hat{H}^{[j]}_{\mathrm{eff}} B^{[j]}\dots B^{[j+s-1]} \quad (7.94)$$

where the effective Hamiltonian is formed from the LR overlaps as in Eq. (7.41). Convergence can be monitored via

$$\left| |\phi\rangle - \hat{H}|\psi\rangle \right|^2 = \mathrm{Tr}\left[\left(A^{[j]}\dots A^{[j+s-1]}\right)^{\dagger} A^{[j]}\dots A^{[j+s-1]}\right] + \langle\psi|\hat{H}^2|\psi\rangle \quad (7.95)$$

$$-2\mathrm{Re}\,\mathrm{Tr}\left[\left(A^{[j]}\dots A^{[j+s-1]}\right)^{\dagger} \hat{H}_{\mathrm{eff}} B^{[j]}\dots B^{[j+s-1]}\right] < \epsilon ,$$

Here, $\langle\psi|\hat{H}^2|\psi\rangle$ can be computed in a manner similar to the variance, and need only be computed once at the beginning of the calculation. We have also assumed that the block of tensors in Eq. (7.95) contains the orthogonality center. Again, this algorithm is often much less costly than the iterative eigenstate search.

We now turn to steps (iii)(d) and (iii)(e) of our algorithm. Step (iii)(d) is usually stated for ordinary vector spaces as

$$|r\rangle = |r\rangle - \alpha_j|v_j\rangle - \beta_{j-1}|v_{j-1}\rangle, \tag{7.96}$$

as $\alpha_j = \langle v_j|r\rangle$ and $\beta_{j-1} = \langle v_{j-1}|r\rangle$ and so this is equivalent to classical Gram–Schmidt orthogonalization. Hence, we could implement step (iii)(d) by using the fitting algorithm Eq. (7.88) to find the MPS closest to $|r\rangle - \alpha_j|v_j\rangle - \beta_{j-1}|v_{j-1}\rangle$. However, we have found that the following algorithm, which is closely related to eMPS, often converges more quickly and also is applicable to step (iii)(e). In our method we look for the optimal MPS $|\phi\rangle$ representing $|\psi\rangle$ but also subject to the constraints that $\langle\phi|\psi_k\rangle = 0$ for some set $\{|\psi_k\rangle\}$. We start by copying the state $|\psi\rangle$ to a variational ansatz $|\phi\rangle$. We then construct overlaps between the state $\langle\phi|$ and $|\psi\rangle$, which we call LR[7] and a set of overlaps between $\langle\phi|$ and $|\psi_k\rangle$, which we call $\mathcal{LR}$. We then sweep through the lattice and make the replacement

$$A^{[j]}\dots A^{[j+s-1]} = \sum\nolimits_{\gamma_1\dots\gamma_{s+1}} L^{[j]i_j}_{\alpha\gamma_1} B^{[j]i_j}_{\gamma_1\gamma_2}\dots B^{[j+s-1]i_j}_{\gamma_s\gamma_{s+1}} R^{[j+s]}_{\gamma_{s+1}\beta}, \tag{7.97}$$

with B the MPS tensors of $|\psi\rangle$ and A the MPS tensors of $|\phi\rangle$. We then apply the projector into the space orthogonal to the $|\psi_k\rangle$ by constructing the set $\{G^{[j]\{\mu\}}\}$, $\mu = 1,\dots,p$, as in Eq. (7.55) and performing

$$A^{[j]}\dots A^{[j+s-1]} = A^{[j]}\dots A^{[j+s-1]} - \mathrm{Tr}\left(G^{[j]\{\mu\}\dagger} A^{[j]}\dots A^{[j+s-1]}\right) G^{[j]\{\mu\}}, \tag{7.98}$$

for $\mu = 1,\dots,p$. Using the fitting algorithm of Eq. (7.88) corresponds to replacing the $\mathcal{LR}$ overlaps, which are between the *variational* state $\langle\phi|$ and the set $\{|\phi_k\rangle\}$, with a set of $\mathcal{LR}$ overlaps between the state $\langle\psi|$ and the set $\{|\phi_k\rangle\}$. Our algorithm, which amounts to fitting followed by modified Gram–Schmidt, uses information about the distance between the variational state and those to be orthogonalized against to determine operations, and hence often converges more quickly and is more stable. Convergence can be monitored by ensuring that $\langle\phi|\psi_k\rangle$ are orthogonal to a precision ϵ via

$$\left|\mathrm{Tr}\left(A^{[\ell]\dagger} F^{[\ell]\{k\}}\right)\right| \le \epsilon. \tag{7.99}$$

If one requires additional truncation of the bond dimension, one can switch to the ordinary fitting algorithm Eq. (7.88) at this point, using a new variational state $|\zeta\rangle$ to fit to $|\phi\rangle$.

We now pause to consider the sources of error in the time-propagation routine. First, because the CFME expansion Eq. (7.83) is only of order δt^{N+1}, the error

[7] In this initialization all of the LR are Kronecker deltas provided that $|\psi\rangle$ has an orthogonality center.

incurred in using this form of the propagator is $\epsilon_{\text{CFME}} = ct_{\text{final}}\delta t^N$, where t_{final} is the final time reached. Thus, as the final time desired becomes longer, smaller time steps should be taken in order to keep the error fixed. The coefficient c can be determined by using this known scaling and decreasing the time step by a constant factor. Factors in the range $2^{1/N}$ to $3^{1/N}$ are practical. This strategy can also be used to devise adaptive time step applications such as those used widely in ordinary differential equation solvers. Next, there is the error ϵ_{Krylov} incurred in the Krylov subspace approximation to the exponential. As discussed above, this error can be minimized by adding more and more Lanczos vectors. This error also involves the time step δt, and so when determining the CFME expansion error constant c one should be careful that $\epsilon_{\text{Krylov}} < \epsilon_{\text{CFME}}$. Finally, there are errors resulting from the variational fitting of MPSs in steps (iii)(b), (iii)(d), and (iii)(e) of the Lanczos algorithm for the matrix exponential. These can be reduced by lowering the variational tolerances, but this is done typically at the expense of a larger bond dimension χ and hence a slower algorithm and more memory usage.

7.9 Simulation Protocol

We are now in a position to devise a complete, generic protocol for the time evolution of a 1D quantum system.

1. *Input*: Input an operator alphabet and a set of FSA rules defining the Hamiltonian MPO. Input the functional forms $\{h^p(t)\}$ of its Hamiltonian parameters, a final time desired t_{final} and a time step δt. Input tolerances $\{\epsilon\}$ for variational ground state search and time evolution.
2. *Initial state preparation*: Find the ground state of the Hamiltonian using variational ground state search from Sect. 7.3. Alternatively, if a different initial state is desired, read in its MPS representation. Set $|\psi(t=0)\rangle$ to be this state.
3. *Measure*: Measure local observables and two-point correlation functions constructed from the OA as well as MPO observables constructed from their own FSA rule sets.
4. *eMPS*: Use eMPS (Sect. 7.4) to find excited states of the Hamiltonian, if desired. Measure properties of the excited states.
5. *tMPS*: Set $t = 0$.

   (a) For $i = s, \dots 1$, construct $\hat{\Omega}_i$ from Eq. (7.84) using the FSA rules and use the Krylov algorithm to apply the matrix exponentials of these operators in succession to $|\psi(t)\rangle$. Set $t = t + \delta t$.
   (b) Measure $|\psi(t)\rangle$ if desired.
   (c) Continue (a) and (b) until t_{final} is reached.

Starting from a base set of rules such as the site, bond, exponential, decaying function, and finite function rules of Sect. 7.2.2, a single implementation can accommodate a vast range of systems based on the particular OA, rules, and quench protocols $\{h^p(t)\}$ used.

7.10 Case Studies

7.10.1 *Case Study: Ising Model in a Transverse Field*

For our first case study we choose the paradigmatic Ising model in a transverse field, with Hamiltonian

$$\hat{H} = -J \sum_{\langle i,j \rangle} \hat{\sigma}_i^z \hat{\sigma}_{i+1}^z - h \sum_i \hat{\sigma}_i^x . \tag{7.100}$$

Here J is the coupling energy, h is a transverse magnetic field, and the $\{\hat{\sigma}_i\}$ are the Pauli spin operators on-site i. We choose this model because its dynamics are amenable to numerically exact study using the time-dependent Bogoliubov–de Gennes formalism (see Appendix for a review) and so we were able to carefully check convergence of our results. The statics have all been converged to eight digits, and the dynamics at all times to at least four digits. Here we refer to convergence of local quantities such as the energy or density of defects. Nonlocal quantities such as the bond entropy will not have this same accuracy, but numerical tests show that the qualitative behavior is unaffected. We begin with a discussion of the statics.

In Fig. 7.6a we demonstrate the gaps from the ground state to the two lowest eigenstates, computed via eMPS. The variances are smaller than the point size in this case. The upper (lower) curve corresponds to even (odd) parity, while the ground state has even parity, where parity is defined as simultaneous inversion of all spins $P = \langle \prod_i \hat{\sigma}_i^x \rangle$. Hence, the relevant gap for discussing the quantum phase transition is in fact the gap to the second excited state, shown in green.[8] Both gaps close at criticality, and this can cause the first excited state returned by eMPS to be a mixture of these two states. This will not affect the energies so long as the variance remains small, but it can affect other observable properties of the states. There are two ways to remove this nuisance. The first, most complex, and most preferable is to use an MPS representation in which the state is explicitly $\mathbb{Z}_2$ invariant [40]. The second is to add a field coupling to the parity $-h_p \prod_i \hat{\sigma}_i^x$ to cause the relevant even-parity subspace to become lower in energy than the odd-parity subspace. This operator is an MPO with bond dimension 1, the MPO equivalent of a product state. The closing of the gap at the known critical point $h = 1$ is linear in $1/L$, indicating a conformally invariant critical theory with dynamical critical exponent $z = 1$.

We venture to determine the central charge of the critical theory by fitting to the Calabrese–Cardy formula in two ways. In the first, we fit to the finite-size formula

$$S_l = \frac{c}{6} \log \left[\frac{2L}{\pi} \sin \left(\frac{\pi l}{L} \right) \right] + a \tag{7.101}$$

[8]The even parity gap for $h > 1$ is in fact twice the demonstrated gap in green, but the essential piece is the closing of the gap at criticality.

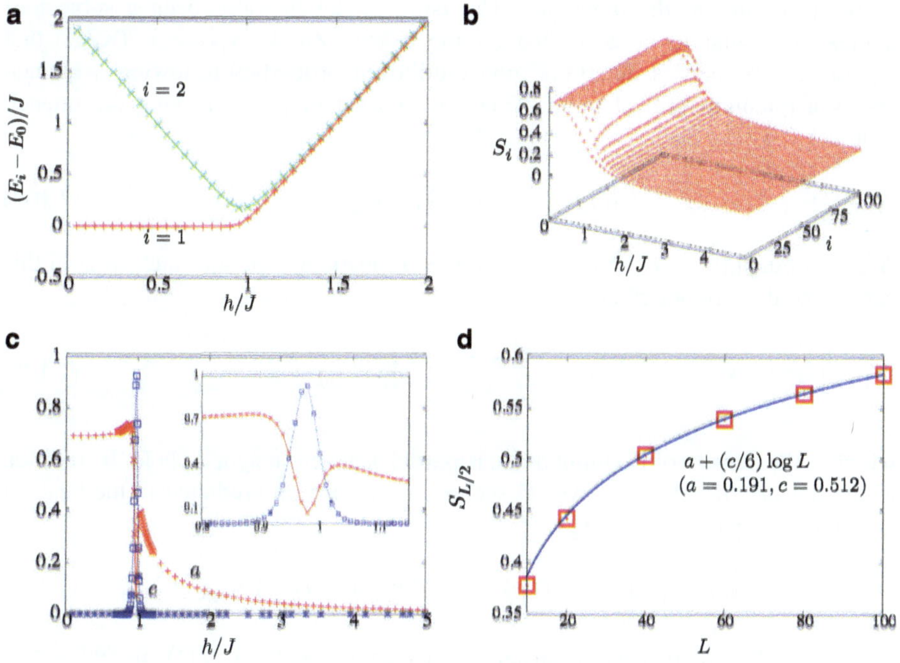

Fig. 7.6 Statics of the Ising model. (**a**) The gaps to the two-lowest-lying eigenstates, computed using eMPS. (**b**) The bond entropy S_i vs. the site index i and magnetic field h for $L = 100$ sites. Note the increased curvature near the critical point $h = 1$. (**c**) The central charge c (*blue line*) and bulk entanglement a (*red line*) extracted from a fit to Eq. (7.101) vs. h for $L = 100$ sites, neglecting 30 sites at both boundaries. The *inset* is a close-up of the critical region. (**d**) The bond entropy of the central site $S_{L/2}$ vs. L at $h = 1.0$ together with a fit extracting the central charge

for fixed L and variable i, and in the second we fix i at $L/2$ and fit $S_{L/2}$ to this formula for various L. Near criticality, the presence of nonzero c indicates a curvature of S_i, while in the gapped phases S_i obeys an area law and is hence a constant apart from finite-size effects. The bond entropy in the bulk approaches the correct limiting values of $\log 2$ as $h \to 0$ and 0 as $h \to \infty$. The first fit, shown in Fig. 7.6a, provides us with a clear indicator of the critical region by the spike in the central charge c. However, the precise determination of c for a finite size system in this case is noisy, likely due to strong finite-size effects. Once we have narrowed down where the critical region is, the second fit, shown in Fig. 7.6b, allows us to extract the anticipated value $c = 1/2$ much more precisely. If the same scaling analysis is attempted at a point which is not the critical point, the bond entropy saturates and $c \to 0$ as $L \to \infty$. We can understand this as a large but finite correlation length ξ. For $L/2 < \xi$, the system appears to be conformal and we see scaling of the bond entropy with L. For $L > \xi$ the bond entropy saturates and this scaling breaks down, indicating that the given region is not critical. We note that in this analysis we have used no properties which are specific to this system, e.g. correlation functions of an order parameter to extract the critical behavior.

We now turn to the dynamics. The Ising model has also been a subject of interest for dynamics as a testbed for the Kibble–Zurek hypothesis (KZH) that equilibrium properties determine nonequilibrium properties following a quench across a quantum critical point. This was studied in [81–83] using the quench protocol

$$h(t)/J = 5\left(1 - \tfrac{t}{\tau}\right), \quad 0 \le t \le \tau. \tag{7.102}$$

A useful quantity for determining how non-adiabatic a particular quench is in this case is the density of defects

$$\rho = \frac{1}{2L}\sum_i \left(1 - \hat{\sigma}_i^z \hat{\sigma}_{i+1}^z\right) \tag{7.103}$$

which is the density of Bogoliubov quasiparticles at zero magnetic field. In addition to the density of defects, universal scaling has also been predicted in the heat, or non-adiabatic part of the energy,

$$Q(t) = \langle \psi(t)|\hat{H}(t)|\psi(t)\rangle - \langle \psi_{\text{g.s.}}(t)|\hat{H}(t)|\psi_{\text{g.s.}}(t)\rangle, \tag{7.104}$$

where $|\psi_{\text{g.s.}}(t)\rangle$ represents the instantaneous ground state of $\hat{H}(t)$. In addition to these quantities, which are amenable to Bogoliubov–de Gennes analysis, we also compute the time-dependent bond entropy. Our results are shown in Fig. 7.7.

We first discuss the heat, as shown in panels (a) and (d). Panel (a) displays the heat as a function of time, and demonstrates a sharp change in the behavior of the system as we pass through the critical point. This is especially true of the longest quenches. In panel (d) we investigate the heat as a function of the quench rate both at the time when h takes on its critical value, $t_c = 4\tau/5$, and at the final time when $h = 0$. The large difference indicates that non-adiabatic processes continue after we have passed from the critical region back into the gapped ferromagnetic region. Thus, the universal scaling of the heat may be difficult to determine if the critical point itself is not known sufficiently well. We now move on to the density of defects as shown in panels (b) and (e). In panel (b) we see that the density of defects at the final time decreases slowly to zero as $\tau \to \infty$; that is, when the quench becomes perfectly adiabatic. This is in accordance with the KZH prediction. Panel (e) demonstrates the large disparity between the density of defects at the critical time and the final time for all but the most rapid of quenches. Finally, in panels (c) and (f) we investigate the bond entropy. In panel (c) we see the bond entropy of the central bond as a function of time. As the quench becomes more adiabatic, the entropy increases more towards the ferromagnetic limiting value of $\log 2$. However, for very slow quenches, the bond entropy reaches this value before the end of the quench and then begins to oscillate. In panel (f) we show the bond entropy as a function of the bond index at the critical time. Bulk curvature such as that seen at criticality in Fig. 7.6d is absent, indicating that the time-evolved state is still quite far from the conformal ground state.

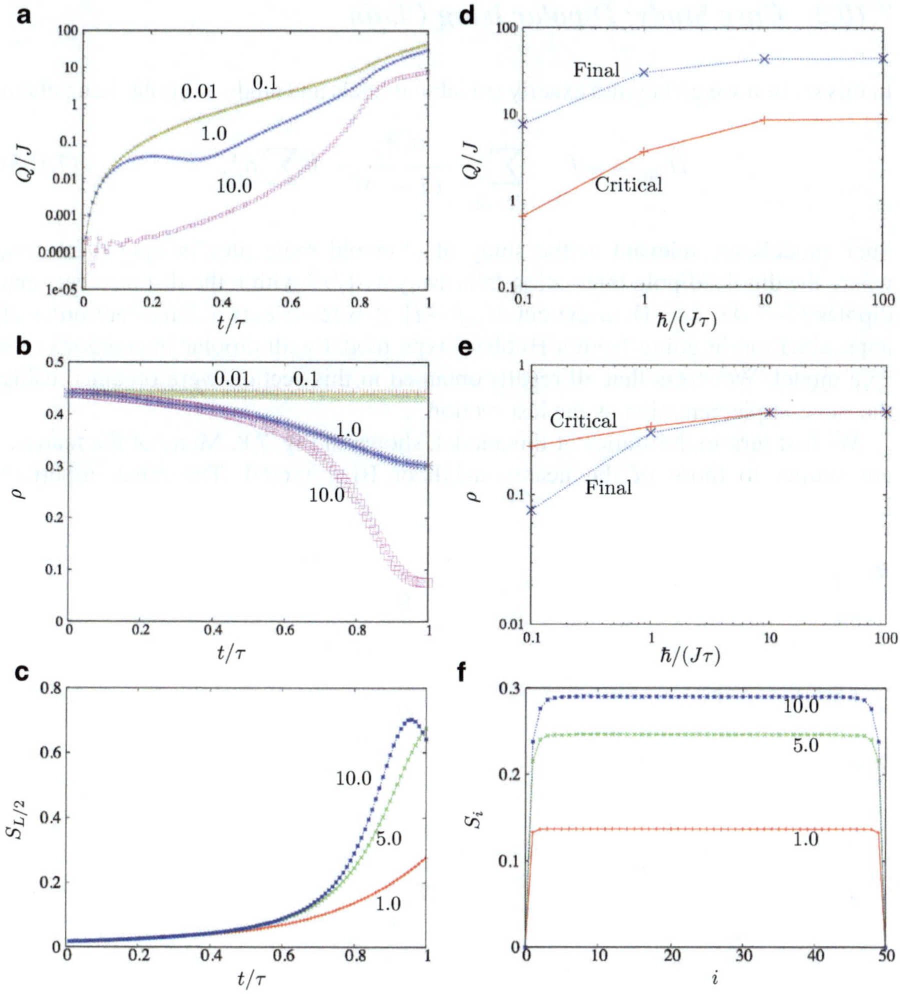

Fig. 7.7 Dynamics of the Ising model. (**a**) The heat as a function of time shows a marked change in behavior as we transition past the critical time $t/\tau = 4/5$. The numbers indicate the value of $\tau J/\hbar$. $\tau J/\hbar = 0.1$ and 0.01 are indistinguishable on the scale of this plot. Both may be considered to be instantaneous. (**b**) The density of defects as a function of time scales to zero as $\tau J/\hbar \to \infty$, in accordance with the KZH. (**c**) The bond entropy of the central bond as a function of time approaches the limiting value log 2 as the quench becomes more adiabatic. For nearly adiabatic quenches, the bond entropy oscillates after the critical point. (**d**) Scaling of the heat in the final and critical stages with the inverse quench time shows marked non-adiabatic processes occurring after passing the critical point. (**e**) Scaling of the density of defects in the final and critical stages with the inverse quench time shows non-adiabatic processes after passing the critical point only for slow quenches. (**f**) Snapshots of the bond entropy at the critical time demonstrate that the system is not generally close to its conformal ground state

7.10.2 Case Study: Dipolar Ising Chain

In this section we go beyond exactly solvable models and study a dipolar Ising chain

$$\hat{H}_{\text{dip}} = -J \sum_{i<j;|j-i|\leq 6} \frac{\hat{\sigma}_i^z \hat{\sigma}_j^z}{(j-i)^3} - h \sum_i \hat{\sigma}_i^x. \tag{7.105}$$

Such models are relevant to the study of ultracold molecules in optical lattices, where the dipole–dipole interaction falls away as $1/r^3$ with r the distance between dipoles [7–9, 48, 84]. Here the cutoff $|j-i| \leq 6$ represents a consistent order of approximation in going from a Hubbard-type model with dipolar interactions to a spin model. We stress that all results obtained in this section were obtained using the same implementation as the last section.

We first turn to the statics of this model, shown in Fig. 7.8. Many of the features are similar to those of the nearest-neighbor Ising model. The most important

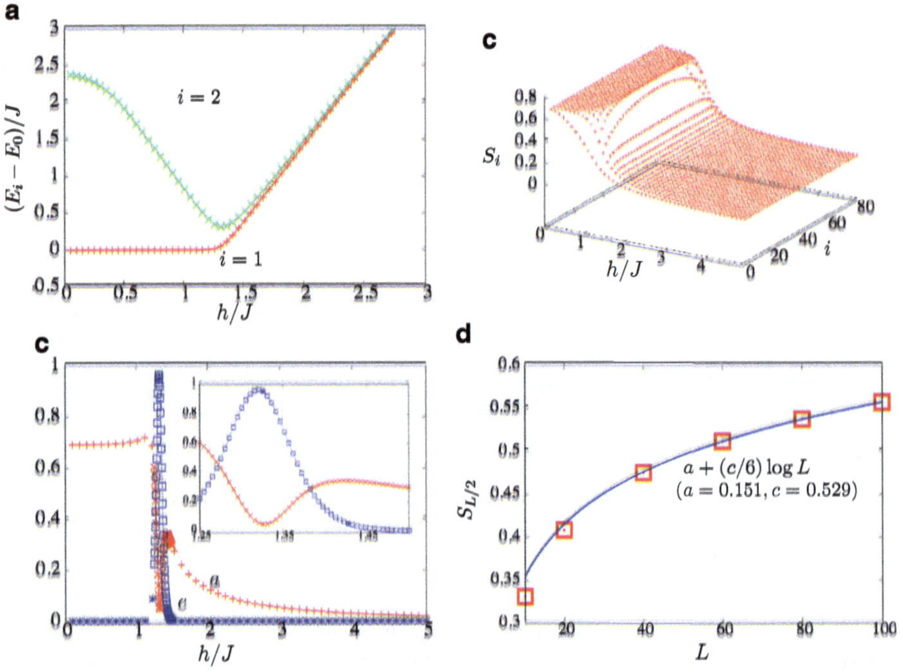

Fig. 7.8 Statics of the dipolar Ising model. (**a**) The gaps to the two-lowest-lying eigenstates, computed using eMPS. Here we see a breakdown of the linear dispersion at small h, indicating interactions between quasiparticles. (**b**) The bond entropy S_l vs. the site index i and magnetic field h for $L = 80$ sites. The point of greatest curvature is shifted towards larger h with respect to the Ising model. (**c**) The central charge c (*blue line*) and bulk entanglement a (*red line*) extracted from a fit to Eq. (7.101) vs. h for $L = 80$ sites, neglecting 30 sites at both boundaries. The *inset* is a close-up of the critical region. (**d**) The bond entropy of the central site $S_{L/2}$ vs. L at $h = 1.362$ together with a fit extracting the central charge

differences are that the critical region is shifted towards larger h with respect to the nearest-neighbor Ising model as seen in panels (b) and (c). This indicates increased stability of the ferromagnetic phase, in accordance with expectations. Using these points as a guide, we determine the critical point to be $h_c = 1.362 \pm 0.01$, as shown by the scaling in panel (d). Additionally, the energy of the first even parity excited state deviates from pure $z = 1$ linear behavior near $h = 0$, indicating interactions between quasiparticles which were noninteracting in the nearest-neighbor Ising model.

We now turn to dynamics, following the same quenching protocol Eq. (7.102) as above. The results are shown in Fig. 7.9. We reiterate that the dynamics of this model cannot be handled by Bogoliubov–de Gennes methods, nor straightforwardly with standard tDMRG/TEBD. The density of defects no longer represents the density of quasiparticles at the critical point, but we compute it for comparison with the results of the nearest-neighbor Ising model. Because of the larger MPO bond dimensions and the more rapid growth of bond dimension for this model, we restrict our analysis

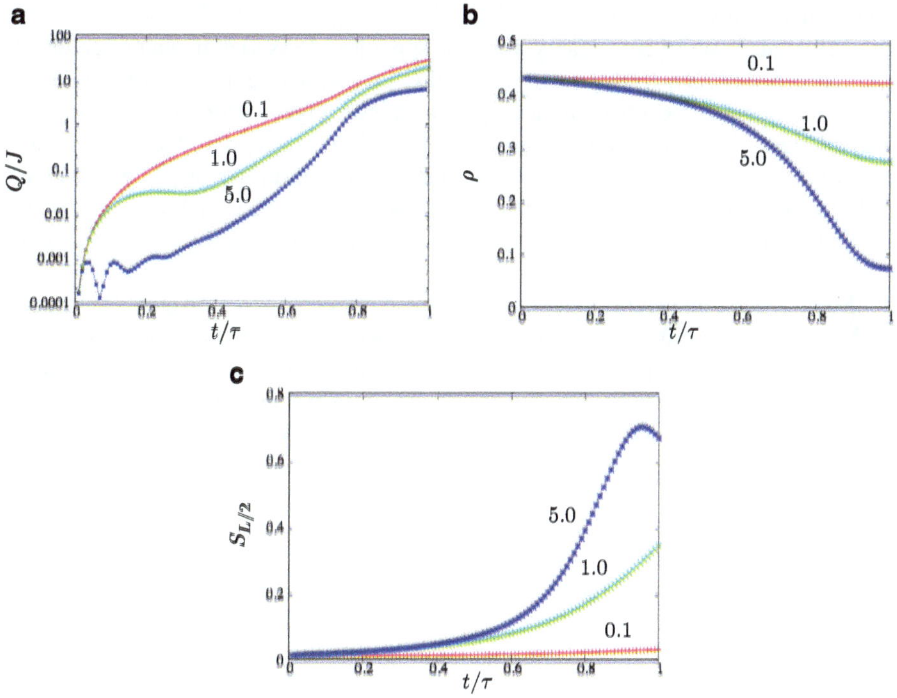

Fig. 7.9 Dynamics of the dipolar Ising model. (**a**) The heat as a function of time displays a slower buildup in post-critical non-adiabatic effects for longer quenches, as in the nearest-neighbor case. (**b**) The density of domain walls as a function of time is comparable to that for the nearest-neighbor case, but no longer has the same interpretation in terms of quasiparticles. (**c**) The bond entropy of the central bond oscillates for the shorter quench time $J\tau/\hbar = 5$ due to the quench passing the critical point sooner than in the nearest-neighbor case

to short times $J\tau/\hbar \leq 5$, though an optimized implementation could reach longer times. The basic features are similar to the dynamics of the nearest-neighbor Ising model. One quantitative difference is that, because the critical point is reached at an earlier time than in the nearest-neighbor Ising model, oscillations in the central bond entropy occur for shorter quench times.

7.11 Conclusions

The power of MPS algorithms over DMRG-based algorithms is most readily apparent when multiple states are involved, as each state may be represented as a separate MPS in the former approach. Because MPSs with a fixed bond dimension do not form a vector space, a set of MPSs carries more information at smaller numerical cost than the same set represented as a multi-state targeted basis in DMRG. We have presented two algorithms, eMPS and tMPS, which use this property to find eigenstates and perform time evolution of strongly correlated 1D quantum systems.

eMPS uses a set of eigenstates stored as separate MPSs to define a projector into the space orthogonal to this set. We use this projector to explicitly orthogonalize a variational state against previously determined eigenvectors in order to find excited states. The explicit MPS representation allows us to store the excited states much more accurately than with standard DMRG, and allows us to ensure global orthogonality between the eigenstates to a desired tolerance. The variance, which is computed exactly using the MPO representation of the Hamiltonian, gives strict error bars on the energies obtained with this procedure.

tMPS avoids the need for an explicit representation of the propagator by using a CFME and then building successive Krylov subspace approximations to the matrix exponentials which appear in the expansion. Each vector in the Krylov subspace is stored as a separate MPS to maximize efficiency. Furthermore, the operators $\hat{\Omega}_i$ have exact representations as MPO with the same bond dimension as the Hamiltonian. Our algorithm eliminates the need for Hamiltonian-specialized implementation of dynamics. Additionally, by carefully accounting for the time dependence of the Hamiltonian with a CFME, the error in our algorithm depends only on commutators of the Hamiltonian with itself at different times and not on its derivatives. As with eMPS, the errors are rigorously accounted for by considering distance functionals with the variational state.

The MPO forms of 1D Hamiltonians can be obtained using a small set of FSA rules such as the site, bond, and finite function rules. Using MPO arithmetic, we can add together the various terms in a Hamiltonian from these rules to form a complete canonical MPO representation. This representation allows for templating of Hamiltonians which depend only on the type of interactions and not on the microscopic constituents of the lattice model. Furthermore, given the time-dependent form of the Hamiltonian parameters, one can use the same template to form the operators $\hat{\Omega}_i$ which appear in tMPS at negligible numerical cost.

We used our algorithms to study both the nearest-neighbor Ising model in a transverse field and a long-range Ising model in a transverse field. By the closing of the gap obtained with eMPS we determined that the critical points of these models were conformal, and so we used the Calabrese–Cardy formula for the bond entropy of conformal systems to locate the critical point and its associated central charge. The known result $h_c = 1$ was verified for the nearest-neighbor case, and the critical point was shifted deeper into the paramagnetic region $h_c = 1.362 \pm 0.01$ for the long-range case, indicating a stabilization of the ferromagnetic phase. We used tMPS to study the dynamics of these models following a linear quench of the transverse field from the paramagnetic phase through the critical point into the ferromagnetic phase. We saw strong non-adiabatic effects in the heat as quenching continued into the ferromagnetic region, scaling of the density of defects consistent with the KZH, and the oscillation of the bond entropy near its limiting ferromagnetic phase value for nearly adiabatic quenches.

This work was supported by the Heidelberg Center for Quantum Dynamics, the Alexander von Humboldt Foundation and the National Science Foundation under Grant PHY-0903457. We also acknowledge the Golden Energy Computing Organization at the Colorado School of Mines for the use of resources acquired with financial assistance from the National Science Foundation and the National Renewable Energy Laboratories. We thank Erman Bekaroglu for discussions and a thorough reading of the manuscript.

Appendix: Time Evolution of Exact Solution of 1D Transverse-Field Quantum Ising Model for Comparison with tMPS

The solution of the statics of the transverse-field quantum Ising model is covered in standard texts [52]. However, for comparison with tMPS we require a description of the dynamics, and so we present the dynamical case here. To find the exact solution of the Ising model,

$$\hat{H} = -J \sum_{\langle i,j \rangle} \hat{\sigma}_i^z \hat{\sigma}_j^z - h \sum_{i=1}^{L} \hat{\sigma}_i^x, \tag{7.106}$$

we affect the Jordan–Wigner transformation

$$\hat{\sigma}_i^x = 1 - 2\hat{c}_i^\dagger \hat{c}_i, \tag{7.107}$$

$$\hat{\sigma}_i^z = -\left(\hat{c}_i + \hat{c}_i^\dagger\right) \prod_{j<i} \left(1 - 2\hat{c}_j^\dagger \hat{c}_j\right), \tag{7.108}$$

where the fermionic operators $\hat{c}_l$ satisfy the anti-commutation relations $\{\hat{c}_i, \hat{c}_j^\dagger\} = \delta_{ij}$, $\{\hat{c}_i, \hat{c}_j\} = \{\hat{c}_i^\dagger, \hat{c}_j^\dagger\} = 0$. This transforms the Ising model into the fermion Hamiltonian

$$\hat{H} = -J \sum_{l=1}^{L-1} \left(\hat{c}_l^\dagger \hat{c}_{l+1}^\dagger + \hat{c}_l^\dagger \hat{c}_{l+1} + \text{h.c.} \right) + 2h \sum_{l=1}^{L} \hat{c}_l^\dagger \hat{c}_l - Lh. \tag{7.109}$$

As this is a quadratic form in fermion operators, it may be diagonalized by a canonical (Bogoliubov) transformation [85, 86]

$$\hat{c}_l = \sum_{k=1}^{L} \left(u_{lk} \hat{\gamma}_k + v_{lk}^\star \hat{\gamma}_k^\dagger \right) \tag{7.110}$$

which provides the set of Bogoliubov–de Gennes equations

$$\epsilon_k \mathbf{u}_k = A\mathbf{u}_k + B\mathbf{v}_k \tag{7.111}$$

$$\epsilon_k \mathbf{v}_k = -B\mathbf{u}_k - A\mathbf{v}_k, \tag{7.112}$$

where $\mathbf{u}_k$ are the elements of $\{u_{ik}, i = 1, \ldots, L\}$ arranged as a vector and likewise for $\mathbf{v}_k$. The matrices A and B are real and tridiagonal, with the nonzero matrix elements $A_{l,l} = 2h$, $A_{l,j} = -J$, $|i - j| = 1$ and $B_{l,l+1} = -B_{l+1,l} = -J$. The pairs (u_{ik}, v_{ik}) with positive energy ϵ_k, $\epsilon_1 \le \epsilon_2 \le \cdots \le \epsilon_L$, and the normalization $|\mathbf{u}_k|^2 + |\mathbf{v}_k|^2 = 1$ define quasiparticle operators

$$\hat{\gamma}_k = u_{ik}^\star \hat{c}_i + v_{ik} \hat{c}_i^\dagger \tag{7.113}$$

which bring the Hamiltonian into the diagonal form

$$\hat{H} = \sum_{k=1}^{N} \epsilon_k \left(\hat{\gamma}_k^\dagger \hat{\gamma}_k - \frac{1}{2} \right). \tag{7.114}$$

Corresponding to every such pair is another pair $(\tilde{u}_{ik}, \tilde{v}_{ik}) = (v_{ik}, u_{ik})$ with $\tilde{\epsilon}_k = -\epsilon_k$ which defines the conjugate operator $\hat{\gamma}_k^\dagger$. Writing $\mathbf{u}_k$ and $\mathbf{v}_k$ together as a composite vector, the Bogoliubov–de Gennes equations take the form of a real symmetric eigenvalue problem of dimension $2L$:

$$\epsilon_k \begin{pmatrix} \mathbf{u}_k \\ \mathbf{v}_k \end{pmatrix} = \begin{pmatrix} A & B \\ -B & -A \end{pmatrix} \begin{pmatrix} \mathbf{u}_k \\ \mathbf{v}_k \end{pmatrix} \tag{7.115}$$

which can be readily solved using standard eigenvalue methods [87].

Evolution under the fermion Hamiltonian Eq. (7.109) does not preserve the number of fermions N_F but it does preserve the fermionic parity $(-1)^{N_F}$. Because the ground state is the Bogoliubov vacuum it contains no fermions, and so the first *accessible* excited state consists of two Bogoliubov excitations, one in each of the lowest two modes. The gap between the ground and first excited states is thus $\epsilon_1 + \epsilon_2$.

We now consider the Heisenberg equations of motion for the fermi operators

$$i\hbar \frac{d\hat{\mathbf{c}}(t)}{dt} = A(t)\,\hat{\mathbf{c}}(t) + B(t)\,\hat{\mathbf{c}}^{\dagger}(t), \tag{7.116}$$

where $A(t)$ and $B(t)$ are the time-dependent generalizations of A and B above and $\hat{\mathbf{c}}(t)$ and $\hat{\mathbf{c}}^{\dagger}(t)$ are the elements of $\{\hat{c}_i(t)\}$ and $\{\hat{c}_i^{\dagger}(t)\}$, respectively, arranged as vectors. Because this equation is linear in the Fermi operators it may be diagonalized with a time-dependent Bogoliubov transformation

$$\hat{c}_i(t) = \sum_{k=1}^{L} \left(u_{ik}(t)\,\gamma_k + v_{ik}^{\star}(t)\,\hat{\gamma}_k^{\dagger} \right), \tag{7.117}$$

where $\mathbf{u}_i(t)$ and $\mathbf{v}_i(t)$ subject to the time-dependent Bogoliubov–de Gennes equations

$$i\hbar \frac{d}{dt} \begin{pmatrix} \mathbf{u}_k(t) \\ \mathbf{v}_k(t) \end{pmatrix} = \begin{pmatrix} A(t) & B(t) \\ -B(t) & -A(t) \end{pmatrix} \begin{pmatrix} \mathbf{u}_k(t) \\ \mathbf{v}_k(t) \end{pmatrix} \equiv \mathcal{H}(t) \begin{pmatrix} \mathbf{u}_k(t) \\ \mathbf{v}_k(t) \end{pmatrix}, \tag{7.118}$$

and $\hat{\gamma}_k$ and $\hat{\gamma}_k^{\dagger}$ diagonalize the Hamiltonian at the initial time. Equivalently, u and v define time-dependent Bogoliubov operators

$$\hat{\gamma}_k(t) = u_{ik}^{\star}(t)\,\hat{c}_i + v_{ik}(t)\,\hat{c}_i^{\dagger} \tag{7.119}$$

such that the time-evolved state $|\psi(t)\rangle$ is the Bogoliubov vacuum of this set, i.e. $\hat{\gamma}_k(t)|\psi(t)\rangle = 0$. To compare with the MPS simulations, we note that the energy at time t is

$$-\frac{1}{2}\sum_k \epsilon_k(t) = -\frac{1}{2}\sum_k \left(\mathbf{u}_k(t)\,\mathbf{v}_k(t)\right) \mathcal{H}(t) \begin{pmatrix} \mathbf{u}_k(t) \\ \mathbf{v}_k(t) \end{pmatrix}. \tag{7.120}$$

Similarly, the density of defects is

$$\rho(t) = \frac{1}{2L}\sum_{i=1}^{L-1} \langle \psi(t)|\left(1 - \hat{\sigma}_i^z \hat{\sigma}_{i+1}^z\right)|\psi(t)\rangle, \tag{7.121}$$

$$= \frac{1}{2L}\sum_{i=1}^{L-1}\left[1 - \sum_{k=1}^{L}\left(v_{i,k}(t) - u_{i,k}(t)\right)\left(u_{i+1,k}^{\star}(t) + v_{i+1,k}^{\star}(t)\right)\right]. \tag{7.122}$$

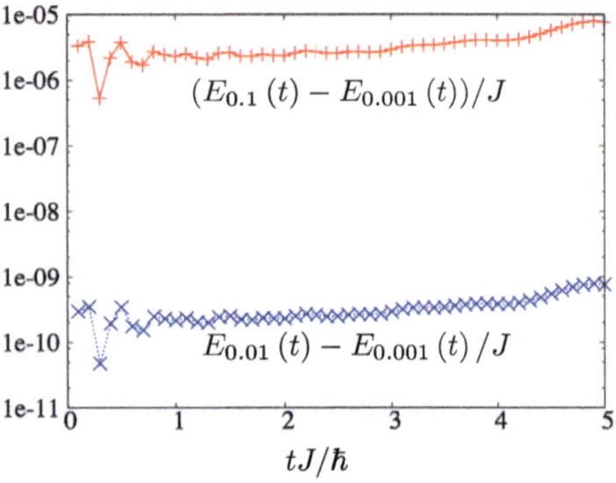

Fig. 7.10 Demonstration of Bogoliubov–de Gennes method convergence. The errors in the energy computed with a given time step δt, $E_{\delta t}$, are shown as a function of time. The ratio of the errors as a function of time is roughly $(0.01/0.1)^4 = 0.0001$, as should be expected for our fourth-order CFME. Errors in the density of defects behave similarly

Time evolution thus reduces to the solution of a $2L \times 2L$ time-dependent matrix differential equation which we solve using a CFME as in Sect. 7.7. Because the dimensions of the system are much smaller than those of the associated MPS problem we are able to take very small time steps, and so the results obtained in the method may be considered to be numerically exact, see Fig. 7.10. Here we demonstrate that the error incurred in the energy as a function of time scales with the infinitesimal time step δt as δt^4 using our fourth-order CMFE. Hence, by decreasing δt, any desired degree of accuracy may be met.

References

1. Wenger, T., Kinoshita, T., Weiss, D.S.: A quantum Newton's cradle. Nature **440**, 900–903 (2006)
2. Trotzky, S., Pollet, L., Gerbier, F., Schnorrberger, U., Bloch, I., Prokof'ev, N.V., Svistunov, B., Troyer, M.: Suppression of the critical temperature for superfluidity near the Mott transition. Nat. Phys. **6**(12), 998–1004 (2010). doi:10.1038/nphys1799. http://dx.doi.org/10.1038/nphys1799 [ISSN 1745-2473]
3. Trotzky, S., Chen, Y.-A., Flesch, A., McCulloch, I.P., Schollwöck, U., Eisert, J., Bloch, I.: Probing the relaxation towards equilibrium in an isolated strongly correlated 1D Bose gas (2011) [arXiv:1101.2659]
4. Ni, K.-K., Ospelkaus, S., de Miranda, M.H.G., Peér, A., Neyenhuis, B., Zirbel, J.J., Kotochigova, S., Julienne, P.S., Jin, D.S., Ye, J.: A high phase-space-density gas of polar molecules. Science **322**, 231–235 (2008)
5. Ospelkaus, S., Ni, K.K., de Miranda, M.H.G., Pe'er, A., Nyenhuis, B., Wang, D., Kotochigova, S., Julienne, P.S., Jin, D.S., Ye, J.: Ultracold polar molecules near quantum degeneracy. Faraday Discuss. **142**, 351–359 (2009)
6. Deiglmayr, J., Grochola, A., Repp, M., Mörtlbauer, K., Glück, C., Lange, J., Dulieu, O., Wester, R., Weidemüller, M.: Formation of ultracold polar molecules in the rovibrational ground state. Phys. Rev. Lett. **101**(13), 133004 (2008). doi:10.1103/PhysRevLett.101.133004
7. Wall, M.L., Carr, L.D.: Hyperfine molecular hubbard hamiltonian. Phys. Rev. A **82**(1), 013611 (2010). doi:10.1103/PhysRevA.82.013611
8. Gorshkov, A.V., Manmana, S.R., Chen, G., Ye, J., Demler, E., Lukin, M.D., Rey, A.M.: Tunable superfluidity and quantum magnetism with ultracold polar molecules. Phys. Rev. Lett. **107**, 115301 (2011). doi:10.1103/PhysRevLett.107.115301. http://link.aps.org/doi/10.1103/PhysRevLett.107.115301

9. Gorshkov, A.V., Manmana, S.R., Chen, G., Demler, E., Lukin, M.D., Rey, A.M.: Quantum magnetism with polar alkali-metal dimers. Phys. Rev. A **84**, 033619 (2011). doi:10.1103/PhysRevA.84.033619. http://link.aps.org/doi/10.1103/PhysRevA.84.033619
10. Polkovnikov, A., Sengupta, K., Silva, A., Vengalattore, M.: Colloquium: Nonequilibrium dynamics of closed interacting quantum systems. Rev. Mod. Phys. **83**, 863–883 (2011). doi:10.1103/RevModPhys.83.863. http://link.aps.org/doi/10.1103/RevModPhys.83.863
11. Poulin, D., Qarry, A., Somma, R., Verstraete, F.: Quantum simulation of time-dependent hamiltonians and the convenient illusion of hilbert space. Phys. Rev. Lett. **106**, 170501 (2011). doi:10.1103/PhysRevLett.106.170501. http://link.aps.org/doi/10.1103/PhysRevLett.106.170501
12. Schollwöck, U.: The density-matrix renormalization group in the age of matrix product states. Ann. Phys. **326**(1), 96–192 (2011). doi:10.1016/j.aop.2010.09.012. http://www.sciencedirect.com/science/article/pii/S0003491610001752 (January 2011 Special Issue) [ISSN 0003-4916]
13. García-Ripoll, J.J.: Time evolution of matrix product states. New J. Phys. **8**(12), 305 (2006). http://stacks.iop.org/1367-2630/8/i=12/a=305
14. Keilmann, T., García-Ripoll, J.J.: Dynamical creation of bosonic cooper-like pairs. Phys. Rev. Lett. **100**, 110406 (2008). doi:10.1103/PhysRevLett.100.110406. http://link.aps.org/doi/10.1103/PhysRevLett.100.110406
15. Eisert, J., Briegel, H.J.: Schmidt measure as a tool for quantifying multiparticle entanglement. Phys. Rev. A **64**, 022306 (2001). doi:10.1103/PhysRevA.64.022306. http://link.aps.org/doi/10.1103/PhysRevA.64.022306
16. Gross, D., Flammia, S.T., Eisert, J.: Most quantum states are too entangled to be useful as computational resources. Phys. Rev. Lett. **102**(19), 190501 (2009). doi:10.1103/PhysRevLett.102.190501
17. Östlund, S., Rommer, S.: Thermodynamic limit of density matrix renormalization. Phys. Rev. Lett. **75**(19), 3537–3540 (1995). doi:10.1103/PhysRevLett.75.3537
18. Rommer, S., Östlund, S.: Class of ansatz wave functions for one-dimensional spin systems and their relation to the density matrix renormalization group. Phys. Rev. B **55**(4), 2164–2181 (1997). doi:10.1103/PhysRevB.55.2164
19. Verstraete, F., Porras, D., Cirac, J.I.: Density matrix renormalization group and periodic boundary conditions: A quantum information perspective. Phys. Rev. Lett. **93**(22), 227205 (2004). doi:10.1103/PhysRevLett.93.227205
20. Pippan, P., White, S.R., Evertz, H.G.: Efficient matrix-product state method for periodic boundary conditions. Phys. Rev. B **81**(8), 081103 (2010). doi:10.1103/PhysRevB.81.081103
21. Rossini, D., Giovannetti, V., Fazio, R.: Stiffness in 1D matrix product states with periodic boundary conditions. J. Stat. Mech. **2011**(10), P05021 (2011)
22. Perez-Garcia, D., Verstraete, F., Wolf, M.M., Cirac, J.I.: Matrix product state representations. Quantum Inf. Comput. **7**, 401 (2007)
23. Auerbach, A.: Interacting Electrons and Quantum Magnetism. Springer, Berlin (1994)
24. Affleck, I., Kennedy, T., Lieb, E.H., Tasaki, H.: Rigorous results on valence-bond ground states in antiferromagnets. Phys. Rev. Lett. **59**, 799–802 (1987). doi:10.1103/PhysRevLett.59.799. http://link.aps.org/doi/10.1103/PhysRevLett.59.799
25. Tu, H.-H., Zhang, G.-M., Xiang, T.: Class of exactly solvable SO(n) symmetric spin chains with matrix product ground states. Phys. Rev. B **78**, 094404 (2008). doi:10.1103/PhysRevB.78.094404. http://link.aps.org/doi/10.1103/PhysRevB.78.094404
26. White, S.R.: Density matrix formulation for quantum renormalization groups. Phys. Rev. Lett. **69**(19), 2863–2866 (1992). doi:10.1103/PhysRevLett.69.2863
27. Hastings, M.B.: An area law for one-dimensional quantum systems. J. Stat. Mech. **2007**(10), P08024 (2007)
28. Verstraete, F., Cirac, J.I.: Matrix product states represent ground states faithfully. Phys. Rev. B **73**(9), 094423 (2006). doi:10.1103/PhysRevB.73.094423
29. Eisert, J., Cramer, M., Plenio, M.B.: Colloquium: Area laws for the entanglement entropy. Rev. Mod. Phys. **82**, 277–306 (2010). doi:10.1103/RevModPhys.82.277. http://link.aps.org/doi/10.1103/RevModPhys.82.277

30. Holzhey, C., Larsen, F., Wilczek, F.: Geometric and renormalized entropy in conformal field theory. Nucl. Phys. B **424**(3), 443–467 (1994). doi:10.1016/0550-3213(94)90402-2. http://www.sciencedirect.com/science/article/pii/0550321394904022 [ISSN 0550-3213]
31. Calabrese, P., Cardy, J.: Entanglement entropy and quantum field theory. J. Stat. Mech. **2004** (06), P06002 (2004). http://stacks.iop.org/1742-5468/2004/i=06/a=P06002
32. Affleck, I., Laflorencie, N., Sørensen, E.S.: Entanglement entropy in quantum impurity systems and systems with boundaries. J. Phys. A **42**(50), 504009 (2009). http://stacks.iop.org/1751-8121/42/i=50/a=504009
33. Calabrese, P., Cardy, J.: Entanglement entropy and conformal field theory. J. Phys. A **42**(50), 504005 (2009). http://stacks.iop.org/1751-8121/42/i=50/a=504005
34. Cardy, J.: Scaling and Renormalization in Statistical Physics. Cambridge University Press, Cambridge (1996)
35. Läuchli, A.M., Kollath, C.: Spreading of correlations and entanglement after a quench in the one-dimensional Bose-Hubbard model. J. Stat. Mech. **2008**(05), P05018 (2008). http://stacks.iop.org/1742-5468/2008/i=05/a=P05018
36. Block, M.S., Mishmash, R.V., Kaul, R.K., Sheng, D.N., Motrunich, O.I., Fisher, M.P.A.: Exotic gapless mott insulators of bosons on multileg ladders. Phys. Rev. Lett. **106**, 046402 (2011). doi:10.1103/PhysRevLett.106.046402. http://link.aps.org/doi/10.1103/PhysRevLett.106.046402
37. Schuch, N., Cirac, I., Verstraete, F.: Computational difficulty of finding matrix product ground states. Phys. Rev. Lett. **100**, 250501 (2008). doi:10.1103/PhysRevLett.100.250501. http://link.aps.org/doi/10.1103/PhysRevLett.100.250501
38. Shi, Y.-Y., Duan, L.-M., Vidal, G.: Classical simulation of quantum many-body systems with a tree tensor network. Phys. Rev. A **74**(2), 022320 (2006). doi:10.1103/PhysRevA.74.022320. http://link.aps.org/abstract/PRA/v74/e022320
39. Golub, G.H., Van Loan, C.F.: Matrix Computations, 3rd edn. Johns Hopkins Studies in Mathematical Sciences. The Johns Hopkins University Press, Baltimore (1996)
40. McCulloch, I.P.: From density-matrix renormalization group to matrix product states. J. Stat. Mech. **2007**(10), P10014 (2007). http://stacks.iop.org/1742-5468/2007/i=10/a=P10014
41. Osborne, T.J.: Efficient approximation of the dynamics of one-dimensional quantum spin systems. Phys. Rev. Lett. **97**, 157202 (2006). doi:10.1103/PhysRevLett.97.157202. http://link.aps.org/doi/10.1103/PhysRevLett.97.157202
42. Crosswhite, G.M., Bacon, D.: Finite automata for caching in matrix product algorithms. Phys. Rev. A **78**, 012356 (2008). doi:10.1103/PhysRevA.78.012356. http://link.aps.org/doi/10.1103/PhysRevA.78.012356
43. Fisher, M.P.A., Weichman, P.B., Grinstein, G., Fisher, D.S.: Boson localization and the superfluid-insulator transition. Phys. Rev. B **40**, 546–570 (1989)
44. Jaksch, D., Bruder, C., Cirac, J.I., Gardiner, C.W., Zoller, P.: Cold bosonic atoms in optical lattices. Phys. Rev. Lett. **81**(15), 3108–3111 (1998). doi:10.1103/PhysRevLett.81.3108
45. Pirvu, B., Murg, V., Cirac, J.I., Verstraete, F.: Matrix product operator representations. New J. Phys. **12**(2), 025012 (2010). http://stacks.iop.org/1367-2630/12/i=2/a=025012
46. Crosswhite, G.M., Doherty, A.C., Vidal, G.: Applying matrix product operators to model systems with long-range interactions. Phys. Rev. B **78**, 035116 (2008). doi:10.1103/PhysRevB.78.035116. http://link.aps.org/doi/10.1103/PhysRevB.78.035116
47. Fröwis, F., Nebendahl, V., Dür, W.: Tensor operators: constructions and applications for long-range interaction systems. Phys. Rev. A **81**, 062337 (2010). doi:10.1103/PhysRevA.81.062337. http://link.aps.org/doi/10.1103/PhysRevA.81.062337
48. Wall, M.L., Bekaroglu, E., Carr, L.D.: Phys. Rev. A **88**, 023605 (2013)
49. Davidson, E.R.: The iterative calculation of a few of the lowest eigenvalues and corresponding eigenvectors of large real-symmetric matrices. J. Comput. Phys. **17**(1), 87–94 (1975). doi:10.1016/0021-9991(75)90065-0. http://www.sciencedirect.com/science/article/pii/0021999175900650 [ISSN 0021-9991]

50. Porras, D., Verstraete, F., Cirac, J.I.: Renormalization algorithm for the calculation of spectra of interacting quantum systems. Phys. Rev. B **73**, 014410 (2006). doi:10.1103/PhysRevB.73.014410. http://link.aps.org/doi/10.1103/PhysRevB.73.014410
51. McCulloch, I.P.: Infinite size density matrix renormalization group, revisited (2008). http://xxx.lanl.gov/abs/0804.2509
52. Sachdev, S.: Quantum Phase Transitions. Cambridge University Press, New York (1999)
53. Alcaraz, F.C., Berganza, M.I., Sierra, G.: Entanglement of low-energy excitations in conformal field theory. Phys. Rev. Lett. **106**, 201601 (2011). doi:10.1103/PhysRevLett.106.201601. http://link.aps.org/doi/10.1103/PhysRevLett.106.201601
54. Bartel, E., Schadschneider, A., Zittartz, J.: Excitations of anisotropic spin-1 chains with matrix product ground state. Eur. Phys. J. B **31**, 209–216 (2003). doi:10.1140/epjb/e2003-00025-7. http://dx.doi.org/10.1140/epjb/e2003-00025-7 [ISSN 1434-6028]
55. Haegeman, J., Pirvu, B., Weir, D.J., Cirac, J.I., Osborne, T.J., Verschelde, H., Verstraete, F.: Variational matrix product ansatz for dispersion relations. Phys. Rev. B **85**, 100408 (2012). doi:10.1103/PhysRevB.85.100408. http://link.aps.org/doi/10.1103/PhysRevB.85.100408
56. Pirvu, B., Haegeman, J., Verstraete, F.: Matrix product state based algorithm for determining dispersion relations of quantum spin chains with periodic boundary conditions. Phys. Rev. B **85**, 035130 (2012). doi:10.1103/PhysRevB.85.035130. http://link.aps.org/doi/10.1103/PhysRevB.85.035130
57. Rigol, M., Dunjko, V., Olshanii, M.: Thermalization and its mechanism for generic isolated quantum systems. Nature (London) **452**, 854 (2008)
58. Rigol, M.: Breakdown of thermalization in finite one-dimensional systems. Phys. Rev. Lett. **103**, 100403 (2009). doi:10.1103/PhysRevLett.103.100403. http://link.aps.org/doi/10.1103/PhysRevLett.103.100403
59. Gogolin, C., Müller, M.P., Eisert, J.: Absence of thermalization in nonintegrable systems. Phys. Rev. Lett. **106**, 040401 (2011). doi:10.1103/PhysRevLett.106.040401. http://link.aps.org/doi/10.1103/PhysRevLett.106.040401
60. Calabrese, P., Cardy, J.: Evolution of entanglement entropy in one-dimensional systems. J. Stat. Mech. **2005**(04), P04010 (2005). http://stacks.iop.org/1742-5468/2005/i=04/a=P04010
61. Mishmash, R.V., Carr, L.D.: Quantum entangled dark solitons formed by ultracold atoms in optical lattices. Phys. Rev. Lett. **103**, 140403 (2009)
62. Mishmash, R.V., Danshita, I., Clark, C.W., Carr, L.D.: Quantum many-body dynamics of dark solitons in optical lattices. Phys. Rev. A **80**, 053612 (2009)
63. Glick, J.A., Carr, L.D.: Macroscopic quantum tunneling of solitons in Bose-Einstein condensates. Phys. Rev. Lett. (2011, under review) [arXiv:1105.5164]
64. Vidal, G.: Efficient classical simulation of slightly entangled quantum computations. Phys. Rev. Lett. **91**(14), 147902 (2003). doi:10.1103/PhysRevLett.91.147902
65. White, S.R., Feiguin, A.E.: Real-time evolution using the density matrix renormalization group. Phys. Rev. Lett. **93**, 076401 (2004)
66. Daley, A.J., Kollath, C., Schollwöck, U., Vidal, G.: Time-dependent density-matrix renormalization-group using adaptive effective Hilbert spaces. J. Stat. Mech. **2004**(04), P04005 (2004). http://stacks.iop.org/1742-5468/2004/i=04/a=P04005
67. Stoudenmire, E.M., White, S.R.: Minimally entangled typical thermal state algorithms. New J. Phys. **12**(5), 055026 (2010). http://stacks.iop.org/1367-2630/12/i=5/a=055026
68. Schachenmayer, J., Lesanovsky, I., Micheli, A., Daley, A.J.: Dynamical crystal creation with polar molecules or rydberg atoms in optical lattices. New J. Phys. **12**(10), 103044 (2010). http://stacks.iop.org/1367-2630/12/i=10/a=103044
69. Schmitteckert, P.: Nonequilibrium electron transport using the density matrix renormalization group method. Phys. Rev. B **70**, 121302 (2004). doi:10.1103/PhysRevB.70.121302. http://link.aps.org/doi/10.1103/PhysRevB.70.121302
70. Kleine, A., Kollath, C., McCulloch, I.P., Giamarchi, T., Schollwöck, U.: Spin-charge separation in two-component bose gases. Phys. Rev. A **77**, 013607 (2008). doi:10.1103/PhysRevA.77.013607. http://link.aps.org/doi/10.1103/PhysRevA.77.013607

71. Bañuls, M.C., Hastings, M.B., Verstraete, F., Cirac, J.I.: Matrix product states for dynamical simulation of infinite chains. Phys. Rev. Lett. **102**, 240603 (2009). doi:10.1103/PhysRevLett.102.240603. http://link.aps.org/doi/10.1103/PhysRevLett.102.240603
72. Müller-Hermes, A., Cirac, J.I., Bañuls, M.C.: Tensor network techniques for the computation of dynamical observables in one-dimensional quantum spin systems. New J. Phys. **14**(7), 075003 (2012). http://stacks.iop.org/1367-2630/14/i=7/a=075003
73. Magnus, W.: On the exponential solution of differential equations for a linear operator. Commun. Pure Appl. Math. **7**(4), 649–673 (1954). doi:10.1002/cpa.3160070404. http://dx.doi.org/10.1002/cpa.3160070404 [ISSN 1097-0312]
74. Blanes, S., Moan, P.C.: Fourth- and sixth-order commutator-free Magnus integrators for linear and non-linear dynamical systems. Appl. Numer. Math. **56**(12), 1519–1537 (2006). doi:10.1016/j.apnum.2005.11.004. http://www.sciencedirect.com/science/article/pii/S0168927405002163 [ISSN 0168-9274]
75. Blanes, S., Casas, F., Oteo, J.A., Ros, J.: The Magnus expansion and some of its applications. Phys. Rep. **470**(56), 151–238 (2009). doi:10.1016/j.physrep.2008.11.001. http://www.sciencedirect.com/science/article/pii/S0370157308004092 [ISSN 0370-1573]
76. Alvermann, A., Fehske, H.: High-order commutator-free exponential time-propagation of driven quantum systems. J. Comput. Phys. **230**(15), 5930–5956 (2011). doi:10.1016/j.jcp.2011.04.006. http://www.sciencedirect.com/science/article/pii/S0021999111002300 [ISSN 0021-9991]
77. Munthe-Kaas, H., Owren, B.: Computations in a free lie algebra. Philos. Trans. R. Soc. Lond. A **357**(1754), 957–981 (1999). doi:10.1098/rsta.1999.0361. http://rsta.royalsocietypublishing.org/content/357/1754/957.abstract
78. Press, W.H., Teukolsky, S.A., Vetterling, W.T., Flannery, B.P.: Numerical Recipes in C: The Art of Scientific Computing. Cambridge University Press, Cambridge (1993)
79. Saad, Y.: Analysis of some Krylov subspace approximations to the matrix exponential operator. SIAM J. Numer. Anal **29**, 209–228 (1992)
80. Hochbruck, M., Lubich, C.: On krylov subspace approximations to the matrix exponential operator. SIAM J. Numer. Anal. **34**, 1911–1925 (1997). doi:10.1137/S0036142995280572. http://epubs.siam.org/sinum/resource/1/sjnaam/v34/i5/p1911_s1
81. Zurek, W.H., Dorner, U., Zoller, P.: Dynamics of a quantum phase transition. Phys. Rev. Lett. **95**(10), 105701 (2005). doi:10.1103/PhysRevLett.95.105701
82. Dziarmaga, J.: Dynamics of a quantum phase transition: Exact solution of the quantum ising model. Phys. Rev. Lett. **95**, 245701 (2005). doi:10.1103/PhysRevLett.95.245701. http://link.aps.org/doi/10.1103/PhysRevLett.95.245701
83. Cincio, L., Dziarmaga, J., Rams, M.M., Zurek, W.H.: Entropy of entanglement and correlations induced by a quench: Dynamics of a quantum phase transition in the quantum Ising model. Phys. Rev. A **75**(5), 052321 (2007). doi:10.1103/PhysRevA.75.052321
84. Carr, L.D., Demille, D., Krems, R.V., Ye, J.: Cold and ultracold molecules: Science, technology, and applications. New J. Phys. **11**, 055049 (2009)
85. Lieb, E., Schultz, T., Mattis, D.: Two soluble models of an antiferromagnetic chain. Ann. Phys. **16**(3), 407–466 (1961). doi:10.1016/0003-4916(61)90115-4. http://www.sciencedirect.com/science/article/pii/0003491661901154 [ISSN 0003-4916]
86. Pfeuty, P.: The one-dimensional Ising model with a transverse field. Ann. Phys. **57**(1), 79–90 (1970). doi:10.1016/0003-4916(70)90270-8. http://www.sciencedirect.com/science/article/pii/0003491670902708 [ISSN 0003-4916]
87. LAPACK package for solution of dense linear algebra problems (2012). http://www.netlib.org/lapack/f

Chapter 8
The Infinite Size Variational Matrix Product State Algorithm

Chapter 7 discusses algorithms using matrix product states (MPSs) as variational ansätze for finding eigenstates and time evolution of 1D quantum systems with a finite number of sites. In this chapter we instead focus on a system which has an infinite number of lattice sites[1] and whose Hamiltonian is spatially homogenous. Because of the translational invariance of the Hamiltonian, the eigenstates may also be chosen to be translationally invariant. However, in a many-body system invariance under translations of a single site may be spontaneously broken, resulting in a system which is only translationally invariant under shifts by q sites. Examples of this type of translational symmetry breaking arise in long-range interacting systems at strong coupling [1, 2], where crystalline states of any rational filling occupy finite regions of the phase diagram. As the broken symmetry is associated with a discrete group, the Mermin–Wagner theorem [3, 4] does not apply and so the system can display true long-range order. This has the unfortunate side effect of inducing very strong finite size effects in simulations with open boundary conditions, where MPS algorithms are most efficient. The algorithm presented in this chapter assumes that the system is invariant under translation by a user-specified number of sites q, and then optimizes this *unit cell* of q sites directly in the limit of an infinite system size. Performing the optimization in this limit removes spurious boundary effects which can completely obscure the properties of the system otherwise. A full set of examples are provided in Sect. 8.4 after the theory of the algorithm has been developed.

An algorithm for optimizing an infinite MPS was present already in the very first DMRG paper [5]. This algorithm, which is known as iDMRG,[2] begins by

[1] Because we are considering spatially discrete systems, the infinities we encounter are always countable.

[2] Many MPS algorithms have infinite counterparts which are denoted by attaching an "i" to the beginning of the name, e.g. iTEBD. This naming scheme was developed well before Apple made it trendy to do so.

M.L. Wall, *Quantum Many-Body Physics of Ultracold Molecules in Optical Lattices*, Springer Theses, DOI 10.1007/978-3-319-14252-4_8

considering a two-site system. The energy is minimized in this configuration, resulting in the MPS[3]

$$|\psi\rangle = \sum_{i_1 j_1} A^{[1]i_1} \Lambda^{[1]} B^{[1]j_1} |i_1 j_1\rangle. \tag{8.1}$$

This minimization, and those that follow, can be performed using the ground state search method presented in Chap. 7. These two-sites now form an environment into which two new sites are embedded. In order to maintain a consistent canonical form for the resulting four-site MPS, the environment is formed from $A^{[1]}$ and $B^{[1]}$ and the two new sites replace $\Lambda^{[1]}$ as the orthogonality center. These two interior sites are optimized by minimizing the energy with the outer two-sites held fixed, resulting in the four-site MPS

$$|\psi\rangle = \sum_{i_1 i_2 j_1 j_2} A^{[1]i_1} A^{[2]i_2} \Lambda^{[2]} B^{[2]j_2} B^{[1]j_1} |i_1 i_2 j_2 j_1\rangle. \tag{8.2}$$

The procedure of inserting two new sites and optimizing them with the others held fixed is repeated, leading at the nth iteration to an MPS of the form

$$|\psi\rangle = \sum_{i_1 \dots i_n j_1 \dots j_n} A^{[1]i_1} A^{[2]i_2} \dots A^{[n]i_n} \Lambda^{[n]} B^{[n]j_n} \dots B^{[2]j_2} B^{[1]j_1} |i_1 \dots i_n j_n \dots j_1\rangle . \tag{8.3}$$

Using this method, White and Huse [6] found many of the properties of the spin-1 antiferromagnetic Heisenberg chain an unprecedented precision of 12 digits. In particular, they verified Haldane's conjecture [7] that this model has a finite gap to excitations, and demonstrated the presence of long-range string order of the correlator

$$g(\ell) = \left\langle \hat{S}_0^z \prod_{k=1}^{\ell-1} (-1)^{\hat{S}_k^z} \hat{S}_\ell^z \right\rangle, \tag{8.4}$$

corresponding to a hidden $\mathbb{Z}_2 \times \mathbb{Z}_2$ order [8].[4]

While iDMRG produced very good results for this case, it is rarely used to study infinite systems because of strong setbacks in the general case. One of the main setbacks is that iDMRG does not produce a translationally invariant wavefunction. That is to say, after a large number of iterations the resulting state Eq. (8.3) is

[3]Here and throughout this section the superscript index in brackets denotes from which iteration of iDMRG the particular tensor was obtained rather than its position in space.

[4]For a physical interpretation of string order and the order parameter $g(\ell)$, see the discussion following Eq. (6.74). While the results given there are for the AKLT state, the AKLT state and the ground state of the spin-1 antiferromagnetic Heisenberg model are in the same quantum phase.

still an open boundary MPS with the central tensors being increasingly accurate representations of the bulk tensors which would be embedded in the center of an infinite chain. This makes it useful as a means to generate an initial ansatz for a finite-size simulation, but not to study the translationally invariant infinite system. As first realized by McCulloch [9], we can identify a translationally invariant unit cell from the tensors obtained by iDMRG by writing the infinite wavefunction in Vidal canonical form as

$$|\psi\rangle = \sum_{\ldots i_k i_{k+1} i_{k+2} i_{k+3},\ldots} \ldots \left(\Lambda^{[A]}\Gamma^{[A]i_k}\Lambda^{[B]}\Gamma^{[B]i_{k+1}}\right)\left(\Lambda^{[A]}\Gamma^{[A]i_{k+2}}\Lambda^{[B]}\Gamma^{[B]i_{k+3}}\right)\ldots \times |\ldots i_k i_{k+1} i_{k+2} i_{k+3}\ldots\rangle\,, \tag{8.5}$$

where the repeat unit has been bracketed out. Translating this result into the A and B language using Eqs. (6.43) and (6.44), we have that the current best guess at the unit cell is

$$A^{[n]i}\Lambda^{[n]}B^{[n]j}\left(\Lambda^{[n-1]}\right)^{-1}. \tag{8.6}$$

With the realization of this repeat unit, we find that the algorithm can produce approximations to an infinite many-body state which is exactly translationally invariant. Explicitly, the infinite state is

$$|\psi\rangle = \sum_{\ldots i_k j_k i_{k+1} j_{k+1}\ldots} \ldots \left(A^{[n]i_k}\Lambda^{[n]}B^{[n]j_k}\left(\Lambda^{[n-1]}\right)^{-1}\right) \left(A^{[n]i_{k+1}}\Lambda^{[n]}B^{[n]j_{k+1}}\left(\Lambda^{[n-1]}\right)^{-1}\right)\ldots \times |\ldots i_k j_k i_{k+1} j_{k+1}\ldots\rangle\,. \tag{8.7}$$

So far we have considered only unit cells which are two-sites in length, as this is the configuration considered in the original iDMRG work. However, the procedure can be applied to unit cells of any length, as required for systems which spontaneously break single-site translational invariance. In order to keep the notation light, we will continue to use the two-site notation for the unit cell. Expressions for larger unit cells may be derived from these by treating i and j as multi-component site indices $\mathbf{i} = \left\{i_1,\ldots,i_{\lfloor q/2\rfloor}\right\}$,[5] $\mathbf{j} = \left\{j_{\lfloor q/2\rfloor+1},\ldots,j_q\right\}$ and then decomposing $A^{[n]\mathbf{i}}$ as a product of left-canonical matrices $A^{[n]i_1}\ldots A^{[n]i_{\lfloor q/2\rfloor}}$ and likewise for $B^{[n]}$. We will use the symbol q to denote the length of the unit cell in what follows. We also note that the computation time scales linearly in the length of the unit cell.

[5]Here $\lfloor n\rfloor$ represents the smallest integer less than or equal to n.

An important practical consequence of McCulloch's unit cell identification is that we can use the optimal tensors from the $(n-1)$ st and nth iteration cycles to construct a guess at the optimal wavefunction of the $(n+1)$ st iteration. The optimal tensor at the nth iteration Eq. (8.3) may be written as

$$\begin{aligned} &A^{[1]i_1} \dots A^{[n]i_n} \Lambda^{[n]} B^{[n]j_n} \dots B^{[1]j_1} \\ &= A^{[1]i_1} \dots \left(A^{[n]i_n} \Lambda^{[n]} B^{[n]j_n} \left(\Lambda^{[n-1]} \right)^{-1} \right) \Lambda^{[n-1]} \dots B^{[1]j_1} . \end{aligned} \tag{8.8}$$

Identifying the unit cell, and recognizing that in the limit of an infinite system the presence of one additional unit cell is inconsequential, we can insert our best guess at the unit cell Eq. (8.6) to obtain a guess at the optimal tensor of the $n+1$th iteration

$$\begin{aligned} &A^{[1]i_1} \dots \left(A^{[n]i_n} \Lambda^{[n]} B^{[n]j_{n+1}} \left(\Lambda^{[n-1]} \right)^{-1} \right) \\ &\left(A^{[n]i_{n+1}} \Lambda^{[n]} B^{[n]j_n} \left(\Lambda^{[n-1]} \right)^{-1} \right) \Lambda^{[n-1]} \dots B^{[1]j_1} \\ &= A^{[1]i_1} \dots A^{[n]i_n} \left(\Lambda^{[n]} B^{[n]j_{n+1}} \left(\Lambda^{[n-1]} \right)^{-1} A^{[n]i_{n+1}} \Lambda^{[n]} \right) B^{[n]j_n} \dots B^{[1]j_1} , \end{aligned} \tag{8.9}$$

where the last line has bracketed out the guess. In this form it is also clear that, for a two-site unit cell, the optimization procedure optimizes the first site and the second site in the unit cell in alternate iterations. That is, the position of the beginning of the unit cell shifts each iteration. For unit cells of larger than two-sites there is greater freedom in how to split the tensor for absorbing into the environment, but practical experience shows that cutting the unit cell in half as in the two-site case works best.[6] This is because this decomposition maintains that the two environments are equally valid representations of their infinite counterparts, avoiding "one-sided" errors.

With these notations, we may now state the infinite-size variational ground state search with MPS (iMPS) algorithm as formulated by McCulloch.

1. *Input*

   Input the matrix product operator (MPO) representation of the Hamiltonian[7] and a sequence of bond dimensions $\{\chi_i\}$, $i = 1, 2, \dots$. Set $\chi = \chi_1$.

[6]For a unit cell with an odd number of sites q, we alternately absorb $\lceil q/2 \rceil$ and $\lfloor q/2 \rfloor$ into the left environment.

[7]See Chap. 7 for a discussion of MPOs.

2. *Initialization*

   (a) Construct a finite-size, open boundary simulation on q sites and find the optimal state for the given χ. Bring into mixed canonical form $A^{[1]}\Lambda^{[1]}B^{[1]}$.
   (b) Absorb $A^{[1]}$ and $B^{[1]}$ into the environment using the caching recursions Eqs. (7.44)–(7.46) discussed in Sect. 7.3. Solve for the ground state of a new q-site cell with this environment, and bring into mixed canonical form $A^{[2]}\Lambda^{[2]}B^{[2]}$. Set $n = 2$.

3. *Iteration*

   (a) Absorb $A^{[n]}$ and $B^{[n]}$ into the environment.
   (b) Initialize a trial guess at the ground state with the new environment $\left(\Lambda^{[n]}B^{[n]j_{n+1}}\left(\Lambda^{[n-1]}\right)^{-1}A^{[n]i_{n+1}}\Lambda^{[n]}\right)$.
   (c) Using the trial guess, find the ground state, and bring into mixed canonical form $A^{[n+1]}\Lambda^{[n+1]}B^{[n+1]}$.
   (d) Check for convergence. If converged, measure desired properties and increment to the next bond dimension χ. If the last χ was computed, exit, otherwise return to 3(a). If convergence is not reached, increment n by one and return to 3(a).

8.1 The Orthogonality Fidelity

If the ground state of the given Hamiltonian has the translational symmetry assumed by the iMPS unit cell ansatz, then it is expected that the iMPS iteration above converges to a fixed point in which the unit cells obtained from concurrent iterations are close in some sense. We can make this intuition precise by introducing the orthogonality fidelity

$$F_{\text{ortho}} = \mathrm{Tr}\sqrt{\sqrt{\rho_R^{[n]}}\rho^{[n-1]}\sqrt{\rho_R^{[n]}}}\,, \tag{8.10}$$

where the reduced density matrices $\rho_R^{[n]}$ and $\rho^{[n-1]}$ are obtained by tracing over everything to the right of the current unit cell at the nth iteration and to the right of the orthogonality center at the $(n-1)$ st iteration, respectively. For a mixed-canonical state, $\rho^{[n]} = \Lambda^{[n]\dagger}\Lambda^{[n]}$. We can compute $\rho_R^{[n]}$ by performing a singular value decomposition $\Lambda^{[n]}B^{[n]} = USV$. $\Lambda_R^{[n]}$ is SV, and $\rho_R^{[n]} = \Lambda_R^{[n]\dagger}\Lambda_R^{[n]}$. Now, using the cyclic properties of the trace for finite matrices, F_{ortho} is the sum of the singular values of $\Lambda_R^{[n]}\Lambda^{[n-1]\dagger}$, which is straightforwardly calculated.

8.2 Orthogonalization of MPSs in the Thermodynamic Limit

Let us now turn to measurements using our fixed point MPS. First, we recall that the translationally invariant unit cell is represented most accurately at the nth iteration as

$$A^{[n]i} \Lambda_n B^{[n]j} \left(\Lambda^{[n-1]}\right)^{-1}, \tag{8.11}$$

or, shifting the position of the beginning of the unit cell, as

$$\left(\Lambda^{[n-1]}\right)^{-1} A^{[n]i} \Lambda^{[n]} B^{[n]j}. \tag{8.12}$$

Considering Eq. (8.11) and shifting the orthogonality center to the edge of the unit cell, we find

$$A^{[n]i} A^{[n]j} \Lambda_R^{[n]} \left(\Lambda^{[n-1]}\right)^{-1}. \tag{8.13}$$

Here, the tensor describing the right half of the unit cell, $\Lambda^{[n]} B^{[n]j}$, has been converted to $A^{[n]j} \Lambda_R^{[n]}$, where $A^{[n]j}$ is left-canonical and different from the tensor $A^{[n]i}$ describing the left half of the unit cell. In order to keep the notation light, we will use A for both of these tensors, and they will be distinguished by the indices and the order in which they appear within the unit cell. The appearance of the matrix $P \equiv \Lambda_R^{[n]} \left(\Lambda^{[n-1]}\right)^{-1}$ indicates a deviation from orthonormality for finite F_{ortho}, which is always the case in numerics.[8] By a deviation from orthonormality, we mean that the right basis of $A^{[n]j}$ is not orthogonal, and also that the transfer matrix has a spectral radius different than 1. Similarly, by extracting $\Lambda_L^{[n]}$ from the SVD of $A^{[n]i} \Lambda^{[n]}$ in Eq. (8.12), one can show that the left basis of $A^{[n]i}$ is also not orthogonal. A means to measure the amount of (non)orthogonality in the bases of these MPS tensors is to measure the expectation of the unit operator between these states. This is readily done within the transfer operator formalism introduced in Sect. 6.5.

We define the transfer operator of a left-canonical unit cell, $T_L(E)$ as

$$T_L(E) = \sum_{ij} P^\dagger A^{j\dagger} A^{i\dagger} E A^i A^j P, \tag{8.14}$$

where P was defined in the last paragraph. Here E is a $\chi \times \chi$ matrix, and the transfer operator T_L is an operator which takes $\chi \times \chi$ matrices to $\chi \times \chi$ matrices. Also, we note that T_L operates on the right of E. Hence, we use the notation $T_L(E)$

[8] Except in extreme cases, for example if the ground state is a product state.

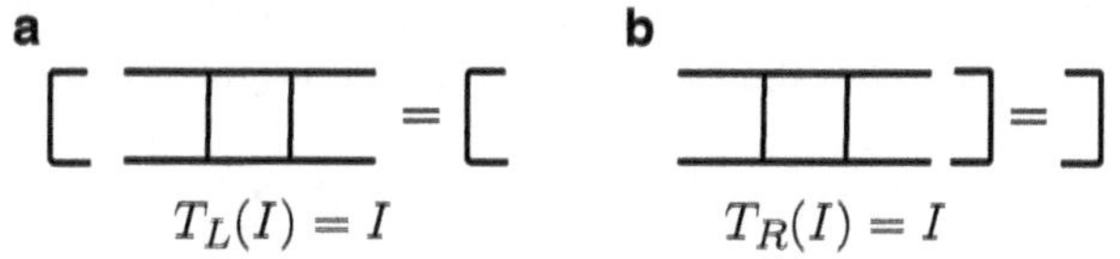

Fig. 8.1 Normalization conditions for infinite MPSs. (**a**) For a left-canonical unit cell, the transfer operator of the unit cell admits the identity as a left eigenmatrix with eigenvalue 1. (**b**) For a right-canonical unit cell, the transfer operator of the unit cell admits the identity as a right eigenmatrix with eigenvalue 1

to avoid confusion with the direction of operation. Also, while naively the operation of the transfer operator would require $\mathcal{O}(\chi^4)$ operations, it can be done in $\mathcal{O}(\chi^3)$ by exploiting its tensor network structure.

The usefulness of the transfer operator is best seen by considering the norm of the translationally invariant wavefunction, which is the contraction of an infinite tensor network. If we assume that we have contracted this network from $-\infty$ to 0 and the result is stored in E, then $T_L(E)$ gives the result of this contraction being carried through one more unit cell. Hence, for an orthonormal state, we would have that $T_L(I) = I$, see Fig. 8.1. Stated equivalently, the identity matrix is a left eigenmatrix of the transfer operator with eigenvalue 1. However, in the present case, $T_L(I) = P^\dagger P$. We can remedy this [10] by solving for the largest eigenmatrix of $T_L(I)$, call it V_L. Here it should be noted that $T_L(I)$ is not symmetric in general. Also, because of the tensor network structure of $T_L(E)$, a sparse eigensolver should be used to find the eigenmatrix corresponding to the largest eigenvalue. In practice we use the routine `dnaupd` from the Arnoldi-based ARPACK [11] package for non-symmetric sparse matrices. Because of the "quadratic" form of $T_L(I)$, V_L is Hermitian and positive definite when the largest eigenvalue is non-degenerate, and hence we can decompose it into $V_L = X^\dagger X$ where X is invertible. While this is suggestive of a Cholesky decomposition [12], Cholesky decomposition becomes unstable when the matrix is nearly singular [13] and so it is advisable to rather perform an eigenvalue decomposition $V_L = U\Lambda U^\dagger$ with U the matrix with eigenvectors of V_L as columns and then $X = \sqrt{\Lambda}U^\dagger$. We then transform the unit cell via a similarity transformation $XA^iA^jPX^{-1}$ whence $T_L(I) = I$, as desired. The case of degenerate maximal eigenvalue will be discussed at the end of this subsection.

Shifting the starting point of the unit cell as in Eq. (8.12) and moving the orthogonality center to the left, we have that the right-canonical unit cell is

$$QB^iB^j\,, \tag{8.15}$$

where $Q \equiv \left(\Lambda^{[n-1]}\right)^{-1}\Lambda_L^{[n]}$. The interpretations of B^i and Λ^L are parallel to that of A^j and Λ^R above. We stress that this unit cell is independent of X, and so all operations here are compatible with the change of basis of the unit cell such that the right basis is orthogonal. For the transfer operator of a right-canonical unit cell

$$T_R(E) = \sum_{ij} QB^i B^j E B^{j\dagger} B^{i\dagger} Q^\dagger, \tag{8.16}$$

we hence have that $T_R(I) = QQ^\dagger$. Proceeding as before, we find the (right) eigenmatrix V_R associated with the largest eigenvalue and decompose it as $V_R = YY^\dagger$ via an eigenvalue decomposition, $V_R = U\Lambda U^\dagger$, $Y = U\sqrt{\Lambda}$. We then transform the unit cell as $Y^{-1}QB^i B^j Y$. If we reinsert all definitions, we find that the properly orthogonalized unit cell may be written as

$$XA^{[n]i}\Lambda^{[n]}B^{[n]j}YY^{-1}\left(\Lambda^{[n-1]}\right)^{-1}X^{-1}. \tag{8.17}$$

This may be brought into the original form of the unit cell

$$A^{[n]i}\Lambda^{[n]}B^{[n]j}\left(\Lambda^{[n-1]}\right)^{-1}, \tag{8.18}$$

by identifying the new set of tensors

$$A^{[n]i} = XA^{[n]i}, \tag{8.19}$$

$$B^{[n]j} = B^{[n]j}Y, \tag{8.20}$$

$$\Lambda^{[n-1]} = X\Lambda^{[n-1]}Y. \tag{8.21}$$

The identification of the tensors Eqs. (8.19)–(8.21) has the benefit of not requiring the inverses of X and Y, but only of $X\Lambda^{[n-1]}Y$. This set of transformations gives us a translationally invariant unit cell with proper left and right orthonormalization ensured. In what follows, when we write, e.g., A_n^i we refer to the orthonormalized tensor and not the raw output from the iMPS iteration.

We can now discuss canonical forms for the translationally invariant MPS state. By writing down a product of several unit cells

$$\ldots A^{[n]i}\Lambda^{[n]}B^{[n]j}\left(\Lambda^{[n-1]}\right)^{-1}A^{[n]i}\Lambda^{[n]}B^{[n]j}\left(\Lambda^{[n-1]}\right)^{-1}A^{[n]i}\Lambda^{[n]}B^{[n]j}\left(\Lambda^{[n-1]}\right)^{-1}\ldots, \tag{8.22}$$

we can identify three different canonical forms paralleling the canonical forms for finite MPSs. The first is fully left-canonical, and is obtained by $A^j = \Lambda^{[n]}B^{[n]j}\left(\Lambda^{[n-1]}\right)^{-1}$:

$$\ldots\left(A^{i_1}A^{j_1}\right)\left(A^{i_2}A^{j_2}\right)\ldots. \tag{8.23}$$

Here, the parentheses denote unit cells, and the ordering of the indices now denotes position in the lattice from left to right rather than the optimization cycle index, compare Eq. (8.3). By the properties of our orthonormalization procedure above, these tensors are truly left-canonical in the sense of Eq. (6.24). The second is fully right-canonical, and is obtained by $B^i = \left(\Lambda^{[n-1]}\right)^{-1} A^{[n]i} \Lambda^{[n]}$:

$$\ldots \left(B^{i_1} B^{j_1}\right) \left(B^{i_2} B^{j_2}\right) \ldots . \tag{8.24}$$

Again, these matrices are all right-canonical in the sense of Eq. (6.33). The mixed-canonical state is obtained by inserting the identity $I = \Lambda^{[n-1]} \left(\Lambda^{[n-1]}\right)^{-1}$ after one of the $\left(\Lambda^{[n-1]}\right)^{-1}$ and using the identifications of A^j and B^i to find

$$\ldots \left(A^{i_1} A^{j_1}\right) \Lambda^{[n-1]} \left(B^{i_2} B^{j_2}\right) \ldots . \tag{8.25}$$

This is the form most useful for expectation values. Here, the parentheses denote the left-canonical and right-canonical unit cells from Eqs. (8.23) and (8.24). The left-canonical unit cell repeats infinitely to the left of $\Lambda^{[n-1]}$, and the right-canonical unit cell repeats infinitely to the right of $\Lambda^{[n-1]}$. We note that the compatibility of Eqs. (8.23)–(8.25) implies that T_R admits $\Lambda^{[n-1]\dagger} \Lambda^{[n-1]}$ as a left eigenmatrix with eigenvalue 1 and similarly T_L admits $\Lambda^{[n-1]} \Lambda^{[n-1]\dagger}$ as a right eigenmatrix with eigenvalue 1. This can also be verified directly.

In the case of a degenerate maximal eigenvalue, the left and right eigenmatrices corresponding to the maximal eigenvalue we find are not unique, but form a basis for the degenerate eigenspace. Hence, the ordering of the numerically obtained eigenmatrices may be such that $\langle k|k\rangle$ is very close to zero, where $\langle k|$ is the kth left eigenmatrix, $|k\rangle$ is the kth right eigenmatrix, and we use the Frobenius inner product. This can cause severe instability in the formation of the properly orthogonalized tensors given in Eqs. (8.19)–(8.21). To avoid this difficulty, we form the Gram matrix of the left and right eigenmatrices

$$M_{kk'} = \langle k|k'\rangle , \tag{8.26}$$

and perform a singular value decomposition $M = USV$. We now unitarily transform the left-and right eigenmatrices as

$$|k\rangle = V_{kk'} |k'\rangle , \tag{8.27}$$

$$\langle k| = U_{k'k} \langle k'| , \tag{8.28}$$

such that their Gram matrix is now the positive diagonal matrix of singular values S. This is effectively a transformation of the left and right eigenmatrices to the well-conditioned subspace. In practice we perform this procedure on any set of Hermitian and positive definite eigenmatrices whose eigenvalues differ in a relative sense by 10^{-4},[9] as a quasi-degeneracy resulting in a large correlation length may be indicative of a true degeneracy that is not converged in χ.

8.3 Expectation Values

We now turn to expectation values. Here we will use the mixed-canonical form Eq. (8.25). Without loss of generality, we assume that the support[10] of our observable operator contains the orthogonality center. On the right of the rightmost operator is an infinite product of right-canonical matrices B. Due to the normalization of the state and the right-canonical condition shown graphically in Fig. 8.1, the result of the infinite contraction of right-canonical matrices becomes $\lim_{N\to\infty} T_R^N(I) = I$. Similarly, on the left of the leftmost operator we have an infinite product of left-canonical matrices A, yielding $\lim_{N\to\infty} T_L^N(I) = I$. Thus, the calculation of expectation values of operators with finite support is obtained by contracting the finite network on which our expectation operators have support and tracing the boundaries, see Fig. 8.2. We will turn to the question of operators with infinite support,[11] whose expectation values diverge for the infinite state, in the next paragraph. When computing two-point correlation functions, e.g., $\langle \hat{n}_0\hat{n}_r\rangle$, we find that best results are obtained when the result is averaged over all possible separations r consistent with translation invariance. Namely, we compute

$$\langle \hat{n}_0\hat{n}_r\rangle = \frac{1}{q}\sum_{i=0}^{q-1}\langle \hat{n}_i\hat{n}_{i+r}\rangle\,, \tag{8.29}$$

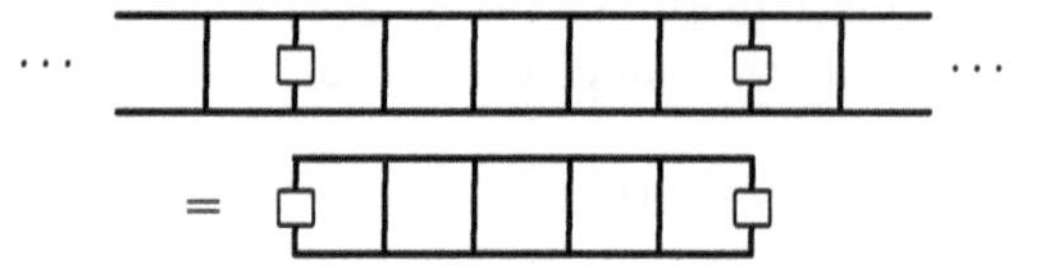

Fig. 8.2 Expectation of a finitely-supported operator within an infinite MPS. Due to the left- and right-canonical conditions shown graphically in Fig. 8.1, the expectation of an operator whose finite support contains the orthogonality center may be evaluated as a finite tensor network contraction using the methods of Sect. 7.5

[9]This value corresponds to a correlation length of roughly 10^4 lattice sites. The reason for this choice is practical; current experimental setups using ultracold gases would have difficulty distinguishing a state with a correlation length this large from a state which is truly long-range ordered. Hence, explicitly breaking symmetries by mixing near-degenerate eigenstates for the benefit of numerical stability will not affect our ability to predict experimental outcomes.

[10]The support of an operator is the region of the lattice on which the operator is nonzero.

[11]That is, expectations of operators which act on all sites in the infinite lattice.

where q is the number of sites in the unit cell and the summation runs over all sites in the unit cell. This accounts for possible breaking of the translational symmetry within the unit cell itself.

We now turn our attention to finding the expectation value of an operator representing an extensive observable, taking the Hamiltonian as a paradigmatic example [14]. This amounts to finding the energy density of the infinite state. We make the assumption that the expectation of the observable grows at most extensively such that it has a finite density in the thermodynamic limit. This holds for all reasonable Hamiltonians, but not, for example, for the square of the Hamiltonian.[12] The Hamiltonian is assumed to be spatially homogenous, and so its MPO representation is specified by a single MPO matrix W. To be concrete, let us consider the MPO representation of the Ising model in a transverse field

$$\hat{H}_{\text{Ising}} = -J \sum_{\langle i,j \rangle} \hat{\sigma}_i^z \hat{\sigma}_j^z - h \sum_i \hat{\sigma}_i^x , \tag{8.30}$$

which is given by

$$W = \begin{pmatrix} \hat{I} & 0 & 0 \\ \hat{\sigma}^z & 0 & 0 \\ -h\hat{\sigma}^x & -J\hat{\sigma}^z & \hat{I} \end{pmatrix} . \tag{8.31}$$

Here $\hat{\sigma}^\nu$ denotes the Pauli matrix along the νth Cartesian direction. This MPO has a bond dimension $\chi_O = 3$.

Assuming the state to be in right-canonical form as in Eq. (8.24), the expectation $\langle \hat{H} \rangle$ becomes an infinitely large contraction of the form

$$\langle \hat{H} \rangle = \text{Tr} \left[\dots \left(\sum_{ii'} B^{i\star} \otimes W^{ii'} \otimes B^{i'} \right) \left(\sum_{jj'} B^{j\star} \otimes W^{jj'} \otimes B^{j'} \right) \dots \right] . \tag{8.32}$$

Comparing with Eq. (6.65) suggests defining a generalized unit cell transfer operator as

$$T_{R;X}(E) = \sum_{iji'j'} X^{ii';jj'} B^{i'} B^{j'} E B^{j\dagger} B^{i\dagger} , \tag{8.33}$$

where $X^{ii';jj'}$ is an operator acting on the local indices of the unit cell. Defining the elements of the MPO representation of our operator acting on the unit cell as

[12] This can be seen by considering the expectation of $\hat{H}^2$ in the ground state, which is $e_0 L^2$ with e_0 the energy density and $L \to \infty$ the number of lattice sites.

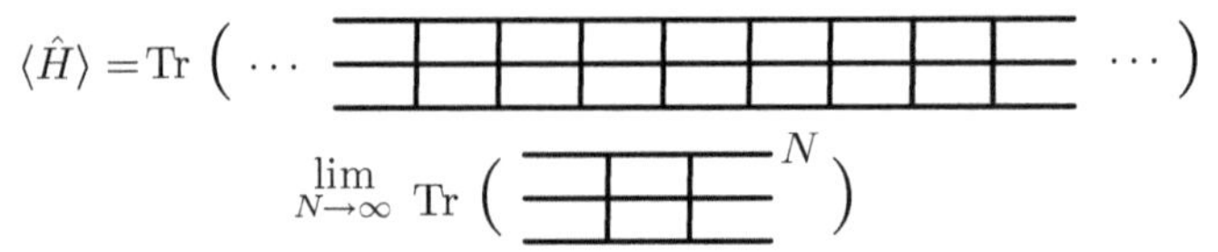

Fig. 8.3 Expectation of an infinitely-supported operator within an infinite MPS. Because of the repeating structure of the unit cell and the homogenous MPO representation, we can identify a repeating tensor network structure which is the generalized transfer operator $T_{R;\mathcal{W}_{\kappa\kappa'}}$. The uncontracted bonds in the center of the expectation correspond to the bond indices of the MPO

$$\mathcal{W}^{ii'jj'}_{\kappa\kappa'} = \sum_{\kappa''} W^{ii'}_{\kappa\kappa''} W^{jj'}_{\kappa''\kappa'}, \tag{8.34}$$

we can write the infinite contraction in terms of the generalized transfer operator Eq. (8.33) as shown graphically in Fig. 8.3.

The general procedure of evaluating the infinite product of transfer operators is similar to the procedure used to normalize infinite MPSs. In the normalization procedure, the restriction that the state have a norm of one implied that the transfer operator had a dominant eigenmatrix with eigenvalue 1. The left- or right-canonical form of the unit cell then allowed us to identify that this dominant eigenmatrix was in fact the identity operator. In the present case, the dominant matrix will have χ_O components, $\left(E_1, \ldots, E_{\chi_O}\right)$, where χ_O is the bond dimension of the MPO. To outline the general structure of the dominant matrices, let us consider multiplying[13] the matrix W in Eq. (8.31) by itself L times to find

$$W^L = \begin{pmatrix} \hat{I}^{\otimes L} & 0 & 0 \\ \hat{I}^{\otimes (L-1)} \otimes \hat{\sigma}^z & 0 & 0 \\ \hat{H}_{\text{Ising}} & -J\hat{\sigma}^z \otimes \hat{I}^{\otimes (L-1)} & \hat{I}^{\otimes L} \end{pmatrix}. \tag{8.35}$$

This is indicative of the general structure of an MPO. In particular, the two halves of the bond term $\hat{\sigma}^z\hat{\sigma}^z$ have a string of identities appended to them to become $\hat{I}^{\otimes(L-1)} \otimes \hat{\sigma}^z$ and $\hat{\sigma}^z \otimes \hat{I}^{\otimes(L-1)}$. In DMRG, these operator strings are referred to as *connection operators*, as they specify how the Hamiltonian connects the L sites acted on by this MPO product to the rest of the system. Furthermore, the lower left element of the MPO is the Hamiltonian on L sites, and the upper left and lower right components are the identity on L sites. Because the only term in the first row is the identity, E_1 is an eigenmatrix of T_R with eigenvalue 1 provided that the unit cell is in right-canonical form. Hence, the first component of the dominant eigenmatrix is the identity matrix.

To find the other components of the dominant matrix, let us consider that we have performed the expectation Eq. (8.32) over n unit cells starting with some boundary

[13] By multiplication, we mean contraction over the bond indices with the matrix-valued elements combined by tensor products, see Sect. 7.2.2.

matrices $\{E_\kappa(0)\}$ and have stored the result in the matrices $\{E_\kappa(n)\}$. Here n denotes over how many unit cells the contraction has been performed. Carrying out the contraction over one further unit cell produces the matrix

$$E_\kappa(n+1) = \sum_{\kappa'} T_{R;\mathcal{W}_{\kappa\kappa'}}(E_{\kappa'}(n)) \, . \tag{8.36}$$

We now focus on the case where all infinite-ranged terms in the Hamiltonian are free of Fermi phases.[14] In these cases any operator appearing on the diagonal of the MPO matrix $\mathcal{W}$ is proportional to the identity. Separating out this part of the MPO explicitly, we have the fixed point relations

$$(I - xT_R)(E_\kappa) = \sum_{\kappa'<\kappa} T_{R;\mathcal{W}_{\kappa\kappa'}}(E_{\kappa'}) \, . \tag{8.37}$$

In Eq. (8.37) we have dropped the arguments n with the understanding that the dominant matrices are fixed points of Eq. (8.37) and are hence independent of n.

The crucial component in solving Eq. (8.37) is that, because of the lower triangular structure and the fact that E_1 is known a priori, the right-hand side of Eq. (8.37) is a known matrix, and hence Eq. (8.37) represents a non-symmetric system of linear equations for E_κ, see Fig. 8.4. For $\kappa \neq \chi_O$, there are two possibilities. The first is that there are no long-range terms, and so $x = 0$. In this case there is no linear system of equations, but rather an equality for the unknown matrix E_κ. The second is that there is a long-range interaction with $x < 1$.[15] In this case we do have to solve a linear system of equations, but the matrix on the left-hand side is nonsingular due to the fact that the spectral radius of T_R is 1. We solve these using the GMRES method [15], a Krylov-subspace based sparse linear solver for non-symmetric matrices. This allows us to take advantage of the tensor network structure when applying T_R so that the solution is $\mathcal{O}(\chi^3)$. In the case of the Ising model with MPO given by Eq. (8.31), all elements with $1 < \kappa < \chi_O$ have $x = 0$, and so the associated $\{E_\kappa(n)\}$ can be solved for explicitly. These matrices do not depend on the initial values $E_\kappa(0)$ provided that $E_1(0) = I$, and so are the fixed points $\{E_\kappa\}$ of Eq. (8.37).

Fig. 8.4 Fixed point relations for the dominant matrices of the Hamiltonian MPO transfer matrix, Eq. (8.37)

[14] This assumption does not hold for long-range tunneling of fermions, for example. Such cases can be dealt with, but are not relevant for this thesis.

[15] $x = 1$ corresponds to uniform long-range interactions, and we know of no physical real-space Hamiltonians involving such terms.

Finally, we turn to $\kappa = \chi_O$. Here we have $x = 1$, and so the left-hand matrix in Eq. (8.37) is right-singular for components of E_{χ_O} along the identity. However, the lower-left element of a product of L MPO matrices is the Hamiltonian on L sites, see Eq. (8.35), and so $E_{\chi_O}(n+1)$ represents the overlap of the Hamiltonian on n unit cells in the basis states of the unit cell MPS decomposition. Hence, the singularity arises from the fact that this expectation can take on any value $e_0 q n$, where e_0 is the energy density, q is the number of lattice sites, and n is a positive integer. That is to say, the result of carrying the contraction through one more unit cell is

$$E_{\kappa}(n+1) = -qe_0 I \delta_{\kappa,\chi_O} + \sum_{\kappa'} T_{R;W_{\kappa\kappa'}}\left(E_{\kappa'}(n)\right) , \tag{8.38}$$

where the constant piece is the energy expectation of the unit cell. We can remove this ambiguity by using the fact that $\Lambda^{[n-1]^\dagger}\Lambda^{[n-1]}$ spans the left null space of the matrix $(I - T_R)$ to find

$$qe_0 = \mathrm{Tr}\left[\Lambda^{[n-1]^\dagger}\Lambda^{[n-1]} \sum_{\kappa'<\chi_O} T_{R;W_{\kappa\kappa'}}\left(E_{\kappa'}\right)\right] . \tag{8.39}$$

This provides us with the energy density. Hence, the relevant fixed-point condition for the dominant matrices is not Eq. (8.37) but

$$(I - xT_R)(E_{\kappa}) = -qe_0 I \delta_{\kappa,\chi_O} + \sum_{\kappa'<\kappa} T_{R;W_{\kappa\kappa'}}\left(E_{\kappa'}\right) , \tag{8.40}$$

which differs from Eq. (8.37) only for $\kappa = \chi_O$. The fact that the energy density takes the form Eq. (8.39) also implies that the right-hand side of Eq. (8.40) is orthogonal to the left null space of the left-hand side matrix. By the fundamental theorem of linear algebra [16], the right-hand side lies completely within the image of the operator on the left-hand side, and so this equation has a consistent solution and not merely a least-squares or pseudo-inverse solution in spite of the fact that the operator on the left-hand side is singular. This construction contrasts with the method taken in [14], in which the extensive part of the observable is not included in Eq. (8.40) and a projection procedure within the GMRES method is used instead. The fact that we can add any multiple of the identity matrix to the dominant matrix E_{χ_O} which solves Eq. (8.40) and still have a solution to Eq. (8.40) corresponds to the freedom in choosing a zero of energy for the infinite system. Using Eq. (8.39) we get the energy per site several orders of magnitude more accurately than we obtain from raw output of the effective Hamiltonian eigenequation used to optimize the iMPS.

8.4 Examples

To demonstrate the iMPS method, we will study the spin-1/2 and spin-1 antiferromagnetic Heisenberg models. Antiferromagnetic Heisenberg models have long been of interest in studies of quantum magnetism, where they represent a minimal Hamiltonian describing particles whose spins wish to be anti-aligned with their neighbors [17]. The spin-1/2 model is exactly solvable via Bethe ansatz [18, 19], and is known to be critical with central charge 1 and z–z correlations decaying asymptotically as

$$\langle \hat{S}_0^z \hat{S}_r^z \rangle \sim (-1)^r \frac{\sqrt{\ln |r|}}{|r|} . \tag{8.41}$$

The z-z correlation function describes how strongly the z-component of a spin at position 0 is correlated with the z-component of a spin r sites away. Experimentally, this information is probed via the dynamical structure factor

$$S(\mathbf{q}, \omega) = L^{-1} \int_{-\infty}^{\infty} dt e^{i\omega t} \sum_{i,j=1}^{L} e^{i\mathbf{q}\cdot(\mathbf{r}_i - \mathbf{r}_j)} \langle \hat{S}_i^z(t) \hat{S}_j^z(0) \rangle , \tag{8.42}$$

where $\hat{S}_i^z(t)$ is a Heisenberg picture operator and L is the number of sites [17]. In one dimension, true long-range ferromagnetic order where $\langle \hat{S}_0^z \hat{S}_r^z \rangle$ approaches a constant as $r \to \infty$ is destroyed by strong fluctuations according to the Mermin–Wagner theorem [3, 4], and so power-law decay represents the strongest allowed degree of correlation. Furthermore, its nearest-neighbor and next-nearest neighbor z-z correlations are known from the Bethe ansatz solution to be $\frac{1}{12}(1 - 4\log 2) \approx -0.14771573$ [19] and $\frac{1}{12}(1 - 16\log 2 + 9\zeta(3)) \approx 0.06067977$ [20]. The spin-1 model has no properties which are amenable to exact computation. However, its properties were of great interest after Haldane [7] conjectured a major difference between integer-spin and half integer-spin antiferromagnetic Heisenberg models. In particular, as opposed to the half-integer case where the system is critical and correlations decay algebraically, integer-spin chains are gapped and non-critical, and z-z correlations decay exponentially as

$$\langle \hat{S}_0^z \hat{S}_r^z \rangle \sim (-1)^r \frac{\exp(-|r|/\xi)}{|r|} . \tag{8.43}$$

DMRG found one of its important early applications for this system in which Haldane's conjecture was verified, and the gap and correlation length ξ were both calculated to high precision [6].

In Fig. 8.5, we show the z-z correlation function of these two models computed from converged iMPS simulations with a range of bond dimensions χ. In the upper plot for the spin-1/2 model, we also plot the nearest-neighbor and next-nearest-neighbor correlations as horizontal lines to accentuate that even MPSs with small

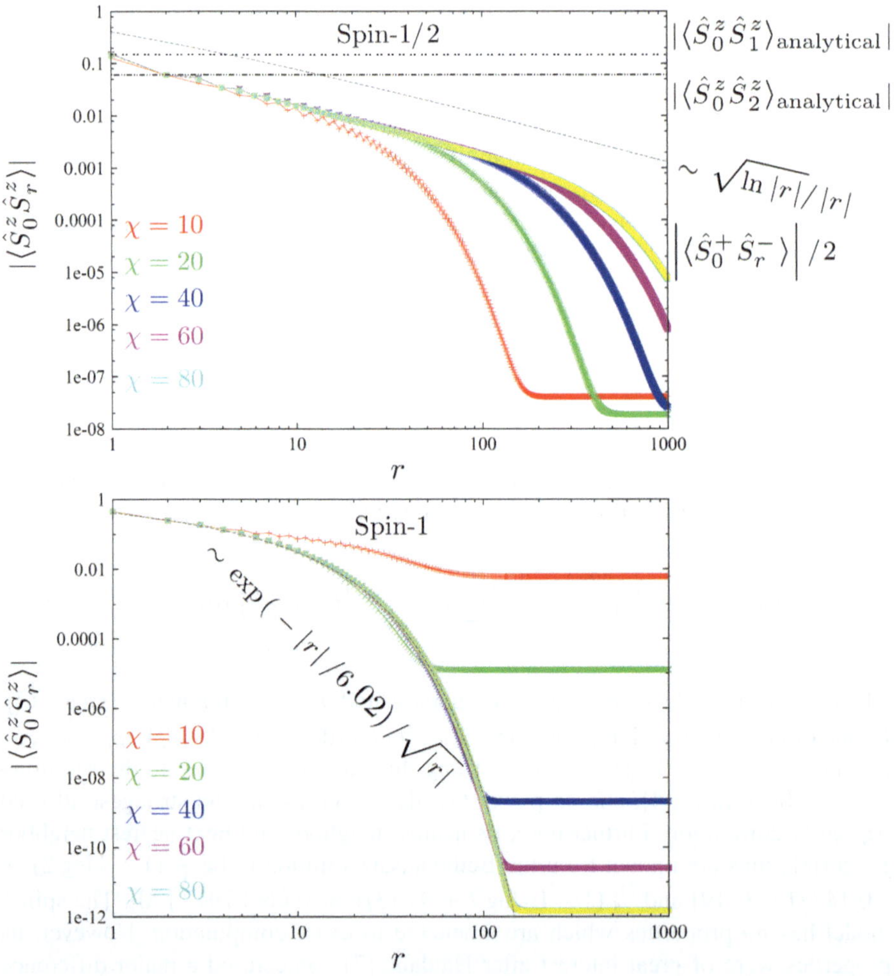

Fig. 8.5 Correlation functions for the spin-1/2 and spin-1 Heisenberg models computed with iMPS. *Top panel*: The *z-z* correlation functions for the spin-1/2 antiferromagnetic Heisenberg model is compared with exactly known results. As the bond dimension is increased, the range over which this function follows the true algebraic decay increases, compare Fig. 6.5. *Bottom panel*: The *z-z* correlation function for the spin-1 antiferromagnetic Heisenberg model is shown together with a fit demonstrating its exponential decay

bond dimension $\chi = 10$ capture the short-distance physics well. The tilted dashed line is a function with the same scaling as the Bethe ansatz result Eq. (8.41). As the bond dimension increases, the power-law behavior of the correlator persists to further distances. Hence, what is more important than the value of the correlator for a single χ is the scaling behavior of the correlator as χ is increased. Plotted in yellow, which lies on top of the light blue line, is the $\langle \hat{S}_0^+ \hat{S}_r^- \rangle/2$ correlator to demonstrate that spin-rotational symmetry is preserved by the simulation. As is always the case

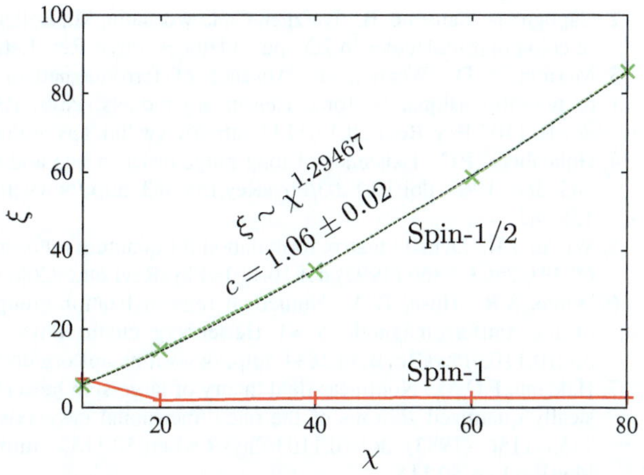

Fig. 8.6 Scaling of the correlation length with the bond dimension χ in iMPS. The spin-1/2 chain obeys the scaling relation Eq. (6.75) between the correlation length and the bond dimension, yielding a central charge in agreement with the analytical prediction. The spin-1 chain shows a saturation of the correlation length with the entanglement cutoff, and is hence non-critical

for MPSs with a finite bond dimension that do not have long-range order, at large enough distances the behavior of the correlator always becomes exponential with a correlation length equal to the largest correlation length of the transfer operator. This behavior may also be compared with the approximation of a power-law by a sum of exponentials shown in Fig. 6.5. In the bottom panel of Fig. 8.5 we show the analogous plot for the spin-1 chain. Here the correlator is exponential, as shown via the fit through the correlator with the largest bond dimension.

In Fig. 8.6, we display the scaling of the largest correlation length obtained from the sub-leading eigenvalue of the transfer operator with the bond dimension. The critical spin-1/2 state shows an increase in the correlation length as the bond dimension increases, and a fit to the prediction Eq. (6.75) yields a central charge in good agreement with the known value. Also shown in this plot is the behavior of the spin-1 chain, in which the saturation of the correlation length to a finite value implies the absence of criticality. Typical values of χ for MPS/DMRG studies on finite lattices are a few hundreds to a few thousands, with values of tens of thousands being reported for quasi-2D systems [21, 22]. In finite-size MPS studies one must first extrapolate $\chi \to \infty$ before one can perform reliable finite-size scaling $L \to \infty$. By instead taking the limit $L \to \infty$ first and performing finite-entanglement scaling, we can obtain very good results using comparatively rather modest values of the bond dimension χ.

References

1. Dalmonte, M., Pupillo, G., Zoller, P.: One-dimensional quantum liquids with power-law interactions: The Luttinger staircase. Phys. Rev. Lett. **105**, 140401 (2010). doi:10.1103/PhysRevLett.105.140401. http://www.link.aps.org/doi/10.1103/PhysRevLett.105.140401

2. Capogrosso-Sansone, B., Trefzger, C., Lewenstein, M., Zoller, P., Pupillo, G.: Quantum phases of cold polar molecules in 2D optical lattices. Phys. Rev. Lett. **104**, 125301 (2010)
3. Mermin, N.D., Wagner, H.: Absence of ferromagnetism or antiferromagnetism in one- or two-dimensional isotropic Heisenberg models. Phys. Rev. Lett. **17**, 1133–1136 (1966). doi:10.1103/PhysRevLett.17.1133. http://www.link.aps.org/doi/10.1103/PhysRevLett.17.1133
4. Hohenberg, P.C.: Existence of long-range order in one and two dimensions. Phys. Rev. **158**, 383–386 (1967) doi:10.1103/PhysRev.158.383. http://www.link.aps.org/doi/10.1103/PhysRev.158.383
5. White, S.R.: Density matrix formulation for quantum renormalization groups. Phys. Rev. Lett. **69**(19), 2863–2866 (1992) doi:10.1103/PhysRevLett.69.2863
6. White, S.R., Huse, D.A.: Numerical renormalization-group study of low-lying eigenstates of the antiferromagnetic S =1 Heisenberg chain. Phys. Rev. B **48**, 3844–3852 (1993) doi:10.1103/PhysRevB.48.3844. http://www.link.aps.org/doi/10.1103/PhysRevB.48.3844
7. Haldane, F.D.M.: Nonlinear field theory of large-spin heisenberg antiferromagnets: Semiclassically quantized solitons of the one-dimensional easy-axis néel state. Phys. Rev. Lett. **50**, 1153–1156 (1983) doi:10.1103/PhysRevLett.50.1153. http://www.link.aps.org/doi/10.1103/PhysRevLett.50.1153
8. Kennedy, T., Tasaki, H.: Hidden $Z_2 \times Z_2$ symmetry breaking in Haldane-gap antiferromagnets. Phys. Rev. B **45**, 304–307 (1992) doi:10.1103/PhysRevB.45.304. http://www.link.aps.org/doi/10.1103/PhysRevB.45.304
9. McCulloch, I.P.: Infinite size density matrix renormalization group, revisited (2008). http://www.xxx.lanl.gov/abs/0804.2509
10. Orús, R., Vidal, G.: Infinite time-evolving block decimation algorithm beyond unitary evolution. Phys. Rev. B **78**, 155117 (2008) doi:10.1103/PhysRevB.78.155117. http://www.link.aps.org/doi/10.1103/PhysRevB.78.155117
11. ARPACK package for sparse solution of eigenvalue problems (1997). http://www.caam.rice.edu/software/ARPACK/
12. Golub, G.H., Van Loan, C.F.: Matrix Computations, 3rd edn. Johns Hopkins Studies in Mathematical Sciences. The Johns Hopkins University Press, Baltimore (1996)
13. Wilkinson, J.H.: Rounding Errors in Algebraic Processes. Prentice-Hall, Englewood Cliffs (1964)
14. McCulloch, I.P., Michel, L.: Schur forms of matrix product operators in the infinite limit (2010). http://www.arxiv.org/abs/1008.4667
15. Saad, Y., Schultz, M.: GMRES: A generalized minimal residual algorithm for solving nonsymmetric linear systems. SIAM J. Sci. Stat. Comput. **7**(3), 856–869 (1986). doi:10.1137/0907058. http://www.epubs.siam.org/doi/abs/10.1137/0907058
16. Strang, G.: The fundamental theorem of linear algebra. Am. Math. Mon. **100**(9), 848–855 (1993) http://www.jstor.org/stable/2324660 [ISSN: 00029890]
17. Auerbach, A.: Interacting Electrons and Quantum Magnetism. Springer, Berlin (1994)
18. Bethe, H.: Zur theorie der metalle. Z. Phys. A **71**, 205–226 (1931). doi:10.1007/BF01341708
19. Hulthén, L.: Über das austauchsproblem eines kristalles. Ark. Mat. Astron. Fys. A **26**, 1 (1938)
20. Takahashi, M.: Thermodynamics of One-Dimensional Solvable Models. Cambridge University Press, Cambridge (1999)
21. Jiang, H.-C., Block, M.S., Mishmash, R.V., Garrison, J.R., Sheng, D.N., Motrunich, O.I., Fisher, M.P.A.: Non-Fermi liquid d-wave metal phase of strongly interacting electrons (2012). http://www.arxiv.org/abs/1207.6608
22. Depenbrock, S., McCulloch, I.P., Schollwöck, U.: Nature of the spin-liquid ground state of the $s = 1/2$ heisenberg model on the kagome lattice. Phys. Rev. Lett. **109**, 067201 (2012). doi:10.1103/PhysRevLett.109.067201. http://www.link.aps.org/doi/10.1103/PhysRevLett.109.067201

Chapter 9
Finite Temperature Matrix Product State Algorithms and Applications

Abstract We review the basic theory of matrix product states (MPS) as a numerical variational ansatz for time evolution, and present two methods to simulate finite temperature systems with MPS: the ancilla method and the minimally entangled typical thermal state (METTS) method. A sample calculation with the Bose–Hubbard model is provided.

9.1 Introduction

The dimension of the Hilbert space for a general many-body system increases exponentially with the system size, severely restricting the system sizes amenable to straightforward numerical study. Several techniques have been developed to deal with this fact, such as the stochastic sampling of the Hilbert space in quantum Monte Carlo techniques and the judicious use of symmetries and sparse matrix structures in exact diagonalizations. The most successful approximate method for 1d systems is the density matrix renormalization group (DMRG) method first pioneered by White [1] (see Chap. 24 for a methodology tailored towards higher-dimensional lattice configurations). Soon after, the theory of matrix product states (MPS) [2, 3] was used to shed light on the amazing success of DMRG [4, 5]. Ideas from quantum information theory, most notably the idea of bipartite entanglement, have led to the development of MPS algorithms which generalize DMRG to time evolution [6, 7], periodic boundary conditions [8], and finite temperature [9, 10]. In this chapter we review algorithms based on MPS for finite temperature simulations and discuss their relevance to studying finite temperature superfluid systems.

Published previously as *Finite Temperature Matrix Product State Algorithms and Applications*, M.L. Wall and L.D. Carr, Chapter in "Quantum Gases: Finite Temperature and Non-Equilibrium Dynamics" (Vol. 1 Cold Atoms Series), N.P. Proukakis, S.A. Gardiner, M.J. Davis and M.H. Szymanska, eds. (Imperial College Press, 2012).

M.L. Wall, *Quantum Many-Body Physics of Ultracold Molecules in Optical Lattices*, Springer Theses, DOI 10.1007/978-3-319-14252-4_9

9.2 Methodology

9.2.1 *Matrix Product States*

A MPS[1] on a lattice with periodic boundary conditions is defined as

$$|\Psi_{\text{mps}}\rangle = \sum_{i_1,i_2,\ldots i_L=1}^{d} \text{Tr}\left(\mathbf{A}^{[1]i_1}\cdots\mathbf{A}^{[L]i_L}\right)|i_1,\ldots,i_L\rangle, \tag{9.1}$$

where the $\mathbf{A}^{[k]i_k}$ are square matrices[2] of dimension χ (the bond dimension), d is the dimension of the Hilbert space spanned by the $\{|i_k\rangle\}$, and L is the number of lattice sites. Let us refer to the set of all MPSs with bond dimension χ as $\mathcal{M}_\chi$. An MPS in $\mathcal{M}_\chi$ contains $Ld\chi^2$ parameters, and so it is clear that any state on a finite lattice can be written as an MPS provided we take the bond dimension to be $\chi_{\text{max}} = d^{\lfloor L/2\rfloor}$. However, the great utility of MPSs is that an MPS with bond dimension $\chi \ll \chi_{\text{max}}$ often provides an excellent approximation to the true state [12] and allows for exponentially more efficient manipulation and calculation of observables than an exact representation.

To visualize MPSs and operations with them, it is useful to introduce the notion of a *tensor network diagram* as in Fig. 9.1. In such a diagram a box represents a tensor, free lines are uncontracted indices, and closed lines are contracted indices. Figure 9.1a shows the state of a many-body system expressed in the basis of the full Hilbert space as an L-index tensor, and Fig. 9.1b shows the same state written as an MPS. The advantage of the MPS representation becomes clear when we compute scalar products such as $\langle\psi|\hat{O}|\phi\rangle$.

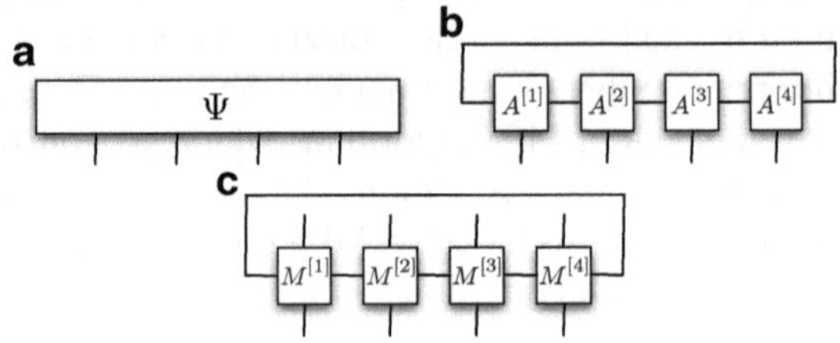

Fig. 9.1 (**a**) Tensor network representation of full four-site wavefunction. (**b**) Tensor network representation of an MPS on four sites. (**c**) Tensor network representation of an MPO on four sites

[1] An MPS is a vector in Hilbert space. The qualifier matrix product refers to the fact that the expansion coefficients in the Fock basis are expressed as products of matrices.

[2] These matrices can be taken to have the same symmetry as the state they represent, e.g., if the state has real coefficients in some basis, then the MPS matrices can be taken to be real. See [11] and the references therein for more details.

Before we discuss how scalar products are efficiently computed, it is advantageous to introduce a *matrix product operator* (MPO) as

$$\hat{O} = \sum_{i_1,\ldots,i_L=1}^{d} \sum_{i'_1,\ldots,i'_L=1}^{d} \mathrm{Tr}\left(\mathbf{M}^{[1]i_1 i'_1} \cdots \mathbf{M}^{[L]i_L i'_L}\right) |i_1,\ldots,i_L\rangle\langle i'_1,\ldots,i'_L|, \tag{9.2}$$

where each of the $\mathbf{M}^{[k]i_k i'_k}$ is a matrix, the dimensions of which are bounded by a fixed number (bond dimension) χ_O. The tensor network representation of an MPO is similar to that of an MPS, but there are two uncontracted indices per tensor corresponding to the bra and ket indices; see Fig. 9.1c. Equivalently, one can think of each element of the matrix $\mathbf{M}^{[k]}$ as being operator-valued, where the operator acts on the space spanned by $\{|i_k\rangle\}$.

Let us now see how to evaluate the scalar product of an operator $\hat{O}$ represented as an MPO between two states $|\psi\rangle$ and $|\phi\rangle$ represented as MPSs. Let us denote the MPO matrices of $\hat{O}$ as $\mathbf{M}$ and the MPS matrices of $|\psi\rangle$ and $|\phi\rangle$ as $\mathbf{A}$ and $\mathbf{B}$, respectively. Then, we have

$$\begin{aligned}\langle\psi|\hat{O}|\phi\rangle &= \sum_{i_1,\ldots,i_L=1}^{d} \sum_{i'_1,\ldots,i'_L=1}^{d} \mathrm{Tr}\left(\mathbf{A}^{[1]i_1\star} \cdots \mathbf{A}^{[L]i_L\star}\right) \\ &\quad \times \mathrm{Tr}\left(\mathbf{M}^{[1]i_1 i'_1} \cdots \mathbf{M}^{[L]i_L i'_L}\right) \mathrm{Tr}\left(\mathbf{B}^{[1]i_1} \cdots \mathbf{B}^{[L]i_L}\right) \qquad (9.3) \\ &= \mathrm{Tr}\left(\left[\sum_{i_1,i'_1=1}^{d} \mathbf{A}^{[1]i_1\star} \otimes \mathbf{M}^{[1]i_1 i'_1} \otimes \mathbf{B}^{[1]i'_1}\right]\right. \\ &\quad \left. \times \cdots \times \left[\sum_{i_L,i'_L=1}^{d} \mathbf{A}^{[L]i_L\star} \otimes \mathbf{M}^{[L]i_L i'_L} \otimes \mathbf{B}^{[L]i'_L}\right]\right) \qquad (9.4) \\ &\equiv \mathrm{Tr}\left(\mathbf{E}_{\mathbf{M}}^{[1]}(\mathbf{A},\mathbf{B}) \cdots \mathbf{E}_{\mathbf{M}}^{[L]}(\mathbf{A},\mathbf{B})\right),\end{aligned}$$

where the last line defines the generalized transfer matrix $\mathbf{E}_{\mathbf{M}}^{[k]}(\mathbf{A},\mathbf{B}) \equiv \sum_{i_k,i'_k=1}^{d} \mathbf{A}^{[k]i_k\star} \otimes \mathbf{M}^{[k]i_k i'_k} \otimes \mathbf{B}^{[k]i'_k}$, which is a $\chi^2\chi_O \times \chi^2\chi_O$ matrix. Naively we would expect that the multiplication of two transfer matrices would require $\mathcal{O}(\chi^6\chi_O^3)$ operations, but the special structure of the transfer matrices allows us to perform such a multiplication in $\mathcal{O}(\chi^5\chi_O^2 d^2)$[3] as

[3]The fact that the boundary matrices of MPSs with open boundary conditions have bond dimension 1 allows us to perform this contraction in $\mathcal{O}(\chi^3\chi_O^2 d^2)$, and recent developments for periodic boundary conditions have reduced the scaling to $\mathcal{O}(\chi^3\chi_O^2 d^2)$ for large systems with only a few relevant correlation lengths [13, 14].

$$\left[\mathbf{E}_{\mathbf{M}}^{[k]}(\mathbf{A},\mathbf{B})\,\mathbf{E}_{\mathbf{M}}^{[k+1]}(\mathbf{A},\mathbf{B})\right]_{[\alpha\gamma\beta],[\alpha'\gamma'\beta']}$$

$$= \sum_{i'=1}^{d}\sum_{\beta=1}^{\chi}\left(\sum_{i=1}^{d}\sum_{\gamma''=1}^{\chi_O}\left(\left[\mathbf{G}_{\mathbf{M}}^{[k]}(\mathbf{A},\mathbf{B})\right]_{[\alpha\gamma\beta],[\alpha'\gamma''\beta']}\right)\mathbf{M}_{\gamma''\gamma'}^{[k+1]ii'}\right)\mathbf{B}_{\beta''\beta'}^{[k+1]i'}, \quad (9.5)$$

where

$$\left[\mathbf{G}_{\mathbf{M}}^{[k]}(\mathbf{A},\mathbf{B})\right]_{[\alpha\gamma\beta],[\alpha'\gamma''\beta']} \equiv \sum_{\alpha''=1}^{\chi}\left[\mathbf{E}_{\mathbf{M}}^{[k]}(\mathbf{A},\mathbf{B})\right]_{[\alpha\gamma\beta],[\alpha''\gamma''\beta'']}\mathbf{A}_{\alpha''\alpha'}^{[k+1]i\,\star}. \quad (9.6)$$

Here the square brackets around indices denote a composite index in the Kronecker representation and parentheses give the order in which the contraction should be performed to ensure the best scaling. In particular, it is essential not to sum over the α'' and β'' indices simultaneously.[4] The tensor network representation of the scalar product procedure is given in Fig. 9.2.

Many operators of interest, such as translationally invariant 1d Hamiltonians, can be easily represented as MPOs with small bond dimension $\chi_O \sim 4$–10 [14, 15], and the MPO representations of more complex operators can be constructed using simple MPO arithmetic [15, 16]. That the MPO form of an operator is optimal for MPS algorithms can be straightforwardly deduced using the tensor network formalism, as the scalar product of an MPO between two MPSs is the most general 1d tensor network that can be efficiently contracted; see Fig. 9.2.

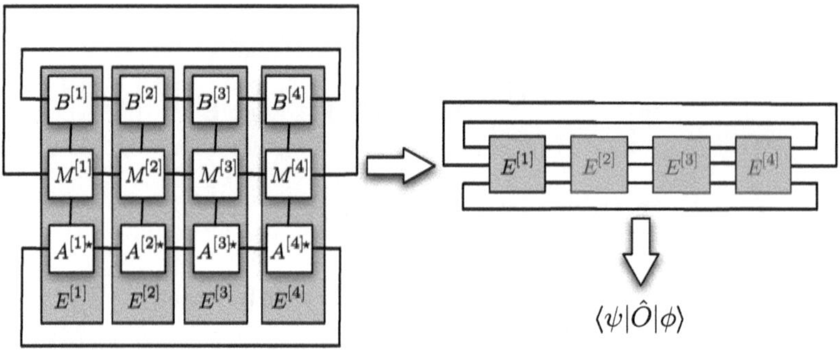

Fig. 9.2 Tensor network representation of the scalar product procedure of Eq. (9.4). The transfer matrices $\mathbf{E}_{\mathbf{M}}^{[k]}(\mathbf{A},\mathbf{B})$ have been abbreviated as $E^{[k]}$ for succinctness

[4] Here and throughout we use Greek indices to denote bond indices and Roman indices to denote physical indices.

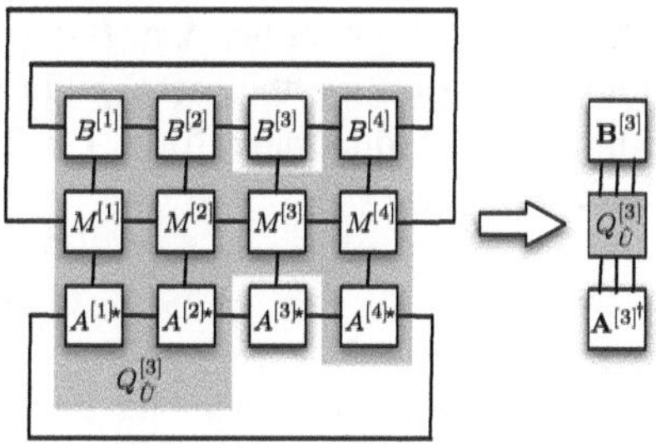

Fig. 9.3 Tensor network representation of the quadratic form representing $\langle\psi|\hat{U}|\phi\rangle$ in Eq. (9.7)

We now turn to the simulation of time evolution using MPSs. The main difficulty of using MPSs is that $\mathcal{M}_\chi$ is not a vector space.[5] Thus, when operators such as the propagator are applied to an MPS we must find the optimal[6] projection into $\mathcal{M}_\chi$ to keep the algorithm efficient. We denote this projection as $\mathcal{P}_\chi$. The optimal MPS $|\psi\rangle \in \mathcal{M}_\chi$ representing the MPS $\hat{U}|\phi\rangle$ is

$$\begin{aligned}\mathcal{P}_\chi\left[\hat{U}|\phi\rangle\right] &= \min_{|\psi\rangle\in\mathcal{M}_\chi}\left||\psi\rangle - \hat{U}|\phi\rangle\right|^2 \\ &= \min_{|\psi\rangle\in\mathcal{M}_\chi}\left[\langle\psi|\psi\rangle + \langle\phi|\hat{U}^\dagger\hat{U}|\phi\rangle - 2\mathcal{R}\left(\langle\psi|\hat{U}|\phi\rangle\right)\right],\end{aligned} \tag{9.7}$$

where $\mathcal{R}(\bullet)$ denotes the real part. Each of the scalar products in Eq. (9.7) may be written as a quadratic form in each of the matrices $\mathbf{A}^{[k]i_k}$, as is demonstrated in the tensor network diagram Fig. 9.3.

Again denoting the matrices in the MPS representation of $|\psi\rangle$ by $\mathbf{A}$ and those of $|\phi\rangle$ by $\mathbf{B}$, the quadratic form of the kth site may be written as

$$\mathcal{Q}^{[k]} = \mathbf{A}^{[k]^\dagger}\mathbf{Q}_{\hat{1}}^{[k]}\mathbf{A}^{[k]} + \mathbf{B}^{[k]^\dagger}\mathbf{Q}_{\hat{U}^\dagger\hat{U}}^{[k]}\mathbf{B}^{[k]} - 2\mathcal{R}\left(\mathbf{A}^{[k]^\dagger}\mathbf{Q}_{\hat{U}}^{[k]}\mathbf{B}^{[k]}\right), \tag{9.8}$$

where $\mathbf{A}^{[k]}$ represents the $d\chi^2$ elements of the $\{\mathbf{A}^{[k]i_k}\}$, arranged as a vector, and the matrices $\mathbf{Q}_{\hat{O}}$ are defined as

$$\left[\mathbf{Q}_{\hat{O}}^{[k]}\right]_{[\alpha i_k\alpha'][\beta i'_k\beta']} = \sum_{\gamma,\gamma'=1}^{\chi_O}\mathbf{M}_{\gamma\gamma'}^{[k]i_k i'_k}\left[\prod_{j\neq k}\mathbf{E}_{\mathbf{M}}^{[j]}(\mathbf{C},\mathbf{D})\right]_{[\alpha\gamma\beta],[\alpha'\gamma'\beta']}, \tag{9.9}$$

[5]This can be seen from the fact that the addition of two MPSs is given by the direct sum of their matrices: $|\psi_C\rangle = |\psi_A\rangle + |\psi_B\rangle \Rightarrow \mathbf{C}^{[k]} = \mathbf{A}^{[k]}\oplus\mathbf{B}^{[k]}$. If the matrices $\mathbf{A}^{[k]}$ and $\mathbf{B}^{[k]}$ have orthogonal bases, then $\dim(\mathbf{C}^{[k]}) = \dim(\mathbf{A}^{[k]}) + \dim(\mathbf{B}^{[k]})$.

[6]By optimal we mean that the overlap is maximal in the two-norm. Although MPSs do not form a vector space, they are embedded in a larger Hilbert space and so this norm is well defined.

where **C** and **D** are either **A** or **B** depending on the quadratic form. The $\mathbf{M}^{i_k i'_k}$ in this final expression are the matrices in the MPO representation of $\hat{O}$. The stationary points of the quadratic form Eq. (9.8) are given by the solution of the linear system[7] [8]

$$\mathbf{Q}_{\hat{1}} \mathbf{A}^{[k]} = \mathbf{Q}_{\hat{U}} \mathbf{B}^{[k]}. \tag{9.10}$$

The algorithmic procedure for time evolution is to sweep back and forth through the lattice, solving Eq. (9.10) at each site until convergence is reached. In practice, it is essential for efficiency not to explicitly form the matrices $\mathbf{Q}_\bullet$, but rather to use iterative methods which require only multiplication by the $\mathbf{Q}_\bullet$ to solve Eq. (9.10). Details on the form of the propagator $\hat{U}$ can be found in [16, 17].

9.2.2 *The Ancilla Method*

At finite temperature, the state of a quantum system is given by the thermal density matrix $\hat{\rho} = e^{-\beta \hat{H}}/Z$. The ancilla method [9, 18] relies on the notion of purification [19] to represent the thermal density matrix as a pure state in an enlarged Hilbert space. Each physical site is augmented with an ancilla which has the same Hilbert space dimension as the physical site. The MPS representation of such a system is

$$|\psi\rangle = \sum_{i_1,\dots,i_L=1}^{d} \sum_{a_1,\dots,a_L=1}^{d} \mathrm{Tr}\left(\mathbf{A}^{[1]i_1 a_1} \cdots \mathbf{A}^{[L]i_L a_L}\right) |i_1 a_1 \cdots i_L a_L\rangle. \tag{9.11}$$

One can think of the combined system as a two-legged ladder, with the physical sites on the lower leg and the ancillae on the upper leg. The purpose of the ancillae is to act as a perfect heat bath which, when traced out, provides the proper thermal density matrix for the physical system. The choice of ancilla for infinite temperature ($\beta = 0$) is simply the normalized purification of the identity

$$|\psi(0)\rangle = \frac{1}{\sqrt{d^L}} \prod_{k=1}^{L} \sum_{i_k,a_k=1}^{d} \delta_{i_k a_k} |i_k a_k\rangle, \tag{9.12}$$

which represents a product of maximally entangled site-ancilla pairs. This state has an MPS representation with bond dimension 1 generated by taking all matrices to

[7] It is important to note that while $\mathbf{Q}_{\hat{1}}$ is the quadratic form representing the scalar product $\langle\psi|\psi\rangle$ it cannot in general be made equal to the identity. The numerical conditioning of this matrix and of the linear system Eq. (9.10) can be improved by suitable choice of "gauge conditions" on the matrices A; see [8].

be $\mathbf{A}_{\alpha\beta}^{[k]i_k a_k} = \delta_{\alpha,1}\delta_{\beta,1}\delta_{i_k a_k}/\sqrt{d}$. The extension to finite inverse temperature β is provided by evolving only the physical sites[8] in imaginary time up to $\beta/2$,

$$|\psi(\beta)\rangle = e^{-\beta\hat{H}/2}|\psi(0)\rangle\,. \tag{9.13}$$

This time evolution can be efficiently performed using the methods of Sect. 9.2.1. Observables are calculated using transfer matrices as above with the additional requirement that the ancilla degrees of freedom are traced over.

The ancilla method is conceptually very simple, and becomes numerically exact for large enough bond dimension. However, because the MPS [Eq. (9.11)] must encode the information of both the system and the bath, it requires a bond dimension $\sim \chi_{\rm gs}^2$ at low temperatures, where $\chi_{\rm gs}$ is the bond dimension required to accurately represent the ground state. Typical values of $\chi_{\rm gs}$ range from 50 to 5,000, making the ancilla method impractical for many systems at very low temperatures.

We conclude this section by remarking that the ancilla method represents a highly idealized heat bath chosen to reproduce the exact thermal density matrix. Many of the current examples of strongly correlated many-body systems, e.g. cold atoms, are very mesoscopic and are in contact with thermal reservoirs which are also mesoscopic. A modification of the ancilla method where the perfect entanglement at infinite temperature is replaced with ancilla–ancilla and ancilla–system couplings in the Hamiltonian can be devised. Alternatively, one can directly simulate master equations by considering matrix product density operators with optimal projections based on the Hilbert–Schmidt distance [8] or matrix product decompositions of superkets with local projections [10].

9.2.3 *Minimally Entangled Typical Thermal States*

A new method for finite temperature MPS simulations has recently been proposed by White [20]. The idea stems from the question "What is a typical wave function of a quantum system at finite temperature?" That is, if we are to measure a quantum system at finite temperature, what "typical" pure states would we find, and with what probabilities? It is clear from the basic tenets of statistical mechanics that any set of typical states $\{|\phi(i)\rangle\}$ must satisfy

$$\sum_i P(i)\,|\phi(i)\rangle\langle\phi(i)| = e^{-\beta\hat{H}}, \tag{9.14}$$

[8]That is, the Hamiltonian only couples physical sites to physical sites, and not ancillae to ancillae or physical sites to ancillae.

where $P(i)$ is the probability of measuring the system to be in state $|\phi(i)\rangle$, and so the expectation of an operator $\hat{A}$ at finite temperature may be written as

$$\langle \hat{A} \rangle = \sum_i \frac{P(i)}{Z} \langle \phi(i)|\hat{A}|\phi(i)\rangle, \tag{9.15}$$

with Z the partition function. From Eq. (9.15), we see that we can calculate observables using an unweighted average of $\langle \phi(i)|\hat{A}|\phi(i)\rangle$, if we choose the $|\phi(i)\rangle$ at random according to their probabilities of being measured $P(i)/Z$. It is easy to generate states satisfying the typicality condition [Eq. (9.14)] simply by taking any orthonormal basis $\{|i\rangle\}$ and defining

$$|\phi(i)\rangle = [P(i)]^{-1/2} \exp\left(-\beta \hat{H}/2\right)|i\rangle, \quad P(i) = \langle i|\exp\left(-\beta \hat{H}\right)|i\rangle. \tag{9.16}$$

We now use the freedom in the choice of the orthonormal basis $\{|i\rangle\}$ to generate typical states with the least amount of spatial entanglement, as these are the states which can be most efficiently represented as MPSs [6, 21]. This amounts to taking the $\{|i\rangle\}$ to be classical product states (CPSs), $|i\rangle = \prod_{k=1}^{L} |i_k\rangle$, where i_k labels the state of site k. The set of $|\phi(i)\rangle$ obtained from this choice of $\{|i\rangle\}$ are called *minimally entangled typical thermal states* (METTS).

The most efficient algorithmic procedure for generating thermal averages using METTS is as follows:

1. Choose a CPS $|i\rangle$ at random.
2. Evolve in imaginary time using the methods of Sect. 9.2.1 to generate the METTS $|\phi(i)\rangle = [P(i)t]^{-1/2} \exp(-\beta \hat{H}/2)|i\rangle$.
3. Compute observables of interest using this METTS and add to the running averages.
4. Randomly select a new CPS $|i'\rangle$ according to the probability $|\langle i'|\phi(i)\rangle|^2$.
5. Repeat from step 2 until converged.

We see that the main loop of this algorithm closely resembles a Monte Carlo iteration with measurement taking the place of the usual configuration updates. However, it does not rely on a rejection method to perform sampling, and so each METTS that is generated can be used to generate statistics. In practice very few ($\sim$100) METTS suffice to obtain the total energy to a relative accuracy of 10^{-5}. For algorithmic details on how to perform the CPS selection to minimize correlations between successive METTS we refer the reader to [16].

This METTS algorithm has many advantages over the ancilla method of the previous section. As we do not have to encode the bath degrees of freedom in our MPS, the bond dimension required to accurately represent each METTS ranges from 1 at infinite temperature to χ_{gs} at very low temperatures. This makes the METTS method more efficient than the ancilla method by a factor of 10^3–10^{10} for typical systems at very low temperatures. Additionally, if the Hamiltonian of interest has a global symmetry, then we can use the fact that the MPS matrices must transform irreducibly to speed up the calculation [15] or find the thermal ensemble corresponding to a fixed quantum number (canonical ensemble). This latter point is relevant to cold atom systems where the total number of atoms is held fixed.[9]

9.3 Validity Issues

It has been shown that MPSs can faithfully represent ground states of 1d gapped Hamiltonians with at most nearest-neighbor interactions with a bond dimension which grows only polynomially in the system size [12]. In higher dimensions this polynomial scaling gives way to an exponential scaling [23], but calculations on 2D systems of width 8–12 are still feasible [24]. Generalizations of MPSs to higher dimensions exist, but are so far limited by poor polynomial scaling of tensor contractions [25–27]. Perhaps the most important quality of MPS methods as compared to other efficient many-body methods, such as quantum Monte Carlo, is that MPS methods work equally well for fermionic or frustrated systems. All of the methods presented here will work equally well for any 1d or quasi-1d physical system.

9.4 Application: Specific Heat of the Hard-Core Extended Bose–Hubbard Model

As an example of how the above methods may be applied to study the behavior of a finite temperature superfluid system, we study the properties of the hard-core extended Bose–Hubbard model

$$\hat{H} = -J \sum_{\langle i,j \rangle} \left(\hat{b}_i^\dagger \hat{b}_j + \text{H.c.} \right) + V \sum_{\langle i,j \rangle} \hat{n}_i \hat{n}_j \tag{9.17}$$

at half filling. This model is known to have a superfluid phase in the XY universality class for $V < 2J$. In Fig. 9.4 we show a typical thermodynamic quantity, the specific heat $C_V = \beta^2(\langle \hat{H}^2 \rangle - \langle \hat{H} \rangle^2)/L$, as a function of temperature and the nearest-neighbor repulsion. Note that computation of $\langle \hat{H}^2 \rangle$ is easily performed when the MPO representation of $\hat{H}$ is known.

[9] The ancilla method can also be used to simulate systems in the canonical ensemble, but the process is complicated by the fact that we need the purification of the constrained infinite temperature density matrix. This purification can be generated using a ground state DMRG-type calculation with a suitably chosen Hamiltonian [22]. The Hamiltonian will contain artificial ancilla–ancilla and ancilla–physical site couplings which are typically highly nonlocal.

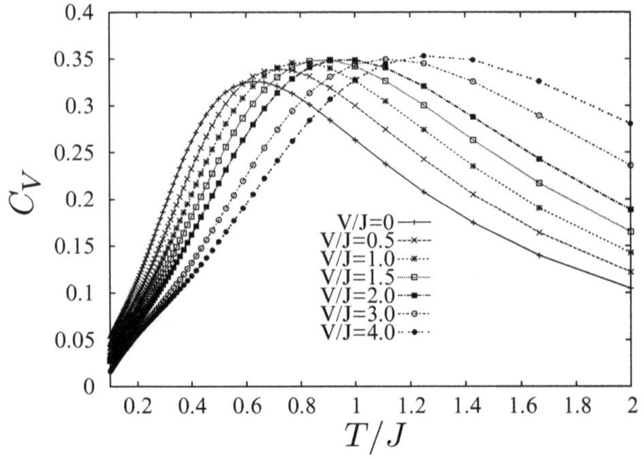

Fig. 9.4 Specific heat of the hard-core extended Bose–Hubbard model on 34 sites for repulsive nearest-neighbor interaction $V/J = 0, 0.5, 1.0, 1.5, 2.0, 3.0, 4.0$. The qualitative behavior of the low temperature specific heat changes as V becomes larger than $2J$ because the system transitions from a gapless superfluid phase into a gapped insulating phase

Acknowledgements We acknowledge useful discussions with Juan José García Ripoll and Miles Stoudenmire. This work was supported by the National Science Foundation under Grant PHY-0903457. MLW thanks the Boulder Summer School for Condensed Matter for stimulating discussions. We also acknowledge the Golden Energy Computing Organization at the Colorado School of Mines for the use of resources acquired with financial assistance from the National Science Foundation and the National Renewable Energy Laboratories.

References

1. White, S.R.: Density matrix formulation for quantum renormalization groups. Phys. Rev. Lett. **69**, 2863 (1992)
2. Klumper, A., Schadschneider, A., Zittartz, J.: Equivalence and solution of anisotropic spin-1 models and generalized t-J fermion models in one dimension. J. Phys. A **24**, L955 (1991)
3. Fannes, M., Nachtergaele, B., Werner, R.F.: Finitely correlated states on quantum spin chains. Commun. Math. Phys. **144**, 443 (1992)
4. Östlund, S., Rommer, S.: Thermodynamic limit of density matrix renormalization. Phys. Rev. Lett. **75**, 3537 (1995)
5. Rommer, S., Östlund, S.: Class of ansatz wave functions for one-dimensional spin systems and their relation to the density matrix renormalization group. Phys. Rev. B **55**, 2164 (1997)
6. Vidal, G.: Efficient simulation of one-dimensional quantum many-body systems. Phys. Rev. Lett. **93**, 040502 (2004)
7. Daley, A.J., Kollath, C., Schollwöck, U., Vidal, G.: Time-dependent density-matrix renormalization-group using adaptive effective Hilbert spaces. J. Stat. Mech. **2004**, P04005 (2004)
8. Verstraete, F., Porras, D., Cirac, J.I.: Density matrix renormalization group and periodic boundary conditions: A quantum information perspective. Phys. Rev. Lett. **93**, 227205 (2004)

9. Verstraete, F., García-Ripoll, J.J., Cirac, J.I.: Matrix product density operators: Simulation of finite-temperature and dissipative systems. Phys. Rev. Lett. **93**, 207204 (2004)
10. Zwolak, M., Vidal, G.: Mixed-state dynamics in one-dimensional quantum lattice systems: A time-dependent superoperator renormalization algorithm. Phys. Rev. Lett. **93**, 207205 (2004)
11. Perez-Garcia, D., Verstraete, F., WolF, M.M., Cirac, J.I.: Matrix product state representations. Quantum Inf. Comput. **7**, 401 (2007)
12. Verstraete, F., Cirac, J.I.: Matrix product states represent ground states faithfully. Phys. Rev. B **73**, 094423 (2006)
13. Pippan, P., White, S.R., Evertz, H.G.: Efficient matrix-product state method for periodic boundary conditions. Phys. Rev. B **81**, 081103 (2010)
14. Pirvu, B., Murg, V., Cirac, J.I., Verstraete, F.: Matrix product operator representations. New J. Phys. **12**, 025012 (2010)
15. McCulloch, I.P.: From density-matrix renormalization group to matrix product states. J. Stat. Mech. **2007**, P10014 (2007)
16. Stoudenmire, E.M., White, S.R.: Minimally entangled typical thermal state algorithms. New J. Phys. **12**, 055026 (2010)
17. García-Ripoll, J.J.: Time evolution of matrix product states. New J. Phys. **8**, 305 (2006)
18. Feiguin, A.E., White, S.R.: Finite-temperature density matrix renormalization using an enlarged Hilbert space. Phys. Rev. B **72**, 220401 (2005)
19. Nielsen, M., Chuang, I.: Quantum Computation and Quantum Information. Cambridge University Press, Cambridge (2000)
20. White, S.R.: Minimally entangled typical quantum states at finite temperature. Phys. Rev. Lett. **102**, 190601 (2009)
21. Vidal, G.: Efficient classical simulation of slightly entangled quantum computations. Phys. Rev. Lett. **91**, 147902 (2003)
22. Feiguin, A.E., Fiete, G.A.: Spectral properties of a spin-incoherent Luttinger liquid. Phys. Rev. B **81**, 075108 (2010)
23. Liang, S., Pang, H.: Approximate diagonalization using the density matrix renormalization-group method: A two-dimensional-systems perspective. Phys. Rev. B **49**, 9214 (1994)
24. White, S.R., Scalapino, D.J.: Pairing on striped $t - t' - J$ lattices. Phys. Rev. B **79**, 220504 (2009)
25. Verstraete, F., Cirac, J.I.: Renormalization algorithms for quantum many-body systems in two and higher dimensions (2004) [arXiv:cond-mat/0407066v1]
26. Schuch, N., Wolf, M.M., Verstraete, F., Cirac, J.I.: Simulation of quantum many-body systems with strings of operators and monte carlo tensor contractions. Phys. Rev. Lett. **100**, 040501 (2008)
27. Evenbly, G., Vidal, G.: Entanglement renormalization in two spatial dimensions. Phys. Rev. Lett. **102**, 180406 (2009)

Part V
Open Source Code and Educational Materials

Chapter 10
Open Source Code Development

The purpose of this chapter is to describe two open source code projects in which the author has been involved. The first, open source time-evolving block decimation (OSTEBD) [1], is a package which was maintained by the Carr theoretical physics research group and heavily modified for flexibility and efficiency by the author. OSTEBD had two releases, one in February of 2009 and the other in October of 2009. The OSTEBD dedicated blog has been viewed uniquely over 2,100 times to date, and OSTEBD has been used in eight publications within the Carr research group [2–9] and six known publications outside of this group [10–15]. The most recent blog statistics, as gathered by http://wordpress.com, are shown in Fig. 10.1. An extensive user's guide was distributed with OSTEBD describing the background of time-evolving block decimation (TEBD), providing exercises to acquaint the user with the package, and manual pages for all routines in the code. The most recent release of OSTEBD, v2.0, may be obtained from the repository at http://sourceforge.net/projects/opentebd/.

In 2010, the code used for the OSTEBD project was significantly altered for stability and speed and merged into the algorithms and libraries for physics simulations (ALPS) package [16], the premier resource for numerical methods for strongly correlated many-body problems. This code was formally included in the v2.0 release of the ALPS package in November 2010. ALPS contains open source implementations of nearly all widely used numerical methods for strongly correlated many-body systems, including exact diagonalization; worm, stochastic series expansion, and quantum Wang–Landau quantum Monte Carlo; dynamical mean field theory; and a static density-matrix renormalization group code. The TEBD software described in this thesis was the first component of the ALPS package to enable the study of dynamics. The ALPS software package is maintained by the ALPS international collaboration presently consisting of 28 researchers. In addition to being used for cutting-edge research world-wide with nearly 250 citations for the first two versions of ALPS at the time of writing [8, 17], ALPS is also used as a pedagogical tool for summer schools. The ALPS collaboration maintains other

M.L. Wall, *Quantum Many-Body Physics of Ultracold Molecules in Optical Lattices*, Springer Theses, DOI 10.1007/978-3-319-14252-4_10

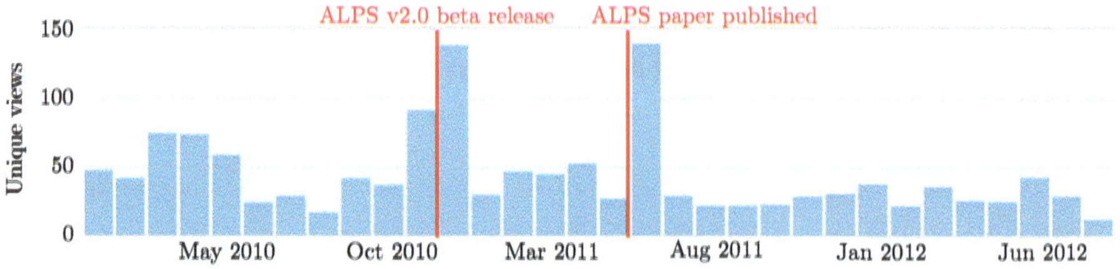

Fig. 10.1 Open source time-evolving block decimation blog statistics. The *red lines* indicate releases of ALPS code and documentation

pedagogical resources such as tutorials and a user's forum. The most recent major release of ALPS at the time of writing of this thesis, ALPS 2.1, is included with the source code CD. Additionally, all documentation for the ALPS TEBD routines which were written solely by the present author, including tutorial exercises, are reprinted in Appendix A.

The OSTEBD manual, which is obtained as part of the code repository, covers the vast majority of that project. Hence, in Sect. 10.1 we provide a brief overview of the package and its capabilities, referring the reader to the manual for more detail. Section 10.1.1 contains information about parallel extensions to OSTEBD which were in alpha version at the time of the v2.0 release. The documentation for the ALPS code is given in Appendix A. It is much less detailed than the documentation for the OSTEBD project. Hence, in Sect. 10.2 we discuss the differences in the ALPS and OSTEBD codes, as well as other parts of the ALPS project which required development by the present author including a Python (http://www.python.org/) front end and integration with the VisTrails workflow provenance system (http://www.vistrails.org/).

10.1 OSTEBD Overview

OSTEBD [1] is a software package, written in Fortran 90, implementing the TEBD algorithm for one-dimensional systems with nearest-neighbor interactions [18, 19]. In addition to real-time propagation, ground states may be found using imaginary time propagation. OSTEBD supports systems of bosons with or without spins, fermions with or without spins, spin systems, and open or periodic boundary conditions [20]. The resulting matrix product state (MPS) representations of the wavefunction are explicitly stored, allowing for checkpointing of long simulations as well as wavefunction-based techniques for detecting quantum phase transitions such as fidelity analysis [6]. Arbitrary one- and two-point correlation functions can be specified by the user, and routines to obtain one and two-body density matrices are included for more complex or quantum-information based measures. Number conservation for particle models or magnetization conservation for spin models is supported [21, 22]. The irreps of the associated on-site Hilbert space are allowed to be degenerate. Both data parallelism and intrinsic parallelism are included in alpha version with the 2.0 release of the code, and are discussed in Sect. 10.1.1. OSTEBD v2.0 contains 21,822 lines of code in its core, i.e., code which does not comprise main files or inputs.

Inputs are provided using the Fortran NAMELIST syntax, which allows for input files to be generated using scripts. For example, the inputs for the `BoseHubbard_ITP.f90` case study are given in the file `BH_ITP.nml` as

```
&SystemSettings
 systemSize=30, maxFilling=5, totNum=30,
     BoundaryCond='O', TrotterOrder=5
&end

&BHParams
 jTunn=1.0, U0=20.0, V0=0.0, mu0=0.0
&end

&ITPParams
 chiMin=15, chiMax=20, convCriterion1=0.00001,
     convCriterion2=0.0000001, stepsForJudge=100,
     dtITP=0.05, maxITPsteps=4000, itpDir='ITPDATA/'
&end
```

The top line gives Hilbert space metadata, including the number of lattice sites, the maximum number of particles allowed on any site, the number of particles, the boundary conditions, and the order of Trotter expansion used for the propagator. The second line gives the parameters of the Bose–Hubbard model. Finally, the last line gives convergence criteria for imaginary time propagation, such as the maximum number of iterations, the imaginary time step, and the maximal bond dimensions. In addition to supporting a wide array of models, the package was designed to be easily modified to suit users' individual needs. The comprehensive manual pages of the OSTEBD manual were part of this approach, as were exercises using the OSTEBD routines which require users to write their own code. For more detail on the educational component of the manual, we refer the reader to Chap. 11.

10.1.1 Parallel Extensions in OSTEBD

Data-parallelism is supported in alpha version in OSTEBD v2.0 via the `PD_Extension`. Here a phase diagram is defined by a range of chemical potentials $\mu_{\min} \leq \mu \leq \mu_{\max}$ with $\mu_{\rm res}$ points or total numbers $N_{\min} \leq N \leq N_{\max}$ with $N_{\rm res}$ points and a range of tunnelings $J_{\min} \leq J \leq J_{\max}$ with $J_{\rm res}$ points. A master node sends points in this phase diagram to worker nodes one at a time, and the worker nodes send observables to the master node when the computation has finished. This automatically scheduled data-parallel paradigm is more efficient than assigning the tasks evenly between all nodes, as points in more highly entangled regions of the phase diagram take more time to compute than less entangled regions, and so nodes which are given only lowly entangled regions may sit idle while others compute highly entangled regions. The code in this extension was used in [6]. For large numbers of tasks, i.e., large $\mu_{\mathrm{res}} J_{\mathrm{res}}$ or $N_{\mathrm{res}} J_{\mathrm{res}}$, the code is $P - 1$ times faster than a single core within a few percent, where P is the number of processors.

`Sites_Parallel_Extension`, also included in alpha version with OSTEBD, is an intrinsically parallel TEBD code using MPI [23]. In the sense of balancing the load evenly and having large granularity, the most naturally parallelizable portion of the TEBD algorithm is the loop over spatial positions during time evolution. The ability to intrinsically parallelize the time evolution sweep in TEBD stems from the fact that in the Vidal canonical form Eq. (6.39), any bipartite splitting can be chosen to be the orthogonality center. Hence, when we use Vidal's suggested Trotterization of the propagator:

$$\begin{aligned} e^{-i\hat{H}\delta t} &= e^{-i\hat{H}_{\mathrm{odd}}\delta t/2\hbar} e^{-i\hat{H}_{\mathrm{even}}\delta t/\hbar} e^{-i\hat{H}_{\mathrm{odd}}\delta t/2\hbar} + \mathcal{O}\left(\delta t^3\right), \\ e^{-i\hat{H}_{\mathrm{odd}}\delta t/2\hbar} &= \prod_{\mathrm{odd}\ l} e^{-i\hat{H}_l\delta t/2\hbar}, \\ e^{-i\hat{H}_{\mathrm{even}}\delta t/\hbar} &= \prod_{\mathrm{even}\ l} e^{-i\hat{H}_l\delta t/\hbar}, \end{aligned} \tag{10.1}$$

we can apply the propagation over even bonds and odd bonds simultaneously at all bipartite splittings. A slab decomposition of the application of propagators across all similar parity bonds would ideally reduce the scaling of the most expensive step by a factor of P, where P is the number of processors.

A two-site operation performed on the two sites l and $l+1$ involves $\lambda^{[l]}$, $\Gamma^{[l]}$, $\lambda^{[l+1]}$, $\Gamma^{[l+1]}$, and $\lambda^{[l+2]}$. Thus, for a given processor to time evolve site l it needs to also own[1] the local tensors of site $l+1$. This means that, in a distributed memory paradigm, the last site that can be time evolved by a given processor is the penultimate one. In order for all processors to possess a current[2] copy of all owned local tensors, they must receive at least one and possibly two sets of local tensors from neighboring sites. Thus, each processor overlaps one Γ tensor and two λ tensors with each of its neighbors. To evenly divide the memory load among all processors, we give each processor the same number of sites to optimize. When the number of processors does not evenly divide the number of sites, we give the first and last processors more sites, as the sites at the end of the chain are more weakly entangled and hence we expect to have better load balancing in this configuration. In addition to the number of Γ tensors, which we denote in the

[1] In this section we use the word own to denote that a processor holds this object in its memory.

[2] By current, we mean time-evolved to the most recent time.

code as `my_local_dim`, we also give each processor a two-component array `my_bounds` which contains the absolute site indices of the first and last Γ tensor that the processor owns.

Initialization of the tensors in the absence of conserved quantities may be done without communication. Similarly, the Hamiltonian, propagators, and all other operators may be initialized independently. In the presence of conserved quantities, the `LabelLeft` and `LabelRight` structures depend on the state of the wavefunction to the left of the given site. Hence, we initialize the state sequentially in processor rank, with each processor sending its penultimate value of cumulative conserved quantity to the next processor.

The parity of the local site indices may not be the same as the parity of the global site indices. That is to say, the first site a processor owns might be the sixth site in the actual chain, and so one must be careful when applying propagators over even or odd bonds to apply them to globally even or globally odd sites. Because updating a local tensor at site l requires owning the local tensors of site $l+1$ and the processors do not own all of the local tensors, after each Trotter sweep we must pass local tensors between processors in order that all processors are properly updated. If the last site that a given processor π updates is one less than the total number of sites that the processor owns, then the next processor will not have updated its first tensor in the same Trotter sweep due to the parity difference. Thus, processor π should send its last updated local tensors to processor $(\pi+1)$, where they will become the first local tensors. In a similar manner, if processor π begins on a site with the same parity as the current Trotter sweep,[3] it should send its first local tensors to processor $(\pi-1)$ where they will become the new last local tensors. Identical analysis applies to the even time step, which completes a Trotter sweep.

We now pause to consider the asymptotic scaling of our algorithm's computation time and communication time. The computational scaling of TEBD is $\mathcal{O}\left[L\frac{t_f}{\delta t}\max\left(d^3\chi^3, d^4\chi^2\right)\right]$, where χ is the bond dimension, d the local dimension, δt the infinitesimal time step, L the number of sites, and t_f the final time. The communication time is proportional to the size of the sent data, which is $\mathcal{O}\left(d\chi^2+\chi\right)$ per time step. In essentially all cases of interest, $\chi \gg d$, and so we will assume that the $\mathcal{O}\left(\chi^3\right)$ scaling dominates. If we now divide the computation time among P processors and have all P processors communicate, we have a computational scaling of $\mathcal{O}\left[\frac{L}{P}\frac{t_f}{\delta t}d^3\chi^3\right]$ and a communication scaling of $\mathcal{O}\left[\frac{t_f}{\delta t}d\chi^2\right]$. We thus expect the best performance when $L \gg P$, $\chi \gg P$, or both, as the computation time is larger than the communication time in these instances. Additionally, parallelization is more efficient for increasing d. The $L \gg P$ condition states that each processor should update several local tensors before communicating, and the conditions involving χ and d state that it is less computationally intensive to send a copy of a tensor than to perform a tensor

[3]Equivalently, processor $(\pi-1)$ ends with parity opposite to the Trotter sweep.

operation such as contraction or decomposition. For large L, the first condition is naturally enforced by the condition $P \leq \lfloor \frac{L}{2} \rfloor$, for if P is larger than this some processors will own less than two local tensors and so two-site operations cannot be performed.

Not all of the expensive operations in TEBD can be parallelized using the above decomposition. The two most expensive non-parallel operations are restoration of canonical form and the computation of two-point correlation functions. The restoration of canonical form must be done in a sweeping fashion for neighboring sites. If canonical form is not restored after every time step but only before measuring observables, then this does not slow down the execution considerably. However, not using a strict canonical form at every time step degrades the quality of a TEBD simulation, as will be discussed in Sect. 10.2.1. Two-point correlation functions $\langle \hat{O}_i \hat{O}_j \rangle$ require communication of a partial overlap between processors when i and j are not both owned by the same processor. However, the amount of information which must be sent is less than in a typical Trotter sweep, and so these operations do not slow down parallel execution considerably.

We quantify the performance of our parallel implementation by three quantities: the speedup, defined as

$$S_P \equiv \frac{T_1}{T_P}, \tag{10.2}$$

where T_P denotes the time it takes the code to run on P processors, the efficiency

$$E_P \equiv \frac{T_1}{PT_P}, \tag{10.3}$$

and the experimentally determined serial fraction

$$\exp_f = \frac{\frac{1}{S_P} - \frac{1}{P}}{1 - \frac{1}{P}}. \tag{10.4}$$

The speedup is said to be ideal if $S_P \approx P$. Likewise, optimal efficiency is $E = 1$. The experimentally determined serial fraction, in an ideal case, should be as small as possible, as its inverse limits the speedup for fixed data as the number of processors increases (by Amdahl's law [24]).

We show these measures for a variety of system sizes L and bond dimensions χ in Fig. 10.2. As expected from the above scaling arguments, simulations with larger L and larger χ perform better in all three measures.

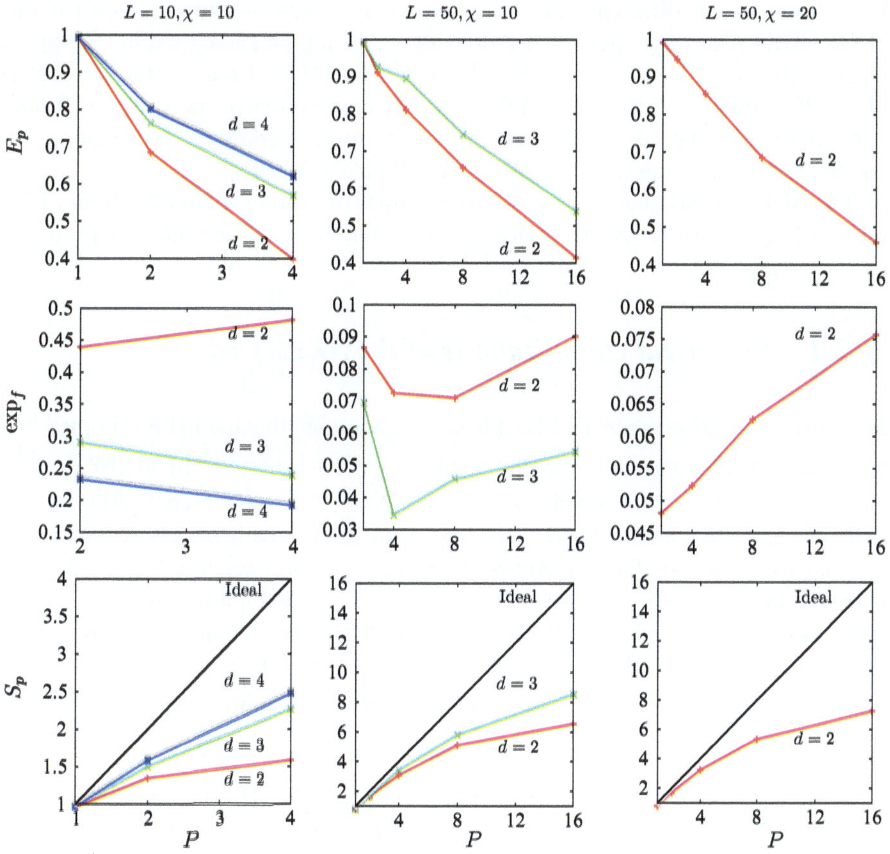

Fig. 10.2 Performance of intrinsically parallel OSTEBD routines. The parallel performance as determined by all three measures increases as L, χ, and d increase, in accordance with the scaling expectations given in the text

10.2 The ALPS TEBD Routines

The ALPS TEBD routines perform similar functions to the OSTEBD routines, performing real or imaginary time evolution of systems of bosons, fermions, or spins. Rather than using different main files or the NAMELIST system of input as in OSTEBD, the ALPS routines instead rely on a Python interface which writes appropriate input files, calls the Fortran routines, and post-processes the output from the Fortran routines. The parts of this front end which are relevant to TEBD are covered in Sect. 10.2.2. The ALPS code is less easily modified to suit user needs than the OSTEBD code due to the requirement of compatibility with the front end, but the author attempted to counteract this by including a wide array of models, measures, and time evolution protocols accessible by the Python interface.

In addition to the different interface, several improvements of the core routines of OSTEBD regarding speed and efficiency were implemented, and are discussed in Sect. 10.2.1. The ALPS TEBD routines can also be utilized via the integration of ALPS with the VisTrails workflow provenance system in part due to Python ports written by the author. Discussions of scientific workflows, provenance, and the ALPS VisTrails interface are relegated to Sect. 10.2.3.

The ALPS TEBD code consists of 10,610 lines of Fortran and on the order of 700 lines of Python written by the author, not including tutorials and other main files.

10.2.1 Numerical Optimizations of the ALPS Code

In two-site operations with TEBD, it is always assumed that the MPS is in the Vidal canonical form, in particular that the Schmidt vectors $\{|\phi_{\alpha}^{[1\dots l-1]}\rangle\}$ and $\left\{|\phi_{\beta}^{[l+2\dots L]}\rangle\right\}$ are orthonormal bases. When this is true, then the tensors on which TEBD operates contain the orthogonality center, and any truncation which occurs represents an optimal truncation of the wavefunction in that it minimizes the 2-norm distance between the truncated state and the true state at that bipartite splitting. When these bases are not orthogonal, the truncation represents an optimal truncation of the particular tensor, but there is not generally any relation between the tensor and the actual wavefunction. After a two-site operation, the Schmidt vectors $\left\{|\phi_{\gamma}^{[1\dots l]}\rangle\right\}$ and $\left\{|\phi_{\gamma}^{[l+1\dots L]}\rangle\right\}$ will be orthogonal by construction, but the same cannot be said of other sets of vectors which change under the course of the operation, for example the set $\left\{|\phi_{\gamma}^{[1\dots l+1]}\rangle\right\}$. This destruction of canonical form always arises from non-unitary operations such as application of an imaginary time propagator, but it also arises from truncation of the bond dimension of a state following application of a unitary operator. The way that OSTEBD accounted for these errors was to explicitly return the state to its canonical form periodically during real and imaginary time evolution. Additionally, in the case that the time step is small, imaginary time propagation is close to unitary, and when truncation is also small the effects of being away from canonical form are not extreme and can be accounted for by extrapolating the time step to zero [25].

A better way to account for the loss of canonical form is to re-order the application of the propagator such that all of the bases which we assume to be orthonormal throughout the course of an update sweep have been made orthonormal by the last operation on them. This can be done by applying operations to successive sites in a directed sweeping motion, e.g. $(l, l+1)$, $(l+1, l+2)$, etc. or $(l+1, l+2)$, $(l, l+1)$, etc. Before computing observables, in which all sites are assumed to be in canonical form, we can return the state to a fully canonical representation by applying the identity operator to two sites at a time in the same forwards and backwards sweeping motion.

In order to apply the propagators in the sweeping fashion as just described, we abandon the use of the Trotter expansions discussed in the OSTEBD manual, and instead use the expansion [26]

$$\exp\left(-i\hat{H}\delta t/\hbar\right) = \prod_{i=1}^{L-1}\exp\left(-i\hat{H}_{i,i+1}\delta t/2\hbar\right)\prod_{i=L-1}^{1}\exp\left(-i\hat{H}_{i,i+1}\delta t/2\hbar\right) + \mathcal{O}\left(\delta t^3\right), \tag{10.5}$$

where $\hat{H}_{i,i+1}$ is the two-site Hamiltonian acting on sites i and $i+1$. The use of this decomposition to explicitly maintain local canonical form is the first of the optimizations implemented in the ALPS code for stability.

Another source of possible numerical error in TEBD as implemented in the OSTEBD package is in the decomposition of the object $\Theta^{ij}_{\alpha\beta}$ representing the two-site wavefunction following application of some operation into the Γ tensors and λ tensors as

$$\Theta_{(\alpha i)(j\beta)} \overrightarrow{\text{SVD}} USV, \tag{10.6}$$

$$\tilde{\lambda}^{[l+1]}_{\alpha_l} = \frac{S_{\alpha_l}}{\sqrt{\sum_{\alpha=1}^{\chi}\left(S_\alpha\right)^2}}, \tag{10.7}$$

$$\tilde{\Gamma}^{[l]i_l}_{\alpha_{l-1}\alpha_l} = U_{(i_l-1)\chi+\alpha_{l-1},\alpha_l}/\lambda^{[l]}_{\alpha_{l-1}}, \tag{10.8}$$

$$\tilde{\Gamma}^{[l+1]i_{l+1}}_{\alpha_l\alpha_{l+1}} = V_{\alpha_l,(i_{l+1}-1)\chi+\alpha_{l+1}}/\lambda^{[l+2]}_{\alpha_{l+1}}. \tag{10.9}$$

Here $\overrightarrow{\text{SVD}}$ indicates performing a singular value decomposition on the left-hand side to yield the right-hand side. In particular, the elements of the tensor λ appearing in Eqs. (10.8) and (10.9) can be very small, of order the machine precision, resulting in loss of precision in the elements of the Γ tensors. To remedy this, we do not store the tensors Γ explicitly, but rather only store products such as $\lambda\Gamma$ or $\Gamma\lambda$ to create left- and right-canonical MPS tensors according to the translation rules Eqs. (6.43) and (6.44). If a single λ tensor is left uncontracted from the remainder of the tensors of the MPS at any time, the resulting MPS decomposition is in mixed canonical form rather than the Vidal canonical form,[4] with the uncontracted λ tensor being the orthogonality center. The splitting of Θ now becomes

$$\Theta_{(\alpha i)(j\beta)} \overrightarrow{\text{SVD}} USV, \tag{10.10}$$

$$\tilde{\lambda}^{[l+1]}_{\alpha_l} = \frac{S_{\alpha_l}}{\sqrt{\sum_{\alpha=1}^{\chi}\left(S_\alpha\right)^2}}, \tag{10.11}$$

[4] See Sect. 6.3 for a discussion of canonical forms for MPSs.

$$(\lambda\Gamma)^{[l]i_l}_{\alpha_{l-1}\alpha_l} = U_{(i_l-1)\chi+\alpha_{l-1},\alpha_l}, \tag{10.12}$$

$$(\Gamma\lambda)^{[l+1]i_{l+1}}_{\alpha_l\alpha_{l+1}} = V_{\alpha_l,(i_{l+1}-1)\chi+\alpha_{l+1}}. \tag{10.13}$$

The bond-centered orthogonality center can now be shifted in the direction of the propagator sweep as discussed in Sect. 6.3. It is remarkable that the mixed canonical form arises naturally when repairing numerical instabilities that arise from the more theoretically appealing Vidal canonical form. This demonstrates the general fact that the theoretically most appealing methods are often not the most numerically stable in practice.

A third optimization is an improved implementation of symmetry-adapted MPSs (see Sect. 6.6) over OSTEBD in the case that all of the on-site irreps are non-degenerate. In this case, the Γ tensors may be written as

$$\Gamma^i_{\alpha\beta} = \Gamma_{\alpha\beta}\delta_{i,q_\beta-q_\alpha+1-q_{i,\min}}, \tag{10.14}$$

where q_i is the quantum number on site i, q_γ is the cumulative quantum number to the left of bond γ, and $q_{i,\min}$ is the minimum allowed value of q_i. Thus, we can store only the matrices $\Gamma_{\alpha\beta}$ rather than the full tensors. In addition to a reduction in storage requirements, this also avoids sums over the local dimension which leads to significant speedup.

Fourth, OpenMP threading [27] over computationally expensive portions of the code is implemented. In particular, the greatest speedup occurs by threading over the SVD of Θ when Abelian symmetries are used. Here the threading is over blocks corresponding to a fixed total charge to the left of the bond on which Θ is centered. This threading displays nearly perfect parallel efficiency because threads operate on blocks independently. Threading is also performed over the formation of Γs from the SVD matrices, the formation of Θ, and propagation of the partial overlap G used for two-point correlation functions, see the OSTEBD manual.

Fifth, in OSTEBD the bond dimension is used as the main truncation parameter, and is fixed by the user without any input from the simulation. A better means to control the entanglement cutoff is to define a tolerance ϵ on the percentage of the singular value norm which can be discarded at a particular two-site operation, and then let this tolerance dynamically define the bond dimension. That is, χ is taken to be the smallest integer satisfying

$$1 - \frac{\sum_{\alpha=1}^{\chi} S_\alpha^2}{\sum_\alpha S_\alpha^2} \leq \epsilon. \tag{10.15}$$

Here, the sum in the denominator runs over all the singular values. The parameter ϵ is represented by `TRUNC_LIMIT` and a safeguard bond dimension is provided by `CHI_LIMIT`. The parameter `ITP_CHIS` is also able to be specified separately for imaginary time propagation. The value of χ now represents a safeguard value to avoid running out of memory, and may be set very large. Provided that this safeguard value of χ is not reached, the simulation remains quasi-exact in the sense that the distance between the state returned by TEBD and the true time evolution obtained from the given initial state is bounded by a known constant.

Finally, it is described in the OSTEBD manual how computing expectations of nonlocal observables expressed as tensor products can be performed more efficiently than a general two-site operator due to the tensor network structure of the contraction. The one place where this was not taken advantage of in OSTEBD was in computing the energy, as the expansions of the two-site Hamiltonians in terms of a separable basis are not generally known. The ALPS routines account for the non-separability of the Hamiltonian $\hat{H}$ which acts on two neighboring sites $|i\rangle|j\rangle$ via the following transformation

$$\left[\hat{H}\right]_{(ij),(i'j')} = \left[\hat{H}_{i,i+1}\right]_{(ii'),(j,j')}, \tag{10.16}$$

$$\left[\hat{H}\right]_{(ii'),(j,j')} \overset{\text{SVD}}{\rightarrow} \sum_{\gamma} U_{(i,i')\gamma} \Sigma_{\gamma} V_{\gamma(j,j')}, \tag{10.17}$$

which implies that we may write the operator as a sum of separable operators as

$$\Rightarrow \left[\hat{H}\right]_{(ij),(i'j')} = \sum_{\gamma} H^{L}_{(i,i')\gamma} H^{R}_{\gamma,(j,j')}, \tag{10.18}$$

where

$$H^{L}_{(i,i')\gamma} = U_{(i,i')\gamma} \Sigma_{\gamma}, \tag{10.19}$$

$$H^{R}_{\gamma,(j,j')} = V_{\gamma(j,j')}. \tag{10.20}$$

Using this decomposition, we can convert any two-site operator into a sum of tensor products of local operators, where the number of terms in the sum is equal to the number of nonzero singular values. The speedup of using this method together with the routines for computing expectations of tensor products versus using the two-site density matrix is striking. This is especially true in systems of bosons with large local dimension where the computation of the energy using the non-tensor-product methods can be the dominant scaling operation. It should also be noted that what we have affected is a decomposition of a rank-4 tensor into a matrix product operator. This same basic decomposition can be applied to turn an operator of any rank into a matrix product operator, see Sect. 7.2.2.

10.2.2 The Python Front End

To discuss the Python front end for the ALPS TEBD code, called *PyALPS*, it is simplest to start with an example, here provided by the TEBD tutorial `tutorial1a.py`. The file begins by loading the required Python modules, here PyALPS and routines for plotting output.

```
import pyalps
import matplotlib.pyplot as plt
import pyalps.plot
```

We now proceed to prepare all of the metadata defining our simulations using an array of Python dictionaries called `parms`

```
parms=[]
count=0
for A in [5.0, 10.0, 15.0, 25.0, 50.0]:
        count+=1
        parms.append({
                'L'                              : 10,
                'MODEL'                          : 'hardcore␣boson',
                'CONSERVED_QUANTUMNUMBERS'       : 'N',
                'N' : 5,
                't'                              : 1.0,
                'V'                              : 10.0,
                'ITP_CHIS' : [20, 30, 35],
                'ITP_DTS' : [0.05, 0.05,0.025],
                'ITP_CONVS' : [1E-8, 1E-8, 1E-9],
                'INITIAL_STATE' : 'ground',
                'CHI_LIMIT' : 40,
                'TRUNC_LIMIT' : 1E-12,
                'NUM_THREADS' : 1,
                'TAUS' : [A, A],
                'POWS' : [ 1.0, 1.0],
                'GS' : ['V', 'V'],
                'GIS' : [10.0, 0.0],
                'GFS' : [0.0, 10.0],
                'NUMSTEPS' : [500, 500],
                'STEPSFORSTORE' : [5, 3],
                'SIMID' : count
              })
```

The parameters which are included in `parms` are discussed in detail in the ALPS documentation, see Appendices A.1 and A.2. Here we only note that what results from these lines is a length-5 array `parms`. Each element of parms is a Python dictionary, which is an unordered set of key:value pairs. The keys in this instance are the strings on the left column, e.g. "L" and "MODEL", and the values form the right column. Thus, `parms[0]['L']` would return 10. Each element of `parms` defines a TEBD simulation.

We now write Fortran-readable files from this simulation metadata with `writeTEBDfiles` and run simulations using this input with `runTEBD` via

```
baseName='tutorial_1a'
#write output files
nmlnameList=pyalps.writeTEBDfiles(parms, baseName)
#run the application
res=pyalps.runTEBD(nmlnameList)
```

We extract information from the outputs of the Fortran code using `loadTimeEvolution` as

```
LEdata=pyalps.load.loadTimeEvolution(pyalps.getResult
   Files(prefix='tutorial_1a'), measurements=
   ['Loschmidt Echo', 'V'])
```

The `measurements` tag specifies which outputs to load. `LEdata` is an array with the same number of elements as `parms`, and contains all of the same simulation metadata together with the specified measurements. This allows for complex evaluations to be performed involving parameters in the Hamiltonian, the system size, or any other input parameters. In the present case, we are interested only in plotting the Loschmidt echo and the parameter V vs. time. We do so by turning these values into ordered (x, y) pairs suitable for a 2D plot with `collectXY`, and then passing these pairs to a plotting front end

```
LE=pyalps.collectXY(LEdata, x='Time',
    y='Loschmidt Echo', foreach=['SIMID'])
for q in LE:
    q.props['label']=r'$\tau=$'+str(q.props['TAUS'][0])

plt.figure()
pyalps.plot.plot(LE)
plt.xlabel('Time $t$')
plt.ylabel('Loschmidt Echo $|< \psi(0)|\psi(t) > |^2$')
plt.title('Loschmidt Echo vs. Time')
plt.legend(loc='lower right')

Ufig=pyalps.collectXY(LEdata, x='Time', y='V',foreach=
   ['SIMID'])
for q in Ufig:
    q.props['label']=r'$\tau=$'+str(q.props['TAUS'][0])

plt.figure()
pyalps.plot.plot(Ufig)
plt.xlabel('Time $t$')
plt.ylabel('V')
plt.title('Interaction parameter $V$ vs. Time')
plt.legend(loc='lower right')
plt.show()
```

The resulting plots are shown as Fig. 11.2 in Chap. 11, where the physical content of these tutorials is discussed in greater detail.

While the actual NAMELIST files which are passed to the Fortran routines are similar to those of OSTEBD above, the Python front end provides a powerful way to automate the writing of these inputs, allows for much easier post-processing of outputs, and also unifies the preparation, execution, and analysis of simulations.

The contributions of the present author to the Python front end were to write the Python routines to write Fortran-readable files for TEBD, run TEBD simulations, and process the outputs of TEBD as a function of time. Since TEBD was the first of the ALPS applications to simulate dynamics, the author also modified several of the other existing ALPS processing routines to accept data which depends on time.

10.2.3 Integration with the VisTrails Workflow Provenance System

The Python front end for ALPS provides a unified way of defining, executing, and analyzing simulations. ALPS v2.0 is also integrated with VisTrails (http://www.vistrails.org/), which is an open-source scientific workflow and provenance management system written in Python. By provenance, we refer to both the steps that need to be followed to produce a specific result (*Prospective provenance*) as well as the steps which were actually taken (*Retrospective provenance*) [28, 29]. Hence, the simulation metadata above, which user ran the simulation, at what time, etc. are all data relevant to the provenance of a particular result. VisTrails caches all of these data in persistent storage such that results are always reproducible without the need to repeat the calculation.

The same tutorial which was used to exemplify the Python front end is shown using VisTrails in Fig. 10.3. In panel (a), the three parts to tutorial 1 are shown as three different workflows, the beige ovals. Under tebd1a, the blue boxes capture provenance information about changes made to the workflow itself. The actual workflow contents of tebd1a are shown in panel (b). Here the large boxes, called *modules* represent subtasks to be performed, with the small boxes on the top corresponding to inputs for a particular task and the boxes on the bottom corresponding to outputs from the task. The overall simulation flow is similar to that using the Python front end: parameters are specified or looped over, Fortran-readable files are written, the simulation is performed, data is extracted and post-processed, and plots are made. With VisTrails, the full provenance information of both how a simulation is to be performed and what the actual values used to achieve a particular result are permanently captured. The contributions of the present author to the VisTrails functionality of ALPS were modules to write files for and run TEBD simulations as well as the post-processing of time series.

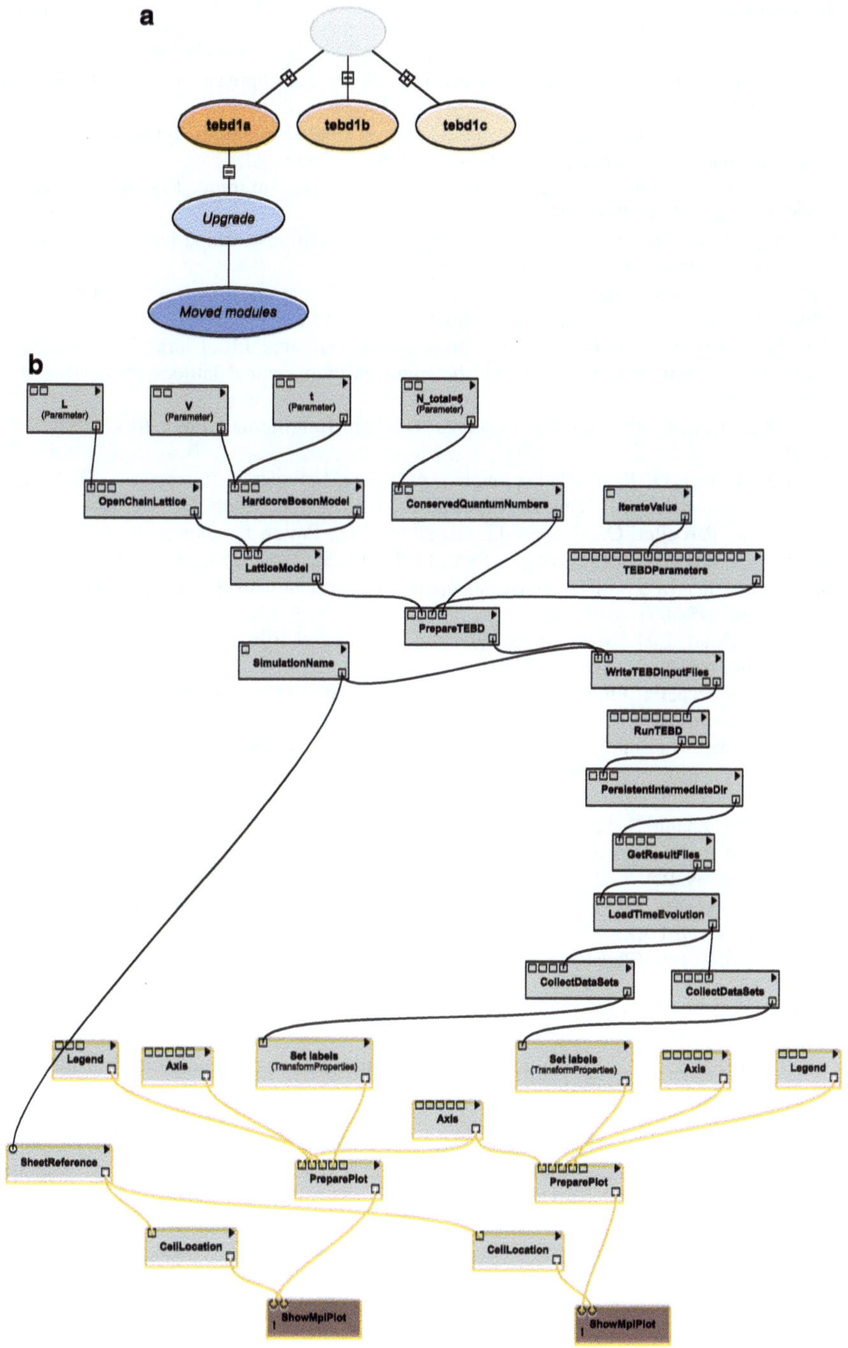

Fig. 10.3 Example VisTrail for ALPS TEBD tutorial 1. (**a**) The workflows for tutorial 1 are represented as the *beige circles*. Changes to the workflow of tebd1a are represented as *blue ovals* below the beige one. (**b**) The contents of the workflow tebd1a. Here *each large box* represents a task performed by PyALPS or an internal VisTrails task

References

1. Open source time-evolving block decimation (2009). http://physics.mines.edu/downloads/software/tebd/
2. Wall, M.L., Carr, L.D.: Emergent timescales in entangled quantum dynamics of ultracold molecules in optical lattices. New J. Phys. **11**, 055027 (2009)
3. Carr, L.D., Oberthaler, M.K.: Spatial dependence of entropy in trapped ultracold Bose gases (2009) [e-print arXiv:0906.4708]
4. Mishmash, R.V., Carr, L.D.: Quantum entangled dark solitons formed by ultracold atoms in optical lattices. Phys. Rev. Lett. **103**, 140403 (2009)
5. Mishmash, R.V., Danshita, I., Clark, C.W., Carr, L.D.: Quantum many-body dynamics of dark solitons in optical lattices. Phys. Rev. A **80**, 053612 (2009)
6. Carr, L.D., Wall, M.L., Schirmer, D.G., Brown, R.C., Williams, J.E., Clark, C.W.: Mesoscopic effects in quantum phases of ultracold quantum gases in optical lattices. Phys. Rev. A **81**, 013613 (2010)
7. Wall, M.L., Carr, L.D.: Hyperfine molecular Hubbard Hamiltonian. Phys. Rev. A **82**, 013611 (2010)
8. Bauer, B., Carr, L.D., Evertz, H.G., Feiguin, A., Freire, J., Fuchs, S., Gamper, L., Gukelberger, J., Gull, E., Guertler, S., Hehn, A., Igarashi, R., Isakov, S.V., Koop, D., Ma, P.N., Mates, P., Matsuo, H., Parcollet, O., Pawlowski, G., Picon, J.D., Pollet, L., Santos, E., Scarola, V.W., Schollwöck, U., Silva, C., Surer, B., Todo, S., Trebst, S., Troyer, M., Wall, M.L., Werner, P., Wessel, S.: The ALPS project release 2.0: open source software for strongly correlated systems. J. Stat. Mech. P05001 (2011)
9. Glick, J.A., Carr, L.D.: Macroscopic quantum tunneling of solitons in Bose-Einstein condensates. [under review, arXiv:1105.5164]
10. Carleo, G., Becca, F., Schiró, M., Fabrizo, M.: Localization and glassy dynamics of many-body quantum systems. Sci. Rep. **2**, 243 (2011)
11. Qian, Y., Gong, M., Zhang, C.: Quantum transport of bosonic cold atoms in double-well optical lattices. Phys. Rev. A **84**, 013608 (2011). doi:10.1103/PhysRevA.84.013608. http://link.aps.org/doi/10.1103/PhysRevA.84.013608
12. Zhou, Z.-W., Zhang, S.-L., Zhou, X.-F., Guo, G.-C., Zhou, X., Pu, H.: Quantum phase transition of Bose-Einstein condensates on a nonlinear ring lattice. Phys. Rev. A **83**, 043626 (2011). doi:10.1103/PhysRevA.83.043626. http://link.aps.org/doi/10.1103/PhysRevA.83.043626
13. Lu, H., Baksmaty, L.O., Bolech, C.J., Pu, H.: Expansion of 1D polarized superfluids: the Fulde-Ferrell-Larkin-Ovchinnikov state reveals itself. Phys. Rev. Lett. **108**, 225302 (2012). doi:10.1103/PhysRevLett.108.225302. http://link.aps.org/doi/10.1103/PhysRevLett.108.225302
14. Qian, Y., Gong, M., Zhang, C.: Many-body Landau-Zener transition in cold atom double well optical lattices. Phys. Rev. A **87**, 013636 (2013)
15. Yin, S., Qin, X., Lee, C., Zhong, F.: Finite-time scaling of dynamic quantum criticality (2012) [http://arxiv.org/abs/1207.1602]
16. ALPS (2012). http://alps.comp-phys.org
17. Albuquerque, A.F., Alet, F., Corboz, P., Dayal, P., Feiguin, A., Fuchs, S., Gamper, L., Gull, E., Gurtler, S., Honecker, A., Igarashi, R., Korner, M., Kozhevnikov, A., Lauchli, A., Manmana, S.R., Matsumoto, M., McCulloch, I.P., Michel, F., Noack, R.M., Pawlowski, G., Pollet, L., Pruschke, T., Schollwock, U., Todo, S., Trebst, S., Troyer, M., Werner, P., Wessel, S.: The ALPS project release 1.3: open-source software for strongly correlated systems. J. Magn. Magn. Mater. **310**, 1187–1193 (2007)
18. Vidal, G.: Efficient classical simulation of slightly entangled quantum computations. Phys. Rev. Lett. **91**, 147902–1–4 (2003)
19. Vidal, G.: Efficient simulation of one-dimensional quantum many-body systems. Phys. Rev. Lett. **93**, 040502 (2004)

20. Danshita, I., Naidon, P.: Bose-Hubbard ground state: extended Bogoliubov and variational methods compared with time-evolving block decimation. Phys. Rev. A **79**, 043601 (2009). doi:10.1103/PhysRevA.79.043601. http://link.aps.org/doi/10.1103/PhysRevA.79.043601
21. Daley, A.J.: Manipulation and simulation of cold atoms in optical lattices. Ph.D. thesis, Leopold-Franzens-Universität Innsbruck (2005)
22. McCulloch, I.P.: From density-matrix renormalization group to matrix product states. J. Stat. Mech. **2007**(10), P10014 (2007). http://stacks.iop.org/1742-5468/2007/i=10/a=P10014
23. Message Passing Interface Forum (2012). http://www.mpi-forum.org/
24. Scott, L., Clark, T., Bagheri, B.: Scientific Parallel Computing. Princeton University Press, Princeton (2005)
25. Orús, R., Vidal, G.: Infinite time-evolving block decimation algorithm beyond unitary evolution. Phys. Rev. B **78**, 155117 (2008). doi:10.1103/PhysRevB.78.155117. http://link.aps.org/doi/10.1103/PhysRevB.78.155117
26. Sornborger, A.T., Stewart, E.D.: Higher-order methods for simulations on quantum computers. Phys. Rev. A **60**, 1956–1965 (1999). doi:10.1103/PhysRevA.60.1956. http://link.aps.org/doi/10.1103/PhysRevA.60.1956
27. Open Multiprocessing (2012). http://openmp.org/wp/
28. Silva, C.T., Freire, J., Callahan, S.P.: Provenance for visualizations: reproducibility and beyond. IEEE Comput. Sci. Eng. **9**(5), 82 (2007). doi:10.1109/MCSE.2007.106
29. Freire, J., Koop, D., Santos, E., Silva, C.T.: Provenance for computational tasks: a survey. Comput. Sci. Eng. **10**(3), 11–21 (2008). doi:10.1109/MCSE.2008.79. http://dx.doi.org/10.1109/MCSE.2008.79 [ISSN 1521-9615]

Chapter 11
Educational Materials

Chapter 10 discusses the author's involvement in the development of open source code implementing variational matrix product state (MPS) algorithms. The present chapter covers educational materials which were distributed with these codes, either as part of the open source packages or within the Carr theoretical physics research group (CTPRG). The educational materials which were disseminated with the open source code are aimed at the level of graduate students performing research in strongly correlated physics, and are designed to enable users to modify the open source codes to meet their own research goals. The materials designed for use by the CTPRG are aimed more to the level of undergraduates who may not have completed a quantum mechanics course and may have no coding experience. Hence, they must strike a balance between being conveying the important ideas of the algorithms without being too technical, and not require coding experience while still enabling the student to perform calculations which are meaningful and exciting. Within the CTPRG several undergraduates at the senior level were able to use these educational materials successfully towards modifying and using the open source codes as part of their senior thesis projects.

11.1 Materials Distributed with the Open Source Packages

The open source time-evolving block decimation (OSTEBD) [1] package includes a manual written by the present author which provides background on the time-evolving block decimation (TEBD) algorithm [2, 3], manual pages for all routines contained in the package, and case studies. Provided with each case study is a main file from which a user can reproduce the figures given in the manual in order to understand how to run the code and post-process its output. The case studies, in order, are a study of the ground state properties of the Bose–Hubbard Hamiltonian, the dynamics of the Bose–Hubbard Hamiltonian following a linear ramp of the

M.L. Wall, *Quantum Many-Body Physics of Ultracold Molecules in Optical Lattices*, Springer Theses, DOI 10.1007/978-3-319-14252-4_11

on-site interaction from the Mott insulator to the superfluid phase and back [4, 5], the ground state properties of spinless fermions, and the dynamics of a domain wall defect in the XX model [6]. At the end of each case study are exercises which require the user to modify the case study base code to extract other observables or perform other tasks. For example, in the Bose–Hubbard case study, the user is required to add a harmonic trap in one of the exercises and study the physics of the trapped system. Several of the exercises, in particular those dealing with dynamics, have the user study the behavior of the simulation as the bond dimension χ, the time step δt, the system size, and the boundary conditions are changed. These exercises help the user gain an intuition for the convergence behavior of typical simulations.

The educational materials distributed with the algorithms and libraries for physics simulations (ALPS) [7–9] TEBD routines take the form of case studies which were posted on the ALPS wiki pages at http://alps.comp-phys.org. These case studies and other ALPS TEBD documentation from the ALPS wiki are included as Appendix A with this thesis. The second tutorial is the more pedagogical, as it uses comparisons with analytically known results to demonstrate the convergence of TEBD simulations. In particular, this tutorial uses the same system as the last case study for OSTEBD in which a single domain wall defect is initialized at the center of a long chain:

$$|\psi\rangle = |\downarrow\downarrow\dots\downarrow\uparrow\dots\uparrow\uparrow\rangle\,. \tag{11.1}$$

At time $t = 0$, this wavefunction is evolved according to the XX model

$$\hat{H}_{\mathrm{XX}} = -\sum_i \left(\hat{S}_i^x \hat{S}_{i+1}^x + \hat{S}_i^y \hat{S}_{i+1}^y\right)\,, \tag{11.2}$$

where $\hat{S}_i^\nu$ is the spin-1/2 operator along the νth Cartesian direction at lattice site i. This Hamiltonian is equivalent to a system of free fermions via a Jordan–Wigner transformation, and so we can solve exactly for the magnetization a distance n from the initial defect position at time t, $M\,(n,t)$, as [10]

$$M\,(n,t) = -\frac{1}{2}\sum_{i=1-n}^{n-1} j_l^2\,(t)\,, \tag{11.3}$$

where $j_l\,(x)$ is the Bessel function of order l. This result also implies that in the limit as $n \to \infty$ and $t \to \infty$, this function approaches as scaling form which depends only on the variable n/t as

$$\lim_{n\to\infty}\lim_{t\to\infty} M\,(n,t) = \phi\left(\frac{n}{t}\right) \equiv -\frac{1}{\pi}\arcsin\left(\frac{n}{t}\right)\,. \tag{11.4}$$

The results comparing these predictions with the outputs of the ALPS TEBD tutorial `tutorial2a.py` are shown in Fig. 11.1. We see very good visual agreement between the numerical results and the predictions for the magnetization in

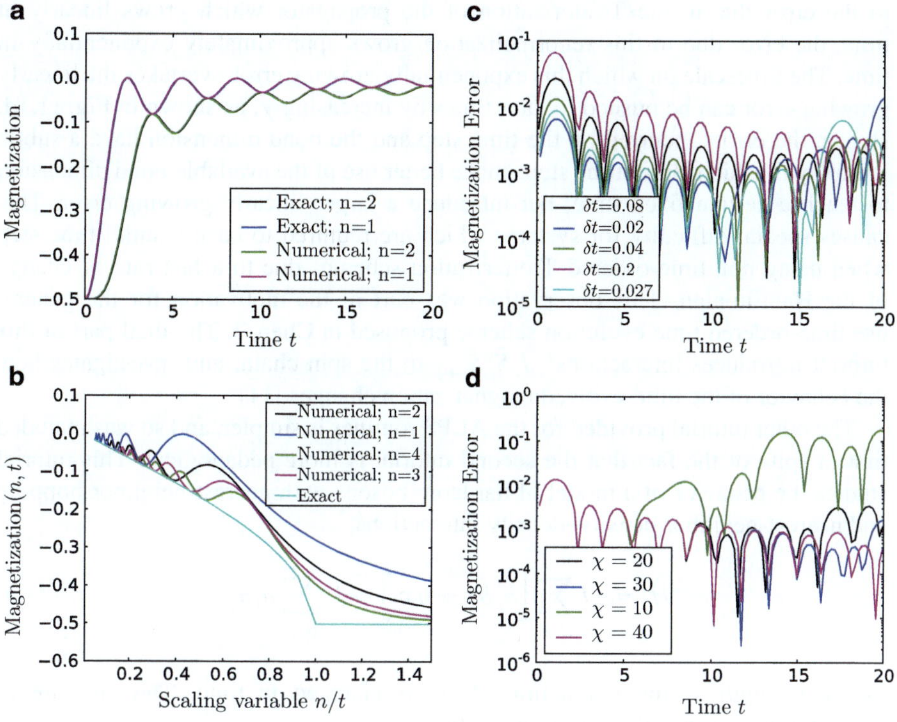

Fig. 11.1 Results of ALPS TEBD tutorial 2. (**a**) The magnetization computed by TEBD for the given situation compared to the prediction Eq. (11.3). (**b**) The scaling limit of the magnetization compared to the scaling function predicted in Eq. (11.4). (**c**) The error in the magnetization as a function of time and the time step δt. (**d**) The error in the magnetization as a function of time and the bond dimension χ

Fig. 11.1a. Figure 11.1b demonstrates that the magnetization does indeed approach the scaling limit Eq. (11.4), with agreement improving for large n and t, as expected. This part of the tutorial also demonstrates the power of the Python front end for performing complex post-processing such as the extraction of scaling forms from the numerical data, see Sect. 10.2.2.

In the next two parts of this tutorial, we examine the deviations of the magnetization with these exact results as we change the two main convergence parameters in TEBD: the time step δt and the bond dimension χ. In Fig. 11.1c we see the deviation of the magnetization as a function of time and the time step δt. At short times, the smallest δt has the smallest error in accordance with our expectations given by the error bounds on the Trotterization of the propagator. However, at long times, we begin to see an exponential growth in the errors for the smallest δt, and this simulation eventually has the largest error of all the simulations. The reason for this is that at each time step we perform a truncation of the bond dimension, and this truncation involves a renormalization of the wavefunction. Hence, in contrast to the error due to the Trotterization of the propagator which grows linearly in time, the error due to this renormalization grows approximately exponentially in time. The timescale on which this exponentially growing error overtakes the linearly growing error can be pushed to later times by increasing χ, as shown in Fig. 11.1d. Hence, the errors incurred by the time step and the bond dimension have a subtle interplay in which larger time steps make better use of the available bond dimension to reach a certain fixed time, but introduce a larger linearly growing error. This causes special difficulty for systems which are required to have a small time step when using non-time-ordered Trotterization schemes due to a fast rate of change of the Hamiltonian. This observation was part of the motivation for the general use time-ordered time evolution scheme proposed in Chap. 7. The final part of this tutorial introduces interactions[1] $J_z\hat{S}_i^z\hat{S}_{i+1}^z$ to the spin chain, and investigates how the behavior of the time-evolved magnetization changes [11].

The other tutorial provided for the ALPS routines is simpler, and so was included first in spite of the fact that the second tutorial is more pedagogical. This tutorial studies the behavior of a model of hardcore bosons with nearest-neighbor hopping and nearest-neighbor density–density interactions,

$$\hat{H} = -t\sum_{\langle i,j\rangle}\left[\hat{b}_i^\dagger\hat{b}_j + \text{h.c.}\right] + V\sum_{\langle i,j\rangle}\hat{n}_i\hat{n}_j\,, \tag{11.5}$$

as the strength of the interactions V/t[2] is changed in time. The measure of adiabaticity we use is the Loschmidt echo

$$L\left(t;\gamma\right) = \left|\langle\psi\left(t\right)|\psi\left(0\right)\rangle\right|^2\,, \tag{11.6}$$

which is the squared overlap of the time-evolved wavefunction with the initial wavefunction. This quantity depends on general on the way in which the parameters are changed in time; this dependence is denoted by the symbol γ.

The first part of the tutorial focuses on quenches from the gapped charge-density wave (CDW) phase to the superfluid phase and back. The exercises explore how difficult it is for such a quench to be adiabatic in the sense that the system returns to its initial configuration, and how this depends on the system size [12–16]. An exercise which represents a possible research project is to extend the simulation to the soft-core boson model with nearest-neighbor interactions in which there exist Mott insulating, CDW, and superfluid phases at unit filling. One can then explore how difficult it is to be adiabatic when quenching between two gapped phases, here the Mott insulator and the CDW. In the next part of the tutorial, the system is held in the superfluid phase for a time τ_{hold} before quenching back to the CDW.

[1] These are interactions in that they provide terms which are quartic in the fermion field operators following a Jordan–Wigner transformation.

[2] In the expression V/t, t denotes the nearest-neighbor tunneling and not the time.

The final part of the tutorial utilizes the capability of the ALPS TEBD routines to also simulate time evolutions described by nonlinear power-law behavior of Hamiltonian parameters. The outputs of the simulations in this tutorial are shown in Fig. 11.2.

11.2 Materials Intended for CTPRG Use

In addition to the OSTEBD manual, which contains a great amount of detail, a smaller work "A gentle introduction to Time Evolving Block Decimation (TEBD)" (henceforth, the TEBD intro) was developed for use by advanced undergraduate and beginning graduate students within the CTPRG. The purpose of this document is to acquaint the student with the basic idea of how MPS algorithms make the many-body problem tractable through restricting entanglement, to provide experience in running and post-processing MPS code, to develop familiarity with the high-performance and parallel computing environments available to CTPRG members, and to enable the student to perform an exciting, cutting-edge calculation regarding quantum phase transitions. This document is included as Appendix B.

The TEBD intro builds upon an earlier set of exercises "Introduction to the Bose–Hubbard Model and Fock State Basis" which provides a basic introduction to strongly correlated lattice models, Fock space, and second quantization. The final exercise in this latter problem set is to write an exact diagonalization program to compute the spectrum of the Bose–Hubbard model on M sites with N particles and at most $d-1$ particles per site. The introductory TEBD exercises begin by having the student explore for what size systems exact diagonalization is numerically practical. For most of the implementations that the author has seen, this is usually 8–10 sites. Then, the notion of the singular value decomposition (SVD) and reduced rank approximations are introduced. Rather than discuss tensor networks in general, it is demonstrated how a matrix represented as its SVD may be applied to vectors much more cheaply than the full matrix representation when the rank of the matrix is small compared to its dimension. Mathematical and programming exercises on the SVD acquaint the user with its properties, and also demonstrate that the rank of a matrix may be difficult to ascertain from its other properties. A physical connection between quantum mechanical states and the SVD is provided by the Schmidt decomposition [17, 18], and the matrices in the mathematical SVD exercises are related to quantum states. Finally, the ideas are put together noting that TEBD performs a truncated Schmidt decomposition at every bipartite splitting, hence restricting the amount of entanglement in the state and allowing for more efficient operations.

At the end of the introductory TEBD materials, some exercises are provided to orient the user with running the OSTEBD code and to introduce quantum phase transitions [6, 19]. The first few exercises consist of comparing the results of small Bose–Hubbard systems with exact diagonalization with and without number conservation. The next exercise requires the student to diagonalize the bosonic

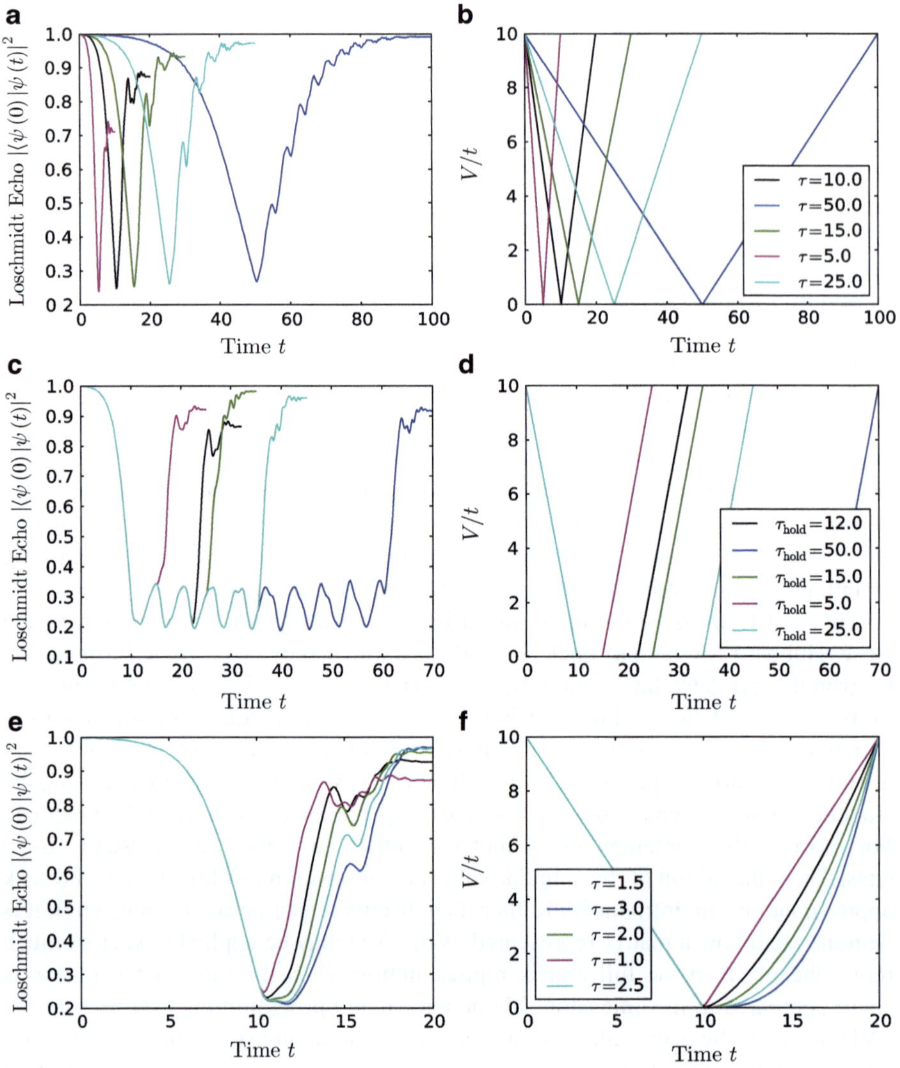

Fig. 11.2 Results of ALPS TEBD tutorial 1. The color scheme in a panel in the *left column* is the same as the neighboring panel in the *right column*. (**a**) The Loschmidt echo for a process which quenches V/t from the CDW phase to the superfluid and back as a function of the linear timescale of the quench τ. (**b**) The form of V/t as a function of time for the simulation in (**a**). (**c**) The Loschmidt echo for a process which quenches V/t from the CDW phase to the superfluid, holds the parameters constant for a time τ_{hold}, and then quenches back as a function of the hold time τ_{hold}. (**d**) The form of V/t as a function of time for the simulation in (**c**). (**e**) The Loschmidt echo for a process which quenches V/t linearly from the CDW phase to the superfluid and nonlinearly quenches back as a function of the power of the return quench. (**f**) The form of V/t as a function of time for the simulation in (**e**)

tight-binding chain via a Fourier transform for both open and periodic boundary conditions and compute the spectrum, the on-site number, and the single-particle density matrix. These results are compared with the open source code. The exercise also acquaints the student with other useful many-body tools such as canonical transformations. The final exercise makes use of the data-parallel capability of OSTEBD (see Sect. 10.1.1) to explore the Mott insulator-superfluid transition in the Bose–Hubbard model. As such, this exercise also serves to introduce students to high-performance and parallel computing environments. The exercise begins with a qualitative treatment of why the ground state must be non-analytic based on strong and weak coupling limits of the depletion, relying on the student's intuition for the single-particle density matrix developed by the tight-binding chain. The critical point is then located numerically via the maximum of the fidelity susceptibility, which gives a system-independent measure of changes in the ground state as a system passes through a quantum critical point [20]. An algebraic scaling function for the position of the fidelity susceptibility maximum as a function of the system size is provided [21], and the student is asked to find the critical point using a scaling analysis for a series of system sizes. Estimates for the appropriate range of parameters are provided by fidelity susceptibility data on small system sizes.

The final educational document to be discussed is "Introduction to MPS Algorithms," (henceforth, the MPS intro) which is included in its most recent version as Appendix C. The MPS intro was developed for future code developers in the CTPRG to write their own small working variational MPS code for the Ising model. This process is guided by introducing a theoretical concept, for example canonical forms for MPSs, and then having the student write code with a specific interface to perform the procedure outlined in theory. A highly stripped-down and de-optimized version of the author's code was provided along with this document. In the present work, this code is reproduced at the end of Appendix C. Each of the procedures of the author's code has the same interface as the procedures outlined in the document, and so students can consult the author's code for implementation hints if they are stuck. The author's code can also be used as a black box to obtain observables for the Ising model for varying system parameters, and so the student can verify his or her own code for correctness. The topics covered are canonical forms, matrix product operators including a section on long-range operators [22–24], caching of effective Hamiltonian overlaps [25], and the formation of the effective Hamiltonian from them, the sparse solution of the effective Hamiltonian eigenproblem using the Lanczos recursion [26, 27], computation of observables, calculation of excited states using eMPS,[3] and TEBD-style time evolution.

[3]See Chap. 7 for a discussion of eMPS.

References

1. Open source time-evolving block decimation (2009). http://physics.mines.edu/downloads/software/tebd/
2. Vidal, G.: Efficient classical simulation of slightly entangled quantum computations. Phys. Rev. Lett. **91**(14), 147902 (2003). doi:10.1103/PhysRevLett.91.147902
3. Vidal, G.: Efficient simulation of one-dimensional quantum many-body systems. Phys. Rev. Lett. **93**(4), 040502 (2004). doi:10.1103/PhysRevLett.93.040502
4. Clark, S.R., Jaksch, D.: Dynamics of the superfluid to Mott-insulator transition in one dimension. Phys. Rev. A **70**, 043612 (2004). doi:10.1103/PhysRevA.70.043612. http://link.aps.org/doi/10.1103/PhysRevA.70.043612
5. Daley, A.J.: Manipulation and simulation of cold atoms in optical lattices. Ph.D. thesis, Leopold-Franzens-Universität Innsbruck (2005)
6. Sachdev, S.: Quantum Phase Transitions. Cambridge University Press, New York (1999)
7. ALPS (2012). http://alps.comp-phys.org
8. Albuquerque, A.F., Alet, F., Corboz, P., Dayal, P., Feiguin, A., Fuchs, S., Gamper, S., Gull, E., Gurtler, S., Honecker, A., Igarashi, R., Korner, M., Kozhevnikov, A., Lauchli, A., Manmana, S.R., Matsumoto, M., McCulloch, I.P., Michel, F., Noack, R.M., Pawlowski, G., Pollet, L., Pruschke, T., Schollwock, U., Todo, S., Trebst, S., Troyer, M., Werner, P., Wessel, S.: The ALPS project release 1.3: Open-source software for strongly correlated systems. J. Magn. Magn. Mater. **310**, 1187–1193 (2007)
9. Bauer, B., Carr, L.D., Evertz, H.G., Feiguin, A., Freire, J., Fuchs, S., Gamper, L., Gukelberger, J., Gull, E., Guertler, S., Hehn, A., Igarashi, R., Isakov, S.V., Koop, D., Ma, P.N., Mates, P., Matsuo, H., Parcollet, O., Pawłowski, G., Picon, J.D., Pollet, L., Santos, E., Scarola, V.W., Schollwöck, U., Silva, C., Surer, B., Todo, S., Trebst, S., Troyer, M., Wall, M.L., Werner, P., Wessel, S.: The ALPS project release 2.0: Open source software for strongly correlated systems. J. Stat. Mech. Theory Exp. **2011**(05), P05001 (2011). http://stacks.iop.org/1742-5468/2011/i=05/a=P05001
10. Antal, T., Rácz, Z., Rákos, A., Schütz, G.M.: Transport in the XX chain at zero temperature: Emergence of flat magnetization profiles. Phys. Rev. E **59**, 4912–4918 (1999). doi:10.1103/PhysRevE.59.4912. http://link.aps.org/doi/10.1103/PhysRevE.59.4912
11. Gobert, D., Kollath, C., Schollwöck, U., Schütz, G.: Real-time dynamics in spin-$\frac{1}{2}$ chains with adaptive time-dependent density matrix renormalization group. Phys. Rev. E **71**, 036102 (2005). doi:10.1103/PhysRevE.71.036102. http://link.aps.org/doi/10.1103/PhysRevE.71.036102
12. Kibble, T.W.B.: Topology of cosmic domains and strings. J. Phys. A Math Gen. **9**, 1387 (1980)
13. Zurek, W.H.: Cosmological experiments in superfluid Helium? Nature **317**, 505–508 (1985)
14. Kibble, T.: Phase-transition dynamics in the lab and the universe. Phys. Today **60**(9), 47 (2007)
15. Damski, B., Zurek, W.H.: Dynamics of a quantum phase transition in a ferromagnetic Bose-Einstein condensate. Phys. Rev. Lett. **99**, 130402 (2007)
16. Saito, H., Kawaguchi, Y., Ueda, M.: Kibble-Zurek mechanism in a quenched ferromagnetic Bose-Einstein condensate. Phys. Rev. A **76**, 043613 (2007)
17. Schmidt, E.: Zur Theorie der linearen und nichtlinearen Integralgleichungen. Math. Ann. **63**, 433–476 (1907). doi:10.1007/BF01449770. http://dx.doi.org/10.1007/BF01449770 [ISSN 0025-5831]
18. Nielsen, M., Chuang, I.: Quantum Computation and Quantum Information. Cambridge University Press, Cambridge (2000)
19. Carr, L.D. (ed.): Understanding Quantum Phase Transitions. CRC press, Boca Raton (2010)
20. Carr, L.D., Wall, M.L., Schirmer, D.G., Brown, R.C., Williams, J.E., Clark, C.W.: Mesoscopic effects in quantum phases of ultracold quantum gases in optical lattices. Phys. Rev. A **81**(1), 013613 (2010). doi:10.1103/PhysRevA.81.013613
21. Buonsante, P., Vezzani, A.: Ground-state fidelity and bipartite entanglement in the Bose-Hubbard model. Phys. Rev. Lett. **98**, 110601 (2007)

22. Crosswhite, G.M., Doherty, A.C., Vidal, G.: Applying matrix product operators to model systems with long-range interactions. Phys. Rev. B **78**, 035116 (2008). doi:10.1103/PhysRevB.78.035116. http://link.aps.org/doi/10.1103/PhysRevB.78.035116
23. Pirvu, B., Murg, V., Cirac, J.I., Verstraete, F.: Matrix product operator representations. New J. Phys. **12**(2), 025012 (2010). http://stacks.iop.org/1367-2630/12/i=2/a=025012
24. Fröwis, F., Nebendahl, V., Dür, W.: Tensor operators: Constructions and applications for long-range interaction systems. Phys. Rev. A **81**, 062337 (2010). doi:10.1103/PhysRevA.81.062337. http://link.aps.org/doi/10.1103/PhysRevA.81.062337
25. Crosswhite, G.M., Bacon, D.: Finite automata for caching in matrix product algorithms. Phys. Rev. A **78**, 012356 (2008). doi:10.1103/PhysRevA.78.012356. http://link.aps.org/doi/10.1103/PhysRevA.78.012356
26. Golub, G.H., Van Loan, C.F.: Matrix Computations, 3rd edn. Johns Hopkins Studies in Mathematical Sciences. The Johns Hopkins University Press, Baltimore (1996)
27. Trefethen, L.N., Bau, D., III.: Numerical Linear Algebra. SIAM, Philadelphia (1997)

Part VI
Conclusions and Appendices

Chapter 12
Conclusions and Suggestions for Future Research

This thesis presents models for the low-energy physics of molecules trapped in optical lattices and simulation methods to elucidate the many-body physics of these models in one dimension. Part II focused on the molecular Hubbard Hamiltonian (MHH), a model for the low-energy physics of $^1\Sigma$ heteronuclear bialkali molecules in an optical lattice [1, 2]. A strong DC electric field applied to these molecules gives rise to resonant dipole–dipole interactions which are long-range, and also allows for tunable access to rotational states other than the lowest rotational level via an AC microwave field. The nuclear spin degrees of freedom couple most strongly to a DC magnetic field, but the coupling of rotational and nuclear spin degrees of freedom provided by nuclear quadrupole interactions implies that the interplay between the rotational and nuclear spin degrees of freedom may be tuned by changing the angle between the electric and magnetic fields. Thus, not only the magnitudes of the parameters of the MHH, but also the number of states involved in the dynamics are amenable to experimental control. Hence, the MHH may be used as a simulator of a quantum complex system, one in which are large number of degrees of freedom are interacting quantum mechanically on a multitude of timescales.

For nearest-neighbor interactions and a single internal state, the MHH admits only superfluid and charge-density wave (CDW) phases [3]. The CDW is characterized by a "checkerboard" pattern in the density which arises as a peak in the structure factor at a wavevector of π/a. For convex long-range interactions, insulating phases appear at every rational filling [4, 5], and so the phase diagram is much richer. While theoretically appealing, these phases may exist only in a very narrow parameter range, and may be destroyed by terms such as tunneling between sites which are not nearest-neighbors and thermal fluctuations. Hence, a consistent analysis would involve carefully enumerating all terms appearing in the many-body Hamiltonian in order of decreasing energy and then performing a self-consistent truncation down to an energy scale set by experimental constraints on timescale. The microscopic

M.L. Wall, *Quantum Many-Body Physics of Ultracold Molecules in Optical Lattices*, Springer Theses, DOI 10.1007/978-3-319-14252-4_12

connections between the few-body physics and the many-body physics presented in this thesis are crucial for this kind of analysis, and hence also for practical guidance of experiments.

When more than one internal state is populated the possibilities for statics become even greater. As discussed in Chaps. 3 and 4, both the tunneling and the interactions depend on the rotational state. This allows for the study of many-body systems in mixtures different quantum phases and possibly far from equilibrium. Furthermore, away from the strong DC field regime studied in Chap. 4, there is a long-ranged interaction term which represents the exchange of a rotational quantum between two molecules in different rotational states [6, 7]. For two internal states $\sigma = \{1, 2\}$, this term has the form

$$\sum_{i<j} E^{i,j} \hat{a}^{\dagger}_{i1} \hat{a}^{\dagger}_{j2} \hat{a}_{j1} \hat{a}_{i2} , \tag{12.1}$$

where i and j are lattice sites, $\hat{a}_{i\sigma}$ destroys a particle in internal molecular state σ at site i, and $E^{i,j}$ obeys a power-law form for large distances $|i - j|$. The *exchange* contribution $E^{i,j}$ has a comparable power-law form to $U^{i,j}$, the direct part of the dipole–dipole interaction, but its magnitude can be tuned independently. The action of this term in the internal degrees of freedom of a molecule is similar to that of *ring-exchange* terms in real space [8]. Such terms have been shown to induce novel Bose metal and gapless Mott insulator phases in ladder systems of hard-core bosons [8, 9] and also drive electronic systems into a non-Fermi liquid phase [10]. The exchange term frustrates the system, and so studies of the MHH with the exchange term must be performed with either exact diagonalization or tensor network methods due to a sign problem in quantum Monte Carlo. Studies of the MHH with all terms accounted for to a self-consistent energetic cutoff, across a wide range of static and dynamical field regimes, and for different molecular species are currently underway [11].

In addition to the statics, the dynamics of the MHH were investigated in this thesis. The particular quench process studied initializes the system in the ground state of the MHH in the absence of an AC field. An AC field is suddenly turned on at a frequency corresponding to a single-molecule rotational resonance. In the absence of couplings between sites, each molecule would undergo coherent Rabi flopping independently of the others. However, when averaged over the many-body wavefunction, the Rabi oscillations of transitions between the single-molecule levels driven on resonance were seen to damp out exponentially with an emergent timescale. This effect was studied for a variety of Hamiltonian parameters near half filling, but a more thorough study of how the emergent timescale depends on the experimental parameters is lacking. Such a study may also shed light on the true microscopic mechanism of quantum dephasing. Is it governed only by equilibrium properties, as suggested by the Kibble–Zurek hypothesis [12–16]? What is the nature of the state as $t \rightarrow \infty$? A final interesting avenue for further research is to study the interplay between the complex many-molecule behavior of quantum dephasing with the complex behavior of a single molecule in the presence of non-collinear electric and magnetic fields.

MHHs may also be derived and studied for more complex molecules using the procedures outlined in this thesis. Near-term extensions include $^2\Sigma$ molecules formed from alkali metals and alkaline earths. These molecules have an unpaired spin giving rise to magnetic dipole moments in addition to electric dipole moments, which may lead to a rich interplay of crystalline and magnetic phases. Longer-term extensions can include molecules in Π or Δ states which have orbital angular momenta. Essentially nothing is known about the possibilities for many-body physics with such molecules.

Part III of this thesis laid the foundations for the Fermi resonance Hamiltonian (FRH) [17]. The FRH is a lattice projection of a two-channel model for a Feshbach resonance between two-component fermions and a bound molecular state. The two-body bound states in the lattice are nontrivial combinations of all Bloch bands from both the open and closed channels. Using a projection method to separate the two-particle Hilbert space into a low energy piece spanned by open channel fermions in the lowest band and a high energy piece spanned by open channel fermions in excited bands and all bands of the closed channel, the relevant high-energy degrees of freedom at low density are identified. The derivation of the FRH consists of re-coupling the low-energy and high energy sectors of Hilbert space at the many-body level, resulting in a multichannel resonance model between fermions in the lowest open channel band and dressed molecules which are nontrivial linear combinations of an infinite number of bands chosen to reproduce the two-body scattering length exactly. The use of the numerically exact lattice solution of the two-body problem leads to novel features of the Hamiltonian, such as diagonal hopping of dressed molecules.

The physics related to the FRH is very much in its infancy. The most immediate question is, what are the many-body features of the FRH? How is the phase diagram affected by diagonal hopping and other features arising from the lattice solution? How do the Hubbard parameters for the FRH behave in confined geometries or geometries which are not separable for a single particle? Furthermore, one can imagine extending the FRH analysis to pairing in higher relative orbital angular momentum, multichannel situations, mass-imbalanced systems, or more realistic inter-channel potentials. A final practical application of the FRH would be to optimize production of Feshbach molecules directly in an optical lattice using the many-body properties of the FRH.

In Parts IV and V of this thesis we devised variational matrix product state (MPS) algorithms for eigenstates and dynamics of generic time-dependent 1D Hamiltonians [18, 19], discussed open source software efforts for MPS algorithms [20], and presented educational materials to aid in the use of the open source implementations to study strongly correlated physics. Our stand-alone implementation of time-evolving block decimation (TEBD), open source TEBD [21], and the TEBD code written for the algorithms and libraries for physics simulations (ALPS) [20, 22, 23] package are both in use by many groups across the world [20, 24–36]. ALPS is also regularly used as a pedagogical tool for learning many-body physics at summer schools. At present, open source implementations of the generic algorithms presented in this thesis are being prepared for merger into the ALPS open source software package in its next release. This will greatly extend the capability of the ALPS package to study the statics and dynamics of a wider range of models, including long-range interactions and general time-dependence. In order to make these tools useful, we also are writing new educational materials to demonstrate the use of these algorithms and their enhanced capabilities over other MPS algorithms for time evolution.

MPS algorithms are very powerful in that they do not depend on the nature of the microscopic constituents of the model under study. Also, MPS methods produce wavefunctions, allowing for the simulation of dynamics and access to a broad range of quantum measures. The flexibility of MPS algorithms makes them especially well suited both to open source implementation and studies of many-body physics with ultracold molecules. Speaking broadly, as more complex molecules approach quantum degeneracy, the number of many-body models and their complexity will also dramatically increase. Searching for emergent phenomena such as quantum order in strongly correlated systems requires exploration of large parameter regimes and careful finite-size or finite-entanglement scaling. Open source tools for modern strongly correlated physics must be adaptable to different physical degrees of freedom, interactions, and dynamical processes. Furthermore, they should support massive parallelization over parameter regimes, and be efficient enough to handle large-scale parameter exploration. Finally, they must be able to calculate a broad range of quantum measures, and contain powerful post-processing tools to extract and manipulate data from large simulations. Many open questions remain regarding MPSs in general. Is it possible to simulate the dynamics of a generic time-dependent infinite system using Krylov methods as outlined for finite systems? How do we improve the numerical conditioning of the higher-dimensional tensor network algorithms so as to make them practical? What is the relevant operator structure, analogous to MPOs in 1D, for higher dimensions? The answer to these questions would provide a "holy grail" of strongly correlated physics: a black box numerical method applicable to any many-body system.

References

1. Wall, M.L., Carr, L.D.: Emergent timescales in entangled quantum dynamics of ultracold molecules in optical lattices. New J. Phys. **11**, 055027 (2009)
2. Wall, M.L., Carr, L.D.: Hyperfine molecular Hubbard Hamiltonian. Phys. Rev. A **82**(1), 013611 (2010). doi:10.1103/PhysRevA.82.013611
3. Kühner, T.D., White, S.R., Monien, H.: One-dimensional Bose-Hubbard model with nearest-neighbor interaction. Phys. Rev. B **61**(18), 12474–12489 (2000). doi:10.1103/ PhysRevB.61.12474
4. Dalmonte, M., Pupillo, G., Zoller, P.: One-dimensional quantum liquids with power-law interactions: The Luttinger staircase. Phys. Rev. Lett. **105**, 140401 (2010). doi:10.1103/ PhysRevLett.105.140401. http://link.aps.org/doi/10.1103/PhysRevLett.105.140401
5. Capogrosso-Sansone, B., Trefzger, C., Lewenstein, M., Zoller, P., Pupillo, G.: Quantum phases of cold polar molecules in 2D optical lattices. Phys. Rev. Lett. **104**, 125301 (2010)

6. Gorshkov, A.V., Manmana, S.R., Chen, G., Ye, J., Demler, E., Lukin, M.D., Rey, A.M.: Tunable superfluidity and quantum magnetism with ultracold polar molecules. Phys. Rev. Lett. **107**, 115301 (2011). doi:10.1103/PhysRevLett.107.115301. http://link.aps.org/doi/10.1103/PhysRevLett.107.115301
7. Gorshkov, A.V., Manmana, S.R., Chen, G., Demler, E., Lukin, M.D., Rey, A.M.: Quantum magnetism with polar alkali-metal dimers. Phys. Rev. A **84**, 033619 (2011). doi:10.1103/PhysRevA.84.033619. http://link.aps.org/doi/10.1103/PhysRevA.84.033619
8. Block, M.S., Mishmash, R.V., Kaul, R.K., Sheng, D.N., Motrunich, O.I., Fisher, M.P.A.: Exotic gapless mott insulators of bosons on multileg ladders. Phys. Rev. Lett. **106**, 046402 (2011). doi:10.1103/PhysRevLett.106.046402. http://link.aps.org/doi/10.1103/PhysRevLett.106.046402
9. Mishmash, R.V., Block, M.S., Kaul, R.K., Sheng, D.N., Motrunich, O.I., Fisher, M.P.A.: Bose metals and insulators on multileg ladders with ring exchange. Phys. Rev. B **84**, 245127 (2011). doi:10.1103/PhysRevB.84.245127. http://link.aps.org/doi/10.1103/PhysRevB.84.245127
10. Jiang, H.-C., Block, M.S., Mishmash, R.V., Garrison, J.R., Sheng, D.N., Motrunich, O.I., Fisher, M.P.A.: Non-Fermi liquid d-wave metal phase of strongly interacting electrons. http://arxiv.org/abs/1207.6608 (2012)
11. Wall, M.L., Bekaroglu, E., Carr, L.D.: Molecular hubbard hamiltonian: field regimes and molecular species. Phys. Rev. A **88**, 023605 (2013). doi:10.1103/PhysRevA.88.023605. http://link.aps.org/doi/10.1103/PhysRevA.88.023605
12. Kibble, T.W.B.: Topology of cosmic domains and strings. J. Phys. A Math Gen. **9**, 1387 (1980)
13. Zurek, W.H.: Cosmological experiments in superfluid Helium? Nature **317**, 505–508 (1985)
14. Kibble, T.: Phase-transition dynamics in the lab and the universe. Phys. Today **60**(9), 47 (2007)
15. Damski, B., Zurek, W.H.: Dynamics of a quantum phase transition in a ferromagnetic Bose-Einstein condensate. Phys. Rev. Lett. **99**, 130402 (2007)
16. Saito, H., Kawaguchi, Y., Ueda, M.: Kibble-Zurek mechanism in a quenched ferromagnetic Bose-Einstein condensate. Phys. Rev. A **76**, 043613 (2007)
17. Wall, M.L., Carr, L.D.: Microscopic model for feshbach interacting fermions in an optical lattice with arbitrary scattering length and resonance width. Phys. Rev. Lett. **109**, 055302 (2012). doi:10.1103/PhysRevLett.109.055302. http://link.aps.org/doi/10.1103/PhysRevLett.109.055302
18. Wall, M.L., Carr, L.D.: Finite temperature matrix product state algorithms and applications. In: Proukakis, N.P., Gardiner, S.A., Davis, M.J., Szymanska, M.H. (eds.) Quantum Gases: Finite Temperature and Non-equilibrium Dynamics. Cold Atoms Series, vol. 1. Imperial College Press, London (2012); e-print http://arxiv.org/abs/1008.4303 (2011)
19. Wall, M.L., Carr, L.D.: Out-of-equilibrium dynamics with matrix product states. New J. Phys. **14**(12), 125015 (2012). http://stacks.iop.org/1367-2630/14/i=12/a=125015
20. Bauer, B., Carr, L.D., Evertz, H.G., Feiguin, A., Freire, J., Fuchs, S., Gamper, L., Gukelberger, J., Gull, E., Guertler, S., Hehn, A., Igarashi, R., Isakov, S.V., Koop, D., Ma, P.N., Mates, P., Matsuo, H., Parcollet, O., Pawłowski, G., Picon, J.D., Pollet, L., Santos, E., Scarola, V.W., Schollwöck, U., Silva, C., Surer, B., Todo, S., Trebst, S., Troyer, M., Wall, M.L., Werner, P., Wessel, S.: The ALPS project release 2.0: Open source software for strongly correlated systems. J. Stat. Mech. Theory Exp. **2011**(05), P05001 (2011). http://stacks.iop.org/1742-5468/2011/i=05/a=P05001
21. Open source time-evolving block decimation (2009). http://physics.mines.edu/downloads/software/tebd/
22. ALPS (2012). http://alps.comp-phys.org
23. Albuquerque, A.F., Alet, F., Corboz, P., Dayal, P., Feiguin, A., Fuchs, S., Gamper, L., Gull, E., Gurtler, S., Honecker, A., Igarashi, R., Korner, M., Kozhevnikov, A., Lauchli, A., Manmana, S.R., Matsumoto, M., McCulloch, I.P., Michel, F., Noack, R.M., Pawlowski, G., Pollet, L., Pruschke, T., Schollwock, U., Todo, S., Trebst, S., Troyer, M., Werner, P., Wessel, S.: The ALPS project release 1.3: Open-source software for strongly correlated systems. J. Magn. Magn. Mater. **310**, 1187–1193 (2007)

24. Wall , M.L., Carr, L.D.: Emergent timescales in entangled quantum dynamics of ultracold molecules in optical lattices. New J. Phys. **11**, 055027 (2009)
25. Carr, L.D., Oberthaler, M.K.: Spatial dependence of entropy in trapped ultracold Bose gases (2009) [e-print arXiv:0906.4708]
26. Mishmash, R.V., Carr, L.D.: Quantum entangled dark solitons formed by ultracold atoms in optical lattices. Phys. Rev. Lett. **103**, 140403 (2009)
27. Mishmash, R.V., Danshita, I., Clark, C.W., Carr, L.D.: Quantum many-body dynamics of dark solitons in optical lattices. Phys. Rev. A **80**, 053612 (2009)
28. Carr, L.D., Wall, M.L., Schirmer, D.G., Brown, R.C., Williams, J.E., Clark, C.W.: Mesoscopic effects in quantum phases of ultracold quantum gases in optical lattices. Phys. Rev. A **81**, 013613 (2010)
29. Wall, M.L., Carr, L.D.: Hyperfine molecular Hubbard Hamiltonian. Phys. Rev. A **82**, 013611 (2010)
30. Glick, J.A., Carr, L.D.: Macroscopic quantum tunneling of solitons in Bose-Einstein condensates. [arXiv:1105.5164]
31. Carleo, G., Becca, F., Schiró, M., Fabrizo, M.: Localization and glassy dynamics of many-body quantum systems. Sci. Rep. **2**, 243 (2011)
32. Qian, Y., Gong, M., Zhang, C.: Quantum transport of bosonic cold atoms in double-well optical lattices. Phys. Rev. A **84**, 013608 (2011). doi:10.1103/PhysRevA.84.013608. http://link.aps.org/doi/10.1103/PhysRevA.84.013608
33. Zhou, Z.-W., Zhang, S.-L., Zhou, X.-F., Guo, G.-C., Zhou, X., Pu, H.: Quantum phase transition of Bose-Einstein condensates on a nonlinear ring lattice. Phys. Rev. A **83**, 043626 (2011). doi:10.1103/PhysRevA.83.043626. http://link.aps.org/doi/10.1103/PhysRevA.83.043626
34. Lu, H., Baksmaty, L.O., Bolech, C.J., Pu, H.: Expansion of 1D polarized superfluids: The Fulde-Ferrell-Larkin-Ovchinnikov state reveals itself. Phys. Rev. Lett. **108**, 225302 (2012). doi:10.1103/PhysRevLett.108.225302. http://link.aps.org/doi/10.1103/PhysRevLett.108.225302
35. Qian, Y., Gong, M., Zhang, C.: Many-body Landau-Zener transition in cold atom double well optical lattices. http://arxiv.org/abs/1110.1653 (2012)
36. Yin, S., Qin, X., Lee, C., Zhong, F.: Finite-time scaling of dynamic quantum criticality. http://arxiv.org/abs/1207.1602 (2012)

Appendix A
Documentation for ALPS v2.0 TEBD Code

This appendix contains the documentation for the TEBD code included in the v2.0 release of the ALPS code, see Chap. 10. At the time of the writing of this thesis, this information was hosted on the ALPS wiki at http://alps.comp-phys.org. Section A.1 is documentation about the code in general, the background of the TEBD algorithm, and explanations of the parameters used as input to the code. The original website for this section is http://alps.comp-phys.org/mediawiki/index.php/Documentation:TEBD. Sections A.2 and A.3 are tutorials on using the ALPS routines. The original websites are http://alps.comp-phys.org/mediawiki/index.php/ALPS_2_Tutorials:TEBD-01_bhquench and http://alps.comp-phys.org/mediawiki/index.php/ALPS_2_Tutorials:TEBD-02_kink, respectively.

A.1 Documentation: TEBD

A.1.1 Time-Evolving Block Decimation

The time-evolving block decimation (TEBD) algorithm is a method for simulating the time evolution of one-dimensional quantum lattice systems governed by a Hamiltonian with at most nearest-neighbor interactions. It is closely related to the density matrix renormalization group (DMRG) method in that both methods operate on a class of states known as matrix product states (MPSs). In addition to real time evolution, imaginary time evolution can also be used to find ground states. Essentially, TEBD consists of two parts: a canonical MPS representation of a many-body state, and a protocol for finding the MPS closest to a state which is acted upon by a two-site operator.

M.L. Wall, *Quantum Many-Body Physics of Ultracold Molecules in Optical Lattices*, Springer Theses, DOI 10.1007/978-3-319-14252-4

The particular implementation of TEBD used in ALPS simulates a series of global parameter quenches of the form $g(t) = g(t_i) + ((t - t_i)/\tau)^p(g(t_f) - g(t_i))$. The timescale τ, power p, initial and final values $g(t_f)$ and $g(t_i)$, and Hamiltonian parameters g of each quench are all amenable to specification by the user. Additionally, because the TEBD method produces wavefunctions, a wide range of observables are available, including entropies, correlation functions, and overlaps between the state at different times.

A.1.2 References

Vidal, G.: Efficient classical simulation of slightly entangled quantum computations. Phys. Rev. Lett. **91**, 147902 (2003)

Vidal, G.: Efficient simulation of one-dimensional quantum many-body systems. Phys. Rev. Lett. **93**, 040502 (2004)

Daley, A.J., Kollath, C., Schollwöck, U., Vidal, G.: Time-dependent density-matrix renormalization-group using adaptive effective Hilbert spaces. J. Stat. Mech. **2004**, P04005 (2004)

A.1.3 TEBD-Specific Parameters

CHI_LIMIT
The maximum bond dimension of the MPS allowed during real time propagation. The default value is 50.

TRUNC_LIMIT
The maximum truncation error allowed for a specific two-site evolution. If the bond dimension corresponding to this truncation is greater than CHI_LIMIT, then CHI_LIMIT is chosen instead. The default value is 10^{-12}.

TAUS
The elements of this vector are the timescales τ of the global quenches.

GS
The elements of this vector are the Hamiltonian parameters g of the global quenches, given as character variables. Note that the elements of this vector may themselves be vectors, which corresponds to quenching several parameters at the same time. If this is so, the corresponding elements of POWS, GIS, and GFS must also be vectors of the same length. Note that TAUS, NUMSTEPS, and STEPSFORSTORE

will not be vectors, as the timescale, number of time steps, and number of steps between outputs are the same for each parameter being quenched.

POWS
The elements of this vector are the powers p of the global quenches.

GIS
The elements of this vector are the initial values of the Hamiltonian parameters g of the global quenches.

GFS
The elements of this vector are the final values of the Hamiltonian parameters g of the global quenches.

CONSERVED_QUANTUMNUMBERS
Quantum numbers conserved by the model of interest. For spin models "Sz" can be conserved, and for particle models "N" can be conserved.

NUMSTEPS
The elements of this vector are the number of time-steps of the global quenches. This implicitly defines the time steps dt of the quenches.

STEPSFORSTORE
The elements of this vector are the number of time-steps between the calculation and output of observables.

INITIAL_STATE
The state used at $t = 0$, before real time propagation begins. Currently, only two values are supported: "kink," which produces a specific initial state to be discussed further in tutorial 2a, and "ground," which calculates the ground state of a specified initial hamiltonian via imaginary time propagation. The default value is "ground". See the tutorials for examples.

ITP_CHIS
The elements of this vector are the maximum bond dimensions used in iterations of imaginary time propagation to find the group state. It is only referenced if INITIAL_STATE is "ground."

ITP_DTS
The elements of this vector are the time steps used in iterations of imaginary time propagation to find the group state. It is only referenced if INITIAL_STATE is "ground."

ITP_CONVS
The elements of this vector are the convergence parameters used in iterations of imaginary time propagation to find the group state. An iteration of imaginary time propagation exits if the maximal difference between singular values at some time interval is less than the convergence parameter. It is only referenced if INITIAL_STATE is "ground."

SIMID
An optional integer input which differentiates a series of simulations and can simplify plotting commands.

NUM_THREADS
The number of OpenMP threads used.

VERBOSE
If set to "true," then the code will output, time values, truncation errors, and other running messages. The default value is "false."

A.2 Tutorials: TEBD-01 bhquench

A.2.1 The Hardcore Boson Model

In this first tutorial we investigate the behavior of the hardcore boson model

$$H = -t\sum_{i=1}^{L-1}(b_i^{\dagger}b_{i+1} + b_i b_{i+1}^{\dagger}) + V\sum_{i=1}^{L-1} n_i n_{i+1} \tag{A.1}$$

as the parameter V is changed in time. It is well known that for large V/t the ground state of the hardcore boson model at half filling is a charge-density wave (CDW) insulator while for small V/t the ground state is a superfluid (SF). It is interesting to consider what happens to the system if we begin in one phase and then dynamically change, or "quench," one of the Hamiltonian parameters t or V such that we are in the other phase. As a simple first foray into the rich physics of quenches, we will consider quenching from one phase to the other and then back into the original phase. A particularly stringent criterion for adiabaticity of such a quench is how close the final state is to the initial state, i.e.

$$L(t;\gamma) \equiv |\langle\psi(t)|\psi(0)\rangle|^2 \tag{A.2}$$

which we call the Loschmidt Echo. Note that the t in this expression is the time and not the hopping parameter t. The parameter γ is meant to convey that this quantity in general depends on the manner in which the system is quenched.

The general structure of a quench in the ALPS TEBD routines is given by the parameterization

$$g(t) = g(t_i) + ((t - t_i)/\tau)^p (g(t_f) - g(t_i)), \tag{A.3}$$

where g is some Hamiltonian parameter. In the present case we will take g to be the interaction parameter V. We will begin our system in the CDW regime with $V/t = 10$, quench to the SF regime where $V/t = 0$, and then quench back to the CDW regime with $V/t = 10$. In the three parts of this tutorial we will investigate (a) the effects of the timescale τ on the Loschmidt echo during a linear quench, (b) the effects of "holding" the system in the SF phase for a time τ_{hold} before returning to the CDW phase, and (c) the effects of changing the power p of the quench function.

A.2.2 *Linear Quench*

First, we will investigate the effects of the quench rate τ on the adiabaticity of a linear quench from the CDW to the SF phase and back.

A.2.2.1 Preparing and Running the Simulation Using Python

To set up and run the simulation in Python we use the script tutorial1a.py. The first parts of this script import the required modules and then prepare the input files as a list of Python dictionaries:

```
import pyalps
import matplotlib.pyplot as plt
import pyalps.plot

parms=[]
count=0
for A in [5.0, 10.0, 15.0, 25.0, 50.0]:
        count+=1
        parms.append({
                'L'                          : 10,
                'MODEL'                      : '
                    hardcore␣boson',
                'CONSERVED_QUANTUMNUMBERS'   : 'N',
                'N' : 5,
                't'                          : 1.0,
                'V'                          : 10.0,
                'ITP_CHIS' : [20, 30, 35],
                'ITP_DTS' : [0.05, 0.05,0.025],
```

```
        'ITP_CONVS' : [1E-8, 1E-8, 1E-9],
        'INITIAL_STATE' : 'ground',
        'CHI_LIMIT' : 40,
        'TRUNC_LIMIT' : 1E-12,
        'NUM_THREADS' : 1,
        'TAUS' : [A, A],
        'POWS' : [1.0, 1.0],
        'GS' : ['V', 'V'],
        'GIS' : [10.0, 0.0],
        'GFS' : [0.0, 10.0],
        'NUMSTEPS' : [500, 500],
        'STEPSFORSTORE' : [5, 3],
        'SIMID' : count
    })
```

Let's go through the TEBD-specific parameters in more detail (see the TEBD documentation, Sect. A.1, for a list of all such parameters). The parameter INITIAL_STATE is set to ground, which means that we begin from the ground state of our Hamiltonian with user-specified parameters. The parameters t and V specify that the initial Hamiltonian parameters $t = 1$ and $V = 10$ are used to find the ground state. In order to find the ground state, TEBD performs evolution in imaginary time. We refer to this step as ITP, and so all parameters containing ITP deal with the ground state properties. The vectors ITP_CHIS, ITP_DTS, and ITP_CONVS are the entanglement cutoff parameters, time steps, and convergence criteria for successive applications of imaginary time propagation. These constitute the main convergence parameters for TEBD, and convergence should always be carefully checked in each parameter. For now, don't worry too much about their actual values, we'll see how errors are controlled in the next set of tutorials.

Now we turn to the real-time propagation parameters. We wish to perform a series of two quenches. First we want to quench the parameter V linearly in time from its initial value 10 to 0. Comparing with the general form of a quench $g(t) = g(t_i) + ((t - t_i)/\tau)^p (g(t_f) - g(t_i))$ we see that this corresponds to $g = V$, $g(t_i) = 10$, $g(t_f) = 0$, $p = 1$, and τ is the free parameter whose effects are to be investigated. Looking at the parameter list, we see that the first elements of the vectors GS, GIS, GFS, and POWS correspond to g, $g(t_i)$, $g(t_f)$, and p, respectively. The first element of the vector TAUS is looped over using the variable A, which means that we will perform a series of simulations with $\tau = 5, 10, 15, 25$, and 50. The second quench is essentially the reverse of the first, with $g = V$, $g(t_i) = 0$, $g(t_f) = 10$, $p = 1$, and τ the same as the first. Comparing with the parameters list, we see that this corresponds to the second elements of the vectors GS, GIS, etc. as above.

Time evolution is simulated by breaking the full propagator approximately into a series of operations which act only on two neighboring sites at a time. The error in using this approximate propagator is second order in the "infinitesimal" time-step dt. TEBD gives a protocol for updating the canonical form of our state after such a two-site operation has been applied. The error in this procedure is controlled by CHI_LIMIT, which is directly related to the amount of spatial entanglement, and TRUNC_LIMIT, which is akin to the TRUNCATION_ERROR in the DMRG routines. The parameter vector NUMSTEPS specifies how many time-steps are taken in performing each quench, which together with τ implicitly defines the time-step dt. The overall error is a nontrivial function of CHI_LIMIT, TRUNC_LIMIT, and NUMSTEPS which will be investigated in the next set of tutorials, so we won't worry about the choice of these much for now. Finally, STEPSFORSTORE determines how many time steps are taken before observables are computed and stored and SIMID is an integer differentiating the simulations with different τ.

We now move on to the actual computation. The lines:

```
baseName='tutorial_1a'
#write output files
nmlnameList=pyalps.writeTEBDfiles(parms, baseName)
#run the application
res=pyalps.runTEBD(nmlnameList)
```

prepare the input files for the TEBD routines and run the simulations for the range of τ specified in the parameters. We now load the Loschmidt Echo and interaction parameter U as functions of time via:

```
#Load the loschmidt echo and V
LEdata=pyalps.load.loadTimeEvolution(pyalps.getResult
 Files(prefix='tutorial_1a'), measurements=
 ['Loschmidt␣Echo', 'V'])
```

Finally, we plot the collected data using:

```
LE=pyalps.collectXY(LEdata, x='Time', y='Loschmidt␣
    Echo',foreach=['SIMID'])
for q in LE:
        q.props['label']=r'$\tau=$'+str(q.props['TAUS'
            ][0])

plt.figure()
pyalps.plot.plot(LE)
plt.xlabel('Time␣$t$')
plt.ylabel('Loschmidt␣Echo␣$|<␣\psi(0)|\psi(t)␣>␣|^2$'
    )
plt.title('Loschmidt␣Echo␣vs.␣Time')
plt.legend(loc='lower␣right')

Ufig=pyalps.collectXY(LEdata, x='Time', y='V',foreach
    =['SIMID'])
for q in Ufig:
        q.props['label']=r'$\tau=$'+str(q.props['TAUS'
            ][0])
```

```
plt.figure()
pyalps.plot.plot(Ufig)
plt.xlabel('Time␣$t$')
plt.ylabel('V')
plt.title('Interaction␣parameter␣$V$␣vs.␣Time')
plt.legend(loc='lower␣right')
plt.show()
```

A.2.2.2 Preparing and Running the Simulation Using Vistrails

To run the simulation in Vistrails open the file tutorial1a.vt and look at the workflow labeled "tutorial1a." Click on "Execute" to prepare the input file, run the simulation and create the output figure.

A.2.2.3 Questions

- How does the behavior of the overlap change as the quench rate decreases?
- Roughly how slowly do you have to perform the quench in order for it to be adiabatic?
- Is it easier or harder for a larger system to be adiabatic? Why?
- Are these properties changed depending on whether the intermediate phase is gapped or not? One can test this by changing from the hardcore boson model to the (softcore) boson Hubbard model, and then quenching from the Mott-Insulating (MI) phase at large U/t and unit filling to the CDW phase with large V. As you quench from the Mott insulating to the CDW phase and back, how difficult is it to be adiabatic?

A.2.3 *Linear Quench with Hold*

In this section we will investigate the effects of "holding" the system in the SF phase for a time τ_{hold} before quenching back to the CDW phase.

A.2.3.1 Preparing and Running the Simulation Using Python

To set up and run the simulation in Python we use the script tutorial1b.py. The first parts of this script import the required modules and then prepare the input files as a list of Python dictionaries:

```
import pyalps
import matplotlib.pyplot as plt
import pyalps.plot

#prepare the input parameters
parms=[]
count=0
for A in [5.0, 10.0, 15.0, 25.0, 50.0]:
        count+=1
        parms.append({
                'L'                              : 10,
                'MODEL'                          : 'hardcore␣boson',
                'CONSERVED_QUANTUMNUMBERS'       : 'N',
                'N' : 5,
                't'                              : 1.0,
                'V'                              : 10.0,
                'ITP_CHIS' : [20, 30, 35],
                'ITP_DTS' : [0.05, 0.05,0.025],
                'ITP_CONVS' : [1E-8, 1E-8, 1E-9],
                'INITIAL_STATE' : 'ground',
                'CHI_LIMIT' : 80,
                'TRUNC_LIMIT' : 1E-12,
                'NUM_THREADS' : 1,
                'TAUS' : [10.0, A, 10.0],
                'POWS' : [1.0, 0.0,1.0],
                'GS' : ['V', 'V', 'V'],
                'GIS' : [10.0,0.0, 0.0],
                'GFS' : [0.0, 0.0, 10.0],
                'NUMSTEPS' : [500, int(A/0.05), 500],
                'STEPSFORSTORE' : [5,5, 3],
                'SIMID' : count
              })
```

Note that in this case we have three quenches as GS, GIS, etc. are all vectors of length three. The second quench keeps the Hamiltonian parameters fixed at $t = 1$, $V = 0$ for a time τ_{hold} before quenching back. We write the input files, run the simulations, get outputs, and plot as above:

```
baseName='tutorial_1b'
#write output files
nmlnameList=pyalps.writeTEBDfiles(parms, baseName)
#run the application
res=pyalps.runTEBD(nmlnameList)

#Load the loschmidt echo and U
LEdata=pyalps.load.loadTimeEvolution(pyalps.
    getResultFiles(prefix='tutorial_1b'), measurements
    =['Loschmidt␣Echo', 'V'])
```

```
LE=pyalps.collectXY(LEdata, x='Time', y='Loschmidt 
    Echo',foreach=['SIMID'])
for q in LE:
        q.props['label']=r'$\tau_{\mathrm{hold}}=$'+str
            (q.props['TAUS'][1])
plt.figure()
pyalps.plot.plot(LE)
plt.xlabel('Time $t$')
plt.ylabel('Loschmidt Echo $|< \psi(0)|\psi(t) > |^2$'
    )
plt.title('Loschmidt Echo vs. Time')
plt.legend(loc='lower right')

Ufig=pyalps.collectXY(LEdata, x='Time', y='V',foreach
    =['SIMID'])
for q in Ufig:
        q.props['label']=r'$\tau_{\mathrm{hold}}=$'+str
            (q.props['TAUS'][1])
plt.figure()
pyalps.plot.plot(Ufig)
plt.xlabel('Time $t$')
plt.ylabel('V')
plt.title('Interaction parameter $V$ vs. Time')
plt.legend()
plt.show()
```

A.2.3.2 Preparing and Running the Simulation Using Vistrails

To run the simulation in Vistrails open the file tutorial1b.vt and look at the workflow labeled "tutorial1b." Click on "Execute" to prepare the input file, run the simulation and create the output figure.

A.2.3.3 Questions

- How does the behavior of the overlap change as the hold time increases?
- Is this behavior monotonic in the hold time? Why or why not?

A.2.4 Nonlinear Quenches

In this section we will investigate the effects of varying the power of the quench away from being linear.

A.2.4.1 Preparing and Running the Simulation Using Python

To set up and run the simulation in Python we use the script tutorial1c.py. The first parts of this script imports the required modules and then prepares the input files as a list of Python dictionaries:

```
import pyalps
import matplotlib.pyplot as plt
import pyalps.plot

#prepare the input parameters
parms=[]
count=0
for A in [1.0, 1.5, 2.0, 2.5, 3.0]:
        count+=1
        parms.append({
                  'L'                           : 10,
                  'MODEL'                       : 'hardcore␣boson',
                  'CONSERVED_QUANTUMNUMBERS'    : 'N',
                  'N' : 5,
                  't'                           : 1.0,
                  'V'                           : 10.0,
                  'ITP_CHIS' : [20, 30, 35],
                  'ITP_DTS' : [0.05, 0.05,0.025],
                  'ITP_CONVS' : [1E-8, 1E-8, 1E-9],
                  'INITIAL_STATE' : 'ground',
                  'CHI_LIMIT' : 40,
                  'TRUNC_LIMIT' : 1E-12,
                  'NUM_THREADS' : 1,
                  'TAUS' : [10.0, 10.0],
                  'POWS' : [1.0, A],
                  'GS' : ['V', 'V'],
                  'GIS' : [10.0, 0.0],
                  'GFS' : [0.0, 10.0],
                  'NUMSTEPS' : [1000, 1000],
                  'STEPSFORSTORE' : [10, 5],
                  'SIMID' : count
                })
```

We then write the input files, run the simulations, get outputs, and plot as above:

```
baseName='tutorial_1c'
#write output files
nmlnameList=pyalps.writeTEBDfiles(parms, baseName)
#run the application
res=pyalps.runTEBD(nmlnameList)

#Load the loschmidt echo and U
```

```
LEdata=pyalps.load.loadTimeEvolution(pyalps.
    getResultFiles(prefix='tutorial_1c'), measurements
    =['V', 'Loschmidt␣Echo'])

LE=pyalps.collectXY(LEdata, x='Time', y='Loschmidt␣
    Echo',foreach=['SIMID'])
for q in LE:
        q.props['label']=r'$\tau=$'+str(q.props['POWS'
            ][1])
plt.figure()
pyalps.plot.plot(LE)
plt.xlabel('Time␣$t$')
plt.ylabel('Loschmidt␣Echo␣$|<␣\psi(0)|\psi(t)␣>␣|^2$'
    )
plt.title('Loschmidt␣Echo␣vs.␣Time␣')
plt.legend(loc='lower␣left')

Ufig=pyalps.collectXY(LEdata, x='Time', y='V',foreach
    =['SIMID'])
for q in Ufig:
        q.props['label']=r'$\tau=$'+str(q.props['POWS'
            ][1])
plt.figure()
pyalps.plot.plot(Ufig)
plt.xlabel('Time␣$t$')
plt.ylabel('U')
plt.title('Interaction␣parameter␣$V$␣vs.␣Time')
plt.legend(loc='lower␣left')
plt.show()
```

A.2.4.2 Preparing and Running the Simulation Using Vistrails

To run the simulation in Vistrails open the file tutorial1c.vt and look at the workflow labeled "tutorial1c". Click on "Execute" to prepare the input file, run the simulation and create the output figure.

A.2.4.3 Questions

- How does the behavior of the overlap change as the power changes?
- Again change from the hardcore boson model to the boson Hubbard model and investigate the dynamics of the MI-CDW transition, this time with a nonlinear quench. Is the behavior different from that of a linear quench?

- The present example uses an asymmetric quench which is linear one side and nonlinear on the other. How is the behavior changed if you make both quenches nonlinear?

A.3 Tutorials: TEBD-02 kink

A.3.1 Evolution of a Domain Wall

In this tutorial we will study the time evolution of an $S = 1/2$ spin chain prepared in a nonequilibrium state. The particular state that we choose is that with all spins to the left of the chain center "down" and all of those to the right of the center "up," $|\downarrow\downarrow \ldots \downarrow\uparrow \ldots \uparrow\uparrow\rangle$. This state can be chosen as the initial state by setting INITIAL_STATE to be "kink". Some exact results are known regarding the evolution of this state under the 1D XX model, which allows for a detailed study of the errors present in TEBD.

A.3.2 Exact Solution for the Case of the XX Model

The time evolution of the kink initial state under the XX model was solved exactly in Phys. Rev. E **59**, 4912 (1999) by a Jordan–Wigner transformation to free fermions. It was found that the expectation value of the magnetization at any site as a function of time can be represented as a sum of Bessel functions, and the magnetization in the limit of long times and large distances from the initial domain wall approaches a scaling form in the variable n/t, where n is the distance from the center and t the time. Explicitly, we have

$$M(n,t) = -\frac{1}{2} \sum_{i=1-n}^{n-1} j_i^2(t) \,, \tag{A.4}$$

$$\lim_{n\to\infty} \lim_{t\to\infty} M(n,t) \to \phi(n/t) = -\frac{1}{\pi} \arcsin(n/t) \,, \tag{A.5}$$

where $M(n,t)$ is the magnetization a distance n from the center and $j_i(t)$ is the Bessel function of order i. In the first part of this tutorial we demonstrate these two results.

A.3.2.1 Preparing and Running the Simulation Using Python

To set up and run the simulation in Python we use the script tutorial2a.py. The first parts of this script import the required modules and prepare the input files as a list of Python dictionaries:

```
import pyalps
import matplotlib.pyplot as plt
import pyalps.plot
import numpy as np
import copy
import math
import scipy.special

#prepare the input parameters
parms = [{
          'L'                            : 50,
          'MODEL'                        : 'spin',
          'local_S'                      : 0.5,
          'CONSERVED_QUANTUMNUMBERS'     : 'Sz',
          'Jxy'                          : 1,
          'INITIAL_STATE' : 'kink',
          'CHI_LIMIT' : 40,
          'TRUNC_LIMIT' : 1E-12,
          'NUM_THREADS' : 1,
          'TAUS' : [20.0],
          'POWS' : [0.0],
          'GS' : ['H'],
          'GIS' : [0.0],
          'GFS' : [0.0],
          'NUMSTEPS' : [500],
          'STEPSFORSTORE' : [2]
        }]
```

The math and scipy.special modules are required to generate the special functions needed to compare with the exact solution. Note that we have chosen POWS to be zero, which corresponds to no quenching at all. Thus, the values of GS, GIS, and GFS are arbitrary, and TAUS and NUMSTEPS give us the total simulation time and the number of time steps, respectively. We write the input files, run the simulation, and get the output as usual:

```
baseName='tutorial_2a'
 nmlname=pyalps.writeTEBDfiles(parms, baseName)
 res=pyalps.runTEBD(nmlname)

 #Get the results of the simulation
```

```
Data=pyalps.load.loadTimeEvolution(pyalps.
    getResultFiles(prefix='tutorial_2a'), measurements
    =['Local␣Magnetization'])
```

We now must postprocess the raw output to compare with the exact solution. To do this we first define empty arrays to hold the postprocessed data

```
#define a dataset numericalSolution to contain the numerical
 result
 numericalResult=[]
 #define a dataset exactSolution to contain the exact
     solution
 exactResult=[]
 #define a dataset scalingForm to contain the scaling form
 scalingForm=[]
```

we then calculate the exact result from the time data, and use the computed values of the magnetization at each site to compare with the exact solution.

```
 #Compute the exact result M(n,t)=<S_n^z>=-(1/2)*sum_{
     i=1-n}^{n-1} j_i(t)^2, where
 # j_i(t) is the Bessel function of order i and compare
     to the numerically obtained result
 for q in Data:
         syssize=q[0].props['L']
         #Assign a label 'Distance' denoting the
             distance from the center n (only do the
             first two sites
         #to avoid cluttering the plot)
         for n in range(1,3):
                 #Create copies of the data for
                     postprocessing
                 numericalCopy=copy.deepcopy(q)
                 exactCopy=copy.deepcopy(q)

                 numericalCopy[0].props['Distance']=n
                 numericalCopy[0].props['SIMID']='
                     Numerical␣at␣n='+str(n)
                 exactCopy[0].props['Distance']=n
                 exactCopy[0].props['SIMID']='Exact␣at␣
                     n='+str(n)

                 #compute the exact result of the
                     manetization n sites from the
                     center
                 loc=0.0
                 for i in range(1-n,n):
```

```
                    loc -=0.5*scipy.special.jn(i,q
                        [0].props['Time'])*scipy.
                        special.jn(i,q[0].props['
                        Time'])
            exactCopy[0].y=[loc]
            #add to the exact dataset
            exactResult.extend(exactCopy)

            #get the numerical result of the
                magnetization n sites from the
                center
            numericalCopy[0].y=[q[0].y[syssize/2+n
                -1]]
            #add to the numerical dataset
            numericalResult.extend(numericalCopy)
```

Next, we calculate the exact scaling function, and then compute magnetization as a function of the scaling variable n/t to compare with the exact solution

```
#compute the scaling form
# \phi(n/t)=-(1/pi)*arcsin(n/t) that M(n,t) approaches
    as n->infinity and t->infinity
# and compare it with the numerically computed values
    of M(n/t)
for q in Data:
        syssize=q[0].props['L']
        #Assign a label 'Distance' denoting the
            distance from the center n (only do the
            first few sites
        #to avoid cluttering the plot)
        for n in range(0,5):
                #Create a copy of the data for
                    postprocessing
                scalingCopy=copy.deepcopy(q)
                scalingCopy[0].props['Distance']=n

                #The first distance contains the exact
                     scaling form \phi(n/t)=-(1/pi)*
                    arcsin(n/t)
                if n==0:
                        scalingCopy[0].props['Time'
                            ]=1.0/scalingCopy[0].props[
                            'Time']
                        scalingCopy[0].y
                            =[-(1.0/3.1415926)*math.
```

```
                    asin(min(scalingCopy[0].
                    props['Time'],1.0))]
                scalingCopy[0].props['SIMID']='
                  Exact'

        #The other distances contain the
          numerical data as a function of the
           scaling variable M(n/t)
        else:
                scalingCopy[0].props['Time']=n
                  /scalingCopy[0].props['Time
                  ']
                scalingCopy[0].y=[scalingCopy
                  [0].y[syssize/2+n-1] ]
                scalingCopy[0].props['SIMID']='
                  Numerical␣at␣n='+str(n)
        #add to the scaling dataset
        scalingForm.extend(scalingCopy)
```

Finally, we plot the exact and numerical results for comparison.

```
#Plot the numerical and exact magnetization for
   comparison
exactMag=pyalps.collectXY(exactResult, x='Time', y='
   Local␣Magnetization',foreach=['SIMID'])
for q in exactMag:
       q.props['label']=q.props['SIMID']
numericalMag=pyalps.collectXY(numericalResult, x='Time
   ', y='Local␣Magnetization',foreach=['SIMID'])
for q in numericalMag:
       q.props['label']=q.props['SIMID']

plt.figure()
pyalps.plot.plot([exactMag, numericalMag])
plt.xlabel('Time␣$t$')
plt.ylabel('Magnetization')
plt.legend(loc='lower␣right')
plt.title('Magnetization␣vs.␣time')

#Plot the scaling form with the numerical data for
   comparison
Scal=pyalps.collectXY(scalingForm, x='Time', y='Local␣
   Magnetization', foreach=['SIMID'])
for q in Scal:
       q.props['label']=q.props['SIMID']
```

```
plt.figure()
pyalps.plot.plot(Scal)
plt.xlabel('Scaling variable $n/t$')
plt.ylabel('Magnetization$(n,t)$')
plt.legend()
plt.xlim(0,1.5)
plt.title('Magnetization scaling function; numerical 
    and exact results')
plt.show()
```

We see that the magnetization agrees very well to visual accuracy, and approaches the exact scaling form in the relevant limit.

A.3.2.2 Preparing and Running the Simulation Using Vistrails

To run the simulation in Vistrails open the file tutorial2a.vt and look at the workflow labeled "tutorial2a." Click on "Execute" to prepare the input file, run the simulation and create the output figure.

A.3.3 Error Analysis of TEBD 1: Time Step Error

We now use the exact solution to compute the error in a TEBD simulation as a function of time. We first investigate the effects of changing the "infinitesimal" time step dt.

A.3.3.1 Preparing and Running the Simulation Using Python

To set up and run the simulation in Python we use the script tutorial2b.py. The first parts of this script import the required modules and prepare the input files as a list of Python dictionaries:

```
import pyalps
import matplotlib.pyplot as plt
import pyalps.plot
import numpy as np
import math
import scipy.special

#prepare the input parameters
parms=[]
count=0
for nsteps in [100, 250, 500, 750, 1000]:
        count+=1
```

```
parms.append({
        'L'                          : 50,
        'MODEL'                      : 'spin',
        'local_S'                    : 0.5,
        'CONSERVED_QUANTUMNUMBERS'   : 'Sz',
        'Jxy'                        : 1,
        'INITIAL_STATE' : 'kink',
        'CHI_LIMIT' : 20,
        'TRUNC_LIMIT' : 1E-12,
        'NUM_THREADS' : 1,
        'TAUS' : [20.0],
        'POWS' : [0.0],
        'GS' : ['H'],
        'GIS' : [0.0],
        'GFS' : [0.0],
        'NUMSTEPS' : [nsteps],
        'STEPSFORSTORE' : [int(math.floor(nsteps/100))],
        'SIMID': count
    })
```

By changing the parameter NUMSTEPS we implicitly change the time step, since the total evolution time TAU is fixed. We now write the input files, run the simulations, and collect data:

```
baseName='tutorial_2b_'
nmlnameList=pyalps.writeTEBDfiles(parms, baseName)
res=pyalps.runTEBD(nmlnameList)

#Get magnetization data
Magdata=pyalps.load.loadTimeEvolution( pyalps.
    getResultFiles(prefix='tutorial_2b'), measurements
    =['Local␣Magnetization'])
```

We now calculate the exact result from the time data, and then calculate the difference between the numerical and the exact result for the magnetization

```
#Postprocessing-get the exact result for comparison
for q in Magdata:
        syssize=q[0].props['L']
        #Get the exact result of M(1,t)=-(1/2)*(j_0(t)
            ^2), where j_0(t) is the 0^{th} order
        # bessel function and M(1,t) is the
            magnetization one site to the right of the
            chain center
        loc=-0.5*scipy.special.jn(0,q[0].props['Time'
            ])*scipy.special.jn(0,q[0].props['Time'])
        #Get the difference between the computed and
            exact results
        q[0].y=[abs(q[0].y[syssize/2+1-1]-loc)]
```

Finally, we plot this magnetization error:

```
#Plot the Error in the magnetization one site to the
    right of the chain center
Mag=pyalps.collectXY(Magdata, x='Time', y='Local␣
    Magnetization', foreach=['SIMID'])
for q in Mag:
        dt=round(q.props['TAUS']/q.props['NUMSTEPS'],3)
        q.props['label']='dt='+str(dt)

plt.figure()
pyalps.plot.plot(Mag)
plt.xlabel('Time␣$t$')
plt.yscale('log')
plt.ylabel('Magnetization␣Error')
plt.title('Error␣in␣the␣magnetization␣vs.␣time')
plt.legend(loc='lower␣left')
plt.show()
```

We see that, for short times, the errors are roughly proportional to dt^2, reflecting the contribution to the error from the trotter breakup of our exponential. At long times, however, the simulations with the smallest dt have errors which become larger than those with larger dt, and eventually the errors blow up! We will have more to say about this behavior in the next section.

A.3.3.2 Preparing and Running the Simulation Using Vistrails

To run the simulation in Vistrails open the file tutorial2b.vt and look at the workflow labeled "tutorial2b." Click on "Execute" to prepare the input file, run the simulation and create the output figure.

A.3.4 Error Analysis of TEBD 2: Entanglement Cutoff Error

We now investigate the effects of changing the entanglement cutoff parameter χ on the errors in the magnetization.

A.3.4.1 Preparing and Running the Simulation Using Python

To set up and run the simulation in Python we use the script tutorial2c.py. The first parts of this script import the required modules and prepare the input files as a list of Python dictionaries:

```
import pyalps
import matplotlib.pyplot as plt
import pyalps.plot
import math
import scipy.special

#prepare the input parameters
parms=[]
count=0
for chi in [10, 20, 30, 40]:
        count+=1
        parms.append({
                    'L'                          : 50,
                    'MODEL'                      : 'spin',
                    'local_S'                    : 0.5,
                    'CONSERVED_QUANTUMNUMBERS'   : 'Sz',
                    'Jxy'                        : 1,
                    'INITIAL_STATE' : 'kink',
                    'CHI_LIMIT' : chi,
                    'TRUNC_LIMIT' : 1E-12,
                    'NUM_THREADS' : 1,
                    'TAUS' : [20.0],
                    'POWS' : [0.0],
                    'GS' : ['H'],
                    'GIS' : [0.0],
                    'GFS' : [0.0],
                    'NUMSTEPS' : [500],
                    'STEPSFORSTORE' : [5],
                    'SIMID': count
                })
```

We now write the input files, run the simulations, collect data, and compute the errors as above

```
baseName='tutorial_2c_'
nmlnameList=pyalps.writeTEBDfiles(parms, baseName)
res=pyalps.runTEBD(nmlnameList)

#Get magnetization data
Magdata=pyalps.load.loadTimeEvolution( pyalps.
    getResultFiles(prefix='tutorial_2c'), measurements
    =['Local␣Magnetization'])

#Postprocessing-get the exact result for comparison
for q in Magdata:
        syssize=q[0].props['L']
```

```
#Get the exact result of M(1,t)=-(1/2)*(j_0(t)
    ^2), where j_0(t) is the 0^{th} order
# bessel function and M(1,t) is the
    magnetization one site to the right of the
    chain center
loc=-0.5*scipy.special.jn(0,q[0].props['Time'
    ])*scipy.special.jn(0,q[0].props['Time'])
#Get the difference between the computed and
    exact results
q[0].y=[abs(q[0].y[syssize/2+1-1]-loc)]
```

Finally, we plot the magnetization error

```
#Plot the Error in the magnetization one site to the
   right of the chain center
Mag=pyalps.collectXY(Magdata, x='Time', y='Local␣
   Magnetization', foreach=['SIMID'])
for q in Mag:
        q.props['label']='$\chi$='+str(q.props['
           CHI_LIMIT'])
plt.figure()
pyalps.plot.plot(Mag)
plt.xlabel('Time␣$t$')
plt.yscale('log')
plt.ylabel('Magnetization␣Error')
plt.title('Error␣in␣the␣magnetization␣vs.␣time')
plt.legend(loc='lower␣left')
plt.show()
```

We see that, for short times, the errors are roughly proportional to dt^2, again reflecting the contribution to the error from the trotter breakup of our exponential. As time increases, however, a cascade of diverging errors ensues. First the simulation with $\chi = 10$ diverges around $t = 5$, then the simulation with $\chi = 20$ diverges around $t = 9$ and so on. This breakdown is due to the fact that the protocol for finding the MPS which best approximates the time-evolved state is approximate when the state becomes highly entangled. This approximation involves a renormalization of the wavefunction, and so the errors accumulate roughly exponentially in time.

This exponential growth of errors also accounts for the failure of the simulations with smaller dt. As dt becomes smaller we must apply the approximate propagation scheme more to reach the same fixed final time, and this means more accumulation of the exponentially growing truncation error. Thus, we must strike a delicate balance between the error incurred by increasing the time step and the error incurred by taking more time steps. All results should be carefully checked for convergence in both dt and χ.

A.3.4.2 Preparing and Running the Simulation Using Vistrails

To run the simulation in Vistrails open the file tutorial2c.vt and look at the workflow labeled "tutorial2c." Click on "Execute" to prepare the input file, run the simulation and create the output figure.

A.3.5 Solution in the Case of the XXZ Model

We saw from the exact solution that the magnetization profile had a well-defined front which expanded ballistically with velocity $v = 1$. The XX model has many special properties and so it is natural to ask if this same magnetization behavior holds under more general conditions. In this part of the tutorial we investigate the effects of adding a $J_z S_i^z S_{i+1}^z$ term to the Hamiltonian, corresponding to the XXZ model. In the limit as this term dominates the spins become frozen in a parallel configuration, and so the initial state becomes an exact eigenstate of the Hamiltonian. The XX terms in the Hamiltonian try to flip the spins, and are responsible for the propagating magnetization wavefront we saw in the pure XX model. As a quantitative measure of the ability of the system to transport spin, we consider the integrated flow of magnetization through the center defined in Phys. Rev. E **71**, 036102 (2005) as

$$\Delta M(t) = \sum_{n>L/2}^{L} (\langle S_n^z(t)\rangle + 1/2).$$

A.3.5.1 Preparing and Running the Simulation Using Python

To set up and run the simulation in Python we use the script tutorial2d.py. The first parts of this script import the required modules and prepare the input files as a list of Python dictionaries:

```
import pyalps
import matplotlib.pyplot as plt
import pyalps.plot
import math
import scipy.special

#prepare the input parameters
parms=[]
count=0
for z in [0.0, 0.3, 0.9, 1.0, 1.1, 1.5]:
        count+=1
        parms.append({
```

```
        'L'                          : 50,
        'MODEL'                      : 'spin',
        'local_S'                    : 0.5,
        'CONSERVED_QUANTUMNUMBERS'   : 'Sz',
        'Jxy'                          : 1,
        'Jz'                            : z,
        'INITIAL_STATE' : 'kink',
        'CHI_LIMIT' : 40,
        'TRUNC_LIMIT' : 1E-12,
        'NUM_THREADS' : 1,
        'TAUS' : [20.0],
        'POWS' : [0.0],
        'GS' : ['H'],
        'GIS' : [0.0],
        'GFS' : [0.0],
        'NUMSTEPS' : [500],
        'STEPSFORSTORE' : [5],
        'SIMID': count
    })
```

Note that we are simulating a range of Jz-couplings. We then write the input files, run the simulation, and get the output as usual:

```
baseName='tutorial_2d'
nmlnameList=pyalps.writeTEBDfiles(parms, baseName)
res=pyalps.runTEBD(nmlnameList)

#Get magnetization data
Magdata=pyalps.load.loadTimeEvolution( pyalps.
    getResultFiles(prefix='tutorial_2d'), measurements
    =['Local␣Magnetization'])
```

From the computed magnetization data we calculate the integrated magnetization as defined above:

```
#Compute the integrated magnetization across the
    center
for q in Magdata:
        syssize=q[0].props['L']
        #Compute the integrated flow of magnetization
            through the center \Delta M=\sum_{n>L/2}^{L}
            (<S_n^z(t)>+1/2)
        #\Delta M= L/4
        loc=0.5*(syssize/2)
        #\Delta M-=<S_n^z(t)> from n=L/2 to L
        q[0].y=[0.5*(syssize/2)+sum(q[0].y[syssize/2:
            syssize])]
```

Finally, we plot the integrated magnetization for the range of Jz couplings simulated.

```
#Plot the integrated magnetization
Mag=pyalps.collectXY(Magdata, x='Time', y='Local␣
    Magnetization', foreach=['Jz'])

plt.figure()
pyalps.plot.plot(Mag)
plt.xlabel('Time␣$t$')
plt.ylabel('Integrated␣Magnetization␣$\Delta␣M(t)$')
plt.title('Integrated␣Magnetization␣vs.␣Time')
plt.legend(loc='upper␣left')
plt.show()
```

We see that for Jz < 1 the integrated magnetization increases roughly linearly in time, and so the magnetization transport is ballistic as in the XX case. For Jz around 1, we see a change in the qualitative behavior to one in which the integrated magnetization eventually saturates.

A.3.5.2 Preparing and Running the Simulation Using Vistrails

To run the simulation in Vistrails open the file tutorial2d.vt and look at the workflow labeled "tutorial2d." Click on "Execute" to prepare the input file, run the simulation and create the output figure.

A.3.5.3 Questions

- The point Jz $= 1$ where the behavior of the integrated magnetization undergoes a distinct qualitative change is the point at which the XXZ model transitions from a critical phase to the Antiferromagnetic phase. However, this phase transition is a priori a low-energy phenomenon, affecting the ground state. Can you deduce how this low energy change affects the dynamical properties of our high-energy initial state?

Appendix B
Educational Materials: A Gentle Introduction to Time Evolving Block Decimation

B.1 Pre-test

Before we begin, I'm assuming that you have finished the "Introduction to the Bose–Hubbard Model and Fock State Basis" problem set.[1] In particular, it is important that you have finished Problem 6 and have a working code which will compute the ground state of an N particle, L site Bose–Hubbard system with local dimension d. The below exercises will test your familiarity with the Bose–Hubbard Hamiltonian and the Fock state basis. If you can't complete them with relative ease, you may want to review the material presented in the other problem set before proceeding.

B.1.1 Exercise 1

Consider a system with three sites occupied by bosons. The Fock space is spanned by the states $|n_1 n_2 n_3\rangle$. What is the overlap of the states $|110\rangle$ and $|010\rangle$ (that is, what is $\langle 110|010\rangle$)? What are $\langle 110|\hat{b}_1|010\rangle$ and $\langle 110|\hat{b}_1^{\dagger}|010\rangle$?

B.1.2 Exercise 2

Now consider the total number operator defined by $\hat{N} \equiv \sum_{i=1}^{3} \hat{n}_i$, where $\hat{n}_i = \hat{b}_i^{\dagger}\hat{b}_i$. What is the action of $\hat{N}$ on the states $|\psi_1\rangle = |231\rangle$, $|\psi_2\rangle = \frac{1}{\sqrt{5}}|022\rangle +$

[1]Note added for thesis version: The only preliminary assumed here is an exact diagonalization code for the Bose-Hubbard model and familiarity with second quantization.

M.L. Wall, *Quantum Many-Body Physics of Ultracold Molecules in Optical Lattices*, Springer Theses, DOI 10.1007/978-3-319-14252-4

$\sqrt{\frac{4}{5}}|101\rangle$, and $|\psi_3\rangle = \sqrt{\frac{1}{3}}|112\rangle + \sqrt{\frac{2}{3}}|022\rangle$. Which of these states (if any) are total number eigenstates such that $\hat{N}|\psi\rangle = N|\psi\rangle$?

B.1.3 Exercise 3

In the Bose–Hubbard Hamiltonian, the tunneling term is

$$-t\sum_{\langle i,j\rangle}\left(\hat{b}_i^\dagger \hat{b}_j + \hat{b}_i \hat{b}_j^\dagger\right) . \tag{B.1}$$

What does the second term in parentheses represent physically? Why is it important mathematically?

B.2 The Limitations of Exact Diagonalization

In the previous problem set you wrote a code to explicitly construct the Bose–Hubbard Hamiltonian, diagonalize it, and thus find the ground state energy and eigenvector. The process of finding the ground state of a many-body hamiltonian in terms of a fixed Fock basis is referred to as *exact diagonalization* (ED). ED is very powerful in that it gives us maximal information about the energy spectrum and eigenstates of the Hamiltonian, and also has the benefit of being very simple to understand and to program. As we shall see, ED is also very limited in the sizes of the systems it can handle.

B.2.1 Exercise 4: The Limitations of Exact Diagonalization

Using your ED code, find the largest Bose–Hubbard system your computer can handle for non-number conserving code before running out of memory. Repeat for the number conserving case.

From this previous exercise, we gather two things. The first is that ED is not going to be of much use in studying systems of more than a few sites. The second is that judicious exploitation of symmetries and their associated conservation laws, such as the conservation of total number in the Bose Hubbard model, drastically improves the performance of many-body simulations.

Time-evolving block decimation (TEBD) is a powerful method to simulate many-body systems such as the Bose–Hubbard model whose computation time and memory usage scale polynomially in the system size (as opposed to exponentially in the system size, as in ED). It is also, however, much more difficult to conceptually understand and to program. In the next few subsections I will provide a crash course in what TEBD is and how it works. Those interested in learning only how to *use* the open source TEBD package should skip to Sect. B.6.

B.3 Preliminaries

B.3.1 Mathematical Preliminaries

As is often the case, before we get to the physics we have to learn some math. In this subsection I present two theorems, the singular value decomposition and the Schmidt decomposition.

Theorem (Singular Value Decomposition). *Let A be an $m \times n$ matrix. Then there exists an $m \times m$ unitary matrix Γ_A, an $n \times n$ unitary matrix Γ_B, and an $m \times n$ positive diagonal (as defined for rectangular matrices) matrix λ such that*

$$A = \Gamma_A \lambda \Gamma_B^T , \tag{B.2}$$

or, stated element-wise,

$$A_{ij} = \sum_{k=1}^{\min(m,n)} [\Gamma_A]_{ik} \lambda_{kk} \left[\Gamma_B^T\right]_{kj} . \tag{B.3}$$

This is referred to as the *singular value decomposition* of A, and the diagonal elements of λ are referred to as the *singular values* of A. Note that the form of the decomposition implies that A has at least one and at most $\min(m,n)$ distinct singular values. The number of nonzero singular values of a matrix is its *rank*.

The most important property of the SVD for our purposes is that the matrix $A^{(l)}$ defined by the matrix elements

$$A_{ij}^{(l)} = \sum_{k=1}^{l} [\Gamma_A]_{ik} \lambda_{kk} \left[\Gamma_B^T\right]_{kj} \tag{B.4}$$

is the closest rank-l matrix to A, meaning that $A^{(l)}$ minimizes the *Frobenius norm*: the sum of the absolute squares of the element-wise difference between the rank-l approximation and the full matrix, $\sum_{ij} \left| A_{ij} - A_{ij}^{(l)} \right|^2$. Thus, the SVD gives us a means to find the best lower-rank approximation to a matrix. The reduction in rank is important because matrices of lower rank require much less storage and allow for more efficient operations. Explicitly, a rank-l approximation to an $m \times n$ matrix is comprised of $ml + nl + l$ numbers whereas the full rank matrix is comprised of mn numbers. For matrices with ranks small compared to their size, $l \ll m, n$, the reduction in storage is enormous.

We saw in the previous paragraph that a reduced rank approximation to a matrix requires less storage than the full rank matrix. Here we also explore how reducing the rank of a matrix improves the *efficiency* of operations performed with it. Consider multiplying an $n \times n$ matrix A to a vector v. The resulting vector Av can be written in indicial notation as

$$[Av]_i = \sum_{k=1}^{n} A_{ik} v_k \,. \tag{B.5}$$

The formation of each element of Av requires n multiplications, and there are n elements of Av, so the formation of the full vector Av requires $\mathcal{O}(n^2)$ operations,[2] where $\mathcal{O}(a)$ is read "of order a."[3] Now consider taking the singular value decomposition of A, and forming a rank-l approximation $A^{(l)}$:

$$A_{ij}^{(l)} = \sum_{k=1}^{l} [\Gamma_A]_{ik} \lambda_{kk} \left[\Gamma_B^T\right]_{kj} \,. \tag{B.6}$$

How many operations are required to form $A^{(l)}v$? Naively, we would write

$$\left[A^{(l)} v\right]_{ij} = \sum_{k=1}^{l} \sum_{j=1}^{n} [\Gamma_A]_{ik} \lambda_{kk} \left[\Gamma_B^T\right]_{kj} v_j \,, \tag{B.7}$$

which seems to involve $\mathcal{O}(n^2 l)$ operations, *increasing* the number of operations by a factor of l! We can be more clever, however, and first multiply v by Γ_B^T to get a vector q

$$q_i = \sum_{k=1}^{n} \left[\Gamma_B^T\right]_{ik} v_k \,. \tag{B.8}$$

Note that q has l elements and this multiplication involves $\mathcal{O}(nl)$ operations. We now multiply each element of q by an element of λ to get

$$q_i = \lambda_i q_i \tag{B.9}$$

which involves $\mathcal{O}(l)$ operations. Finally, we multiply q by Γ_A to get

$$\left[A^{(l)} v\right]_i = \sum_{j=1}^{l} [\Gamma_A]_{ij} q_j \,, \tag{B.10}$$

---

[2] By operation we mean an elementary operation such as an addition, multiplication, subtraction, or division. In computer science language, where we deal with floating point representations of numbers, such an operation is called a floating-point operation, or FLOP.

[3] For more information on this notation, see the wikipedia page for Big-O notation.

which again involves $\mathcal{O}(nl)$ operations. Thus, our total operation count is $nl+l+nl$, which is $\mathcal{O}(nl)$! For $l \ll n$, the reduction in the number of operations is again enormous.

B.3.1.1 Exercise 5: General Matrix–Vector Multiply Using the SVD

How many operations does it take to multiply an $m \times n$ matrix A to an n dimensional vector v both with and without the singular value decomposition?

B.3.1.2 Exercise 6: Simple Examples of the SVD

Compute the singular value decompositions of the matrices

$$c_1 = \begin{pmatrix} 1 & 0 \\ 0 & 0 \end{pmatrix}, \quad c_2 = \begin{pmatrix} 0 & 1/\sqrt{2} \\ -1/\sqrt{2} & 0 \end{pmatrix}, \quad c_3 = \begin{pmatrix} 1/2 & 1/2 \\ 1/2 & 1/2 \end{pmatrix} \tag{B.11}$$

numerically. The simplest way to do so is using *Matlab* or *Mathematica*, where the commands are `[U,S,V]=svd(A)` and `{U,S,V}=SingularValueDecomposition[A]`,[4] respectively. Those who wish to use Fortran or C++ can use the lapack routine `dgesvd`. What are the ranks of these matrices? Could you have guessed their ranks simply by looking at them?

The singular value decomposition is a general mathematical statement about a way to rewrite any arbitrary matrix. The next theorem, the Schmidt decomposition, gives a more physical understanding of what the singular value decomposition *means* in the context of a tensor product space.

Theorem (Schmidt Decomposition). *Let $|\psi\rangle$ be a state vector in the $d_A d_B$ dimensional Hilbert space $\mathcal{H}_A \otimes \mathcal{H}_B$.*[5] *Then there exist vectors $\{|\phi_\alpha^A\rangle\}$ and $\{|\phi_\alpha^B\rangle\}$ and scalars χ_S and $\{\lambda_\alpha\}$ such that*

$$|\psi\rangle = \sum_{\alpha=1}^{\chi_S} \lambda_\alpha |\phi_\alpha^A\rangle \otimes |\phi_\alpha^B\rangle\,, \tag{B.12}$$

$$1 \le \chi_S \le \min(d_A, d_B)\,, \tag{B.13}$$

$$\lambda_1 \ge \lambda_2 \ge \cdots \ge \lambda_{\chi_S} > 0\,, \tag{B.14}$$

$$\sum_\alpha \lambda_\alpha^2 = 1\,. \tag{B.15}$$

---

[4] Note in *Mathematica* that it returns U, S, and V such that $A = USV^T$

[5] To be concrete, you could consider $\mathcal{H}_A$ to be the Hilbert space of the first site and $\mathcal{H}_B$ to be the Hilbert space of the second site of a two-site system. Or you could consider $\mathcal{H}_A$ to represent the internal states of some particle and $\mathcal{H}_B$ to represent the internal states of another particle.

This is referred to as the *Schmidt decomposition* of $|\psi\rangle$. χ_S is referred to as the *Schmidt rank*, and the $\{\lambda_\alpha\}$ are referred to as the *Schmidt coefficients*. We shall prove this theorem to elucidate the connection between the Schmidt rank and the singular value decomposition.

Proof. Let $\{|j_A\rangle\}$ and $\{|n_B\rangle\}$ be two orthonormal bases of $\mathcal{H}_A$ and $\mathcal{H}_B$ of dimension d_A and d_B, respectively. The most general decomposition of $|\psi\rangle$ in this basis is

$$|\psi\rangle = \sum_{j=1}^{d_A}\sum_{n=1}^{d_B} c_{jn}|j_A\rangle \otimes |n_B\rangle . \tag{B.16}$$

Applying the singular value decomposition to the coefficient matrix C such that

$$c_{jn} = \sum_{i=1}^{\min(d_A,d_B)} [\Gamma_A]_{ji}\, \lambda_{ii} \left[\Gamma_B^T\right]_{in} , \tag{B.17}$$

we have

$$|\psi\rangle = \sum_{j=1}^{d_A} \sum_{i=1}^{\min(d_A,d_B)} \sum_{n=1}^{d_B} [\Gamma_A]_{ji}\, \lambda_{ii} \left[\Gamma_B^T\right]_{in} |j_A\rangle \otimes |n_B\rangle . \tag{B.18}$$

If we now define

$$|\phi_i^A\rangle \equiv \sum_{j=1}^{d_A} [\Gamma_A]_{ji}\, |j_A\rangle , \quad |\phi_i^B\rangle \equiv \sum_{n=1}^{d_B} \left[\Gamma_B^T\right]_{in} |n_B\rangle , \quad \lambda_i \equiv \lambda_{ii} , \tag{B.19}$$

we have

$$|\psi\rangle = \sum_{i=1}^{\chi_S} \lambda_i |\phi_i^A\rangle \otimes |\phi_i^B\rangle , \tag{B.20}$$

as was to be proven. We can identify χ_S generally as the number of nonzero singular values of the decomposition matrix, with $1 \le \chi_S \le \min(d_A, d_B)$. We also note that the Schmidt decomposition may be easily performed numerically using the singular value decomposition of the coefficient matrix C.

What has the Schmidt decomposition done for us? The Schmidt decomposition of $|\psi\rangle$ provides us with unique orthonormal bases for the two subsystems A and B such that $|\psi\rangle$ can be written as a superposition of tensor products with the least possible number of terms. In addition, the sizes of the Schmidt coefficients λ_i give the "importance" of the particular state $|\phi_i^A\rangle \otimes |\phi_i^B\rangle$ in the representation of $|\psi\rangle$, and the Schmidt rank tells us how difficult it is to write $|\psi\rangle$ as a tensor product, loosely speaking.

These statements can be made more precise and physically meaningful by introducing the quantum mechanical idea of *entanglement*, as is done in the next subsection.

B.3.2 Entanglement

We again consider a system with two parts, which we refer to as a *bipartite* system. A general ket $|\psi\rangle$ in this system can be written as

$$|\psi\rangle = \sum_{n,m} c_{nm}|n\rangle \otimes |m\rangle, \tag{B.21}$$

where the $\{|n\rangle\}$ and $\{|m\rangle\}$ form complete orthonormal bases for the first and second parts, respectively. We see that this is not, in general, a tensor product $|\phi_1\rangle \otimes |\phi_2\rangle$ of kets in the subspaces. States in the composite space which are not tensor products of kets from the subspaces are said to be *entangled*. For example, if we examine a two-qubit[6] system, a general ket can be written

$$|\psi\rangle = c_{11}|+\rangle \otimes |+\rangle + c_{12}|+\rangle \otimes |-\rangle + c_{21}|-\rangle \otimes |+\rangle + c_{22}|-\rangle \otimes |-\rangle \tag{B.22}$$

whereas general normalized kets of the subsystems can be written

$$|\phi_1\rangle = a_1|+\rangle + b1_1|-\rangle \tag{B.23}$$

$$|\phi_2\rangle = a_2|+\rangle + b_2|-\rangle \tag{B.24}$$

$$\Rightarrow |\phi_1\rangle \otimes |\phi_2\rangle = a_1a_2|+\rangle \otimes |+\rangle + a_1b_2|+\rangle \otimes |-\rangle + b_1a_2|-\rangle \otimes |+\rangle + b_1b_2|-\rangle \otimes |-\rangle \tag{B.25}$$

$|\psi\rangle = |\phi_1\rangle \otimes |\phi_2\rangle$ iff $c_{11}c_{22} = c_{12}c_{21}$, which is certainly not the general case.

We saw above that the Schmidt rank gave us a measure of how difficult it was to accurately represent our state as a tensor product. From the above considerations, we see that states with a higher Schmidt rank are *more entangled*. By truncating at a fixed Schmidt rank, we find the best approximation to our state in a Hilbert space that truncates the amount of entanglement allowed. This statement is very key, so I'm going to write it out in bold so that you remember it:

> By performing the Schmidt decomposition on the state of a bipartite system and truncating the Schmidt rank, we find the best approximation to the state in a Hilbert space with restricted entanglement.

The efficient approximation of a state in a space with restricted entanglement is the key idea of TEBD. In the next subsection we will compare the above mathematical ideas with the more tangible idea of image compression.

---

[6] A *qubit* is a general two-state system, for example the internal states of a spin-1/2 particle. Here $|+\rangle$ represents one of the two states and $|-\rangle$ the other.

B.3.2.1 Exercise 7: Studying Entanglement via the Schmidt Decomposition

We can now give a physical interpretation to the purely mathematical Exercise 6. Write down the coefficient matrices of the states

$$|\psi_1\rangle = |+\rangle \otimes |+\rangle \tag{B.26}$$

$$|\psi_2\rangle = \frac{1}{\sqrt{2}}\left(|+\rangle \otimes |-\rangle - |-\rangle \otimes |+\rangle\right) \tag{B.27}$$

$$|\psi_3\rangle = \frac{1}{2}\left[|+\rangle \otimes |+\rangle + |+\rangle \otimes |-\rangle + |-\rangle \otimes |+\rangle + |-\rangle \otimes |-\rangle\right] \tag{B.28}$$

$$\tag{B.29}$$

in the form

$$C = \begin{pmatrix} c_{11} & c_{12} \\ c_{21} & c_{22} \end{pmatrix}, \tag{B.30}$$

where c_{ij} is given as in Eq. (B.22). You should recover the matrices of Exercise 6. Using their singular value decompositions, find the Schmidt ranks. Which of these states are entangled according to the Schmidt rank? For those states that are not entangled, can you find a way to write them as a tensor product? How might you use the singular value decomposition to accomplish this?

B.4 Conceptual Basis of TEBD

The essential fact that allows for the efficiency and success of TEBD is that the singular values of reduced density matrices,[7] when arranged in non-increasing fashion, have an approximately exponential decay. This means that a rank-χ approximation to the reduced density matrix, $\hat{\rho}^{(\chi)}$, formed from the singular value decomposition as in Eq. (B.4) with $\chi \ll \chi_S$ will provide an excellent approximation to the true reduced density matrix. We can understand this fact using an analogy to image compression.

Consider the JPEG image shown in Fig. B.1a. We can represent this picture as a 300×416 array of pixels, requiring 124,800 words of storage. If we look at the singular values of this pixel array, we find the distribution shown in Fig. B.1b, namely that the singular values decay roughly exponentially. This implies that the

[7]The *reduced density matrix* is a general way of dealing with subsystems of quantum mechanical systems. For our purposes, it takes the place of the coefficient tensor when the system has more than two parts (e.g., more than two lattice sites).

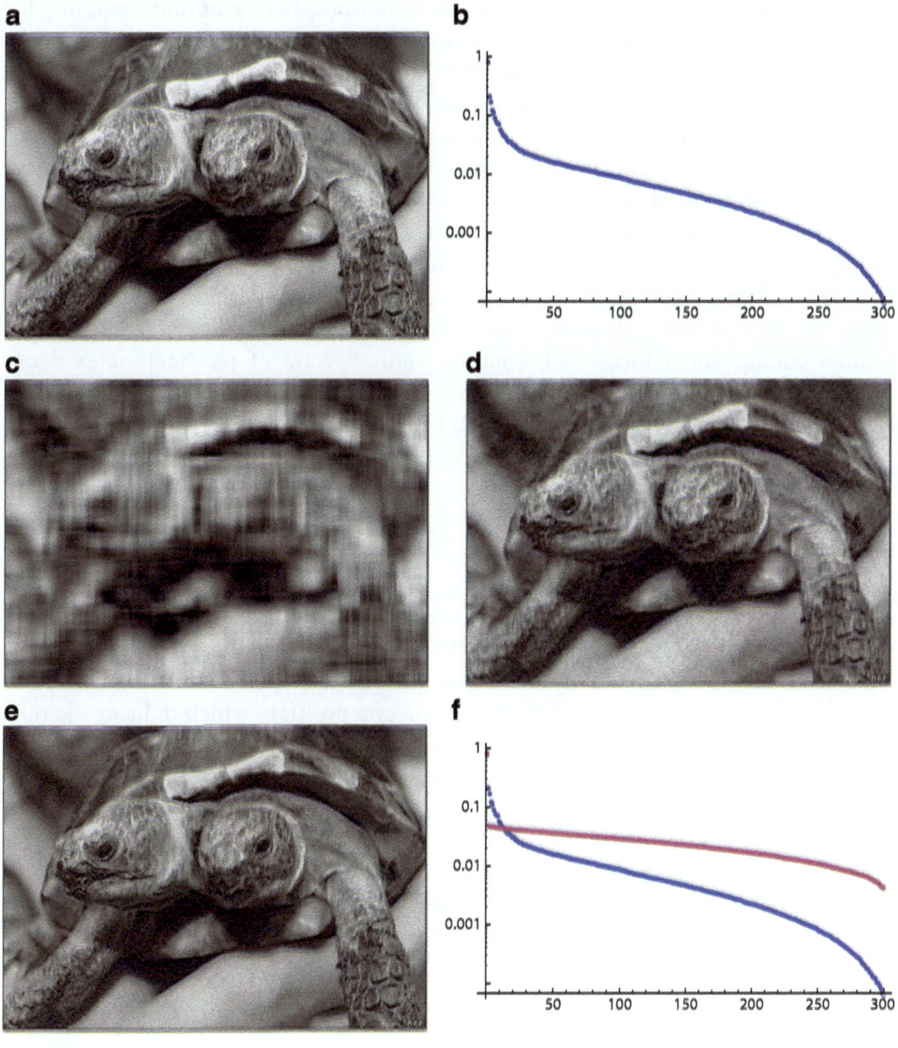

Fig. B.1 SVD representation of an image. (**a**) Original image; (**b**) log plot of singular values; (**c**) 10 singular values; (**d**) 50 singular values; (**e**) 100 singular values; (**f**) log plot of singular values: general pixel array (*pink*), physical pixel array (*blue*)

best approximation to the image given by the SVD will be excellent even if we keep only a fraction of the total singular values. To see this in action, examine Fig. B.1c–e, which show the best approximation to the image for various numbers of singular values. Keeping 100 singular values gives an excellent approximation to the original image with only a quarter of the data storage. If we consider that the original image was a JPEG which itself was compressed down from a $\sim 3000 \times 3000$ pixel

(10 Megapixel) raw array, we have benefitted enormously by carefully sampling the parts of "image space" that are the most important via the SVD.

Why were we able to represent the image with so little data? The reason is that a physical images are a very special subset of all 2D pixel arrays; they have a great deal of structure and regularity. If we consider instead the most probable image of the same size as our original image—one consisting of random pixel values—we get the singular value scaling shown in Fig. B.1f. In this case the singular values are all roughly the same, and our SVD compression would yield a miserable approximation to the original image if we were to use only a quarter of the singular values. In the same manner typical physical states in Hilbert space, including the ground state, have limited entanglement as quantified by *some* entanglement measure. This limited entanglement means the singular value spectra of physical states decay exponentially more quickly than a general state, enabling them to be "entanglement compressed" by algorithms such as TEBD.

B.5 TEBD in Summary

Before going on to use TEBD, let's pause to reiterate what TEBD is and why it works where ED fails. TEBD is a method which gives the best approximation to the state of a many-body system by truncating the amount of entanglement allowed in the state. This is done by the singular value decomposition, which reduces the rank of the matrices carrying the state information. We have seen that reducing the rank of a matrix allows it to be stored more compactly, and operations to be done more efficiently. In the case of TEBD, the rank of the matrices we keep is *exponentially smaller* than the size of the matrices they approximate. The exponential decrease in storage and exponential increase in efficiency of operations are what allow us to use TEBD where ED fails. For all of the gory details on TEBD, you can consult the User's guide for the open source package, located on the downloads page.

B.6 Using OSTEBD

To begin, download the most recent version of the open source TEBD package from the downloads page (unless you have a more recent version from another place). After decompressing the `.tar.gz` file, go into the `Case_Studies` directory, and then into `Bose_Hubbard_Wrapper`. Open a terminal in this directory. Type `make ITP`. If you see `Execute_ITP` in this folder after a while, the case study has compiled. Otherwise, it didn't compile for some reason. You should be sure that you have gfortran as well as lapack installed and try again.

Once you have compiled successfully, open the file `BoseHubbard_ITP.nml` This is the input file, which sets the parameters for a specific TEBD job. You will see three lines, denoted by `SystemSettings`, `BHParams`, and `ITPParams`. The parameters in `SystemSettings` define the properties of the system as a whole. Specifically, they are

1. `systemSize`, the number of lattice sites.
2. `maxFilling`, the maximum number of bosons allowed per lattice site.
3. `totNum`, the total number of bosons for number conserving code. If `totNum`=0, the non-number conserving code is used.
4. `BoundaryCond`, the boundary conditions used. It can be either "O" for open boundary conditions or "P" for periodic boundary conditions.
5. `trotterOrder`, the order of the trotter expansion of the propagator. It can be either 2 or 5.

In very recent versions you may also have `numThr`, which specifies the number of threads used for OpenMP parallelization. Unless you know what you are doing, set this to be 1. The parameters in `BHParams` represent the parameters in the Bose–Hubbard model. Specifically they are

1. `jTunn`, the tunneling energy t.
2. `U0`, the on-site interaction energy U.
3. `V0`, the nearest-neighbor interaction energy V.
4. `mu0`, the chemical potential μ. Note that the chemical potential sets the total number when the non-number conserving code is used, but only gives an overall (unimportant) energy shift when the number conserving code is used.

The parameters contained in `ITPParams` determine the convergence and output properties of the TEBD run. Specifically, they are

1. `chiMin`, the minimum value of the entanglement cutoff parameter χ used to calculate the ground state.
2. `chiMax`, the maximum value of the entanglement cutoff parameter χ used to calculate the ground state.
3. `convCriterion1`, the convergence criterion for the first ITP.[8] attempt.
4. `convCriterion2`, the convergence criterion for the first ITP attempt.
5. `stepsForJudge` determines how many ITP steps occur before convergence is checked for.
6. `dtITP`, the "infinitesimal" time-step used for ITP.
7. `maxITPsteps`, the maximum number of ITP steps allowed.
8. `itpDir`, the directory where the output will be written. Note that this directory must exist when you run the program or it will exit!

After setting the parameters you want in the `BoseHubbard_ITP.nml` file, type `./Exectute_ITP` to run the TEBD program.

[8] ITP is short for *imaginary time propagation*. Imaginary time propagation replaces $t = -it$ in the Schrödinger equation, which turns it into a diffusion equation. The highest energy eigenmodes decay the most quickly during imaginary time evolution, and so in the limit of long times imaginary time propagation gives us the ground state.

B.6.1 Exercise 8: Using TEBD: Non-number Conserving Version

Set `systemSize=4`, `totNum=0`, `BoundaryCond='O'`, `trotterOrder=2`, `chiMin=chiMax=10`, and `maxFilling` to 3. Then set `jTunn=1.12`, `U0=0.72`, `V0=0.0`, and `mu0=0.5`. Run the code as described above, and note the total number and total energy. Using your ED code, calculate the ground state energy and total number of particles for the same parameters and without number conservation and compare them to the TEBD predictions.

B.6.2 Exercise 9: Using TEBD: Number Conserving Version

Set `systemSize=4`, `totNum=4`, `BoundaryCond='O'`, `trotterOrder=2`, `chiMin=chiMax=10`, and `maxFilling` to 3. Then set `jTunn=1.12`, `U0=0.72`, `V0=0.0`, and `mu0=0.0`. Run the code as described above, and note the total energy. Using your ED code, calculate the ground state energy with number conservation and compare it to the TEBD prediction. Also note the large increase in efficiency of the number conserving codes vs. non-number conserving codes.

I hope the last two exercises give you some confidence that the OSTEBD package works. However, TEBD is hardly useful for four sites where ED still works and is likely faster. In the next exercise we will study the analytically solvable *tight-binding* limit of the Bose–Hubbard model using TEBD for large numbers of sites where ED fails.

B.6.3 Exercise 10: The Tight-Binding Chain

Consider the Bose–Hubbard model in the limit $U \to 0$, $V \to 0$ and a fixed number of particles:

$$\hat{H} = -t \sum_i \left(\hat{b}_i^\dagger \hat{b}_{i+1} + \hat{b}_i \hat{b}_{i+1}^\dagger \right) . \tag{B.31}$$

This Hamiltonian, known as the *tight-binding* Hamiltonian, represents noninteracting bosons which are free to hop between sites on a lattice. It is the many-body generalization of the particle in an infinite well problem from single-particle quantum mechanics, and can be solved similarly. We can write the kth eigenstate for a single particle as

$$|\psi_k\rangle = \sum_{i=1}^{L} \psi_{ik} \hat{b}_i^\dagger |0\ldots0\rangle, \tag{B.32}$$

where the ψ_{ik} are complex scalars to be determined and $|0\ldots0\rangle$ is the Fock state with no particles in any site (the Fock vacuum). Inserting this solution into the Schrödinger equation $\hat{H}|\psi\rangle = E|\psi\rangle$, we have the recursion relation

$$-t\,(\psi_{i-1,k} + \psi_{i+1,k}) = E_k \psi_{ik} \tag{B.33}$$

together with appropriate boundary conditions. For open boundary conditions, we have

$$\psi_{0k} = 0\,,\ \psi_{L+1,k} = 0 \tag{B.34}$$

and for periodic boundary conditions we have

$$\psi_{1k} = \psi_{L+1,k}\,. \tag{B.35}$$

Let us now define the ansatz

$$\psi_{jk} = \begin{cases} Ae^{ikj} + Be^{-ikj} & \text{OBC} \\ Ae^{ikj} & \text{PBC} \end{cases}\,. \tag{B.36}$$

Inserting this ansatz into the above recursion relation gives a set of algebraic equations for the ψ_{ik} and the energy E_k. Solve these equations and determine the single-particle eigenfunctions and energies for both open and periodic boundary conditions. Then, normalize these solutions such that $\langle\psi_k|\psi_k\rangle = 1$.

We note that the operator

$$\hat{s}_k^\dagger = \sum_{i=1}^{L} \psi_{ik} \hat{b}_i^\dagger \tag{B.37}$$

creates an eigenstate of the tight-binding Hamiltonian with quantum number k. Since the particles do not interact, the solution for N particles with quantum numbers $k_1, k_2 \ldots k_N$ is simply

$$|\psi\rangle = \mathcal{N} \hat{s}_{k_N}^\dagger \ldots \hat{s}_{k_2}^\dagger \hat{s}_{k_1}^\dagger |0\ldots0\rangle\,, \tag{B.38}$$

where $|0\ldots0\rangle$ is again the Fock vacuum and

$$\mathcal{N} = \frac{1}{\sqrt{\prod_k n_k!}} \tag{B.39}$$

is a normalization factor accounting for the indistinguishability of bosons. The energy of this state is $E = E_{k_1} + E_{k_2} + \cdots + E_{k_N}$. This completes the solution of the N-particle tight-binding Hamiltonian. For bosons, the lowest energy state is the one where all particles are in the lowest energy single-particle eigenstate. Using the above results, find what the expectation value of the number operator in the ground state is at each site for both open and periodic boundary conditions.[9] As a slight generalization, also compute the *single-particle density matrix* (SPDM) defined by

$$\rho_{ij} = \langle \mathrm{g.s.}|\hat{b}_i^\dagger \hat{b}_j|\mathrm{g.s.}\rangle \,. \tag{B.40}$$

Using some system size which is much too large for ED (say, 25), study the tight-binding Hamiltonian using TEBD for both open and periodic boundary conditions. In particular, compare the results for the energy and the on-site number expectation values. To compare with your predictions for the on-site number expectation values, look in the directory `ITPDATA` for the file ending in `localmeasures.dat`. The full filename will be in the form `BH_L$N@ChiXjTunnP.PPUQQ.QQ*BClocalmeasures`, where `$` is the number of lattice sites, `@` the number of particles, `X` the chi value and so on. On the first line of this file are the on-site number expectation values, listed in order of increasing site index.

B.6.4 Exercise 11: Fidelity Susceptibility and the Superfluid-Mott Insulator Quantum Phase Transition

The previous exercise allowed us to go beyond where ED could, but the results were still amenable to analytic computation. Here we go beyond the reach of analytic techniques to study a strongly interacting Bose–Hubbard system, and investigate its *quantum phase transition*. By a quantum phase transition we mean a transition between two different states of matter (characterized by different symmetries) which is driven by quantum fluctuations at zero temperature.[10] Specifically, in the Bose–Hubbard model the quantum phase transition occurs because of fluctuations in the on-site number as we change the ratio t/U of the hopping strength to the interaction.

[9]There's an easy way and a hard way to do this. The easy way involves representing the number operator $\hat{n}_i = \hat{b}_i^\dagger \hat{b}_i$ in terms of the $\hat{s}_k$ using Eq. (B.37) and then taking the expectation value using Eq. (B.38). The hard (but fun) way is to expand the solution Eq. (B.38) in terms of $\hat{b}_i$ using Eq. (B.37) and then take the expectation value of $\hat{n}_i = \hat{b}_i^\dagger \hat{b}_i$.

[10]This is in contrast to classical phase transitions, which are driven by thermal fluctuations, and thus necessarily occur at nonzero temperatures.

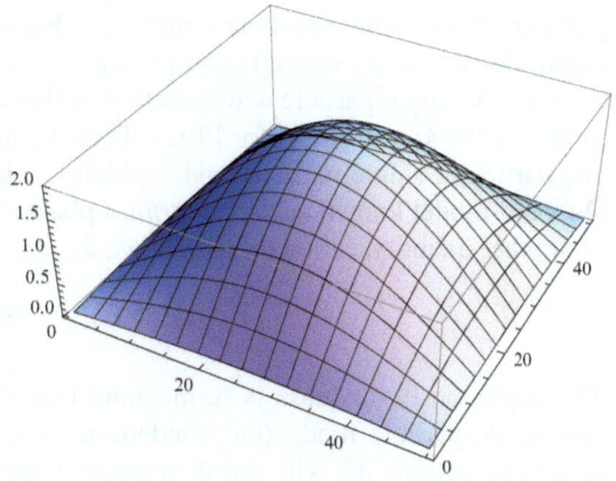

Fig. B.2 Single particle density matrix, noninteracting system of $L = 50$ lattice sites and $N = 50$ particles

To discover where the quantum phase transition is in the Bose–Hubbard model, let us return to the SPDM from the last exercise, and focus on the case of OBC and $N = L$ (as many particles as lattice sites).[11] You should have found that the SPDM in the non-interacting OBC case had the form

$$\rho_{ij} = \frac{2N}{L+1} \sin\left(\frac{\pi i}{L+1}\right) \sin\left(\frac{\pi j}{L+1}\right) \tag{B.41}$$

which is plotted in Fig. B.2. What does this quantity tell us? Notice that the elements of the SPDM can be interpreted as the overlap of the two wavefunctions $\hat{b}_j |\text{g.s.}\rangle$ and $\hat{b}_i |\text{g.s.}\rangle$ Thus, we can interpret the elements of the SPDM as the probability amplitude that if we remove a particle from the ground state at site i we will find it missing at site j, averaged over all the other $N - 1$ particles (with the "diagonal" elements $i = j$ just being the number on-site i, of course). For this to be nonzero with $i \neq j$ implies that the particles must be delocalized. In the present case the particles are highly delocalized, and this gives rise to mostly nonzero SPDM matrix elements. Secondly, because this matrix is Hermitian it may be diagonalized and written in terms of its eigenvalues $\mathcal{N}_\nu$ and eigenvectors $\phi_\mu(i)$ as

$$\rho_{ij} = \sum_\mu \mathcal{N}_\mu \phi_\mu(i)^\star \phi_\mu(j) \ . \tag{B.42}$$

The eigenfunctions $\phi_\mu(i)$ behave in many respects like single-particle wavefunctions (although they do not, in general, diagonalize the single-particle part of the full many-body Hamiltonian) and we can interpret the eigenvalues $\mathcal{N}_\mu$ as the number of

[11] For good reason, as this quantum phase transition happens to occur only for integer filling $N/L = 1, 2, \ldots$.

particles in the "single-particle state" ϕ_μ.[12] For our noninteracting case there is one eigenvalue $\mathcal{N}_1 = N$, with all others being zero. The corresponding eigenfunction ϕ_1 is an actual single-particle wavefunction in this case, and corresponds to the $k = 1$ mode (or the $k = 0$ mode for PBC). Thus we have that all of the particles occupy one particular single-particle mode, which is a hallmark of the phase of the Bose–Hubbard model known as the *superfluid* phase. We can express this result in terms of another quantity, called the *depletion*, as

$$D = 1 - \mathcal{N}_1/N \, . \tag{B.43}$$

The depletion in this case is 0, meaning that all the particles are condensed into one single-particle mode (the condensate is not at all depleted). When we add small interactions $\mathcal{N}_1$ will shrink somewhat, giving rise to nonzero depletion, and the eigenfunction ϕ_1 will no longer be a true single-particle wavefunction, but the overall behavior will be qualitatively the same (most of the elements of ρ_{ij} will be nonzero, there will be one SPDM eigenvalue that is $\mathcal{O}(N)$, etc.).

Let us now focus on the opposite limit of the Bose–Hubbard model, where we let the hopping $t \to 0$ and have U nonzero. The only surviving term for a fixed number of particles is

$$\hat{H} = \frac{U}{2}\sum_i \hat{n}_i\left(\hat{n}_i - 1\right) . \tag{B.44}$$

This Hamiltonian does not couple the different sites together, and so its ground state is a Fock state. In particular, for $N = L$, the ground state is the Fock state with one particle at each lattice site $|11\dots1\rangle$.[13] What does the SPDM look like in this case? A trivial calculation gives

$$\rho_{ij} = \delta_{ij} \, , \tag{B.45}$$

where δ_{ij} is the Kronecker delta. Loosely speaking, there is *no* probability of finding a removed particle missing from anywhere besides the exact spot it was removed from, and so we see that the particles are now highly localized! Also, we see that all of the eigenvalues of the SPDM are 1, giving the depletion

$$D = 1 - 1/N \, . \tag{B.46}$$

[12] For more information on the physical content of the eigenvalues/vectors of the SPDM and their connection with Bose-Einstein condensation, I recommend A.J. Leggett's highly readable book "Quantum Liquids," which is available in the library.

[13] We arrive at this conclusion by noting that occupation of any site by two-particles costs an energy U more than if those two-particles occupied separate sites.

In the limit as $L \to \infty$ and $N \to \infty$ with N/L fixed (the thermodynamic limit) where a true quantum phase transition occurs, the depletion becomes 1, meaning that there are no particles in a condensed state (the condensate is fully depleted). The resulting phase is known as the *Mott Insulator*. If the transition from the superfluid to Mott insulator occurs at a finite value of t/U, then we expect the depletion to be zero below this value and nonzero above it. It is clear that the depletion is thus not a smooth (meaning infinitely continuously differentiable) function of t/U, and it is this singularity in the depletion (or one of its derivatives) that signals the quantum phase transition.

To "see" the quantum phase transition using TEBD, we could calculate the ground state for a series of lattices with increasing $L = N$ and a range of t/U, compute the depletion for each case, and try to find the point where it goes to zero as $L \to \infty$. However, this would require us to know ahead of time that the depletion is the quantity signaling the quantum phase transition. In TEBD we have the option of taking a more general approach, which I will outline here.

The ground states on either side of a quantum phase transition are very different from one another, and the non-analyticity at the transition point disallows us from smoothly "connecting" one type of ground state to another. Thus, we expect that the overlap of the ground state on one side of the critical point with the ground state on the other side should go to zero in the thermodynamic limit. This motivates the definition of the *fidelity*

$$f\left(\frac{t}{U},\delta\frac{t}{U}\right) \equiv \left|\langle \text{g.s.}\left(\frac{t}{U}\right)|\text{g.s.}\left(\frac{t}{U}+\delta\frac{t}{U}\right)\rangle\right| \tag{B.47}$$

which is the overlap of the ground state at t/U with the ground state at $t/U + \delta t/U$. For small $\delta t/U$ this quantity should drop sharply only right near a quantum critical point, and does not require us to know anything about the system at hand. To remove the dependence on the step size $\delta t/U$ it is actually better to work with the *fidelity susceptibility*

$$\chi_f\left(\frac{t}{U}\right) = \frac{2}{L\left(\delta\frac{t}{U}\right)^2}\left(1 - f\left(\frac{t}{U},\delta\frac{t}{U}\right)\right) \tag{B.48}$$

which can be shown to be independent of $\delta\frac{t}{U}$. This quantity diverges at the quantum critical point, and will be the basis of our TEBD study.

In order to be able to properly extrapolate finite size results to the infinite size limit, we will need to run TEBD for several numbers of sites L. In addition, because the prediction of where the quantum critical point is will change as L changes, we will need to simulate several t/U for each L. This amounts to running many simulations, and you will run out of time and/or patience if you try and do it sequentially on a desktop computer! Thus, in this exercise we will use the high performance resources available here at Mines to run TEBD simulations in parallel.

The TEBD code we will use distributes a group of tasks (here computation of the ground state for a range of t/U) among a group of processors, and then compiles the results into a single file.

Log in either to Ra or Mio (Mio is preferred) and go to the carr group directory.[14] Go to the Case studies folder and then enter the `MPI_BH_Wrapper` directory. Type `make FS`, which compiles the main program `BoseHubbard_FS_MPI.f90`. This program calculates the ground state of the Bose–Hubbard model for $N = L$ and a user-specified range of t/U in parallel, and saves the states to disk as they are generated. Once all states have been generated, the routine reads them in and computes the fidelity susceptibility. We can use this routine to study the behavior of the fidelity susceptibility as L gets progressively larger. Also, because we save the states we generate, we can progressively refine our result for the fidelity susceptibility by gradually increasing the amount of entanglement allowed (controlled by χ) or by making the imaginary time step smaller.

The NAMELIST parameter file is `BH_FS.nml`. The parameters are:

1. `systemSize`, same meaning as before.
2. `maxFilling`, same meaning as before.
3. `BoundaryCond`, same meaning as before.
4. `Jmin`, the minimum value of t/U considered.
5. `jMax`, the maximum value of t/U considered.
6. `jres`, the number of t/U points from `Jmin` to `jMax`, inclusive.
7. `chiOld`, the old value of χ, used when reading in previously generated states.
8. `chiIn`, the value of χ for which you want to generate the fidelity susceptibility.
9. `convCriterion`, the convergence criterion for ITP.
10. `stepsForJudge`, same meaning as before.
11. `dtOld`, the old value of δt, used when reading in previously generated states.
12. `dtIn`, the value of δt for which you want to generate the fidelity susceptibility.
13. `maxITPsteps`, same meaning as before.
14. `FSDir`, the directory where the fidelity susceptibility data is stored.
15. `statesDir`, the directory where the state information is stored.

For the first run at a given `systemSize`, `chiOld` and `chiIn` should be the same. If more accuracy is desired after this first run, `chiOld` should be set to the value of χ you previously generated and `chiIn` should be set to the value you want to output at the end of the calculation. Identical reasoning applies to `dt*`. The output file in the directory `FSDir` ending in `FS.dat` has as its columns the values of t/U and χ_f. These can be plotted using your favorite program (e.g., gnuplot, *Matlab*, or *Mathematica*). See Fig. B.3 for example fidelity susceptibility curves to get you started on appropriate parameters.

[14]On Ra, this is /lustre/scratch/projects/carrgroup/. We will get such a directory started on Mio once it is fully operational.

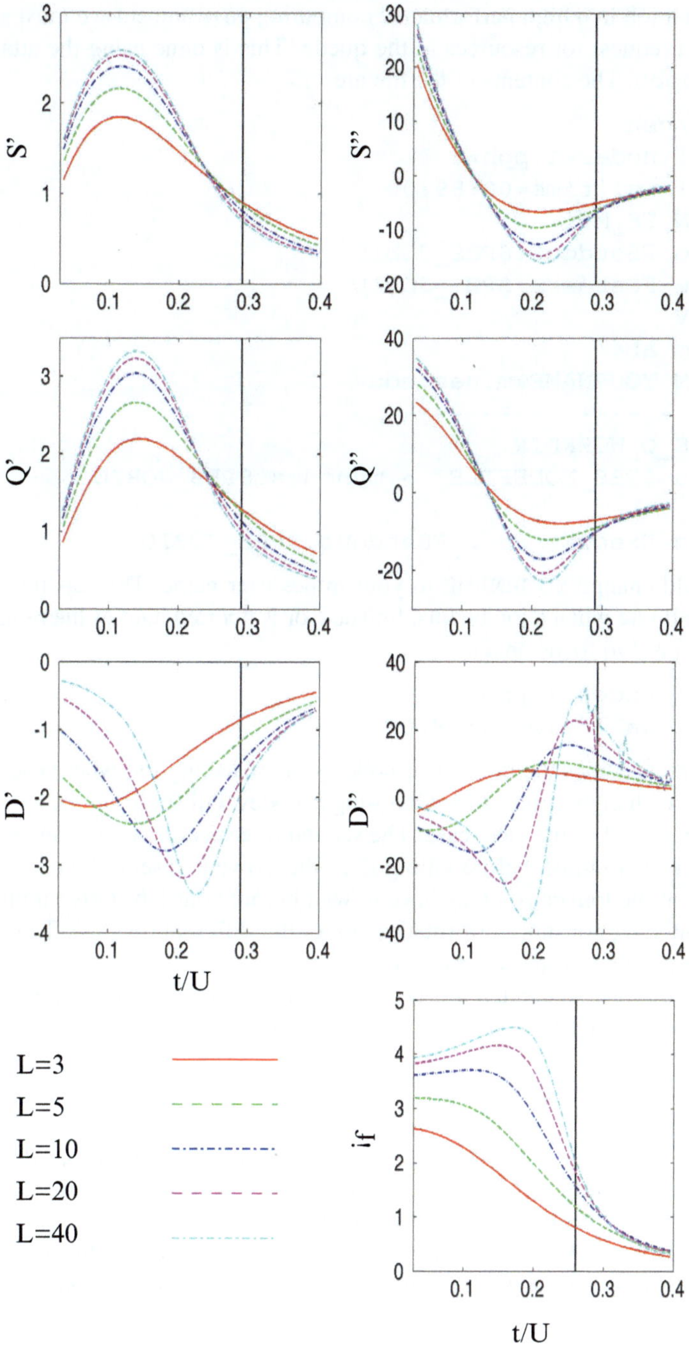

Fig. B.3 Fidelity susceptibility for (*bottom to top*) $L = 3, 5, 10, 20, 40$

To run a job in a high performance computing environment we must submit our job with a request for resources to the queue. This is done using the attached PBS file `tebdpbs`. The contents of the file are

```
#!/bin/csh
#PBS -l nodes=1:ppn=8
#PBS -l walltime=00:59:00
#PBS -N FS_MPI
#PBS -o FSstdout.$PBS_JOBID
#PBS -e FSstderr.$PBS_JOBID
#PBS -V
#PBS -m abe
#PBS -M YOURNAME@mines.edu
#-----------------------
cd $PBS_O_WORKDIR
sort -u $PBS_NODEFILE  > mynodes.$PBS_JOBID

mpiexec Execute_FS > FSstdout.$PBS_JOBID
```

You should change `YOURNAME` to your mines user name. This has the supercomputer email you when a job begins, finishes, or is aborted. Out of the remainder, the lines you need to focus on are

```
#PBS -l nodes=1:ppn=8
#PBS -l walltime=00:59:00
```

The first specifies the number of nodes you are asking for, where each node is a computer with eight processors. Choosing n nodes will speed up the calculation by a factor of $8n-1$ in the ideal case. The second line specifies how long the computer gives your job to finish before killing it. In the above it is set to 59 min. The longer you ask for, the longer you may have to wait before your job starts running (see the GECO webpage for more information about the different queues). To submit your job to the queue, type `msub tebdpbs`.

After you have generated χ_f for a series of L values, plot the locations of the maxima versus the lattice size and fit to a function of the form

$$L = f\left(\zeta_{\max L}\right) = C\left|\zeta_{\max L} - \zeta_{\max\infty}\right|^{-\eta} \tag{B.49}$$

with $C, \eta > 0$ as fit parameters, $\zeta = t/U$, $\zeta_{\max L}$ the value of ζ where χ_f is a maximum for a particular L, and $\zeta_{\max\infty}$ a fit parameter estimating the location of the quantum phase transition in the thermodynamic limit. Note that this calculation was only first done in 2007![15] Quantum phase transitions are an exciting and very current area of research, and powerful new methods such as TEBD are key to advancing understanding in the field.

[15]The relevant paper is *Phys. Rev. Lett.* **98** 110601, but don't look at it until you have a prediction for $\zeta_{\max\infty}$!

Appendix C
Educational Materials: Introduction to MPS Algorithms

The goal of this document is to give a series of successive steps leading to a working variational matrix product state (MPSs) code for the Ising model. New in v3: sections marked with an asterisk (∗) represent optimizations which should be skipped on first reading, but are important for efficient MPS code.

C.1 MPSs and Their Canonical Forms

The definition of a MPS is

$$|\mathrm{MPS}\rangle = \sum_{i_1,\dots,j_L=1}^{d} \mathrm{Tr}\left(A^{i_1}\dots A^{i_L}\right)|i_1\dots i_L\rangle\,, \tag{C.1}$$

where L is the number of lattice sites on a 1D chain, i_j is a *physical index* denoting the state of the jth lattice site, and d is the on-site dimension (which I will also call the local dimension). Because each A^{i_j} with i_j fixed is a matrix we use the terminology MPS. However, each A in fact has three elements: $A^{i}_{\alpha\beta}$. We will refer to an object with greater than two indices generically as a tensor, and the number of indices will be its rank. The dimensions of the spaces indexed by α and β are referred to as *bond dimensions*, with the bond dimension of an MPS being the maximum bond dimension of the matrices A. We will generically give a bond dimension the symbol χ. Simple arguments using the Schmidt decomposition show that the left and right bond dimensions in the case of open boundary conditions satisfy

$$\begin{aligned}\chi_\alpha &\le \min\left(d^{j-1},d^{L-j+1}\right)\,,\\ \chi_\beta &\le \min\left(d^{j},d^{L-j}\right)\end{aligned} \tag{C.2}$$

M.L. Wall, *Quantum Many-Body Physics of Ultracold Molecules in Optical Lattices*, Springer Theses, DOI 10.1007/978-3-319-14252-4

on the jth site, as can be easily proved by Schmidt decomposition. Note in particular that this implies that the furthest left and furthest tensors $A^{[1]}$ and $A^{[L]}$[1] are max $(\chi_\alpha, d, \chi_\beta) = (1, d, d)$ and $(d, d, 1)$, respectively. We will focus on the case of open boundaries for the remainder of this document. The case of periodic boundary conditions is considerably more complicated and should really be considered a separate algorithm altogether.

We plan to use MPS as a variational ansatz for finding the ground state of the Ising model. As such, we will need to define an MPS structure and initialize it with random numbers.

C.1.1 Step 1: MPS Structure

Define a `tensor` [2] structure which is an allocatable array with three indices. This structure will hold the tensors $A^i_{\alpha\beta}$. Next, define an `MPS` structure which is an array of `tensors`. For the first part of this problem set, real `tensors` and `MPSs` will suffice.

C.1.2 Step 2: MPS Initialization

Write a subroutine `AllocateMPS(psi,L,bondD,d)` which allocates an `MPS` `psi` on `L` sites with maximum bond dimension `bondD` and local dimension `d`. The left and right bond dimensions of each tensor A should obey the conditions Eq. (C.2). Also write a routine `DeallocateMPS` which deallocates the `MPS` structure. Finally, write a routine `CreateRandomMPS(psi,L,bondD,d)` which not only allocates an `MPS` as in `AllocateMPS`, but also assigns each element a random number $\in [-1, 1]$.

As discussed in Schollwöck,[3] the imposition and maintenance of canonical forms is absolutely essential to the efficiency of a variational MPS program. The canonical form we will use most often is *Mixed canonical form* at site k, in which all tensors to the left of k satisfy the left orthogonality conditions

$$\sum_{\alpha,i} A^{i}_{\alpha\beta}{}^{\star} A^{i}_{\alpha\beta'} = \delta_{\beta,\beta'} \tag{C.3}$$

[1]The notation $A^{[j]}$ means the tensor $A^{i_j}_{\alpha\beta}$ on-site j. When all four indices are required, we will use $A^{[j]i_j}_{\alpha\beta}$

[2]I will use `typewriter face` whenever I refer to something which is to be coded.

[3]By which I mean Prof. Ulrich Schollwöck's major review *The density-matrix renormalization group in the age of MPSs*, Annals of Physics **326**, 96 (2011).

and all tensors to the right of k satisfy the right orthogonality conditions

$$\sum_{i,\beta} A^i_{\alpha\beta} A^i_{\alpha'\beta}{}^\star = \delta_{\alpha,\alpha} . \tag{C.4}$$

The site k itself satisfies no such orthogonality conditions, but we note that $\mathrm{Tr}_{i_1,\dots,i_L}\left(|\psi\rangle\langle\psi|\right) = \langle\psi|\psi\rangle = \sum_{\alpha,\beta,i_k} A^{[k]i_k}_{\alpha\beta} A^{[k]i_k}_{\alpha\beta}{}^\star$, and so this site carries all information about the norm of the state. We will refer to the site k in mixed canonical form as the *orthogonality center* of the MPS.

We can always choose the orthogonality conditions Eqs. (C.3) and (C.4) to hold because of the so-called *gauge freedom* inherent in the MPS representation. This freedom refers to the fact that if we consider any two MPS tensors $A^{[i]}$ and $A^{[i+1]}$, the same MPS results if we replace these by $\tilde{A}^{[j]i_j}$ and $\tilde{A}^{[j+1]i_{j+1}}$, where

$$\tilde{A}^{[j]i_j} \equiv A^{[j]i_j} X , \tag{C.5}$$

$$\tilde{A}^{[j+1]i_{j+1}} \equiv X^{-1} A^{[j+1]i_{j+1}} , \tag{C.6}$$

and X is any invertible matrix. In practice we implement the orthogonality conditions Eqs. (C.3) and (C.4) as in Fig. C.1 and Fig. C.2, respectively.[4]

You should take a minute to convince yourself of why this works, recalling that the matrices U and V of the singular value decomposition are unitary. Furthermore, you should pay close attention to groupings of indices and note that *two* tensors are changed at a time.

1. Reshape $A^{[j]i}_{\alpha\beta} \to A_{[\alpha i],\beta}$.
2. Perform the singular value decomposition (SVD) on this matrix to obtain $A_{[\alpha i]\beta} \to U_{[\alpha i]\gamma} S_\gamma V_{\gamma\beta}$.
3. Replace $A^{[j]i}_{\alpha\beta} = U_{[\alpha i]\beta}$.
4. Combine $(SV)_{\gamma\beta} = S_\gamma V_{\gamma\beta}$.
5. Contract SV into the A tensor to the right of the one just replaced $A^{[j+1]i}_{\alpha\beta} = (SV)_{\alpha\gamma} A^{i[j+1]}_{\gamma\beta}$.

Fig. C.1 Algorithm to put site j into left-canonical form

[4]The notation $[\alpha i]$ refers to the *Kronecker product* of the indices α and i. The product index $\xi = [\alpha i]$ runs from 1 to $d_\alpha d_i$, and an explicit representation is $\xi = (\alpha - 1)\, d_i + i$.

1. Reshape $A^{[j]i}_{\alpha\beta} \to A_{\alpha[i\beta]}$.
2. Perform an SVD on this matrix to obtain $A_{\alpha[i\beta]} \to U_{\alpha\gamma}S_{\gamma}V_{\gamma[i\beta]}$.
3. Replace $A^{[j]i}_{\alpha\beta} = V_{\alpha[i\beta]}$.
4. Combine $(US)_{\alpha\gamma} = U_{\alpha\gamma}S_{\gamma}$.
5. Contract US into the A tensor to the left of the one just replaced $A^{[j-1]i}_{\alpha\beta} = A^{[j-1]i}_{\alpha\gamma}(US)_{\gamma\beta}$.

Fig. C.2 Algorithm to put site j into right canonical form

C.1.3 Step 3: Canonical Form

Write a routine `OrthogonalizeMPS(psi,k,kl,kr)` which puts the MPS `psi` into mixed canonical form with the orthogonality center at site `k`. The optional arguments `kl` and `kr` denote the furthest left and furthest right sites which are altered. The default values are 1 and L, respectively. Test this routine by explicitly checking the relations Eqs. (C.3) and (C.4) at each site. Do tensors in left-canonical form obey any sort of right-canonical form or vice versa? What about the orthogonality center—does it have any structure that you can discern?[5] Then, write a routine `OrthonormalizeMPS(psi,k)` which shifts the orthogonality center of `psi` to site `k` and then normalizes the state by replacing $A^{[k]i}_{\alpha\beta} \to A^{[k]i}_{\alpha\beta}/\sqrt{\langle\psi|\psi\rangle}$, where $\langle\psi|\psi\rangle = \sum_{\alpha,\beta,i_k} A^{[k]i_k}_{\alpha\beta} A^{[k]i_k\star}_{\alpha\beta}$ as shown above. The svd may be performed using the LAPACK routine `dgesvd`.

OK, you now have an MPS in a form amenable to the variational procedure. Let's now move on to constructing the Hamiltonian in an appropriate way.

C.2 Matrix Product Operators

Just as an MPS can be viewed as a sum over tensor products of state-valued matrices:

$$|\mathrm{MPS}\rangle = \sum_{i_1\ldots i_L} \mathrm{Tr}\left(\mathcal{A}^{i_1}\otimes\cdots\otimes\mathcal{A}^{i_L}\right), \tag{C.7}$$

$$\mathcal{A}^{i_j}_{\alpha\beta} \equiv A^{i_j}_{\alpha\beta}|i_j\rangle \tag{C.8}$$

[5] These are not trick questions….

we can define an object which is like an MPS but is operator-valued instead of state-valued

$$\hat{H}_{\text{MPO}} = \sum_{i_1 \dots i_L} \text{Tr}\left(\mathcal{W}^{i_1 i_1'} \otimes \dots \otimes \mathcal{W}^{i_L i_L'}\right), \tag{C.9}$$

$$\mathcal{W}_{\alpha\beta}^{i_j i_j'} \equiv W_{\alpha\beta}^{i_j i_j'} |i_j\rangle\langle i_j'| \,. \tag{C.10}$$

Such an object is called a *matrix product operator* (MPO). As with an MPS, the i_j indices are physical, referring to actual sites in the lattice, and the indices implicit in the matrix product are called bond indices. The reason that MPOs are useful is that they map MPSs to MPSs. To see this, note that an MPO-MPS product is

$$\hat{H}|\psi\rangle = \text{Tr}\left(\tilde{A}^{i_1} \dots \tilde{A}^{i_L}\right)|i_1 \dots i_L\rangle, \tag{C.11}$$

$$\tilde{A}_{[\alpha k][\beta k']}^{i_j} \equiv \sum_{i_j'} A_{\alpha\beta}^{i_j'} W_{kk'}^{i_j i_j'} \,. \tag{C.12}$$

We see that, generically, the action of an MPO on an MPS increases the bond dimension of that MPS by the multiplicative factor χ_H, the bond dimension of the MPO. Because MPOs are operators, the relevant norm for them is the Frobenius norm $\langle\hat{H}|\hat{H}\rangle = \sqrt{\text{Tr}\left(\hat{H}^\dagger\hat{H}\right)}$. However, this norm scales as d^L, and so we cannot hope to orthogonalize the matrices of an MPO in the same way that we did MPSs without running into serious numerical issues concerning precision. Luckily, because of the structure of physical operators such as many-body Hamiltonians it is possible to explicitly construct the MPOs in a canonical form in which all of the (matrix-valued!) matrices $W_{kk'}$ are lower triangular. For the case of a generic nearest-neighbor Hamiltonian

$$\hat{H} = \sum_{\langle i,j\rangle}\sum_{\alpha=1}^{p} J_\alpha \hat{O}_{i\alpha}^{(1)}\hat{O}_{j\alpha}^{(2)} + \sum_i \sum_{\beta=1}^{q} h_\beta \hat{O}_{i\beta}^{(0)}, \tag{C.13}$$

where α denotes the different nearest-neighbor terms and β denotes the different on-site terms (each characterized by different operators $\hat{O}$), the MPO representation is

$$W^{[1]} = \left(\sum_{\beta=1}^{q} h_\beta \hat{O}_{1\beta}^{(0)} \; J_1\hat{O}_{11}^{(1)} \; \dots \; J_p\hat{O}_{1p}^{(1)} \; I\right), \tag{C.14}$$

$$W^{[j]} = \begin{pmatrix} I & 0 & \dots & 0 & 0 \\ \hat{O}_{j1}^{(2)} & 0 & \dots & 0 & 0 \\ \vdots & \vdots & \ddots & \vdots & \vdots \\ \hat{O}_{jp}^{(2)} & 0 & \dots & 0 & 0 \\ \sum_{\beta=1}^{q} h_\beta \hat{O}_{j\beta}^{(0)} & J_1\hat{O}_{j1}^{(1)} & \dots & J_p\hat{O}_{jp}^{(1)} & I \end{pmatrix}, j \neq 1, L, \tag{C.15}$$

$$W^{[L]} = \begin{pmatrix} I \\ \hat{O}^{(2)}_{L1} \\ \vdots \\ \hat{O}^{(2)}_{Lp} \\ \sum_{\beta=1}^{q} h_\beta \hat{O}^{(0)}_{L\beta} \end{pmatrix} . \tag{C.16}$$

To be very concrete, in the case of the Ising model

$$\hat{H} = -J \sum_{\langle i,j \rangle} \hat{\sigma}_i^z \hat{\sigma}_j^z - h \sum_i \hat{\sigma}_i^x , \tag{C.17}$$

and so $p = 1$, $q = 1$, $J_\alpha = -J$, $h_\beta = -h$, $\hat{O}^{(1)}_{i\alpha} = \hat{O}^{(2)}_{i\alpha} = \hat{\sigma}_i^z$, $\hat{O}^{(0)}_{i\beta} = \hat{\sigma}_i^x$. The MPO in this case is

$$W^{[1]} = \begin{pmatrix} -h\sigma^x & -J\hat{\sigma}^z & I \end{pmatrix} , \tag{C.18}$$

$$W^{[j]} = \begin{pmatrix} I & 0 & 0 \\ \hat{\sigma}^z & 0 & 0 \\ -h\hat{\sigma}^x & -J\hat{\sigma}^z & I \end{pmatrix} , j \neq 1 , L ,$$

$$W^{[L]} = \begin{pmatrix} I \\ \hat{\sigma}^z \\ -h\hat{\sigma}^x \end{pmatrix} .$$

We note that the MPO has a bond dimension of 3.[6] The reader should convince themselves that the MPO form given reproduces the correct Hamiltonian. A simple check is to compute the two-site hamiltonian using only the boundary operators and then the three-site Hamiltonian with an additional matrix between them, etc. with the understanding that the matrices that are elements of the matrices W are multiplied according to the tensor product. This method also suggests a straightforward way to prove the consistency by induction for those who wish to be more thorough.

C.2.1 Step 4: MPO Structure

Introduce a `matrix` structure which is an allocatable array with two indices. Then, define an `MPOm` structure which is an allocatable array of matrices with two indices. This represents the on-site MPO tensors $W^{ii'}_{kk'}$, with i and i' being the indices of the on-site operator. Finally, define an `MPO` structure which is an allocatable array of `MPOms`.

[6]The general case given above has a bond dimension of $p + 2$, *independent* of the number of on-site operators.

C.2.2 Step 5: MPO Representation of the Ising Hamiltonian

Write a subroutine `IsingMPO(H,L,Jz,hx)` which creates an `MPO H` on `L` sites containing the MPO representation of the Ising model with J = `Jz` and h = `hx` according to Eq. (C.18). Additionally, write a routine `DeallocateMPO(H)` which deallocates the `MPO H` which was allocated by the `IsingMPO` routine.

C.2.3 * Long-Range Interactions with MPO

Just as exponentially decaying correlations are naturally supported by an MPS with a constant bond dimension, interactions with exponentially decaying weight may be easily expressed as an MPO with a constant (independent of system size) bond dimension. As an example, a long-range Ising model

$$\hat{H} = -J \sum_{i<j} e^{-\xi|i-j-1|} \hat{\sigma}_i^z \hat{\sigma}_j^z - h \sum_i \hat{\sigma}_i^x \tag{C.19}$$

can be expressed as an MPO as

$$W^{[1]} = \begin{pmatrix} -h\sigma^x & -J\hat{\sigma}^z & I \end{pmatrix}, \tag{C.20}$$

$$W^{[j]} = \begin{pmatrix} I & 0 & 0 \\ \hat{\sigma}^z & e^{-\xi} & 0 \\ -h\hat{\sigma}^x & -J\hat{\sigma}^z & I \end{pmatrix}, j \neq 1, L,$$

$$W^{[L]} = \begin{pmatrix} I \\ \hat{\sigma}^z \\ -h\hat{\sigma}^x \end{pmatrix}.$$

Power-law decaying functions can be approximated by using sums of exponentials in the following manner. We have some function $f(i-j-1)$ and we want to approximate it by the sum of k weighted exponentials $\sum_{i=1}^k \alpha_k \beta_k^{i-j-1} = \sum_{i=1}^k \alpha_k e^{(i-j-1)\log \beta_k}$. We do so in the least squares sense, that is we find the α_k and β_k such that

$$\left| f(x) - \sum_{i=1}^{k} \alpha_k \beta_k^x \right|^2 \tag{C.21}$$

is minimized in the desired (integer valued) domain of x. This can be done with standard numerical packages such as MINPACK. My own implementation in Python follows:

```
from math import sqrt
import numpy as np
from scipy.optimize import leastsq
def fittoexpSum(func,L,maxnterms,tol):
    """Fit a sum of exponentials \sum_n a_n b_n**{x-1}
        to the function func
        across the range [1:L]. maxnterms is the
            maximum number of exponentials allowed
        and tol is the tolerance used to obtain the
            actual number of terms."""
    x=np.linspace(1,L,L)
    #function evaluated at x-points
    y= func(x)
    fail=True
    for n in range(1,maxnterms+1):
        #residual function
        resid = lambda p, x, y : sumexp(p,x)-y
        #initial guess
        p0=np.zeros(2*n)
        if n==1:
            for i in range(2*n):
                p0[i]=0.1
        else:
            #use old values to refine guess
            p0[:2*(n-1)]=p
            p0[2*(n-1):]=0.1
        p,cov,infodict,mesg,ier=leastsq(resid, p0,args
            =(x,y),ftol=tol,gtol=tol, maxfev=100000,
            full_output=1,warning=True)
        if np.dot(infodict['fvec'],infodict['fvec'])<
            tol:
            print 'resid',np.dot(infodict['fvec'],
                infodict['fvec'])
            fail=False
            break
    if fail :
        raise Exception("Unable␣to␣converge␣
            decayingFunction␣to␣the␣desired␣tolerance␣
            with␣the␣given␣number␣of␣terms!␣␣Try␣
            increasing␣maxnterms␣or␣decreasing␣tol!")
    return p

def sumexp(p,x):
    """Return y=\sum_n a_n b_n^{r-1}."""
```

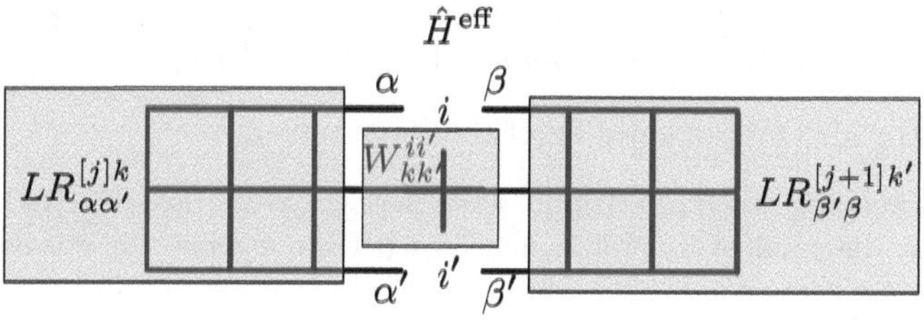

Fig. C.3 Construction of the effective hamiltonian in tensor network diagram representation

```
val=np.double(0.0)
nterms=len(p)/2
for n in range(nterms):
    val=val+p[2*n]*(p[2*n+1]**(x-1))
return val
```

The resulting MPO has the same overall format as the exponential Ising MPO, with the border terms and the diagonals nonzero, but now there are k such nonzero terms instead of just 1. The number of terms which must be kept depends on the particular form of f and the domain of x.

C.3 Construction of the Effective Hamiltonian

We now have the state and Hamiltonian in the form appropriate for variational MPS studies. We now wish to variationally optimize the MPS tensors A one at a time until a global optimum has been reached. We optimize the state in the usual way, by changing its parameters such that the total energy functional

$$E\left[|\psi\rangle\right] = \frac{\langle\psi|\hat{H}|\psi\rangle}{\langle\psi|\psi\rangle} \tag{C.22}$$

is minimized. As shown in Schollwöck, minimization with respect to a single tensor A chosen to be the orthogonality center leads to the linear eigenequation

$$\sum_{\alpha',i',\beta'} \hat{H}^{\mathrm{eff}}_{[\alpha i\beta][\alpha' i'\beta']} A^{i'}_{\alpha'\beta'} = \lambda A^{i}_{\alpha\beta}\,, \tag{C.23}$$

where the effective Hamiltonian $\hat{H}^{\mathrm{eff}}$ is defined by the tensor network diagram shown in Fig. C.3. We call this object the effective Hamiltonian because it gives the energies when all tensors except those at site j are held fixed, and so it incorporates nonlocal effects of the many-body state when choosing the minimum energy configuration at a particular site.

1. Create a random MPS with fixed bond dimension, normalize it, and shift the orthogonality center to some user-specified site j. Set $k = j$
2. Create the MPO representation of the Hamiltonian.
3. (Begin sweeping procedure: right sweep part 1) Create the effective Hamiltonian at site k, diagonalize it, and replace A with the eigenvector corresponding to the lowest eigenvalue.
4. Move the orthogonality center one site to the right. $k = k + 1$
5. Repeat 3 and 4 until site L is the orthogonality center.
6. (sweeping procedure: left sweep) Create the effective Hamiltonian at site k, diagonalize it, and replace A with the eigenvector corresponding to the lowest eigenvalue.
7. Move the orthogonality center one site to the left. $k = k - 1$
8. Repeat 6 and 7 until site 1 is the orthogonality center.
9. (sweeping procedure: right sweep part 2) Repeat 3 and 4 until site j is the orthogonality center.
10. Test for convergence (to be discussed) and either repeat from 3 or exit the loop.

Fig. C.4 Pseudocode to find the ground state variationally

The overall algorithm for variationally finding the ground state is outlined in Fig. C.4.

From Fig. C.3, we see that construction of the effective Hamiltonian at each site requires $\mathcal{O}(L)$ operations, as we have to contract over all MPS tensors not currently being optimized. Because a complete sweep runs over L sites, this implies that the overall algorithm scales as $\mathcal{O}(L^2)$, which is unacceptably slow. We now turn to how to generate the effective Hamiltonian using $\mathcal{O}(L)$ time. We do so by introducing an array of $L+1$ `tensors`, which we call *LR*. This object is defined for orthogonality center ℓ via the relations

$$LR^{[1]k}_{\alpha\alpha'} = \delta_{k,1}\delta_{\alpha,1}\delta_{\alpha',1} \tag{C.24}$$

$$LR^{[j]k}_{\alpha\alpha'} = \sum_{i,i',k',\gamma,\gamma'} A^{[j-1]i\star}_{\gamma\alpha} LR^{[j-1]k'}_{\gamma\gamma'} A^{[j-1]i'}_{\gamma'\alpha'} W^{[j-1]ii'}_{k'k}\,, \quad 2 \le j \le \ell \tag{C.25}$$

$$LR^{[L+1]k}_{\beta'\beta} = \delta_{k,1}\delta_{\beta',1}\delta_{\beta,1} \tag{C.26}$$

$$LR^{[j]k}_{\beta'\beta} = \sum_{i,i',k',\gamma,\gamma'} A^{[j]i'}_{\beta'\gamma'} LR^{[j+1]k'}_{\gamma'\gamma} A^{[j]i\star}_{\beta\gamma} W^{[j]ii'}_{kk'}\,, \quad \ell+1 \le j \le L\,. \tag{C.27}$$

Very special attention should be paid to the ordering of indices in each term. While the order of indices may seem arbitrary, the particular choice shown here will turn out to be beneficial for later optimization. We now note that the effective Hamiltonian at site j can be constructed from these overlaps as

$$\hat{H}^{\text{eff}}_{[\alpha i \beta][\alpha' i' \beta']} = \sum_{k,k'} LR^{[j]k}_{\alpha\alpha'} W^{ii'}_{kk'} LR^{[j+1]k'}_{\beta'\beta}\,, \tag{C.28}$$

as shown in Fig. C.3. Once the present site has been variationally optimized and the orthogonality center shifted to the next site, we update the jth element of the LR array using Eq. (C.25) if we are sweeping to the right (steps 3/4 or 9) or Eq. (C.27) if we are sweeping to the left (step 6/7). Note also that we must initialize the LR array for the beginning orthogonality center j.

C.3.1 Step 6: LR Array

Create a routine `InitializeLR(LR,H,psi,k)` which initializes the $L+1$-dimensional array of `tensors` `LR` according to the above relations for orthogonality center `k`. Also, write a routine `UpdateLR(LR,H,psi,k,sense)` which updates the $k+1$th element of `LR` according to Eq. (C.25) if the integer `sense` is greater than 0 and which updates the kth element of `LR` according to Eq. (C.27) if `sense` is less than 0.[7] These routines allow us to "recycle" old values of `LR` and keep the algorithm scaling as $\mathcal{O}(L)$.

C.3.2 Step 7: Effective Hamiltonian

Create a routine `EffectiveHamiltonian(j,Heff,H,LR)` which creates the `matrix` `Heff` at site `j` according to Eq. (C.28). Create another routine `MinimizeSite(j,psi,H, LR, energy)` which creates the effective Hamiltonian at site `j`, diagonalizes it,[8] outputs the lowest eigenvalue as `energy`, and replaces `psi` with the eigenvector corresponding to the lowest energy (be careful with indices).

---

[7]Note the asymmetry between which element we update based on the direction of the sweep. This arises from the fact that we use $LR^{[j]}$ and $LR^{[j+1]}$ to update site j.

[8]Because the effective Hamiltonian is real and symmetric (prove this to yourself) you can use the LAPACK routine DSYEV.

C.3.3 Step 8: Putting It All Together

Write a routine `FindGroundState(H,psi,k)` which takes as input an `MPO` Hamiltonian `H`, a random `MPS` `psi` with orthogonality center `k` and variationally optimizes it to the ground state of `H`. We do this in several parts. First, call `InitializeLR(LR,H,psi,k)` and set the logical flag `converged` to `.FALSE.`, `energySave=10000.0`, and `numsweeps=1`. We then have the following psuedocode recursion:

```
DO WHILE(.not.converged)
    sense=1
    DO j=k,L-1
        CALL MinimizeSite(j,psi,H,LR,energy)
        CALL OrthogonalizeMPS(psi,j+1,j,j+1)
        CALL UpdateLR(LR,H,psi,j,sense)
    END DO
    sense=-1
    DO j=L,2,(-1)
        CALL MinimizeSite(j,psi,H,LR,energy)
        CALL OrthogonalizeMPS(psi,j-1,j-1,j)
        CALL UpdateLR(LR,H,psi,j,sense)
    END DO
    sense=1
    DO j=1,kin-1
        CALL MinimizeSite(j,psi,H,LR,energy)
        CALL OrthogonalizeMPS(psi,j+1,j,j+1)
        CALL UpdateLR(LR,H,psi,j,sense)
    END DO
    IF(ABS(energySave-energy)<0.01) converged=.true.
    IF(numsweeps==10) EXIT
    IF(converged) EXIT
    numsweeps=numsweeps+1
    energySave=energy
END DO
```

Compare with Fig. C.3. After convergence is reached, print the ground state energy and whether it converged to the screen. Write a main program which creates a random MPS, creates the MPO representation of the Ising Hamiltonian with user-specified J_z and h_x, and then finds the ground state energy. Compare your results with the results of the attached code. The main program is `IsingGSMain.f90` which is compiled as `make IsingGSMain` and run as `./Execute_IsingGSMain`.

C.3.4 * Sparse Eigensolvers: The Lanczos Algorithm

Looking again at Fig. C.3, we see that the application of the effective Hamiltonian onto some MPS tensor A may be broken into three pieces:

$$F^{k'}_{\alpha i\beta} = \sum_{\beta'} A^{[j]i}_{\alpha\beta'} LR^{[j+1]k'}_{\beta'\beta}, \quad G^{k}_{\alpha i\beta} = \sum_{k'i'} W^{ii'}_{kk'} F^{k'}_{\alpha i'\beta}, \quad \left[\hat{H}_{\mathrm{eff}} A\right]^{i}_{\alpha\beta} = \sum_{\alpha' k} LR^{[j]k}_{\alpha\alpha'} G^{k}_{\alpha i\beta}. \tag{C.29}$$

The relative scalings of these contractions are $\chi^3 d\chi_O$, $\chi_O^2 d^2\chi^2$, and $\chi^3 d\chi_O$, where χ is the bond dimension of the MPS, d the local dimension, and χ_O the bond dimension of the MPO. On the other hand, the direct construction of the effective Hamiltonian, Eq. (C.28), scales as $\chi^4\chi_O^2 d^2$, which is *much* slower than a single Hamiltonian-tensor multiply, and the diagonalization of the Hamiltonian scales as $\chi^6 d^3$, which is incomparably slower than either of these operations when $\chi \gg 1$, as is typically the case. Luckily, there exist eigensolvers for one or a few of the extremal eigenvalues which require only Hamiltonian-tensor multiplies. The *Lanczos algorithm* is an example of such an algorithm. It may be stated as follows[9]:

```
input psi, a normalized vector (tensor)
input H, an effective Hamiltonian in the LR/W format
input epsilon, a tolerance for convergence

k=1
lanc=psi
v=H*psi
alpha(k)=psi*v
v=v-alpha(k)*psi
beta(k)=sqrt(v*v)
do until convergence
        temp=psi
        psi=v/beta(k)
        v=-beta(k)*temp
        v=v+H*psi
        k=k+1
        alpha(k)=psi*v
        v=v-alpha(k)*psi
        beta(k)=sqrt(v*v)
        solve alpha/beta system and test for
            convergence
end
```

[9]A great deal more can be said about the origins and interpretation of the Lanczos algorithm. See, for example, Trefethen and Bau's book "Numerical Linear Algebra" or Golub and Van Loan's book "Matrix Computations." Here, for the sake of compactness, we just outline the algorithm.

```
!generate eigenvector
psi=eigvec(1)*lanc
v=0
v=H*lanc
alpha(k)=lanc*v
v=v-alpha(k)*lanc
beta(k)=sqrt(v*v)
do i=1,k-1
        temp=lanc
        lanc=v/beta(k)
        v=-beta(k)*lanc
        psi=psi+eigvec(i+1)*lanc
        v=v+H*lanc
        k=k+1
        alpha(k)=lanc*v
        v=v-alpha(k)*lanc
        beta(k)=sqrt(v*v)
end
renormalize psi
```

where $*$ for two tensors is the inner product of the Hilbert space $\langle A, B\rangle \equiv A * B = \sum_{\alpha i\beta} A^{i\star}_{\alpha\beta} B^{i}_{\alpha\beta}$ and for Hamiltonian-tensor multiplies is the three-step process Eq. (C.29). The section `solve alpha/beta system` at the kth step means to solve for the lowest eigenvalue/eigenvector pair of the symmetric tridiagonal matrix with `alpha(1:k)` on the diagonals and `beta(1:k-1)` on the offdiagonals. `evec(i)` in the above represents the ith component of this eigenvector. Specialized routines for tridiagonal problems exist in lapack. *Exact* convergence occurs when one of the `beta(k)` is zero, to within numerical tolerance. Convergence within a specified tolerance ε, in the sense that $\left|\hat{H}_{\mathrm{eff}} A - \lambda A\right| < \varepsilon$, occurs when $|e_k \beta_{k-1}| < \varepsilon$, where e is the eigenvector of the alpha/beta matrix at the kth iteration and β is `beta`.

The Lanczos algorithm often converges in a constant (independent of χ) number of iterations on the order of a few tens or hundreds. This leads to an algorithm which scales as χ^3 overall, *much* faster than the χ^6 scaling of the direct algorithm. The Lanczos algorithm should thus be considered an essential part of any efficient MPS program.

C.4 Observables

The simplest observables to compute are those at a single site: $\langle \hat{O}^j \rangle$. If that site is the orthogonality center, then we have

$$\langle \hat{O}^j \rangle = \sum_{\alpha,\beta,i,i'} A^{[j]i\star}_{\alpha\beta} \hat{O}^j_{ii'} A^{[j]i'}_{\alpha\beta} . \tag{C.30}$$

To test these routines, compute $\sum_i \langle \hat{\sigma}^i_z \rangle$ and demonstrate that it is 0 (by symmetry) and compute $\sum_i \langle \hat{\sigma}^i_x \rangle$ and show it is $-\partial \langle \hat{H} \rangle / \partial h = -\left[E\left(h+\delta h\right) - E\left(h-\delta h\right) \right] / 2\delta h + \mathcal{O}\left(\delta h^2\right)$. To compute the expectation of a general operator expressed as an MPO you can use the LR recursion

$$LR^{[L+1]k}_{\beta'\beta} = \delta_{k,1}\delta_{\beta',1}\delta_{\beta,1} \tag{C.31}$$

$$LR^{[j]k}_{\beta'\beta} = \sum_{i,i',k',\gamma,\gamma'} A^{[j]i'}_{\beta'\gamma'} LR^{[j+1]k'}_{\gamma'\gamma} A^{[j]i\star}_{\beta\gamma} W^{[j]ii'}_{kk'} , \quad j = 1 \ldots L , \tag{C.32}$$

the expectation is $LR^{[1]1}_{11}$ (note that this tensor is $1 \times 1 \times 1$).

C.5 Excited States

To compute the nth excited state, we again minimize the energy functional

$$E\left[|\psi\rangle\right] = \frac{\langle \psi | \hat{H} | \psi \rangle}{\langle \psi | \psi \rangle} \tag{C.33}$$

but this time subject to the constraints $\langle \psi | \varphi_\gamma \rangle = 0$, $\gamma = 1, \ldots, n$, where $|\varphi_\gamma\rangle$ is the γth lowest energy state already obtained by the ground state procedure above or the current procedure. We note that each one of the constraints is a *linear form* in each of the tensors $A^{[\ell]\star}$. Thus, we can represent the constraints as[10]

$$\langle \psi | \varphi_\gamma \rangle = \sum_{\alpha i \beta} A^{[\ell]i\star}_{\alpha\beta} \mathbb{B}^{[\ell](\gamma)i}_{\alpha\beta} = 0 , \tag{C.34}$$

where

$$\mathbb{B}^{[\ell](\gamma)i}_{\alpha\beta} \equiv \frac{\partial}{\partial A^{[\ell]\star}} \langle \psi | \phi_\gamma \rangle = LR^{[\ell]\gamma}_{\alpha\alpha'} B^{[\ell](\gamma)i}_{\alpha'\beta'} LR^{[\ell+1]\gamma}_{\beta'\beta} . \tag{C.35}$$

The LR overlaps used in this expression are

$$LR^{[1]\gamma}_{\alpha\alpha'} = \delta_{k,1}\delta_{\alpha,1}\delta_{\alpha',1} \tag{C.36}$$

---

[10]Note that the bases of $\mathbb{B}$ are the same as the bases of A, as can be clearly seen from Fig. C.5.

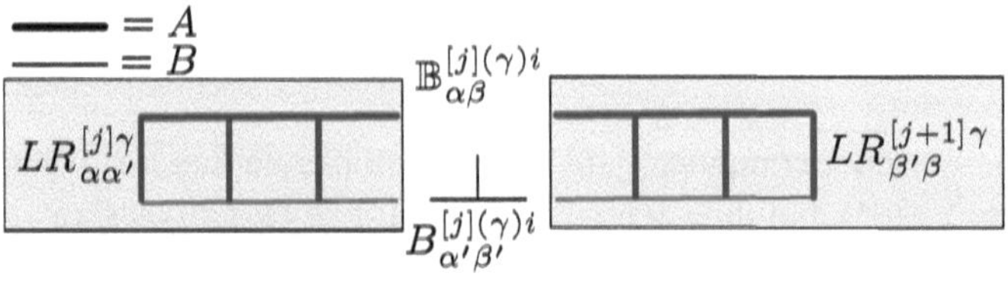

Fig. C.5 Construction of linear forms for projection in tensor network diagram representation. The *thick lines* correspond to contractions over the bond dimensions of A and the *thin lines* are contractions over the bond dimensions of B

$$LR^{[j]\gamma}_{\alpha\alpha'} = \sum_{i,\beta,\beta'} A^{[j-1]i\star}_{\beta\alpha} LR^{[j-1]\gamma}_{\beta\beta'} B^{[j-1](\gamma)i}_{\beta'\alpha'}\,, \quad 2 \le j \le \ell \tag{C.37}$$

$$LR^{[L+1]\gamma}_{\beta'\beta} = \delta_{k,1}\delta_{\beta',1}\delta_{\beta,1} \tag{C.38}$$

$$LR^{[j]\gamma}_{\beta'\beta} = \sum_{i,\alpha,\alpha'} B^{[j](\gamma)i}_{\beta'\alpha'} LR^{[j+1]\gamma}_{\alpha'\alpha} A^{[j]i\star}_{\beta\alpha}\,, \quad \ell+1 \le j \le L\,, \tag{C.39}$$

and $B^{[\ell](\gamma)}$ is the on-site tensor at site ℓ for state $|\varphi_\gamma\rangle$. The tensor network diagram representing these objects is shown in Fig. C.5. We use the $\mathbb{B}^{(\gamma)}$ to construct projectors into the subset of states orthogonal to the n lowest states

$$\mathcal{P}^{[\ell]} = 1 - \sum_{\gamma\gamma'} \mathbb{B}^{[\ell](\gamma)} \left(\mathcal{N}^{-1}\right)_{\gamma\gamma'} \mathbb{B}^{[\ell](\gamma')\dagger}, \tag{C.40}$$

where $\mathcal{N}^{-1}$ is the inverse of the Gram matrix

$$\mathcal{N}_{\gamma\gamma'} = \sum_{\alpha i\beta} \mathbb{B}^{[\ell](\gamma)i\star}_{\alpha\beta} \mathbb{B}^{[\ell](\gamma')i}_{\alpha\beta}\,. \tag{C.41}$$

The Gram matrix inverse is required for the projector to be idempotent: $\mathcal{P}^2 = \mathcal{P}$. While the n lowest states are all orthogonal the $\mathbb{B}$ are not guaranteed to be orthogonal, and may even be linearly dependent.[11] Thus, instead of the true inverse we use the Moore–Penrose pseudo-inverse of the (Hermitian) Gram matrix.

$$\left(\mathcal{N}^{-1}\right)_{\gamma\gamma'} = \sum_{\beta=1}^{N_S} V_{\gamma\beta} \frac{1}{\lambda_\beta} V^{\star}_{\gamma'\beta}, \tag{C.42}$$

[11]One can see that if the number of states desired is larger than the local Hilbert space of a single MPS tensor then the vectors *must* be linearly dependent just by the dimension of the space.

where V is the matrix with the eigenvectors of $\mathcal{N}$ as columns and N_S is the number of eigenvalues greater than a set tolerance (to avoid overflow).[12] By defining the new set of tensors

$$\mathbb{C}^{[\ell](\gamma)i}_{\alpha\beta} = \sum_{\gamma'} \frac{1}{\sqrt{\lambda_\gamma}} V_{\gamma'\gamma} \mathbb{B}^{[\ell](\gamma')i}_{\alpha\beta} \tag{C.43}$$

the projector becomes

$$\mathcal{P}^{[\ell]}_{[\alpha i\beta],[\alpha' i'\beta']} = \delta_{[\alpha i\beta],[\alpha' i'\beta']} - \sum_{\gamma} \mathbb{C}^{[\ell](\gamma)i}_{\alpha\beta} \mathbb{C}^{[\ell](\gamma)i'\star}_{\alpha'\beta'} . \tag{C.44}$$

We compute the projected effective Hamiltonian as $\mathcal{P}^{[\ell]\dagger} \hat{H}_{\text{eff}} \mathcal{P}^{[\ell]}$. The total algorithm proceeds much like the ground state search algorithm except that we also initialize the $\mathbb{B}$ overlaps LR using Eqs. (C.37) and (C.39), compute the projectors at each site to compute and diagonalize the projected effective Hamiltonian, and update the $\mathbb{B}$ LR after each optimization.[13] As a final note, clearly 0 is a valid eigenvalue of the projected Hamiltonian with eigenvector 0. Thus, if the next excited state energy is positive, then the algorithm will fail to converge trying to force the state to be zero. This difficulty can be avoided by shifting the spectrum of the Hamiltonian to be negative definite.[14]

C.5.1 * Sparse Solution of the Projected Eigenproblem

The above method can also easily be recast in a form amenable to the Lanczos algorithm. Every Hamiltonian-tensor multiply is replaced by projection, Hamiltonian-tensor multiply, and then projection again, where projection is

$$A^{i}_{\alpha\beta} = A^{i}_{\alpha\beta} - \sum_{\gamma} \left[\sum_{\alpha'\beta' i'} A^{i'}_{\alpha'\beta'} \mathbb{C}^{[\ell](\gamma)i'\star}_{\alpha'\beta'} \right] \mathbb{C}^{[\ell](\gamma)i}_{\alpha\beta} . \tag{C.45}$$

[12]A good choice for the tolerance is $n\|\mathcal{N}\|_2\varepsilon = n\sqrt{\max\lambda}\varepsilon$ where ε is the machine epsilon, n is the linear dimension of the matrix, and $\sqrt{\max\lambda}$ is the square root of the largest eigenvalue. Note that because $\mathcal{N}$ is a Gram matrix it is guaranteed to be positive semidefinite and have at least one positive eigenvalue.

[13]Note that we don't need to shift the orthogonality center of the Bs, as all of them get contracted each time we form the $\mathbb{B}$s.

[14]Note that the largest eigenvalue of $\hat{H}$ can be found by solving for the "ground state" of $-\hat{H}$.

C.6 Time Evolution with TEBD

In this section we discuss the simplest method for time evolution of an MPS: Time-evolving block decimation (TEBD). TEBD involves two parts: (1) the approximation of the full propagator by a series of two-site propagators for short times and (2) representation of a two-site operator times an MPS as an MPS. There are many ways to accomplish the first task which fall generally under the heading of Suzuki–Trotter expansions. A simple expression which fits our purposes nicely is

$$\mathcal{U} = \exp\left(-i\hat{H}\delta t\right) = \prod_{j=L-1}^{1} \exp\left(-i\hat{H}^{(j)}\delta t/2\right)\prod_{j=1}^{L-1}\exp\left(-i\hat{H}^{(j)}\delta t/2\right) + \mathcal{O}\left(\delta t^3\right) \tag{C.46}$$

$$= \prod_{j=L-1}^{1} \mathcal{U}^{(j)} \prod_{j=1}^{L-1} \mathcal{U}^{(j)} + \mathcal{O}\left(\delta t^3\right), \tag{C.47}$$

where $\hat{H}^{(j)}$ is a *two-site* Hamiltonian acting on-sites j and $j+1$ whose precise form will be determined shortly. We first note that this restriction implies that TEBD is restricted to nearest-neighbor Hamiltonians. As was discussed in the MPO section, the MPO of any Hamiltonian with only nearest-neighbor and local interactions has nonzero entries only on the "border" of the MPO matrices. We construct the Hamiltonians in Eq. (C.46) as

$$\hat{H}^{(1)}_{[ij][i'j']} = \sum_{k=1}^{\chi_O-1}\left(W^{[1]}\right)^{ii'}_{1k}\left(W^{[2]}\right)^{jj'}_{k1} + \frac{1}{2}\left(W^{[1]}\right)^{ii'}_{1\chi_O}\left(W^{[2]}\right)^{jj'}_{\chi_O 1}, \tag{C.48}$$

$$\begin{aligned}\hat{H}^{(\ell)}_{[ij][i'j']} =& \frac{1}{2}\left(W^{[\ell]}\right)^{ii'}_{11}\left(W^{[\ell+1]}\right)^{jj'}_{11} + \sum_{k=2}^{\chi_O-1}\left(W^{[\ell]}\right)^{ii'}_{1k}\left(W^{[\ell+1]}\right)^{jj'}_{k1} \\ &+ \frac{1}{2}\left(W^{[\ell]}\right)^{ii'}_{1\chi_O}\left(W^{[\ell+1]}\right)^{jj'}_{\chi_O 1}, \quad 2\le \ell \le L-2\end{aligned} \tag{C.49}$$

$$\hat{H}^{(L-1)}_{[ij][i'j']} = \frac{1}{2}\left(W^{[L-1]}\right)^{ii'}_{11}\left(W^{[L]}\right)^{jj'}_{11} + \sum_{k=2}^{\chi_O}\left(W^{[L-1]}\right)^{ii'}_{1k}\left(W^{[L]}\right)^{jj'}_{k1}. \tag{C.50}$$

Note that, with these definitions, $\hat{H} = \sum_{j=1}^{L-1}\hat{H}^{(j)}$. The factors of $1/2$ in the local terms with $k = 1$ or χ_O account for the fact that $\hat{H}^{(\ell)}$ and $\hat{H}^{(\ell+1)}$ both contribute a local term at site ℓ, and the boundary Hamiltonians account for the fact that only $\hat{H}^{(1)}$ and $\hat{H}^{(L-1)}$ contain local terms for the first and last site, respectively. With these Hermitian matrices in hand we can easily find the exponentials appearing in Eq. (C.46) by performing the eigenvalue decomposition $H = V\lambda V^{\dagger}$, where

V is the matrix with the eigenvectors of H as columns and λ is a diagonal matrix with the eigenvalues on the diagonal. The matrix exponential $\exp(-iH\delta t)$ is $V\exp(-i\lambda\delta t)\,V^\dagger$ in the same notation.

Our task of applying the full propagator $\mathcal{U}$ to a site $|\psi\rangle$ expressed as an MPS has been reduced to computing the application of a two-site operator to an MPS. We do this in the following manner. We first define a new `fourtensor` structure which has four indices. We then fuse two tensors A and B into a fourtensor T as

$$T_{\alpha ij\beta} = \sum_\gamma A^i_{\alpha\gamma} B^j_{\gamma\beta} \,. \tag{C.51}$$

The operation of a two-site operator $\mathcal{U}$ on this object is

$$\tilde{T}_{\alpha ij\beta} = \sum_{i',j'} \mathcal{U}_{[ij][i'j']} T_{\alpha i'j'\beta} \,. \tag{C.52}$$

Finally, a fourtensor may be split into two tensors by first reshaping it into a $d_\alpha d_i \times d_j d_\beta$ matrix:

$$T_{[\alpha i][j\beta]} = T_{\alpha ij\beta} \tag{C.53}$$

performing an SVD

$$T_{[\alpha i][j\beta]} = U_{[\alpha i]\gamma} S_\gamma V_{\gamma[j\beta]} \tag{C.54}$$

and then unpacking the SVD matrices

$$A^i_{\alpha\gamma} = U_{[\alpha i]\gamma}\,, \quad B^j_{\gamma\beta} = S_\gamma V_{\gamma[j\beta]} \,. \tag{C.55}$$

As written, if either A or B was the orthogonality center then B is the new orthogonality center. One can make A the orthogonality center by contracting S into A instead of B:

$$A^i_{\alpha\gamma} = U_{[\alpha i]\gamma} S_\gamma\,, \quad B^j_{\gamma\beta} = V_{\gamma[j\beta]} \,. \tag{C.56}$$

The growth of the number of singular values is typically exponential in the number of time steps, and this causes the bond dimensions of the tensors A and B to also grow exponentially. We truncate the bond dimension optimally by truncating the singular values. In simulations it is often better to impose a condition on the decay of the singular values rather than to impose a condition on the bond dimension itself. That is, we want to discard at most ϵ of the norm of the state at each two-site operation. This defines the new bond dimension χ_{new} implicitly as

$$1 - \sum_{\alpha=1}^{\chi_{\text{new}}} \frac{S_\alpha^2}{\sum_\beta S_\beta^2} < \epsilon \,. \tag{C.57}$$

Note that the sum in the denominator runs over *all* values of S, and so the sum of the squares is normalized to 1. Once we have χ_{new}, we define $S = S(1 : \chi_{\text{new}})/\sqrt{\langle S(1 : \chi_{\text{new}})|S(1 : \chi_{\text{new}})\rangle}$ and the γ summations in Eqs. (C.55) and (C.56) run from 1 to χ_{new}.

The procedure for TEBD with a time-independent Hamiltonian is thus as follows:

1. Shift the orthogonality center to the first site. $t = 0$. If the MPS structure is real, transfer to a complex MPS representation (be careful in your complex code that you have implemented all conjugates properly—this is an easy mistake to make!).
2. Construct the two-site Hamiltonians $\hat{H}^{(j)}$ as in Eqs. (C.48)–(C.50) and exponentiate them to find the two-site propagators $\hat{\mathcal{U}}^{(j)} = \exp\left(-i\hat{H}^{(j)}\delta t\right)$.
3. Starting from the first site, $j = 1$, fuse sites j and $j + 1$ into a fourtensor T.
4. Act with $\hat{\mathcal{U}}^{(j)}$ on T, and split into new updated $A^{[j]}$ and $A^{[j+1]}$ using Eq. (C.55).
5. Repeat 3 and 4 up to and including $j = L - 1$.
6. Starting from $j = L - 1$ fuse sites j and $j + 1$ into a fourtensor T.
7. Act with $\hat{\mathcal{U}}^{(j)}$ on T, and split into new updated $A^{[j]}$ and $A^{[j+1]}$ using Eq. (C.56).
8. Repeat 5 and 6 down to and including $j = 1$.
9. $t = t + \delta t$
10. Compute any desired observables. If t is greater than or equal to the maximum time desired, exit. Otherwise, return to step 3.

This completes the discussion of TEBD in the MPS language. It turns out to be a much simpler algorithm than the ground state search.

C.7 Source Code

C.7.1 Module: LinearAlgebraBackend.f90

```
MODULE LinearAlgebraBackend
! Purpose:  basic derived types, random numbers, lapack
     wrappers
IMPLICIT NONE
!real vector
TYPE vector
     REAL(KIND=8), POINTER :: v(:)
END TYPE vector
!real matrix
TYPE matrix
     REAL(KIND=8), POINTER :: m(:,:)
END TYPE matrix
```

```
!real 3-tensor
TYPE tensor
    REAL(KIND=8), POINTER :: t(:,:,:)
END TYPE tensor

CONTAINS
SUBROUTINE Seed_Init()
!Purpose:Randomly seed the random number generator
   using the intrinsic system clock
IMPLICIT NONE
INTEGER :: seedsize !Size of seed-returned by intrinsic
     procedure
INTEGER :: clock !Clock time
INTEGER, DIMENSION(:), ALLOCATABLE :: seed !Seed passed
     to intrinsic procedure
INTEGER :: i !dummy integer

CALL RANDOM_SEED(size=seedsize) !Ask for the size of
   the seed
ALLOCATE(seed(seedsize)) !Allocate the seed using the
   given seed size
CALL SYSTEM_CLOCK(COUNT=clock) !Get the clock time
FORALL(i=1:seedsize)
    seed(i)=clock+17*(i-1)
END FORALL
CALL RANDOM_SEED(PUT=seed) !Seed the random number
   generator using the given seed
END SUBROUTINE Seed_Init

REAL(KIND=8) FUNCTION Rand_Num(min,max)
!Purpose:Generate a random number in the range [min,max
   )
IMPLICIT NONE
REAL(KIND=8), INTENT(IN) :: min, max !Bounds
REAL(KIND=8) :: dumrand !Dummy real

CALL RANDOM_NUMBER(dumrand) !Get a random number in the
     range [0,1)
Rand_Num=(max-min)*dumrand+min !Put the number in the
   proper range
END FUNCTION Rand_Num

SUBROUTINE SSvd(U,S,V,A)
```

```
!Purpose: Perform an SVD on A or A^T, whichever is more
    efficient
IMPLICIT NONE
REAL(KIND=8), INTENT(INOUT) :: A(:,:)
REAL(KIND=8), ALLOCATABLE :: B(:,:)
TYPE(matrix) :: U, V, VT
TYPE(vector) :: S
INTEGER :: m, n

m=SIZE(A,1)
n=SIZE(A,2)

IF(m.ge.n) THEN
ALLOCATE(U%m(m,n),S%v(n),V%m(n,n))
    CALL DGESVD_Wrapper(U%m,S%v,V%m,A)
ELSE
    ALLOCATE(U%m(m,m),S%v(m),VT%m(n,m))
    ALLOCATE(B(n,m))
    B=TRANSPOSE(A)
    CALL DGESVD_Wrapper(VT%m,S%v,U%m, B)
    DEALLOCATE(B)
    ALLOCATE(V%m(m,n))
    V%m=TRANSPOSE(VT%m)
    U%m=TRANSPOSE(U%m)
    DEALLOCATE(VT%m)
END IF
END SUBROUTINE SSvd

SUBROUTINE DGESVD_Wrapper(U,S,V,A)
!Purpose: Perform an economy SVD on A using DGESVD
IMPLICIT NONE
REAL(KIND=8), INTENT(INOUT) :: A(:,:), U(:,:), V(:,:)
REAL(KIND=8), INTENT(INOUT) :: S(:)
REAL(KIND=8) :: work(5*SIZE(A,1))
INTEGER :: info

CALL DGESVD('S','A',SIZE(A,1), SIZE(A,2), A, SIZE(A,1),
   S, U, SIZE(A,1), V, SIZE(A,2),&
        work, 5*SIZE(A,1), info)
END SUBROUTINE DGESVD_Wrapper

SUBROUTINE DSYEV_Wrapper_vecs(A, eigvals)
!Purpose: Diagonalize the real symmetric matrix A using
    dsyev.
```

```
! Return the eigvals and eigvecs.
IMPLICIT NONE
REAL(KIND=8), INTENT(INOUT) :: A(:,:)
REAL(KIND=8), INTENT(IN) :: eigVals(:)
REAL(KIND=8), ALLOCATABLE :: work(:)
INTEGER :: info

ALLOCATE(work(3*SIZE(A,1)-1))
CALL DSYEV('V','U',SIZE(A,1),A,SIZE(A,1),eigVals,work
    ,3*SIZE(A,1)-1,info)
DEALLOCATE(work)
END SUBROUTINE DSYEV_Wrapper_vecs
END MODULE LinearAlgebraBackend
```

C.7.2 Module: MPSOps.f90

```
MODULE LinearAlgebraBackend
! Purpose: basic derived types, random numbers, lapack
    wrappers
IMPLICIT NONE
!real vector
TYPE vector
    REAL(KIND=8), POINTER :: v(:)
END TYPE vector
!real matrix
TYPE matrix
    REAL(KIND=8), POINTER :: m(:,:)
END TYPE matrix
!real 3-tensor
TYPE tensor
    REAL(KIND=8), POINTER :: t(:,:,:)
END TYPE tensor

CONTAINS
SUBROUTINE Seed_Init()
!Purpose:Randomly seed the random number generator
   using the intrinsic system clock
IMPLICIT NONE
INTEGER :: seedsize !Size of seed-returned by intrinsic
    procedure
INTEGER :: clock !Clock time
INTEGER, DIMENSION(:), ALLOCATABLE :: seed !Seed passed
    to intrinsic procedure
INTEGER :: i !dummy integer
```

```
CALL RANDOM_SEED(size=seedsize) !Ask for the size of
    the seed
ALLOCATE(seed(seedsize)) !Allocate the seed using the
    given seed size
CALL SYSTEM_CLOCK(COUNT=clock) !Get the clock time
FORALL(i=1:seedsize)
     seed(i)=clock+17*(i-1)
END FORALL
CALL RANDOM_SEED(PUT=seed) !Seed the random number
    generator using the given seed
END SUBROUTINE Seed_Init

REAL(KIND=8) FUNCTION Rand_Num(min,max)
!Purpose:Generate a random number in the range [min,max
    )
IMPLICIT NONE
REAL(KIND=8), INTENT(IN) :: min, max !Bounds
REAL(KIND=8) :: dumrand !Dummy real

CALL RANDOM_NUMBER(dumrand) !Get a random number in the
     range [0,1)
Rand_Num=(max-min)*dumrand+min !Put the number in the
    proper range
END FUNCTION Rand_Num

SUBROUTINE SSvd(U,S,V,A)
!Purpose: Perform an SVD on A or A^T, whichever is more
     efficient
IMPLICIT NONE
REAL(KIND=8), INTENT(INOUT) :: A(:,:)
REAL(KIND=8), ALLOCATABLE :: B(:,:)
TYPE(matrix) :: U, V, VT
TYPE(vector) :: S
INTEGER :: m, n

m=SIZE(A,1)
n=SIZE(A,2)

IF(m.ge.n) THEN
ALLOCATE(U%m(m,n),S%v(n),V%m(n,n))
     CALL DGESVD_Wrapper(U%m, S%v, V%m, A)
ELSE
     ALLOCATE(U%m(m,m),S%v(m),VT%m(n,m))
     ALLOCATE(B(n,m))
```

```
    B=TRANSPOSE(A)
    CALL DGESVD_Wrapper(VT%m, S%v ,U%m, B)
    DEALLOCATE(B)
    ALLOCATE(V%m(m, n))
    V%m=TRANSPOSE(VT%m)
    U%m=TRANSPOSE(U%m)
    DEALLOCATE(VT%m)
END IF
END SUBROUTINE SSvd

SUBROUTINE DGESVD_Wrapper(U,S,V,A)
!Purpose: Perform an economy SVD on A using DGESVD
IMPLICIT NONE
REAL(KIND=8), INTENT(INOUT) :: A(:,:), U(:,:), V(:,:)
REAL(KIND=8), INTENT(INOUT) :: S(:)
REAL(KIND=8) :: work(5*SIZE(A,1))
INTEGER :: info

CALL DGESVD('S','A',SIZE(A,1), SIZE(A,2), A, SIZE(A,1),
   S, U, SIZE(A,1), V, SIZE(A,2),&
        work, 5*SIZE(A,1), info)
END SUBROUTINE DGESVD_Wrapper

SUBROUTINE DSYEV_Wrapper_vecs(A, eigvals)
!Purpose: Diagonalize the real symmetric matrix A using
     dsyev.
! Return the eigvals and eigvecs.
IMPLICIT NONE
REAL(KIND=8), INTENT(INOUT) :: A(:,:)
REAL(KIND=8), INTENT(IN) :: eigVals(:)
REAL(KIND=8), ALLOCATABLE :: work(:)
INTEGER :: info

ALLOCATE(work(3*SIZE(A,1)-1))
CALL DSYEV('V','U',SIZE(A,1),A,SIZE(A,1),eigVals,work
   ,3*SIZE(A,1)-1,info)
DEALLOCATE(work)
END SUBROUTINE DSYEV_Wrapper_vecs
END MODULE LinearAlgebraBackend
```

C.7.3 Module: MPOOps.f90

```
MODULE MPOOps
! Purpose: MPOs and Overlaps
```

```
USE LinearAlgebraBackend
USE MPSOps
IMPLICIT NONE

!Matrix product operator matrix
TYPE MPOm
     TYPE(matrix), POINTER :: Op(:,:)
END TYPE MPOm
!Matrix product operator
TYPE MPO
     TYPE(MPOm), POINTER :: W(:)
END TYPE MPO
CONTAINS

SUBROUTINE IsingMPO(H,L,Jz,hx)
!Purpose: Define the matrix product operator form of
    the transverse field Ising Hamiltonian on L sites
IMPLICIT NONE
TYPE(MPO) :: H
INTEGER :: L
REAL(KIND=8) :: Jz, hx
TYPE(matrix) :: sx, sz, id
INTEGER :: k, kp, i
INTEGER :: bd

!bond dimension of Ising MPO is 3
bd=3
!define I, \sigma_z, and \sigma_x
ALLOCATE(id%m(2,2))
id%m=0.0_8
id%m(1,1)=1.0_8
id%m(2,2)=1.0_8

ALLOCATE(sz%m(2,2))
sz%m=0.0_8
sz%m(1,1)=1.0_8
sz%m(2,2)=-1.0_8

ALLOCATE(sx%m(2,2))
sx%m=0.0_8
sx%m(1,2)=1.0_8
sx%m(2,1)=1.0_8

ALLOCATE(H%W(L))
```

```
!leftmost MPO is a row vector
ALLOCATE(H%W(1)%Op(1,3))
DO k=1,1
     DO kp=1,bd
          ALLOCATE(H%W(1)%Op(k,kp)%m(2,2))
          H%W(1)%Op(k,kp)%m=0.0_8
     END DO
END DO
H%W(1)%Op(1,1)%m=id%m
H%W(1)%Op(1,2)%m=-Jz*sz%m
H%W(1)%Op(1,3)%m=-hx*sx%m

!rightmost MPO is a column vector
ALLOCATE(H%W(L)%Op(3,1))
DO k=1,bd
     DO kp=1,1
          ALLOCATE(H%W(L)%Op(k,kp)%m(2,2))
          H%W(L)%Op(k,kp)%m=0.0_8
     END DO
END DO
H%W(L)%Op(1,1)%m=-hx*sx%m
H%W(L)%Op(2,1)%m=sz%m
H%W(L)%Op(3,1)%m=id%m

DO i=2,L-1
     !intermediate MPOs are lower triangular matrices
     ALLOCATE(H%W(i)%Op(3,3))
     DO k=1,bd
          DO kp=1,bd
               ALLOCATE(H%W(i)%Op(k,kp)%m(2,2))
               H%W(i)%Op(k,kp)%m=0.0_8
          END DO
     END DO
     H%W(i)%Op(1,1)%m=id%m
     H%W(i)%Op(2,1)%m=sz%m
     H%W(i)%Op(3,1)%m=-hx*sx%m
     H%W(i)%Op(3,2)%m=-Jz*sz%m
     H%W(i)%Op(3,3)%m=id%m
END DO

DEALLOCATE(id%m,sz%m,sx%m)
END SUBROUTINE IsingMPO

SUBROUTINE DeallocateMPO(H)
```

```
!Purpose: deallocate an MPO
IMPLICIT NONE
TYPE(MPO) :: H
INTEGER :: L
INTEGER :: i, k, kp

L=SIZE(H%W)
DO i=1,L
    DO k=1,SIZE(H%W(i)%Op,1)
        DO kp=1,SIZE(H%W(i)%Op,2)
            DEALLOCATE(H%W(i)%Op(k,kp)%m)
        END DO
    END DO
    DEALLOCATE(H%W(i)%Op)
END DO
DEALLOCATE(H%W)
END SUBROUTINE DeallocateMPO

SUBROUTINE InitializeLR(LR,H,psi, k)
!Purpose: Initialize the overlaps used to define the
    effective Hamiltonian
! this routine assumes that the first site is the
    orthogonality center
IMPLICIT NONE
TYPE(tensor), POINTER :: LR(:)
TYPE(MPO) :: H
TYPE(MPS) :: psi
INTEGER :: L, i, kin
INTEGER, INTENT(IN), OPTIONAL :: k

kin=1
IF(PRESENT(k)) kin=k
L=SIZE(psi%A)
ALLOCATE(LR(L+1))
ALLOCATE(LR(1)%t(1,1,1))
LR(1)%t=1.0_8
ALLOCATE(LR(L+1)%t(1,1,1))
LR(L+1)%t=1.0_8
DO i=L,kin+1,(-1)
    CALL RightMPOOverlap(LR(i+1),LR(i),H%W(i),psi%A(i))
END DO
DO i=2,kin
    CALL LeftMPOOverlap(LR(i-1),LR(i),H%W(i-1),psi%A(i
        -1))
```

```
END DO
END SUBROUTINE InitializeLR

SUBROUTINE RightMPOOverlap(Ru,R,W,A)
!Purpose: propagate the leftmoving overlap used to
    define the effective Hamiltonian
IMPLICIT NONE
TYPE(tensor) :: Ru, R
TYPE(MPOm) :: W
TYPE(tensor) :: A
REAL(KIND=8), ALLOCATABLE :: temp1(:,:,:,:), temp2
    (:,:,:,:)
INTEGER :: i, ip, alphap, beta,betap, k, kp
INTEGER :: da,dap, db,dbp, d, dk, dkp

!first contraction: temp1(alpha,i,k,beta)= A_{alpha, i,
     alphap} Ru_{alphap,k,beta}
da=SIZE(A%t,1)
d=SIZE(A%t,2)
dk=SIZE(Ru%t,2)
db=SIZE(Ru%t,3)
dap=SIZE(Ru%t,1)
ALLOCATE(temp1(da,d,dk,db))
temp1=0.0_8
DO alphap=1,dap
    DO k=1,dk
        DO beta=1,db
            temp1(:,:,k,beta)=temp1(:,:,k,beta)+A%t
                (:,:,alphap)*Ru%t(alphap,k,beta)
        END DO
    END DO
END DO
!second contraction: temp2(alpha,i,k,beta)= W_{k kp}^{i
     ip} temp1(alpha,ip,kp,beta)
da=SIZE(temp1,1)
d=SIZE(temp1,2)
dk=SIZE(W%Op,1)
dkp=SIZE(temp1,3)
db=SIZE(temp1,4)
ALLOCATE(temp2(da,d,dk,db))
temp2=0.0_8
DO i=1,d
    DO ip=1,d
        DO k=1,dk
```

```
            DO kp=1,dkp
                temp2(:,i,k,:)=temp2(:,i,k,:)+W%Op(k,kp
                    )%m(i,ip)*temp1(:,ip,kp,:)
            END DO
        END DO
    END DO
END DO
DEALLOCATE(temp1)
!third contraction: R_{alpha, k, beta}=temp2(alpha,i,k,
    betap)*Aconj_{beta, i, betap}
da=SIZE(temp2,1)
d=SIZE(temp2,2)
dk=SIZE(temp2,3)
db=SIZE(A%t,1)
dbp=SIZE(temp2,4)
ALLOCATE(R%t(da,dk,db))
R%t=0.0_8
DO i=1,d
    DO betap=1,dbp
        DO beta=1,db
            R%t(:,:,beta)=R%t(:,:,beta)+temp2(:,i,:,
                betap)*A%t(beta,i,betap)
        END DO
    END DO
END DO
DEALLOCATE(temp2)
END SUBROUTINE RightMPOOverlap

SUBROUTINE LeftMPOOverlap(Lb,L,W,A)
!Purpose: propagate the rightmoving overlap used to
    define the effective Hamiltonian
IMPLICIT NONE
TYPE(tensor) :: Lb, L
TYPE(MPOm) :: W
TYPE(tensor) :: A
REAL(KIND=8), ALLOCATABLE :: temp1(:,:,:,:), temp2
    (:,:,:,:)
INTEGER :: i, ip, alpha,alphap, betap, k, kp
INTEGER :: da,dap, db,dbp, d, dk, dkp

!first contraction: temp1(alpha,i,k,beta)= Lb_{alpha,k,
    betap} A_{betap, i, beta}
da=SIZE(Lb%t,1)
d=SIZE(A%t,2)
```

```
dk=SIZE(Lb%t,2)
db=SIZE(A%t,3)
dbp=SIZE(Lb%t,3)
ALLOCATE(temp1(da,d,dk,db))
temp1=0.0_8
DO betap=1,dbp
     DO k=1,dk
          DO alpha=1,da
               temp1(alpha,:,k,:)=temp1(alpha,:,k,:)+Lb%t(
                    alpha,k,betap)*A%t(betap,:,:)
          END DO
     END DO
END DO
!second contraction: temp2(alpha,i,k,beta)= W_{kp k}^{i
      ip} temp1(alpha,ip,kp,beta)
da=SIZE(temp1,1)
d=SIZE(temp1,2)
dk=SIZE(W%Op,2)
dkp=SIZE(temp1,3)
db=SIZE(temp1,4)
ALLOCATE(temp2(da,d,dk,db))
temp2=0.0_8
DO i=1,d
     DO ip=1,d
          DO k=1,dk
               DO kp=1,dkp
                    temp2(:,i,k,:)=temp2(:,i,k,:)+W%Op(kp,k
                         )%m(i,ip)*temp1(:,ip,kp,:)
               END DO
          END DO
     END DO
END DO
DEALLOCATE(temp1)
!third contraction: L_{alpha, k, beta}=temp2(alphap,i,k
     ,beta) * A_{alphap, i, alpha}
da=SIZE(A%t,3)
dap=SIZE(temp2,1)
d=SIZE(temp2,2)
dk=SIZE(temp2,3)
db=SIZE(temp2,4)
ALLOCATE(L%t(da,dk,db))
L%t=0.0_8
DO i=1,d
     DO alphap=1,dap
```

```
        DO alpha=1,da
            L%t(alpha,:,:)=L%t(alpha,:,:)+temp2(alphap,
                i,:,:)*A%t(alphap,i,alpha)
        END DO
    END DO
END DO
DEALLOCATE(temp2)
END SUBROUTINE LeftMPOOverlap

SUBROUTINE UpdateLR(LR,H,psi,k, sense)
!Purpose: After updating site k and shifting the
    orthogonality center, update the overlaps
! sense tells whether the sweep is moving rightwards (
    sense>0) or leftwards (sense<0)
IMPLICIT NONE
TYPE(tensor), POINTER :: LR(:)
TYPE(MPO) :: H
TYPE(MPS) :: psi
INTEGER, INTENT(IN) :: k, sense

!rightmoving-add an L
IF(sense.gt.0) THEN
    DEALLOCATE(LR(k+1)%t)
    CALL LeftMPOOverlap(LR(k),LR(k+1),H%W(k),psi%A(k))
!leftmovingmoving-add an R
ELSE
    DEALLOCATE(LR(k)%t)
    CALL RightMPOOverlap(LR(k+1),LR(k),H%W(k),psi%A(k))
END IF
END SUBROUTINE UpdateLR
END MODULE MPOOps
```

C.7.4 Module: EffHamiOps.f90

```
MODULE EffHamiOps
! Purpose: Construction and spectrum of Effective
    Hamiltonian
USE LinearAlgebraBackend
USE MPSOps
USE MPOOps
IMPLICIT NONE

CONTAINS
```

```
SUBROUTINE FormEffectiveHamiltonian(j,Heff,H,LR)
!Purpose: Construct the effective Hamiltonian at site j
INTEGER :: j
TYPE(matrix) :: heff
TYPE(MPO) :: H
TYPE(tensor), POINTER :: LR(:)
INTEGER :: da, dap, d, dk ,dkp, db, dbp
INTEGER :: alpha, beta, alphap, betap, i,ip, k,kp, fi,
    si

!Heff(k)_{[alpha, i, beta], [alphap, ip, betap]}= LR(j)
    (alpha,k,alphap)*W_{k kp}^{i ip}*LR(j+1)(betap,kp,
    beta)
da=SIZE(LR(j)%t,1)
dap=SIZE(LR(j)%t,3)
db=SIZE(LR(j+1)%t,3)
dbp=SIZE(LR(j+1)%t,1)
d=SIZE(H%W(j)%Op(1,1)%m,1)
dk=SIZE(H%W(j)%Op,1)
dkp=SIZE(H%W(j)%Op,2)

ALLOCATE(Heff%m(da*d*db,dap*d*dbp))
Heff%m=0.0_8
DO alpha=1,da
DO i=1,d
DO beta=1,db
   fi=(alpha-1)*d*db+(i-1)*db+beta
   DO alphap=1,dap
   DO ip=1,d
   DO betap=1,dbp
      si=(alphap-1)*d*dbp+(ip-1)*dbp+betap
      DO k=1,dk
      DO kp=1,dkp
         Heff%m(fi,si)=Heff%m(fi,si)+LR(j)%t(alpha,k
            ,alphap)*H%W(j)%Op(k,kp)%m(i,ip)*LR(j+1)
            %t(betap,kp,beta)
      END DO
      END DO
   END DO
   END DO
   END DO
END DO
END DO
END DO
```

```
END SUBROUTINE FormEffectiveHamiltonian

SUBROUTINE MinimizeSite(j,psi,H,LR,energy)
!Purpose : Minimize the energy with all sites but j
    held fixed
IMPLICIT NONE
INTEGER, INTENT(IN) :: j
TYPE(MPS) :: psi
TYPE(MPO) :: H
TYPE(tensor), POINTER :: LR(:)
REAL(KIND=8) :: energy
TYPE(matrix) :: Heff
TYPE(vector) :: eigvals
INTEGER :: alpha,i,beta, da, d, db

CALL FormEffectiveHamiltonian(j,Heff,H,LR)
ALLOCATE(eigvals%v(SIZE(Heff%m,1)))
CALL DSYEV_Wrapper_vecs(Heff%m, eigvals%v)
energy=eigvals%v(1)

da=SIZE(psi%A(j)%t,1)
d=SIZE(psi%A(j)%t,2)
db=SIZE(psi%A(j)%t,3)

DO alpha=1,da
    DO i=1,d
        DO beta=1,db
            psi%A(j)%t(alpha,i,beta)=Heff%m((alpha-1)*d
                *db+(i-1)*db+beta,1)
        END DO
    END DO
END DO
DEALLOCATE(Heff%m, eigvals%v)
END SUBROUTINE MinimizeSite

SUBROUTINE FindGroundState(H,psi, k)
!Purpose :Find the ground state psi (represented as an
    MPS) of the hamiltonian H (represented as an MPO)
! variationally. psi should contain the initial guess
    on input. k is an optional parameter telling
! where to start the sweep
INTEGER, INTENT(IN), OPTIONAL :: k
INTEGER :: kin
INTEGER :: j,L
```

```
TYPE(MPO) :: H
TYPE(MPS) :: psi
REAL(KIND=8) :: energy, energySave, tenergy
TYPE(tensor), POINTER :: LR(:)
INTEGER :: sense, numsweeps
LOGICAL :: converged

kin=1
IF(PRESENT(k)) kin=k

L=SIZE(psi%A)
converged=.false.

!set up overlaps
CALL InitializeLR(LR,H,psi, kin)
numsweeps=1
energySave=100000.0_8

DO WHILE(.not.converged)
    sense=1
    DO j=kin,L-1
        CALL MinimizeSite(j,psi,H,LR,energy)
        CALL OrthogonalizeMPS(psi,j+1,j,j+1)
        CALL UpdateLR(LR,H,psi,j,sense)
    END DO
    sense=-1
    DO j=L,2,(-1)
        CALL MinimizeSite(j,psi,H,LR,energy)
        CALL OrthogonalizeMPS(psi,j-1,j-1,j)
        CALL UpdateLR(LR,H,psi,j,sense)
    END DO
    sense=1
    DO j=1,kin-1
        CALL MinimizeSite(j,psi,H,LR,energy)
        CALL OrthogonalizeMPS(psi,j+1,j,j+1)
        CALL UpdateLR(LR,H,psi,j,sense)
    END DO
    IF(ABS(energySave-energy).lt.0.01_8) converged=.
        true.
    IF(numsweeps==10) converged=.true.
    IF(converged) EXIT
    numsweeps=numsweeps+1
    energySave=energy
END DO
```

```
IF(ABS(energySave-energy).gt.0.01_8) THEN
    PRINT *, 'Failed␣to␣converge␣in',numsweeps,'sweeps.
        ␣␣Energy:',energy, '␣with␣\Delta␣E=',ABS(
        energySave-energy)
ELSE
    PRINT *, 'Converged␣in',numsweeps,'sweeps.␣␣Energy:
        ',energy, '␣with␣\Delta␣E=',ABS(energySave-
        energy)
END IF
END SUBROUTINE FindGroundState
END MODULE EffHamiOps
```

C.7.5 Main Program: OrthTestMain.f90

```
PROGRAM OrthTestMain
! Purpose: Find the ground state of the Ising model
    variationally
USE LinearAlgebraBackend
USE MPSOps
USE MPOOps
USE EffHamiOps
IMPLICIT NONE

TYPE(MPS) :: psi
TYPE(matrix) :: M
INTEGER :: L, bondD, d
INTEGER :: k, a, i, b, bp ,ap

d=2 !fixed by the spin-1/2 constraint
L=10 !number of lattice sites-variable
bondD=6 !bond dimension-variable
k=3
!Create a random MPS
CALL CreateRandomMPS(psi,L,bondD,d)
!normalize it and put into canonical form
CALL OrthonormalizeMPS(psi,k)
DO k=1,L
    PRINT *, 'site␣k=',k
    ALLOCATE(M%m(SIZE(psi%A(k)%t,3),SIZE(psi%A(k)%t,3))
        )
    M%m=0.0_8
    DO b=1,SIZE(psi%A(k)%t,3)
    DO bp=1,SIZE(psi%A(k)%t,3)
        DO a=1,SIZE(psi%A(k)%t,1)
```

```
          DO i=1,SIZE(psi%A(k)%t,2)
          Mm(b,bp)=Mm(b,bp)+psi%A(k)%t(a,i,b)*psi%A
              (k)%t(a,i,bp)
          END DO
       END DO
       IF(ABS(Mm(b,bp)).gt.1E-4) PRINT *, '␣␣␣left␣o.
           ␣c.␣b␣bp',b, bp,Mm(b,bp)
   END DO
   END DO
   DEALLOCATE(Mm)
   PRINT *, ''
   ALLOCATE(Mm(SIZE(psi%A(k)%t,1),SIZE(psi%A(k)%t,1))
       )
   Mm=0.0_8
   DO a=1,SIZE(psi%A(k)%t,1)
   DO ap=1,SIZE(psi%A(k)%t,1)
       DO b=1,SIZE(psi%A(k)%t,3)
          DO i=1,SIZE(psi%A(k)%t,2)
          Mm(a,ap)=Mm(a,ap)+psi%A(k)%t(a,i,b)*psi%A
              (k)%t(ap,i,b)
          END DO
       END DO
       IF(ABS(Mm(a,ap)).gt.1E-4) PRINT *, '␣␣␣right␣o
           .␣c.␣a␣ap',a, ap,Mm(a,ap)
   END DO
   END DO
   DEALLOCATE(Mm)
   PRINT *, ''
   PRINT *, ''
END DO
CALL DeallocateMPS(psi)
END PROGRAM OrthTestMain
```

C.7.6 Main Program: IsingGSMain.f90

```
PROGRAM IsingGSMain
! Purpose: Find the ground state of the Ising model
    variationally
USE LinearAlgebraBackend
USE MPSOps
USE MPOOps
USE EffHamiOps
IMPLICIT NONE
TYPE(MPS) :: psi
```

```
TYPE(MPO) :: IsingH
REAL(KIND=8) :: Jz, hx !Ising couplings
INTEGER :: L, bondD, d
INTEGER :: k

d=2 !fixed by the spin-1/2 constraint
L=100 !number of lattice sites-variable
bondD=8 !bond dimension-variable
Jz=1.0_8
hx=1.0_8

!Create a random MPS
CALL CreateRandomMPS(psi,L,bondD,d)
!normalize it and put into canonical form
CALL OrthonormalizeMPS(psi,1)
!Create the MPO rep of the Ising Hamiltonian
CALL IsingMPO(IsingH,L,Jz,hx)
!Find ground state variationally
CALL FindGroundState(IsingH,psi, 1)
!clean up
CALL DeallocateMPS(psi)
CALL DeallocateMPO(IsingH)
END PROGRAM IsingGSMain
```

MIX
Papier aus verantwortungsvollen Quellen
Paper from responsible sources
FSC® C105338

If you have any concerns about our products,
you can contact us on
ProductSafety@springernature.com

In case Publisher is established outside the EU,
the EU authorized representative is:
Springer Nature Customer Service Center GmbH
Europaplatz 3, 69115 Heidelberg, Germany

Printed by Libri Plureos GmbH
in Hamburg, Germany